8

REEDS MARINE ENGINEERING AND TECHNOLOGY

GENERAL ENGINEERING KNOWLEDGE
FOR MARINE ENGINEERS

Revised by Paul A Russell

Leslie Jackson

Thomas D Morton

REEDS

LONDON • OXFORD • NEW YORK • NEW DELHI • SYDNEY

REEDS
Bloomsbury Publishing Plc
50 Bedford Square, London, WC1B 3DP, UK

BLOOMSBURY, REEDS, and the Reeds logo are trademarks of Bloomsbury Publishing Plc

First published in Great Britain 1966
Second edition 1971
Third edition 1978
Fourth edition 1986
Fifth edition 2013
This sixth edition published 2018

A catalogue record for this book is available from the British Library

Library of Congress Cataloguing-in-Publication data has been applied for.

ISBN: PB: 978-1-4729-5273-8; eBook: 978-1-4729-5272-1; ePDF: 978-1-4729-5271-4

2 4 6 8 10 9 7 5 3 1

Typeset in Myriad Pro 10/14 by Newgen KnowledgeWorks Pvt. Ltd., Chennai, India
Printed and bound in Great Britain by CPI Group (UK) Ltd, Croydon CR0 4YY

Bloomsbury Publishing Plc makes every effort to ensure that the papers used in the manufacture of our books are natural, recyclable products made from wood grown in well-managed forests. Our manufacturing processes conform to the environmental regulations of the country of origin.

To find out more about our authors and books visit
www.bloomsbury.com and sign up for our newsletters

CONTENTS

PREFACE

The object of this book is to prepare students for the General Engineering Knowledge part of the Certificates of Competency for marine engineering officers, issued by flag state administrations, of which the Maritime Coastguard Agency (MCA) is responsible in the United Kingdom. The engineering certificates issued by the different flag states are designed against the criteria determined by the International Maritime Organisation (IMO). These are detailed in Chapter III of the Standards of Training, Certification and Watchkeeping for Seafarers (STCW). The latest edition of STCW includes the 2010 Manila Amendments, which are also included in this edition of *General Engineering Knowledge*. This edition also includes the most up-to-date information relating to the requirements of IMO's MARine POLution regulation, MARPOL Annex VI, which sets out the agenda for reducing the environmental impact of ships.

The text is intended to cover the groundwork required for examinations at the different levels of Engineering Officer of the Watch, Second Engineering Officer and Chief Engineering Officer. The engineering principles involved are the same but a much more detailed answer will be expected at the higher levels. It is extremely important for the students preparing for the Officer of the Watch examination to concentrate on safety procedures and the practices of marine engineering. It should also be remembered that a good set of comprehensive technical manuals is an invaluable on-board resource for all engineering staff.

This publication can now also be considered as more than an examination guide and will be useful to Superintendent engineers wishing to have a general guide to the latest trends from which they can seek more detail.

Engineering knowledge is delivered via several different academic pathways in the United Kingdom. These are the Scottish Qualifications Authority's (SQA) Maritime Studies Qualification (MSQ) through Higher National Diplomas (HND) to foundation degrees and full honours degrees. The drawings are still intended to have direct relevance to the examination requirements but it is left to each student to practise his/her own versions. I have found it particularly useful to use an artist's sketch pad, fill it with relevant drawings and practise them so that they can be reproduced as required in the examination.

Preparation for any examination is not easy and requires hard work; this is difficult with a subject as complex as marine engineering knowledge and one text book cannot cover all the possible questions. As a guide it is suggested that the student completes his/her research first and then attempts each question in the book in turn, basing answers on either a good descriptive sketch and writing or a description covering about one-and-a-half pages of A4 paper. The aim should be to complete each question in half an hour.

MATERIALS

Manufacture of Iron and Steel

Iron ores are the basic material used in the manufacture of the various steels and irons in current use. In its natural state iron ore may contain many impurities and vary considerably in its iron content. Some of the more important iron ores are listed below:

1. Hematite 30–65% iron content approximately.
2. Magnetite 60–70% iron content approximately.
3. Taconite 25–30% iron by weight

The most common 'Hematite' used to be mined in many places around the world but mining is now mainly confined to Australia, Brazil, Canada, China, India, Russia, South Africa, Ukraine, Venezuela and the USA. It is also now more common for iron and steel to be recycled.

Iron ores are not usually fed directly into the blast furnace in their natural or mined condition, they are prepared first. The preparation may consist of some form of concentrating process (e.g. washing out the earthy matter) followed by a crushing, screening and sintering process.

The crushing produces even-sized lumps and dust or fines. The fines are separated from the lumps by screening and then they are mixed with coal or tar dust and sintered. Sintering causes agglomeration of the fines and coal dust, and also causes removal of some of the volatiles. The sinter along with the unsintered ore is fed into the blast furnace as part of the charge (or burden); the remainder of the charge is principally coke – which serves as a fuel – and limestone which serves as a flux. Preparation of the iron ores in this way leads to a distinct saving in fuel and a greater rate of iron production.

Iron production

In the most up-to-date blast furnaces the charge is subjected to intense heat, the highest temperature is normally just above the pressurised air entry points, (sometimes up to 1,800°C). The specially designed section to create the pressurised air is known as the Tuyeres.

The following are some of the reactions which take place in a blast furnace:

1. At the bottom, carbon + oxygen = carbon dioxide.
2. At the middle, carbon dioxide + carbon = carbon monoxide.
3. At the top, iron oxide + carbon monoxide = iron + carbon dioxide.

At position 3 the iron that is produced from this oxidation – reduction action – is a spongy mass that gradually falls to the furnace bottom, melting as it falls and taking into the solution carbon, sulphur, manganese, etc. as it goes. The molten iron is collected in the hearth of the furnace, with the slag floating upon its surface. Tapping (the removal of the iron) of the furnace takes place about every 6 hours, the slag being tapped more frequently. When tapped the molten iron runs from the furnace through sand channels into sand pig beds (hence the term *pig iron*) or it is led into tubs, which are used to supply the iron in a molten condition to converters or Open Hearth furnaces for steel manufacture. Pig iron is very brittle and has little use, an analysis of a sample is given below:

Combined carbon	0.5%	Manganese	0.5%
Graphite	3.4%	Phosphorus	0.03%
Silicon	2.6%	Sulphur	0.02%

Converting Iron to Steel

Bessemer process

This is the first mass-production steel-making process (circa 1847) where a blast of air is blown through a charge of molten pig iron contained in a Bessemer converter.

The refining sequence can be followed by observing the appearance of the flames discharging from the converter, since the air will bring about oxidation of the carbon, etc. After pouring the charge, a mixture of iron, carbon (usually in the form of coke) and manganese is added to adjust the carbon content, etc., of the steel.

The principal difference between the Open Hearth and the Bessemer steels of similar carbon content is brought about by the higher nitrogen content in the Bessemer steel and is also partly due to the higher degree of oxidation with this process. This leads to a greater tendency for embrittlement of the steel due to strain-aging in the finished product. Typical nitrogen contents are: Bessemer steel 0.015% approximately, Open Hearth steel 0.005% approximately.

Open hearth process

In this process (circa 1865) a broad shallow furnace is used to support the charge of pig iron and scrap steel. Pig iron content of the charge may constitute 25–75% of the total, which may vary in mass – depending upon furnace capacity – between 10 and 50 tons. Scrap steel is added to reduce melting time if starting from cold.

Fuel employed in this process may be enriched blast furnace gas (blast furnace gas may contain 30% CO after cleaning) which melts the charge by burning across its surface. Reduction of carbon content is achieved by oxidation, and this may be assisted by adding a pure iron oxide ore to the charge. Other impurities are reduced either by oxidation or absorption in the slag.

At frequent intervals samples of the charge are taken for analysis and when the desired result is obtained the furnace is tapped. See Table 1.1 for analysis of metal and slag in a basic open hearth furnace.

Modern processes

Various modern steel-making processes have been developed and put into use, some extensively. These include the Linz-Donawitz (L.D.), Kaldo, Rotor and Spray processes.

The L.D. method of steel manufacture – the letters are the initials of twin towns in Austria, Linz and Donawitz – uses a converter similar in shape to the old Bessemer, mounted on trunnions to enable it to be swung into a variety of desired positions.

Table 1.1 *Analysis of metal and slag found in a basic open hearth furnace*

	Constituent	When melted %	After 6–20 hours Finished steel %
Metal	Carbon	1.1	0.55
	Silicon	–	0.1
	Sulphur	0.04	0.03
	Phosphorus	0.4	0.03
	Manganese	–	0.6
Slag	Silica	19.5	–
	Iron oxide	5.6	–
	Alumina	1.2	–
	Manganous oxide	8.7	–
	Lime	50.0	–
	Magnesia	5.0	–
	Phosphorus	9.0	–
	Sulphur	0.2	–

Figure 1.1 is a diagrammatic arrangement of the L.D. converter (1952). Scrap metal and molten iron from the blast furnace would be fed into the converter, which would then be turned to the vertical position after charging. A water-cooled oxygen lance would then be lowered into the converter and oxygen at a pressure of up to 11 bar approximately would be injected at high speed into the molten iron causing oxidation. After refining, the lance is withdrawn and the converter is first tilted to the metal pouring position and finally to the slag pouring position.

If the metal is of low phosphorus content, only oxygen is used; if, however, it is high in phosphorus, powdered lime is injected with the oxygen and the blow is in two parts, the process being interrupted in order to remove the high phosphorus-content slag.

The Kaldo and Rotor processes have not found the same popularity as the L.D., even though they are similar in that they use oxygen for refining. They both use converters that are rotated and the process is slower and more expensive.

BISRA (British Iron and Steel Research Association 1944–1971) have developed a process in which the molten iron running from the blast furnace was subjected to jets of high-speed oxygen that spray the metal into a container. This gives rapid refining since the oxygen and the metal intimately mix. The main advantages with this system

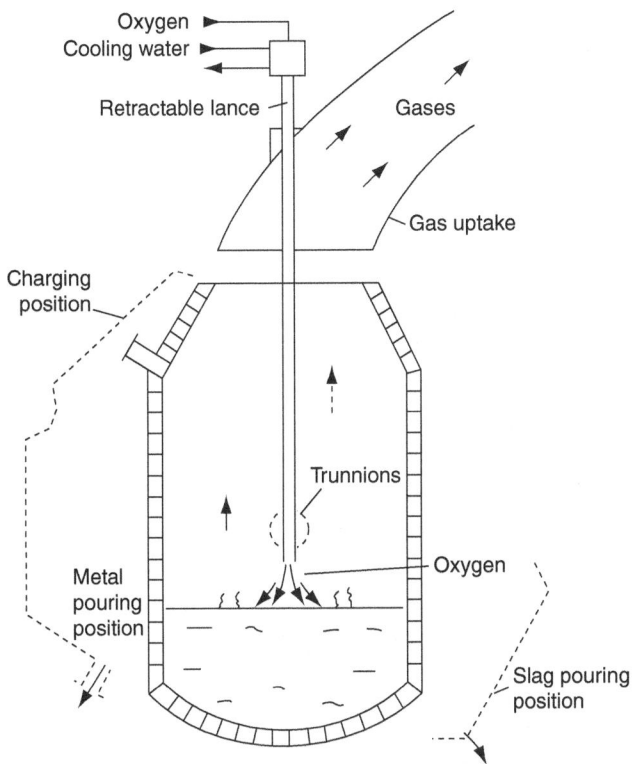

▲ **Figure 1.1** *L.D. process*

were that the intermediate stage of carrying the molten metal from the blast furnace to steel-making plant is eliminated, and the steel production rate is increased as a consequence.

Open Hearth furnaces were modernised by the fitting of oxygen lances in their roofs. This speeds up the steel production and gives the process similar characteristics to the L.D. process. By the 1990s all the older methods were superseded by the Basic Oxygen Process (BOP), which is a refinement of the L.D. process of steel conversion.

Acid and basic chemical processes

When pig iron is refined by oxidation a slag is produced. Depending upon the nature of the slag one of two types of processes is employed. If the slag is siliceous it is the acid process, if it is high in lime content the basic process is used. Hence the furnace lining that is in contact with the slag is made of siliceous material or basic material according to the nature of the slag, thus avoiding the standard chemical reaction:

ACID + BASE = SALT + WATER

Low-phosphorus pig irons are usually rich in silicon, this produces an *acid* slag, silica charged, which would react with a *basic* lining, hence silica bricks are used, which are *acidic*.

High-phosphorus pig iron requires an excess of lime added to it in order to remove the phosphorus. The slag formed will be rich in lime, which is a basic substance that would react with a silica brick lining. Hence a basic lining must be used, for example, oxidised dolomite (carbonates of lime and magnesia).

Both acid and basic processes can be operated in the Open Hearth, Bessemer, L.D. and Electric Arc furnaces, etc. The modern process uses computers to analyse spectroscopic readings that are taken at regular intervals. These readings measure the energy that is emitted from the furnace and from the analysis the composition can be changed to ensure the quality of the final product. This action has reduced the time and energy required for the process and has also increased the quality of the final product. Electric arc furnaces are being used increasingly to melt scrap iron and steel.

Electric Arc Furnace (EAF)

The electric arc furnace is relatively simple to understand. A high voltage is set up across two electrodes and the high temperature (up to 1600°C) 'arc' that is generated melts the metal that is within the furnace. Most of the material loaded into the EAF is recycled steel, although pig iron is added on occasion to correct any chemical imbalance. Flux is added to gather up the impurities that are then removed through a 'taphole'.

Cast Iron

Cast iron is produced by re-melting pig iron in a cupola (a small type of blast furnace) wherein the composition of the iron is suitably adjusted. The cooling process is very important for the quality of cast iron. Fast cooling produces a fine-grain structure while slow cooling produces a coarse-grain structure, uneven cooling produces poor-quality cast iron. The fluidity and low shrinkage on solidification of this material makes it suitable for complicated castings; other properties include machinability, wear resistance and high compressive strength. The careful use of computerised systems improves the quality of the products throughout each step in the process. The moulds can be made more accurately from CAD models, the machines pouring the liquid metal can be controlled more accurately and the quality of the final casting can be checked using computer-controlled laser measuring equipment.

Simple Metallurgy of Iron, Cast Iron and Steel and the Effect of Carbon

Carbon is a non-metallic substance with an atomic number of 6 and an atomic weight of 12.0107. It has the chemical symbol 'C' and readily forms a very large number of compounds such as carbon dioxide and carbon monoxide. In its amorphous or pure carbon form it is used in things such as paints and dry cell batteries and when it is formed into a crystalline structure it can be either a lubricant such as graphite or a cutting tool made from diamonds. The difference between the two is the way that the carbon structure is linked together. The graphite has its crystalline structure arranged in a layered form where the atoms are linked together in strong two-dimensional layers allowing the layers to move over each other. The diamond, on the other hand, has the carbon linked in a very strong three-dimensional structure, giving the diamond both its hardness and brittle characteristics.

Pure iron (ferrite) is soft and ductile with considerable strength, but when carbon is added to the iron in increasing concentrations, the strength and the hardness of the combination increases. Iron mixed with 0.4% carbon could be as much as twice the strength of pure iron, and with a 1.0% mix the result might be as much as three times the strength. Higher concentrations of carbon start to introduce brittleness for very little increase, or possibly a decrease, in strength.

Another interesting feature of iron is that its internal structure changes, as it is heated. At room temperature and up to about 900°C the atoms are arranged in a 'Body Centred Cubic' framework. Heating above 900°C sees the iron adopt a 'Face Centred Cubic' framework. This feature means that at the higher temperature, more carbon can combine with the iron due to the extra space created by the different microstructure of the face centred framework. Cooling the iron/carbon mix at different rates then produces different properties of the final 'room temperature' product. Cooling slowly will allow more carbon to 'de-couple' itself from the iron than will be the case where rapid cooling is employed.

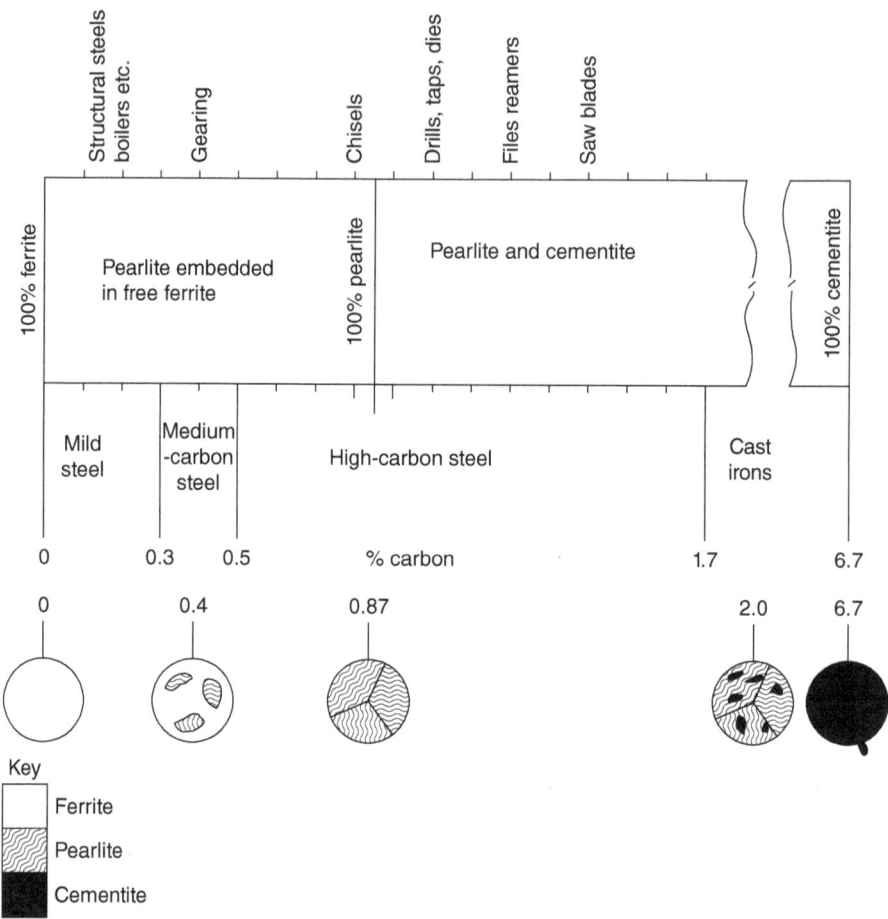

▲ **Figure 1.2** *Microstructure variation with increasing carbon content*

This compound of iron and carbon called iron carbide or cementite (Fe3C) lies side by side with ferrite in laminations to form a structure called pearlite, so called because of its mother-of-pearl appearance. As more carbon is added to the iron, more iron carbide and hence more pearlite is formed, with a reduction in the amount of free ferrite. When the carbon content is approximately 0.9% the free ferrite no longer exists and the whole structure is composed of pearlite alone. Further increases in carbon to the iron produces free iron carbide with pearlite reduction.

The steel range terminates at approximately 2% carbon content and the cast iron range commences. Carbon content for cast iron may vary from 2% to 4%. This carbon may be present in either the form of cementite or graphite (combined or free carbon) depending upon certain factors, one of which is the cooling rate. Grey or malleable cast iron is composed of pearlite and graphite and can be easily machined. Pearlite and cementite gives white cast iron, which is brittle and difficult to machine and hence is not normally encountered in Marine work. Figure 1.2 analyses the above in diagrammatic form.

Killed Steel

During the manufacturing process and while the steel is still in its liquid state, it ends up containing dissolved oxygen. The oxygen combines with some of the carbon to form bubbles of carbon monoxide that can then be seen coming to the surface of the steel. De-oxidants such as aluminium and/or silicon can be added to the mix to remove the oxygen and stop the bubbling effect.

The terms given to this type of steel are Killed, Semi-Killed and Rimmed. The name depends upon how much of the oxygen has been removed. The killed steel has virtually all the oxygen removed and is used where a high strength and resistance to impact is required. Semi-killed steels are used for general structural steel products and rimmed steel is used where a good surface finish is desirable, such as with the steel wheels on a motorcar.

The grades of steel common in shipbuilding are EH36, DH36 and AH36. These steels will come to the shipbuilder with a certificate to show that the steel has been tested to the class requirements. The 'Mill Test Certificate' (MTC) should show that the steel has also been inspected by people independent from the manufacturing process.

Properties of Materials

The choice of a material for use as an engineering component depends upon the conditions under which it will be employed. Conditions could be simple or complex and therefore in choosing a material the design engineer requires detailed knowledge of a material's mechanical properties as well as things such as electrical conductivity and magnetic properties. Increasingly, the materials selected will be a combination of different materials that when combined will result in material with properties better suited to its intended use than any of the parent materials. An alloy is the name given to a mixture of metals, for example, steel can be alloyed with nickel to give the resultant metal, stainless steel, anti-corrosive properties. The principal mechanical properties of most interest are given in Figure 1.3.

Other properties that may have to be considered depending upon the use of the material include: corrosion resistance, electrical conductivity, thermal conductivity and resistance to expansion when heated.

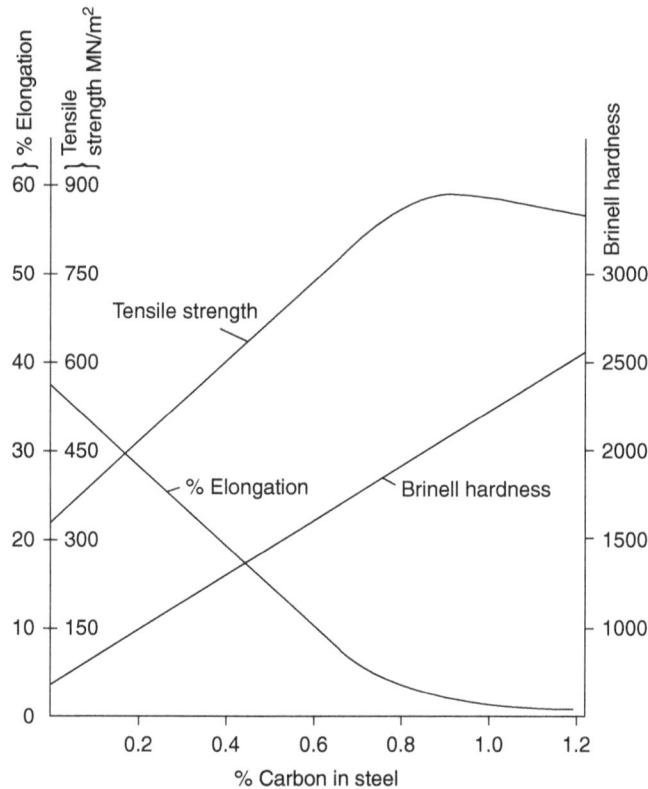

▲ **Figure 1.3** *Diagram showing effect upon mechanical properties by increase in carbon content*

When considering the suitability of a material for a particular task there is very often a compromise to be reached. There are also several factors to take into account when working out the properties, advantages and disadvantages of materials for particular components. For example, ship-side valve, safety valve spring, etc. At the design stage engineers should consider the following:

1. Working conditions for the component, for example will it be erosive, corrosive or have fatigue, stresses and/or thermal shock etc. imposed upon it.
2. Shape and method of manufacture, for example casting, forging, machining, drawing etc.
3. Repairability, for example can it be brazed, welded, metal-locked, easily replaced etc.
4. Cost and/or what is the return on investment.

Ductility	Is that property of a material that enables it to be drawn easily into wire form. The percentage elongation and contraction of area, as determined from a tensile test, are a good practical measure of ductility.
Brittleness	Could therefore be defined as lack of ductility (an extreme example is glass). A brittle metal does not have a 'plastic' region; it will deform slightly from the original shape but will then fracture.
Crack resistance	Is a measure of the metal's ability to remain intact when subjected to a high frequency of load fluctuations
Malleability	Is a property similar to ductility. If a material can be easily beaten or rolled into plate form it is said to be malleable.
Elasticity	This is a measure of a material's ability to return to the original dimensions following deformation by a load (one highly elastic material is rubber).
Plasticity	If none of the strain in a stressed material disappears upon removal of the stress the material is plastic.
Hardness	Is a measure of a material's resistance to deformation of its surface due to impact, erosion or wear.
Strength	The greater the load that can be carried the stronger the material.
Toughness	A material's ability to sustain variable load conditions without failure is a measure of a material's toughness or tenacity. Materials could be strong and yet brittle but a material that is tough has strength and resilience.

Therefore the answer for a ship-side valve, for example a sea water suction, would include the following:

1. Working conditions: corrosive, erosive, little variation in temperature, relatively low stresses, possibility of impact. Material required should be hard, corrosion resistant and with a relatively high impact value.
2. Shape and method of manufacture: relatively intricate shape, therefore the component would most probably be cast.
3. Material could be spheroidal graphitic cast iron, cast steel or phosphor bronze.
4. Increasingly expensive, easier to repair, increasing in corrosion resistance and impact value. In a well-designed valve, the internal parts might be easy to change but the proximity of the valve to the ship's side means that if the vessel is not in drydock then the hole on the outside would need to be blocked.

Testing of Materials

The designer will complete an array of ever increasingly complex calculations in order to determine the forces that will be acting upon all the different part of a ship. They will then want to choose different materials to fit their design criteria so that the ship will function as intended. A variety of destructive and non-destructive tests are carried out upon materials to determine their suitability for use in different engineering situations. The results are recorded and are often presented in tabular form. The most common tests are explained over the next few pages.

Tensile test

This test is carried out to ascertain the strength and ductility of a material.

A simple tensile testing machine is shown in Figure 1.4. The specimen is held in self-aligning grips and is subjected to a gradually increasing tensile load; the beam must be maintained in a floating condition by movement of the jockey weight as the oil pressure to the straining cylinder is increased. An extensometer fitted across the specimen gives extension readings as the load is applied. Modern, compact, tensile testing machines using mainly hydraulic pressure are more complex and difficult to reproduce for examination purposes. For this reason the authors have retained this simple machine. With values of load plotted against the corresponding extension, a nominal stress–strain curve can be drawn; the actual stress–strain curve is drawn for comparison purposes on the same diagram. The difference is due to the fact that the values of stress in the nominal diagram are calculated using the original cross-sectional area of the specimen when in actual fact the cross-sectioned area of the specimen is reducing as the specimen is extended.

Specimens may be round or rectangular in cross-section, the gauge length being formed by reducing the cross-section of the centre portion of the specimen. This reduction must be gradual as rapid changes of section can affect the result. The relation – gauge length to cross-sectional area of specimen – is important, otherwise varying values of percentage elongation may result for the same material. The following formula attempts to standardise this relationship in the United Kingdom:

Gauge length = 4√ Cross-sectional area

During the tensile test the specimen is broken, following this the broken ends are fitted together and the distance between reference marks and the smallest diameter are measured. Maximum load and load at yield are also determined. From these values the following can be calculated:

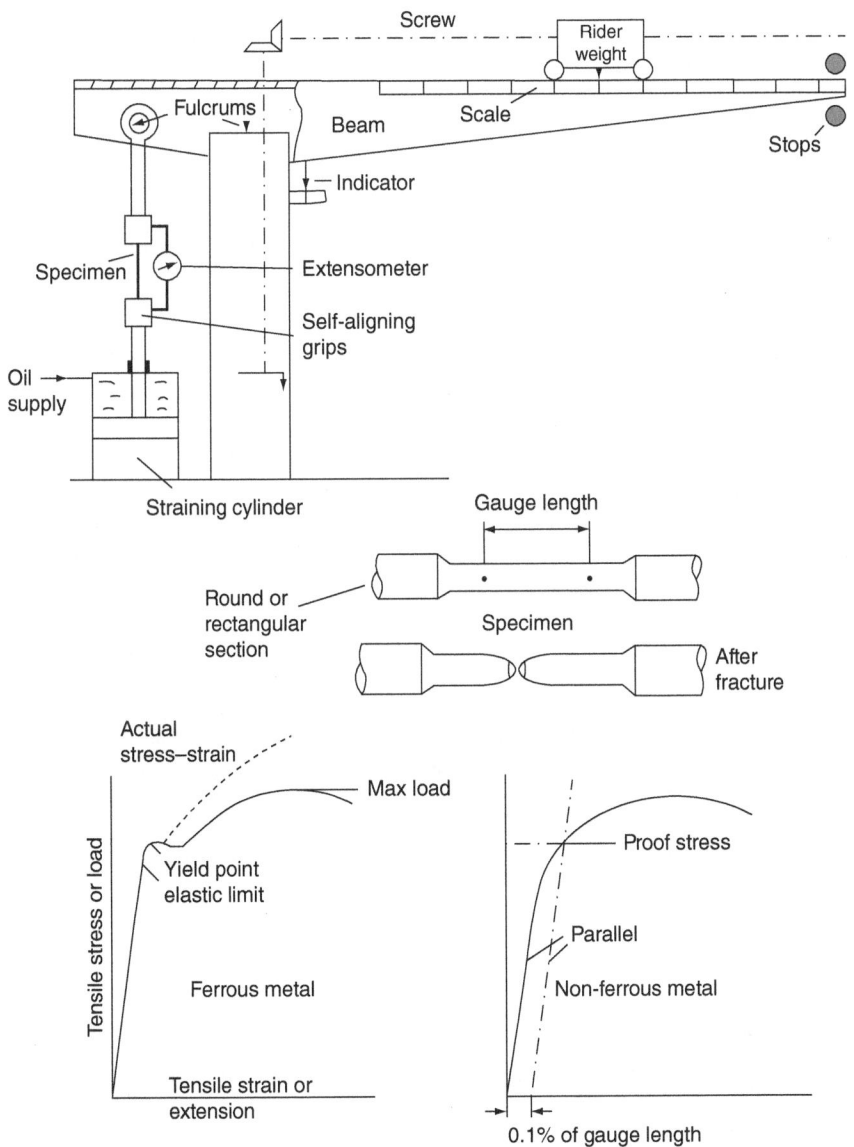

▲ **Figure 1.4** *Nominal stress–strain diagram*

$$\text{Percentage elongation} = \frac{\text{Final length} - \text{Original length}}{\text{Original length}} \times 100$$

$$\text{Percentage contraction of area} = \frac{\text{Original area} - \text{Final area}}{\text{Original area}} \times 100$$

$$\text{Ultimate tensile stress (UTS)} = \frac{\text{Maximum load}}{\text{Original cross-sectional area}}$$

$$\text{Yield stress} = \frac{\text{Yield load}}{\text{Original cross-sectional area}}$$

Percentage elongation and percentage contraction of area are measures of a material's ductility. UTS is a measure of a material's strength. Yield stress gives an indication of departure from an approximate linear relationship between stress and strain. It is the stress that will produce some permanent set in the material, for example, when water or fire tubes in a boiler are expanded into the holes in the face plate.

Factor of safety – this is defined as the ratio of working stress allowed to the ultimate stress, hence:

$$\text{Factor of safety} = \frac{\text{UTS}}{\text{Working stress}} \text{ and is always greater than unity}$$

Components that are subjected to fatigue and corrosion fatigue under normal working conditions are given higher factors of safety than those subjected to static loading, for example tail end shafts are at 12 or above, while boiler stays are about 7 to 8.

Hooke's law states that stress is proportional to strain if the material is stressed within the elastic limit.

$$\therefore \text{Stress} \propto \text{Strain}$$
$$\text{or Stress} = \text{Strain} \times \text{a constant}$$

The constant is given the symbol E and is called the Young's modulus or the modulus of elasticity.

$$\therefore \frac{\text{Stress}}{\text{Strain}} = E$$

The modulus of elasticity of a material is an indication of stiffness and resilience. As E increases then stiffness increases. By way of a simple explanation, we could consider two identical simply supported beams, one of cast iron the other of steel, each carrying a central load W. The deflection of a beam loaded in this way is given by

$$\delta = \frac{WL^3}{48EI}$$

where δ = deflection of beam under the load W, L = length of the beam, I = second moment of area of section and E = modulus of elasticity of the material.

Since the beams are identical $\delta \alpha \dfrac{1}{E}$

i.e. $\delta \times E$ = a constant

E for steel is greater than E for cast iron, hence, δ for steel is less than δ for cast iron. Therefore the steel is stiffer than cast iron and for this reason as well as strength, less steel is required in a structure than would be in the same structure built of cast iron.

0.1% Proof stress

For non-ferrous metals and some alloy steels no definite yield point is exhibited in a tensile test (see Figure 1.4). In this case the 0.1% proof stress may be used for purposes of comparison between metals. With reference to the graph (Figure 1.4), a point A is determined and a line AB is drawn parallel to the lower portion of the curve. Where this line AB cuts the curve, the stress at that point is read from the graph. This stress is called the 0.1% proof stress, that is, the stress required to give a permanent set of approximately 0.1% of the gauge length.

Hardness test

The hardness of a material determines its resistance to surface deformation due to impact or wear (Figure 1.5). There are several tests that can be employed to determine hardness, the most important of which are:

- Rockwell hardness test
- Brinell hardness test

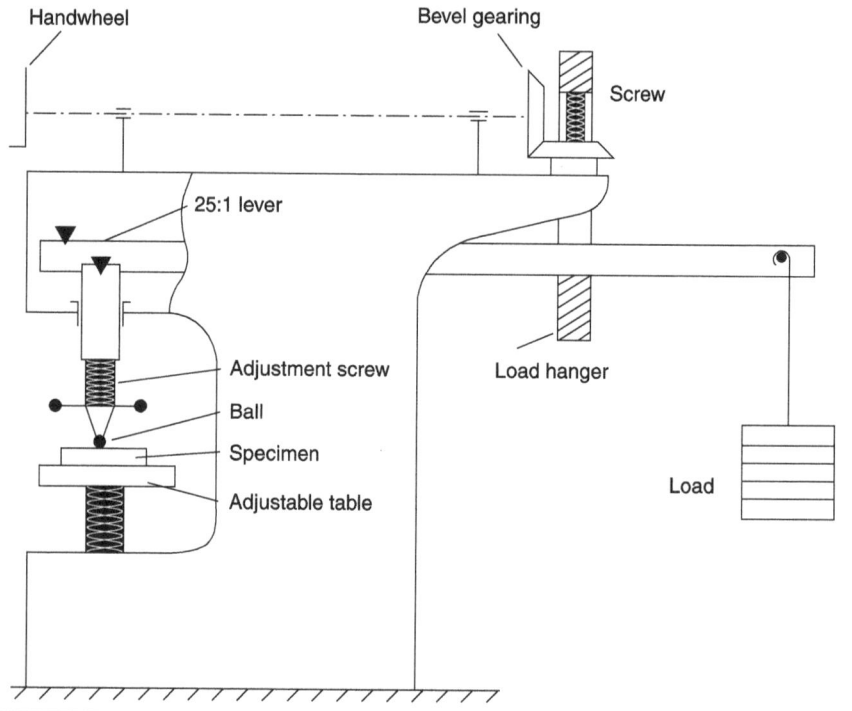

▲ **Figure 1.5** *Brinell hardness testing machine*

- Vickers hardness test
- Microhardness test

Brinell test: This test consists of indenting the surface of a metal by means of a 10mm diameter hardened steel ball under load. The Brinell number is a function of the load applied and the area of indentation, thus:

$$\text{Brinell number} = \frac{\text{Load in Newtons}}{\text{Area of indentation in mm}^2}$$

Only the diameter of the indentation is required and this is determined by a low-powered microscope with a sliding scale. Tables have been compiled to avoid unnecessary calculations in ascertaining the hardness numeral. Loads normally employed are 30,000 N for steels, 10,000 N for copper and brasses and 5,000 N for aluminium. Duration of application of the load is usually 15 seconds. (Industry is still using the old system of calculating Brinell numbers, that is, load in kilograms/area of indentation in mm². Hence, their Brinell numbers will be less by a factor of 10.)

Table 1.2 *Typical hardness values*

Material	Brinell number	VPN
Brass	600	600
Mild steel	1,300	1,300
Grey cast iron	2,000	2,050
White cast iron	4,150	4,370

Vickers hardness test: The surface of the metal under test is indented by a diamond square-based pyramid and the Vickers pyramid number (VPN) is determined by dividing the area of indentation into the load applied with the measurements being taken with a precision microscope. This test is also suitable for extremely hard materials, giving accurate results, whereas the Brinell test's reliability is doubtful above 6,000 Brinell. Table 1.2 gives some typical values.

The *microhardness test* uses the Vicker's testing procedure but uses highly accurate instruments for testing very hard metals.

Impact test

This test is useful for determining differences in materials due to heat treatment, working and casting, which would not be otherwise indicated by the tensile test. It does not give accurately a measure of a material's resistance to impact.

A notched test piece is gripped in a vice and is fractured by means of a swinging hammer (Figure 1.6). After the specimen is fractured the hammer arm engages with a pointer that is carried for the remainder of the swing of the arm. At the completion of the hammer's swing the pointer is disengaged and the reading indicated by the pointer is the energy given up by the hammer in fracturing the specimen. Usually three such tests are carried out upon the same specimen and the average energy to fracture is the impact value.

By notching the specimen the impact value is to some extent a measure of the material's notch brittleness or ability to retard crack propagation. From the practical standpoint this may be clarified to some extent: where changes of section occur in loaded materials (e.g. shafts, bolts, etc.) stress concentration occurs and the foregoing test measures the material's resistance to failure at these discontinuities. It also follows that an impact test reflects the ability of a material to absorb energy without fracturing. Materials that are tougher will have more ability to withstand an impact than more brittle materials. Think of the result of a sharp impact on glass and on rubber. The outcome will probably be very different in each case.

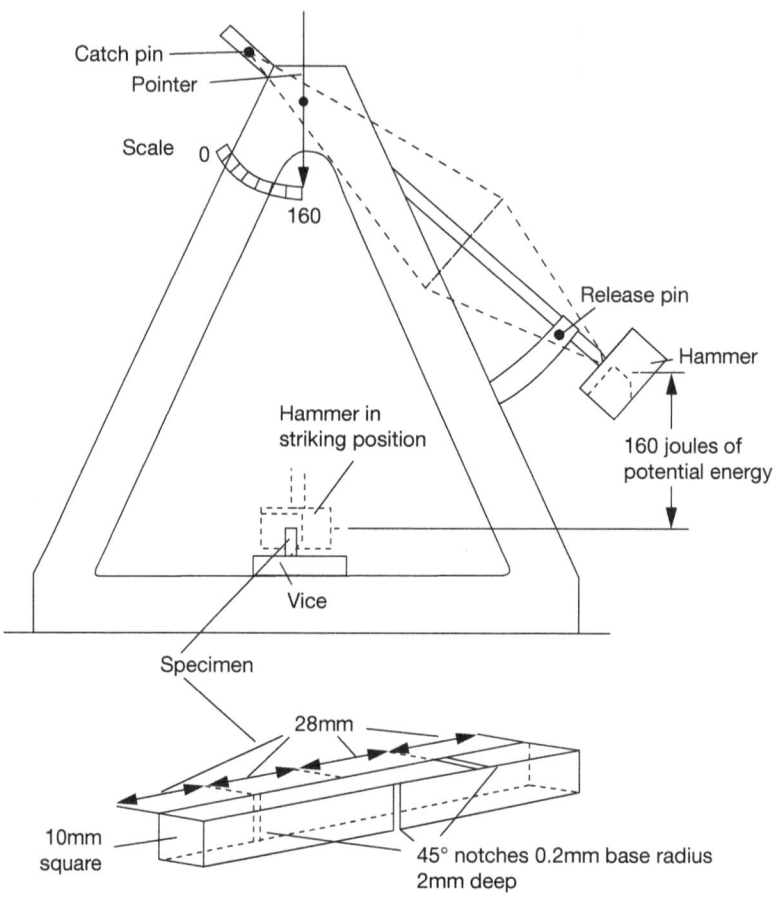

Catch pin
Pointer
Scale 0
160
Release pin
Hammer
Hammer in
striking position
160 joules of
potential energy
Vice
Specimen
28mm
10mm
square
45° notches 0.2mm base radius
2mm deep

▲ **Figure 1.6** *IZOD impact machine*

Table 1.3 gives some typical IZOD values for different materials; considerable variation in IZOD values can be achieved by suitable treatment and alteration in composition.

Charpy V Notch uses a different hammer-and-vice arrangement. The IZOD machine can be converted into a Charpy V Notch machine where the specimen is placed

Table 1.3 *Typical IZOD values*

Material	IZOD Value (Joules)	Uses
18/8 Stainless steel	136	Turbine blades
0.15 C, 0.5 Mn steel	54	General-purpose mild steel
Steel grey iron (annealed)	16	Camshafts, gear wheels
Grey cast iron	Up to 3	Cylinders, valves

horizontally upon two parallel stops between which the hammer swings and breaks the specimen.

The advantage to be gained by this method is that the specimens can be very quickly set up in the machine. Hence impact values for specimens at different temperatures can be accurately obtained. These tests were used extensively before the advent of 'fracture mechanics theory' but nowadays they are not used so much. *Brittle fracture* is a fracture in which there is no evidence of plastic deformation prior to failure. It can occur in steels whose temperature has been lowered, and the steel undergoes a transition. Figure 1.7 illustrates the considerable drop in impact value for mild steel as it passes through the transition range of temperature.

Factors that affect the transition temperature are as follows:

1. The elements – carbon, silicon, phosphorus and sulphur raise the temperature at which the fracture occurs and nickel and manganese lower the temperature.
2. Grain size – the smaller the grain size the lower the transition temperature, therefore grain refinement can be beneficial.
3. Work hardening – this appears to increase transition temperature.
4. Notches – possibly occurring during assembly, for example weld defects or machine marks. Notches can increase tendency to brittle fracture.

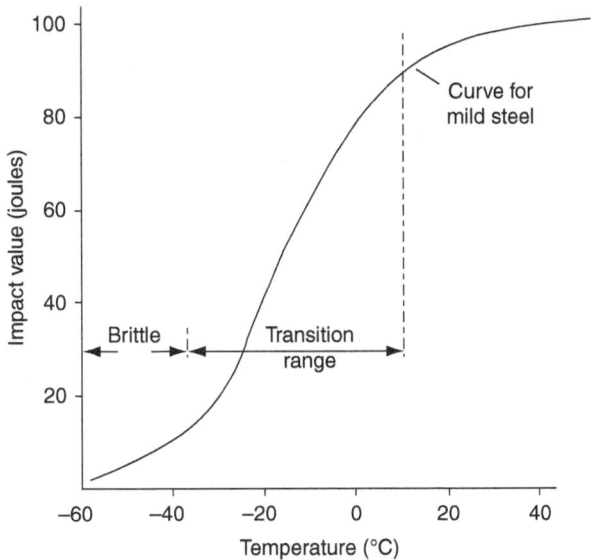

▲ **Figure 1.7** *Temperature-related impact values for mild steel*

Obviously transition temperature is an important factor in the choice of materials for the carriage of low-temperature cargoes, for example LPG and LNG carriers. A typical stainless steel used for containment would be as follows: 18.5% chrome, 10.7% nickel, 0.03% carbon, 0.75% silicon, 1.2% manganese UTS 560 MN/m^2, 50% elongation, Charpy V Notch 102 joules at $-196°C$. The issue has become important as the classification societies look for ways in which to arrest brittle cracking in Ultra Large container vessels by altering the design, especially that of the hatches.

Creep test

Creep may be defined as the slow plastic deformation of a material under a constant stress. A material may fail under creep conditions at a much lower stress and elongation than would be ascertained in a straight tensile test. Hence tests have to be conducted to determine a limiting creep stress with small creep rate.

The creep test consists of applying a fixed load to a test piece that is maintained at a uniform temperature. The test is a long-term one and a number of specimens of the same material are subjected to this test simultaneously, all at different stresses but at the same temperature. In this way the creep rate and limiting stress can be determined; these values depend upon how the material is going to be employed. Some permissible values are given in Table 1.4. Creep test results for all materials are given at working temperature.

Figure 1.8 shows a typical creep curve for a metal. To obtain the minimum uniform creep rate V (i.e. the slope of the line AB) it is necessary that the test be conducted for long enough in order to reach the second stage of creep. Hence, for a time t greater than that covered by the test, the total creep or plastic strain is given approximately by $\varepsilon_p = \varepsilon_0 + Vt$, where ε_p is the plastic strain that would be expected at the end of the first stage, this is important to the designer when considering tolerances, t is the time usually in hours.

Table 1.4 *Creep test results*

Component	Creep rate (m/mh)	Time of test (h)	Maximum strain
Turbine discs	10^{-9}	10^5	0.0001
Steam pipes, boiler tubes	10^{-7}	10^5	0.003
Superheater tubes	10^{-6}	20×10^3	0.02

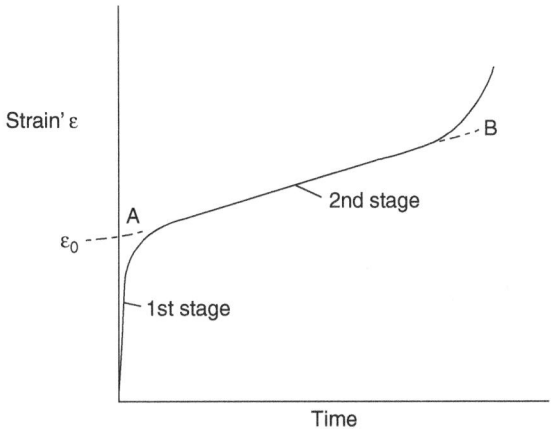

▲ **Figure 1.8** *Creep curve*

Fine-grained materials creep more readily than coarse-grained ones because of their greater amorphous metal content, that is, the structureless metal between the grains.

Fatigue test

Fatigue may be defined as the failure of a material due to repeatedly applied stress. The stress required to bring about such a failure may be much less than that required to break the material in a tensile test.

In this test a machine that can give a great number of stress reversals in a short duration of time is employed. The test is carried out on similar specimens of the same material at different stresses and the number of stress reversals to fracture is noted for each stress, normally 20 million reversals of stress would not be exceeded if failure did not occur. The results are plotted on a graph (Figure 1.9) from which a limiting fatigue stress (fatigue limit) can be ascertained. It is usual, since the number of stress reversals will be high, to condense the graph by taking logarithms of the stress and number of reversals to give a log S – log N curve.

Materials have varying fatigue limits. The limit can be increased by suitable treatment, use of alloy steels, etc. It can be reduced due to 'stress raisers' and changes of section, oil holes, fillets, etc. Environment alters the limit; if it is corrosive the limit could be reduced by about a third.

Figure 1.10 shows the different types of stress that a component could be subjected to in practice.

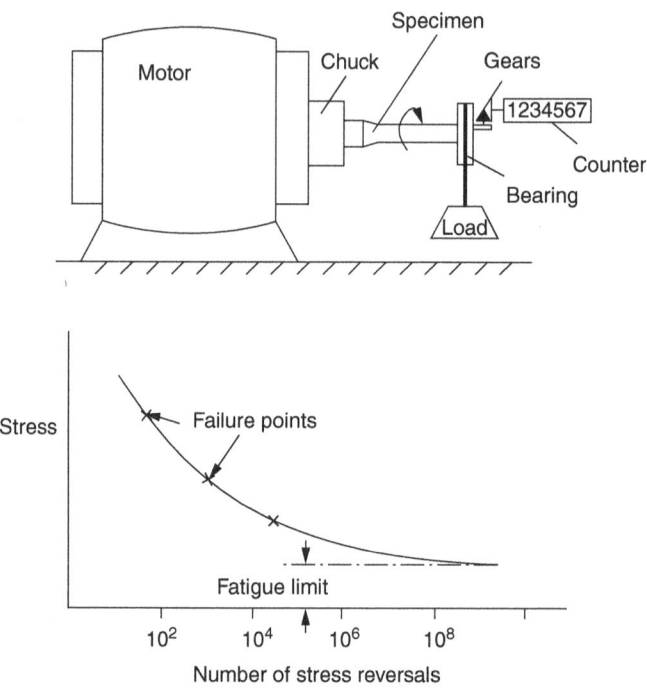

▲ **Figure 1.9** *Fatigue testing*

What is of greatest importance is undoubtedly the *range* of stress, which governs the life of the component. It has been found that if the range of stress passes through zero this can have the effect of lowering the life span for the same stress range.

A fatigue failure is normally easily recognisable one portion of the fracture will be discoloured and relatively smooth, while the other portion will be clean and also fibrous or crystalline depending upon the material. The former part of the fracture contains the origin point of failure, the latter is caused by the sudden failing of the material.

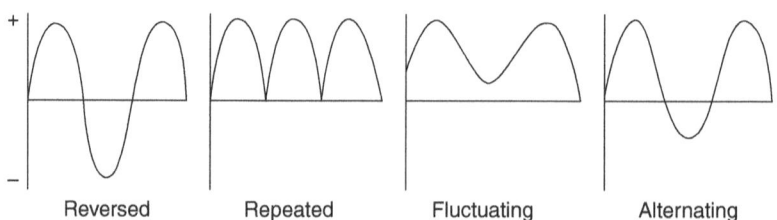

▲ **Figure 1.10** *Types of stress reversal*

Reversed stress	Stress range is symmetrical about zero stress line, for example propeller or centrifugal pump shaft.
Repeated stress	Component is stressed and then completely unloaded, for example gear teeth, cam.
Fluctuating stress	Component is stressed, either compressive or tensile, but stress range does not pass through zero, for example tie bolts, bottom end bolts.
Alternating stress	Stress range passes through zero stress line hence it changes from tensile to compressive, but is asymmetrical about the zero stress line, for example piston rod in double-acting engine or pump, crankweb in a diesel engine.

Bend test

This is a test that is carried out on boiler plate materials and consists of bending a straight plate specimen through 180° around a former. For the test to be satisfactory, no cracks should occur at the outer surface of the plate (see Figure 1.11).

Non-destructive tests

Apart from tests that are used to determine the dimensions and physical or mechanical characteristics of materials, the main non-destructive tests are those used to locate defects. This area of engineering is growing in importance as it becomes an integral

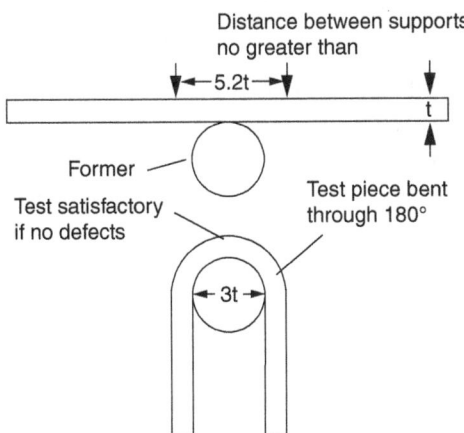

▲ **Figure 1.11** *Bend test*

part of the Condition Based Monitoring systems that are now being used to evaluate the condition of machinery plant life.

Methods of detecting surface defects

For example cracks.

1. A visual examination, including the use of a microscope, hand lens, borescope (inside machinery) and drones to identify possible cracking in places that are difficult to reach, such as at the top corners of very large cargo holds.

2. Liquid penetrant testing. Penetrant liquids must have a low viscosity in order to find their way into fine cracks. A dye is then used to highlight the liquid that is left in the fine cracks.

    (a) *Oil and whitewash*. This is one of the oldest and simplest of the penetrant tests. The oil is first applied to the metal and then the metal surface is wiped clean. Whitewash or chalk is then painted or dusted over the metal and oil remaining in the cracks will discolour the whitewash or chalk. Paraffin oil is frequently used because of its low viscosity and the component may be alternately stressed and unloaded to assist in bringing oil to the surface.

    (b) *Fluorescent penetrant*. This is wiped or sprayed over the metal surface, which is then washed, dried and inspected under near ultraviolet light. A developer may be used to act as a blotter to cause re-emergence of the penetrant, so that it can be iridised at the surface.

    (c) *Red dye penetrant*. This is probably the most popular of the penetrant methods because of its convenience. Three aerosol cans are supplied: red dye penetrant, cleaner and developer. Components must be thoroughly cleaned and degreased, then the red dye is applied by a spray. Excess dye is removed by hosing with a jet of water, or cleaner is sprayed on and then wiped off with a dry cloth. Finally, a thin coating of white developer is applied and when it is dry the component is examined for defects. The red dye stains the developer almost immediately but further indication of defects can develop after 30 minutes or more. Precautions that must be observed: (1) use of protective gloves, (2) use of aerosols only in well-ventilated places, (3) no naked lights, the developer is inflammable.

3. Electro-magnetic crack detection. A magnetic field is applied to the component under test, and wherever there is a surface or a subsurface defect, flux leakage will occur. Metallic powder applied to the surface of the component will accumulate at the defect to try to establish continuity of the magnetic field. This will also occur if there is anything non-metallic in the metal at or just below the surface. Magnetic flux leakage methods and eddy current testing are other methods that can be used.

Methods of detecting defects within a material

1. Suspend the component and strike it sharply with a hammer to hear if it rings true.

2. Radiography. This can be used for the examination of welds, forgings and castings: X-rays or γ-rays, which can penetrate up to 180 mm of steel, pass through the metal and impinge upon a photographic plate or paper to give a negative. Due to the variation in density of the metal, the absorption of the rays is non-uniform, hence giving a shadow picture of the material – it is like shining light through a semi-transparent material. X-rays produced in a Coolidge tube give quick results and a clear negative. Radioactive material (e.g. Cobalt 60), which emits γ-rays, does

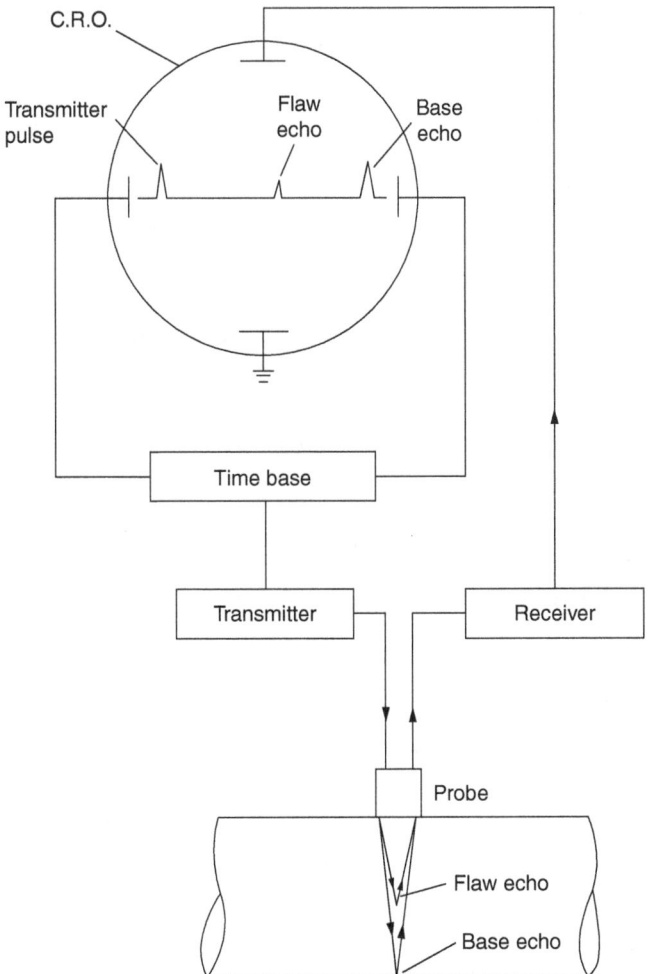

▲ **Figure 1.12** *Ultrasonic testing*

not give a picture as rapidly as the X-rays. However, to compensate for its slowness, it is a compact and simple system.

3.   Ultrasonics. With ultrasonics we do not have the limitations of metal thickness to consider as we have with radiographic testing, high-frequency sound waves reflect from internal interfaces of good metal and defects, and these reflected sound waves are then displayed on to the screen of a cathode ray oscilloscope. Size and position of a defect can be ascertained, and it can also be used for checking material thicknesses, for example, a probe could be passed down a heat exchanger tube (see Figure 1.12).

A portable, battery-operated, hand-held cylindrical detector with a cable to a set of headphones can be used to detect leakages, for example in a vacuum, air lines, superheated steam, air conditioning, etc. A recent application of ultrasonics is testing condensers.

A generator placed inside the condenser 'floods' it with ultrasound. By using a head set and probe, tube leakage can be homed in on. Where a pinhole exists sound 'leaks' through and where a tube is thinned it vibrates like a diaphragm transmitting the sound through the tube wall.

Thermography is where a temperature 'map' of a structure or a set of connections is examined. In the case of electrical equipment the current situation can also be compared with the previous readings to look for defects. Defects in the furnace walls of refractory can also be examined by using the same techniques. Recording the thermal signature of electrical junctions is now a much more common form of testing for developing defects in electrical equipment.

Treatment of Metals

Hardening and tempering

In the process of converting ice into dry, saturated steam by supplying heat, two distinct changes of state occur: from solid to liquid and from liquid to a vapour. When iron is heated up to its melting point two similar events occur where there is heat absorption but no change in temperature. The temperatures at which these events occur are called 'critical points' and these are of great importance. At these critical points, considerable changes of internal structure take place and therefore different physical processes and changes to the original material are possible if these opportunities are created by different techniques.

With steels, these changes in the internal structure of the iron at the critical points also affect the carbon which is present in the form of iron carbide. At the upper critical temperature range 720–900°C in the solid state (the range is due to the variable carbon content), the iron structure formed has the ability to dissolve the iron carbide into solution, forming a new structure. If at this stage the steel is suddenly quenched in water the iron carbide will remain in solution in the iron, but the iron's structure will have reverted to its original form. This completely new structure, which has been brought about by heating and then rapidly cooling the steel, is called 'Martensite' – a hard, needle-like structure consisting of iron supersaturated with carbon, which is basically responsible for hardening steels.

If steel of approximately 0.4% carbon content were heated to a temperature above its upper critical (about 800°C see Figure 1.13) and was then suddenly cooled by quenching, its Brinell hardness numeral would be increased from approximately 2,000 to 6,000. In this condition the steel would be fully hardened, that is, fully martensitic. Choosing a temperature lower than the above but not lower than 720°C (lower critical) and then quenching will produce a partly hardened steel having a Brinell numeral between 2,000 and 6,000.

Hardening material in this way produces internal stresses and also makes the material brittle. To relieve the stresses and restore ductility without loss of hardness or toughness, the material is tempered.

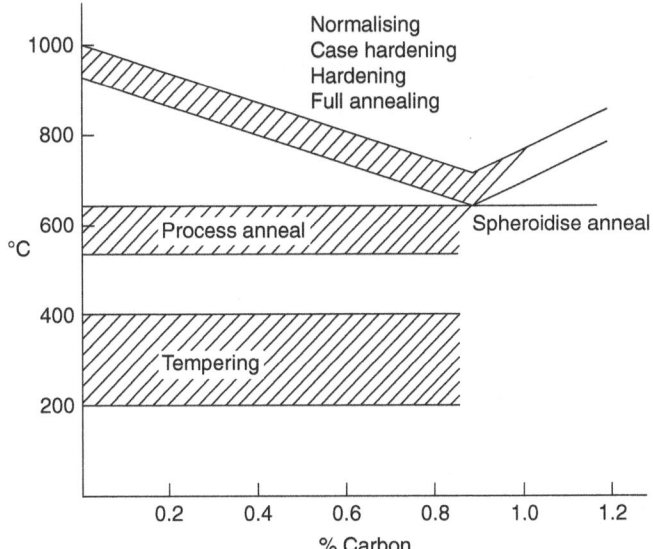

▲ **Figure 1.13** *Treatment diagram*

Tempering consists of heating the material to about 250°C, retaining this temperature for a duration of time (this depends upon the mass and the degree of toughness required) and then quenching or cooling in air.

The combination of hardening and tempering is frequently employed with steels and alloy steels, and a wide range of properties is available thereby. Components such as drills, chisels, punches, saws, reamers and other tools are invariably subjected to the above process.

Straight carbon steels whose carbon content is below 0.2% are *not* usually subjected to hardening and tempering processes. The reason could be attributed to the smaller quantity of Martensite that would be produced.

Annealing and normalising

The object of annealing is either to grain refine, induce ductility, stress relieve or a combination of these. Castings, forgings, sheets, wires and welded materials can be subjected to an annealing process. This process consists of heating the material to a predetermined temperature, possibly allowing it to soak at this temperature, then cooling it in the furnace at a controlled rate. For full annealing and normalising, the temperature for carbon steels is usually 30–40°C above the upper critical temperature. Essentially, the difference between full annealing and normalising is that in the case of the annealing process the material is cooled slowly in the furnace whereas for normalising the material is cooled in still air out of the furnace.

These processes of full annealing and normalising are mainly used on castings since they will usually have variation in size and shape of grain. The casting is heated to about 40°C above the upper critical temperature and held at this temperature until it is uniform in temperature throughout. Then it is cooled. This produces a uniform, grained structure (re-crystallisation temperature about 500°C) with increased ductility, for example, a 0.5% carbon steel casting could have its percentage elongation increased from 18% in the as-cast condition to about 35% in the annealed condition.

The more rapid cooling of the casting that occurs with normalising gives a better, closer-grained, surface finish.

If steel has been cold worked in manufacture the ferrite grains and the pearlite are distorted. Recrystallisation of the ferrite and increased ductility is brought about by employing an annealing procedure known as process annealing. This is similar in most respects to full annealing except that it is conducted at a temperature between 500°C and 650°C, which is below the lower critical temperature.

Blackheart process

For high-carbon castings, for example, 2.5% C content, the Blackheart process may be used to produce a softer, ductile and more easily machined component that would be similar mechanically to cast steel.

The castings are placed in air tight (to prevent burning), heat-resistant metal containers and heated up to 1,000°C. They are kept at this annealing temperature for up to 160 hours or so depending upon material analysis. The prolonged heating causes breakdown of the cementite to give finely divided 'temper carbon' in a matrix of ferrite, which has a black appearance – hence Blackheart.

Work hardening

If a metal is cold worked it can develop a surface hardness. Shot blasting is a method of producing surface hardness that consists of hitting the surface of a component with many hardened steel balls. Expansion and contraction of copper piping, used for steam, can also lead to a hardness and brittleness that has to be removed by annealing. Lifting tackle such as shackles, chains, etc., can develop surface hardness and brittleness due to cold working, hence they have to be annealed at regular intervals (as laid down by the MCA working in close co-ordination with the Health and Safety Executive).

What actually happens is that the work forces cause dislocations to be set up in the crystal latticework (i.e. the geometric arrangement of the metal atoms) of the metal, and in order to remove these dislocations considerable force is required. This considerable force is the evidence of work hardening, since it is the force necessary to dent the surface of the material.

Case hardening

This is sometimes referred to as 'pack carburising'. The steel component to be case hardened is packed in a box, which may be made of fire clay, cast iron or a heat-resisting nickel–iron alloy. Carbon-rich material such as charred leather, charcoal, crushed bone and horn or other material containing carbon is the packing medium, which would encompass the component. The box is then placed in a furnace and raised in temperature to above 900°C. The surface of the component will then absorb carbon forming an extremely hard case. Depth of case depends upon two main factors: the length of time and the carbonaceous material employed. Actual case depth with this process may vary between 0.8 mm and 3 mm requiring between 2 and 12 hours to achieve, for these limits.

Gudgeon pins and other bearing pins are examples of components that may be case hardened. They would possess a hard outer case with good wearing resistance and a relatively soft inner core that retains the ductility and toughness necessary for such components.

Nitriding

In this process the steel component is placed in a gas-tight container through which ammonia gas (NH_3) is circulated. Container and component are then raised in temperature to approximately 500°C. Nitrides are then formed in the material, at and close to the surface, which considerably increases the surface hardness. A nitride is an element combined with nitrogen; usually nitride promoting elements are present in the steel such as aluminium, chromium, vanadium or molybdenum. Actual depth of the hard case is not as great with this process as compared to case hardening: 0.0125–0.05 mm in 5–24 hours for nitriding, compared with 0.8–3mm in 2–12 hours for case hardening. An essential difference from case hardening is the more gradual changeover from the hardened to unhardened section of the metal, thus reducing the risk of the two dissimilar metals parting company.

Flame hardening

This process is used for increasing the surface hardness of cast irons, steels, alloy cast irons and alloy steels. With the increase in surface hardness there is a high improvement in wear resistance.

To flame harden a component (e.g. gear teeth), an oxy-acetylene torch is used to preheat the surface of the metal to a temperature between 800°C and 850°C. A water spray closely following the oxy-acetylene torch quenches the material thereby inducing hardness. Care in the operation of this process is essential, and overheating must be prevented.

Induction hardening

This is a method of surface hardening steels by the use of electrical energy. In Figure 1.14, a high-frequency AC electromagnetic field is shown heating up the surface of the components to be hardened by hysteresis and eddy currents, after surface heating is quenched.

Hysteresis loss is heat energy loss caused by the steel molecules behaving like tiny magnets that are reluctant to change their direction or position with each alteration of electrical supply thus creating molecular friction.

Eddy currents are secondary electrical currents caused by the presence of a primary current nearby. The resistance of the steel molecules to the passage of eddy currents generates heat.

Important points regarding induction hardening are as follows:

1. Time of application of electrical power governs depth to which heat will penetrate.
2. It reduces time of surface hardening to seconds – that is, the process is *very fast*.
3. Rapid heating and cooling produces a fine-grained martensitic structure.
4. Due to speed of operation no grain growth or surface decarburization occurs.
5. There is no sharp division between case and core.

Components that are induction hardened include gudgeon pins and gear pinions.

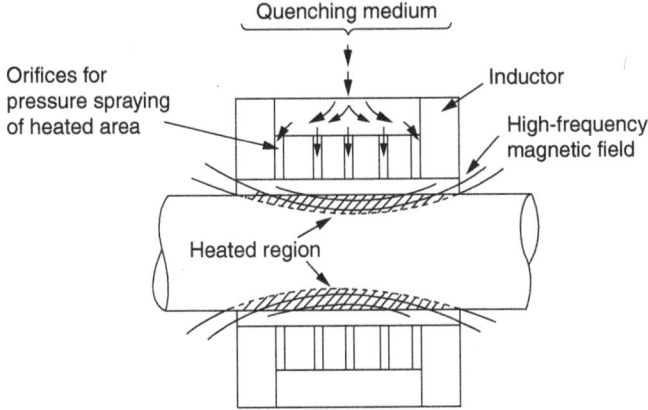

▲ **Figure 1.14** *Induction hardening*

Spheroidising anneal

Spheroidising of steels is accomplished by heating the steel to a temperature between 650°C and 700°C (below lower critical line) when the pearlitic cementite will become globular. This process is employed to soften tool steels so that they may be easily drawn and machined. After shaping, the material is heated for hardening and the globules or spheres of cementite will be dissolved. Refining of the material prior to spheroidising may be resorted to in order to produce smaller *globules*.

Forming of Metals

Sand casting

A mould is formed in high-refractory sand by a wooden pattern whose dimensions are slightly greater than the casting to allow for shrinkage. To ensure a sound casting the risers have to be carefully positioned to give good ventilation.

Defects that can occur in castings are as follows:

1. Shrinkage cavities.
2. Blowholes caused by ineffective venting and dissolved gases in steels that have not been killed (de-oxidised – see page 9).
3. Oxidation.
4. Impurities.

Sand casting is slow and expensive and would only be used if the metal and casting shape are unsuitable for other techniques. However, sand is useful as it is resistant to change when it comes into contact with the high temperatures of the molten metal. Moisture content of the sand is key – too much and the steam produced could affect the casting.

Traditionally the pattern for a mould was made from wood. Modern materials such as polyurethane can be used to make solid objects that can be placed in the sand to form the final shape of the casting. Upon completion of the casting the sand is removed (Figure 1.15) Rapid prototyping together with 3D printers can be used to produce patterns quickly and therefore this process is suited to low-volume production runs for a specific casting.

A shape punched into an aluminium plate can be used for automated 'long run' productions of castings using the 'Shell Mould Casting' technique.

Die casting

Used mainly for aluminium and zinc base alloys. The molten metal is either poured in under gravity or high pressure – hence gravity and pressure die casting. This process gives a fine-grained uniform structure meaning that complicated shapes can be cast and the mould can be re-used. The process lends itself to a high degree of automation making it more suitable for high-volume production runs (Figure 1.15).

Centrifugal casting

A metal cylindrical mould is rotated at speed about its axis, in either the horizontal or vertical plane, and molten metal is poured in. Centrifugal action throws the molten metal radially out on to the inner surface of the mould to produce a uniform, close-grained – due to chilling effect of mould – non-porous cylinder.

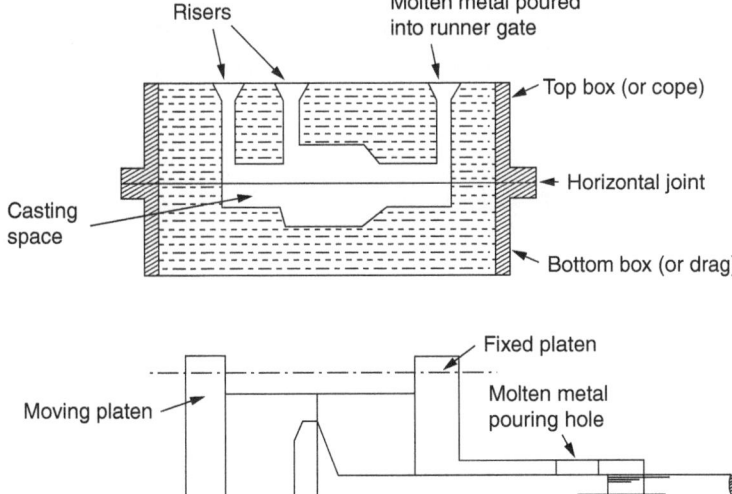

▲ **Figure 1.15** *Die casting*

Such a casting process can be used for pipes and tubes, cylinder liners and some piston rings that are cut from a machined centrifugally cast cylinder. Where the final product is a solid, such as a pulley or the wheel of a gantry crane, the process is known as Semi-Centrifugal Casting. This is due to the difference in the force at the centre of the rotating disc and the force at the edge of the disc.

Forging

This is the working and shaping of hot, or cold, metal by mechanical or hand processes. Smaller sections of metal are hammered by hand and shaped with tools called swages. During the process the original course, as cast, structure of the metal is broken down to form a finer-grained structure with the impurities distributed into a fibrous form.

Items that are forged using heavy machines include connecting rods, crankshafts, upset ends of shafts and boiler stays, etc.

Cold working

Is the cutting, bending, squeezing and pulling of metal through dies to form different shapes such as wires and tubes, cold rolling of plate, expansion of the ends of tubes in boilers and heat exchangers so that they form a tight fit in the end plates, caulking of plates.

Elements in Irons and Steels

The following normally occur naturally in the iron ore from which the steels etc., are originally made.

Manganese

This element, which is found in most commercial irons and steels, is used as an alloying agent to produce steels with improved mechanical properties. Manganese is partly dissolved in the iron and partly combines with the cementite. If the manganese content is high enough, martensite, with its attendant hardness and brittleness, will be formed in the steel even if the steel is slow cooled. For this reason the manganese content will not normally exceed 1.8% although one heat-treated steel known as Hadfields manganese steel contains 12–14% manganese.

Silicon

Tends to prevent the formation of cementite and produce graphite. In steels it increases strength and hardness but reduces ductility. As a graphitiser it is useful in cast irons, tending to prevent the formation of white cast iron and forms instead graphitic cast iron. The quantity of silicon in an iron or steel may vary between 0.5% and 3.5%.

Sulphur

Reduces strength and increases brittleness. It can cause 'hot shortness', that is, it is liable to crack when hot. Normally the sulphur content in finished iron or steel does not exceed 0.1%.

Phosphorus

This also causes brittleness and reduction of strength but it increases fluidity and reduces shrinkage, which are important factors when casting steels and irons. It can produce 'cold shortness', that is, liable to crack when cold worked. Normally the phosphorus content does not exceed 0.3%.

Effect of Alloying Elements

Nickel

This element increases strength and erosion resistance. It does not greatly reduce ductility until 8% nickel is reached. A low- to medium-carbon steel with 3–3.75% nickel content is used for connecting rods, piston and pump rods, etc. Nickel forms a finer-grained material.

Chromium

Increases grain size induces hardness and improves resistance to erosion and corrosion. This element is frequently combined with nickel to produce stainless steels and irons that are used for such items as turbine blades, pump rods and valves.

Molybdenum

Used to increase strength, especially employed for increasing strength at high temperatures, which is one reason why it is used for superheater tubes, turbine rotors, etc. Another reason for its use is its action in removing the possibility of embrittlement occurring in those steels that are prone to it, for example, nickel–chrome steels.

Vanadium

Increases strength and fatigue resistance. Used in conjunction with molybdenum for boiler tube materials.

Other alloying elements include tungsten, which induces self-hardening properties and is used for heat-resisting steels, for example: machine tools; copper that improves corrosion resistance; and cobalt that is used as a bond in stellite alloys. Manganese and silicon, discussed earlier, are also employed as alloying agents.

Non-ferrous Metals

Copper

This material is used extensively for electrical fittings as it has good electrical conduction properties. It is also used as the basis for many alloys and as an alloying agent. If copper is cold worked, its strength and brittleness will increase, but some restoration of ductility can be achieved by annealing. Hence, in this way, a wide range of physical properties are available.

Brass

Brasses are basically an alloy of copper and zinc, usually with a predominance of copper. When brasses are in contact with corrosive conditions, for example atmospheric salt or in salt water, they may dezincify (removal of the zinc phase) leaving a porous, spongy mass of copper. To prevent dezincification, an inhibitor is added to the brass. One such inhibitor is arsenic, of which only a small proportion is employed. Brasses have numerous uses, decorative and purposeful. Marine uses include: valves, bearings, condenser tubes, etc. Alloying elements such as tin, aluminium and nickel are frequently employed to improve brasses. With these elements the strength and erosion resistance of brasses can be greatly improved.

Bronze

Bronze is basically an alloy of copper and tin, but the term bronze is frequently used today to indicate a superior type of brass. It resists the corrosive effect of sea water, has considerable resistance to wear, and is used for these reasons for many marine fittings. With the addition of other alloying elements, its range of uses becomes extensive. Manganese in small amounts increases erosion resistance and forms manganese bronze (propeller brass). Phosphorus, used as a deoxidiser, prevents formation of troublesome tin oxides, improves strength and resistance to corrosion, and provides an excellent hard, glassy bearing surface. Aluminium and zinc give aluminium bronze and gunmetal respectively, which are suitable materials for casting.

Aluminium

This material is progressively supplanting other materials in use for specific items in the marine industry. It resists atmospheric corrosion and its specific gravity is about one-third that of steel. In the pure state its strength is low, but by alloying and by mechanical and thermal treatment its strength can be raised to equal and even surpass that of steel without great loss of ductility. In this form it is used extensively for structural work.

Copper–nickel Alloys

Cupro-nickel alloys have considerable strength, and resistance to corrosion and erosion. The 80/20 or 70/30 cupro-nickels are used for condenser tubes as they strongly resist the attack of estuarine and sea waters. Monel metal is a well-known alloy composed of approximately two-thirds nickel with the remainder being principally copper. It is used for turbine blades, pump rods and impellers, scavenge valves and superheated steam valves. Monel metal retains its high strength at high temperatures. With the addition of 2–4% aluminium, forming a material known as 'K' monel, it can be temper-hardened, thus its strength can be increased still further without detracting from its other properties.

White metals

White metal bearing alloys may be either tin or lead-based materials containing antimony and copper or antimony alone. Tin-based white metals are sometimes referred to as 'Babbitt metals', after Sir Isaac Babbitt who patented them originally. These metals are the most commonly used of the white metals because of (1) their good bearing surface and (2) their uniform microstructure.

The use of copper in a white metal ensures uniform distribution of the hard cuboids of the inter-metallic compound of antimony and tin within the soft, tin-rich matrix. Coefficient of friction for a white metal bearing when lubricated is approximately 0.002. The melting point of white metal varies with composition but is approximately between 200°C and 300°C (see Table 1.5).

Modern engine design incorporates the 'Thin-Shell' type bearings. These are a thin-walled steel shell lined with a thin layer of 'white metal'. This design gives superior performance

Table 1.5 *Composition of white metal*

Composition %				Uses
Tin	Antimony	Copper	Lead	
86	8.5	5.5	–	Heavy duty, high temperatures
78	13	6	3	Normal loading
–	20	–	80	Normal to low load bearings (magnolia metal relatively cheap)

over the thick white metal bearings of old. The steel gives strength to the shell and a layer of copper between the steel and the white metal improves the heat flow, enabling the whole structure to carry more load and run cooler than the older designs.

Titanium

Ideal where resistance to erosion and impingement-corrosion are the more important requirements. It is virtually completely resistant to corrosion in sea water, and only under exceptional conditions of erosion would the protective oxide film be damaged. When alloyed with about 2% copper a moderate increase in strength results. Used in heat exchangers, usually of the plate variety.

Non-metallic Materials

Plastics (polymers)

Most are organic materials, synthetic and natural, consisting of combinations of carbon with hydrogen, oxygen, nitrogen and other substances. Dyes and fillers can be added to give colour and alter properties. Some of the fillers used are glass fibre for strength, asbestos fibre to improve heat resistance, and mica for reducing electrical conductivity. Polymers can be plastic, rigid or semi-rigid, or elastomeric (rubber-like).

These are some of their general properties. (1) Good thermal resistance. Most can be blown to give cellular materials of low density, which is useful for thermal insulation and also stops the spread of fire. (2) Good electrical resistance. (3) Unsuitable for high temperatures. Since they are hydro-carbons they will contribute to fires producing smoke and possibly toxic fumes (PVC releases hydrogen chloride gas). (4) Good corrosion resistance.

Some polymers and other materials in common use are as follows.

Nitrile. Used in place of rubber, unaffected by water, paraffin, gas oil and mineral lubricating oil. Can be used for tyres in hydraulic systems (see Pilgrim nut) anti-vibration mountings, jointing etc.

PTFE. Unaffected by dry steam, water, oils and a considerable range of chemicals. Low friction, used for water-lubricated bearings, gland rings, jointing tape etc.

Expoxy resin. Pourable epoxy resin that cures at room temperature is unaffected by sea water and oils etc. It is extremely tough, solid and durable and is used for chocking engines, winches, pumps, etc. Hence no machining of base plates or foundations, simplified alignment retention, reduced time and cost.

Rubber. Attacked by oils and steam, unaffected by water. Used for fresh and salt water pipe joints and water-lubricated bearings. In a highly vulcanised state it is called ebonite, which is used for bucket rings in feed pumps.

Asbestos. Unaffected by steam, petrol, paraffin, fuel oils and lubricants. In the presence of water it needs a waterproof binder. This used to be a near universal jointing and packing material. However, the severe carcinogenic safety hazard means that no asbestos is used in modern ships, and any older vessels that had asbestos fitted should now have had it removed.

Cotton. Unaffected by water and oils, used as a framework to give strength to rubber and produce rubber insertion jointing, also used in packing.

Silicon nitride. Used as seals in place of bronze wear rings in sea water pumps. UTS 700 MN/m^2, greatly resistant to erosion, inert chemically and galvanically (latter is important in salt water pumps).

No attempt has been made to cover all the materials used in marine engineering as the range is wide and complex, but materials for components discussed elsewhere in the book, not covered in this chapter, will be dealt with as necessary at the appropriate point.

Composites

When two or more different materials combine to form a material superior to the original, they are known as composites. This generic term has now become synonymous with glass or carbon fibre, where the fibres carry the bulk of the load and the resin is there to bind the fibres together. The different types of fibre can be grouped into organic, inorganic and metallic.

The manufacturing techniques are important when the raw material is turned into an actual object. The resin provides chemical and thermal properties, gives chemical resistance and transfers the load to and between the fibres.

The reinforcing fibre determines the mechanical properties, provides mechanical strength, disperses the load and gives dimensional stability to the completed structure. The fibre can also be cut or chopped, continuous, or continuous and inter-woven; each will provide different strength characteristics.

The final raw product can then be made into components and objects ranging from parts of cars and aeroplanes to marine pump impellers and strum boxes. A range of techniques are used from compression moulding and vacuum-induced moulding to machining solid blocks.

The manufacture of boats and yachts with hulls made from GRP or composite material, up to about 50–60 m, is now a regular occurrence, with a limited number being built in excess of this length.

The technique known as vacuum infusion manufacturing has greatly assisted the ability of manufacturers to build larger structures. The strength of any material relates to the consistency of the internal chemical composition. With molten metals, they are carefully tested so that the process controllers know when the chemistry inside of a furnace is correct.

When manufacturing GRP or other composite materials, achieving an even chemical make-up to give a structure that is the same throughout is more difficult. The quality control of each step is very important and even then there could be slight differences in the overall make-up of a hull, for example.

The vacuum infusion technique makes use of the idea that if the matted fibres are positioned in the correct place within a mould, and then the mould is placed in a sealed bag, the whole structure can then be placed under a vacuum. When the resin is introduced through small tubes it will then make its way into all of the very small gaps and holes between the fibres.

This process makes for a product that is much more consistent throughout the structure. There is much less chance of air entering with the resin then there is if the fibre layers are laid out and the resin applied by hand. The process still needs very careful supervision but it is much cleaner and safer for the controllers.

Welding

Welding processes may be thought about as being divided into two main groups: pressure welding and non-pressure welding.

Any welding process that requires pressure, to join the metal together, is generally referred to as a forge welding process and these processes do not usually require a filler metal or flux as the original metal in the different parts are being forced to fuse together with the pressure to which they are being subjected. The parts to be welded, however, should be clean and free from grease, etc.

The oldest form of forge welding was completed within the blacksmith's forge. The process consists of heating the metal components to be welded in a blacksmith's fire until the parts to be united are plastic; then the parts of the components are removed from the heat source and hammered together to form a union.

Resistance welding is a more modern example of another forge welding process. Current and pressure are supplied to the parts being welded but no filler metal or flux is required. The heat that is generated in order to form the weld depends upon (1) the square of the current supplied, (2) the metal to be welded and the contact resistance, (3) the time of application of current and pressure. Examples of resistance welding are studs welded to decks or to boiler tubes in water tube boilers (Table 1.6).

Welding processes that do not require any pressure are often referred to as fusion welding processes. Fusion welding processes require a filler metal and often a flux is used. The most popular and most convenient form of fusion welding is the electric arc welding process, sometimes called the manual metal arc (MMA) welding process.

The application of welding to shipbuilding is almost entirely restricted to fusion welding especially in the form of metallic arc welding, and on board the ship gas welding and cutting is also a very important and useful maintenance and repair tool.

Welding safety

The safety of the staff on board modern ships is obviously very important, especially as some of the activities can be potentially hazardous. Welding is one of those activities that can go very wrong and cause major problems if precautions are not taken by knowledgeable staff.

Any proposed work must be discussed at the safety briefing and the necessary 'permits to work' must also be issued. The ship's safety management system is also available to guide staff on the specific company procedures necessary.

General

The most obvious hazard, which is common across the different welding techniques, is due to heat and sparks. Students will be able to see that these have the ability to start a fire if action is not taken to guard against them.

It is fairly obvious that all combustible material must be removed from the area surrounding the 'hot work' and fire extinguishing equipment kept ready nearby. However, it is less obvious to check the other side of metal bulkheads and deck heads especially if welding is being undertaken close to one of these features. Fires have also been started by sparks falling into the bilges where oil could be present.

People also become complacent about wearing protective clothing; however, when welding, at all times, protective clothing is important. In this case protection is needed against the heat causing burns, which means protective gloves and possibly additional heat-resistant clothing over ordinary work clothes.

The welding process gives off fumes, the composition of which depends upon the method of welding taking place, the filler rods being used and the composition of the metals being welded. Nitrous oxide, for example, is one of the main gases present in welding fumes. The fumes are released as the filler rod and the flux coating are vaporised by the electric arc. The exact composition of the rods or wire should be available on the product information sheets that accompany the products when they are delivered to the ship.

The important rule for any welding is to complete the process in a well-ventilated area. If this is not possible, such as inside a small tank, then the welder must be provided with suitable fresh air breathing equipment or Self Contained Breathing Apparatus (SCBA).

The welding equipment keeps out oxygen as much as possible by surrounding the melting metal with a form of protective coating. When the welded metals are cooling the protection forms a hard, brittle outer shell. When this shell, called slag, is removed, by using a chipping hammer, there is a possible danger of sharp pieces flying towards the eyes and causing harm if they are unprotected. The eye protection for this task involves using 'clear' goggles as opposed to the darkened glasses that are used for the actual welding.

Safety specific to the MMA welding process

The ultraviolet light given off by the manual metal arc (MMA) welding process is much more intense than with the gas welding process, and therefore it is vital to use the correct standard of darkened glass in the welding mask. It is also very important to check that the eye protection is in good condition: with no damage, carries the correct international quality standard and its use is understood by the person undertaking the welding. It's important to remember that not only is the welder vulnerable to eye damage from arc welding, but people passing by are also susceptible if they happen to look at the electrode just as the welder strikes an arc.

Electric shock is also a danger when using arc welding equipment. It is important that the equipment and surrounding areas be kept dry.

Table 1.6 *Composition and qualities of common metals found in marine equipment*

Material	Composition %	Treatment	UTS MN/m²	0.1% PS MN/m²	Fatigue limit MN/m²	Elongation %	Modulus of elasticity kN/mm²	Brinell hardness numeral	Uses
Admiralty brass	70 Cu 29 Zn 1 Sn	Annealed	340	75		70		650	Condenser tubes and tube plates. Arsenic added to prevent dezmcifcation
		Cold worked	590	430		10		1,750	
Aluminium	Nearly pure	Annealed	59		31	60	14	150	As a base metal for many aluminium alloys. Electrical fittings
Aluminium brass	76 Cu 22 Zn 2 A1	Annealed	370	105		70	21	650	Condenser tubes and tube plates. Improved resistance to erosion with addition of aluminium
		Cold worked	610	460		8		1,750	
Brass	70 Cu 30 Zn	Annealed	320	85	114	67.5	20	620	General-purpose brass Bearing liners, etc.
		Cold worked	460	380	152	19.5			
Cast iron (grey)	3.25 C. 2.25 Si 0.65 Mn	Sand cast	310		138	0	23	2,500	Cylinder heads, pistons, etc.
Copper	Nearly pure	Annealed	217	46	66	60	21	420	As a base metal for many alloys. Electrical fittings

Material	Composition	Condition							Applications
Cupro-nickel	70 Cu 30 Ni	Annealed	355	105		45		800	Cooler and condenser tubes where good resistance to erosion and corrosion is required
		Cold worked	650	540		5		1,750	
Gun metal	88 Cu 10 Sn 2 Zn	Sand cast	295	124		16	18.6	850	Pump liners, valves (good casting properties)
Monel metal	68 Ni 29 Cu Fe and Mn	Annealed	540	210		45	36	1,200	Pump impellers, valves, turbine blading scavenge, pump valves
		Cold worked	730	570		20		2,200	
Muntz metal	60 Cu 40 Zn	Hot rolled	370	105		40	20	750	General-purpose brass
Phosphor bronze	95 Cu 5 Sn approx. Small amount of P.	Cold worked	710	640	186	5.5		1,880	An excellent bearing alloy Develops hard glassy surface in use
Stainless iron	0.08 C 13.5 Cr 0.15 Ni	Annealed	480				37	1,400	Turbine nozzles and blading
Stainless steel	18 Cr 8 Ni 0.12 C	Softened	460	260		30	30	1,700	Valves, turbine blading
Wrought iron	0.02 C. 0.02 Si 0.05 P 1.0 Slag 0.01 S.	Hot rolled	310	200	186	30	40	1,000	Decorative

The wire's connections and equipment must be checked for damage before any welding is undertaken. Faulty equipment could lead to overheating, electric shock and/or further damage to the equipment.

Safety specific to the gas welding process

As with electric arc welding, eye protection is extremely important in both gas welding and cutting. The problem is that there is always the temptation to discard the use of goggles when using the gas equipment thinking that the light given off from the gas welding process can be viewed with the naked eye. This is not the case and eyesight damage is bound to happen with prolonged exposure to the gas welding process.

Careful attention must be paid to assembling the gas welding equipment. There are two separate systems that are kept apart until the final flame at the end of the torch. A combustible gas, usually acetylene, is used in one system and oxygen in the other. The two systems are both colour coded and given incompatible fittings, so that an acetylene hose cannot be used on an oxygen fitting. Hoses, connections and equipment must be checked for damage before use. The thread on the connections are arranged to couple up in different directions. The oxygen has a conventional 'right'-handed system while the combustible gas has the less common 'left'-handed thread system. Care must be taken not to force the nut from one system on to the fitting of another. If by some outside chance and by using considerable force this were accomplished, the threads would be stripped off and leakage would occur.

Electric arc welding

Electric arc welding is sometimes known as 'stick' welding and with this form of welding an electric arc is set up or 'struck' between the electrode, which may also serve as the filler metal, and the metal to be welded. The heat that is generated causes the electrode to melt and the molten metal is transferred from the electrode to the plate (Figure 1.16).

If the electrode is bare, the arc tends to wander and is therefore difficult to control. Also, the arc stream is open to contamination from the atmosphere and this results in a porous, brittle weld. To avoid these defects, flux-coated electrodes are generally used.

The flux coating melts at a higher temperature than the electrode metal core and thus the coating protrudes beyond the core during welding. This gives better stability, control and concentration of the arc. The coating also shields the arc and the molten metal pool from the atmosphere by means of the inert gases given off as it vaporises.

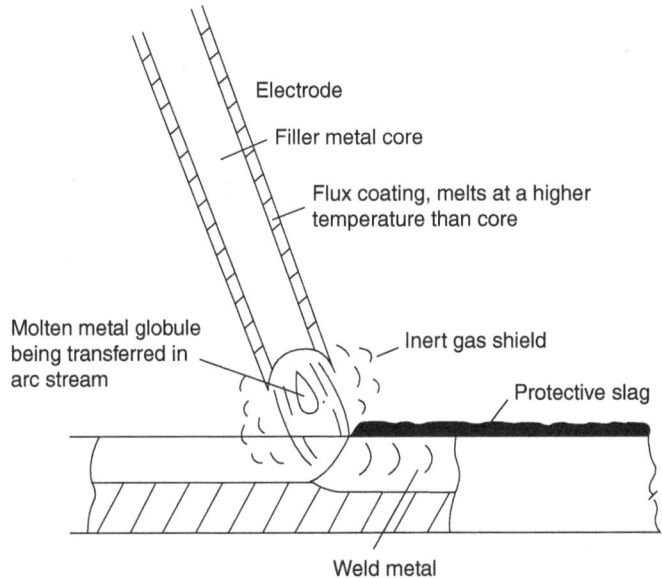

Electrode

Filler metal core

Flux coating, melts at a higher
temperature than core

Molten metal globule
being transferred in
arc stream

Inert gas shield

Protective slag

Weld metal

▲ **Figure 1.16** *Section through electric arc welding*

Silicates, formed from the coating, form a slag upon the surface of the hot metal and this protects the hot metal from the atmosphere as it cools. Also due to the larger contraction of the slag than the metal as cooling is taking place, the slag is easily removed.

Electric arc welding may be done using DC or AC supply. About 50 open circuit volts are required to strike the arc when DC is used, and about 80 volts when AC is used.

The advantages of AC supply that make it more popular than DC are as follows:

1. More compact plant.
2. Less plant maintenance required.
3. Higher efficiency than DC plant.
4. Initial cost is less for similar-capacity plants.

The disadvantages of AC supply are as follows:

1. Higher voltage is used, hence greater shock risk.
2. More difficult to weld cast iron and non-ferrous metals.

Figure 1.17 gives an indication of the ideal weld and also some of the imperfections that may occur on the surface or internally to the weld and adjacent metal.

The defects are generally due to mal-operation of the welding equipment and for this reason welders should be tested regularly and their welding examined for defects. Some of the defects with causes are listed below:

1. *Overlap*: this is caused by an overflow, without fusion, of weld metal over the parent metal. the defect can usually be detected by a magnetic crack detector.

2. *Undercut*: this is a groove or channel along the toe of the weld caused by wastage of the parent metal, which could be due to too high a welding current or low welding speed.

3. *Spatter*: globules or particles of metal scattered on or around the weld. this may be caused by too high a current or voltage making the metal splash or splatter.

4. *Blowhole*: this is a large cavity caused by entrapped gas.

5. *Porosity*: a group of small gas pockets.

6. *Inclusion*: any slag or other entrapped matter is an inclusion defect. surface to be welded must be free from foreign matter, for example grease, oil, millscale, metal chipping, etc. during welding the slag must not be allowed to get in front of the molten metal or it may become entrapped. also when welding is interrupted for changing of electrode or when another run is to be laid, the already deposited metal should be allowed to cool, the slag should then be chipped and brushed off.

7. *Incomplete root penetration*: this is a gap caused by failure of the weld metal to fill the root. this may be due to a fast welding speed or too low a current.

8. *Lack of fusion*: this could occur between weld metal and parent metal, between different layers of weld metal or between contact surfaces of parent metal. it could be caused by incorrect current or voltage, dirt or grease, etc.

Most of the surface defects that occur in welding can be removed by grinding but internal defects, which can be detected by radiographic or ultrasonic methods, necessitate repeating the operations.

Inspection of welding should be carried out during welding as well as after, since the defects if discovered early mean savings in material and labour costs.

During welding by the metal arc process some of the points to be observed are rate of electrode consumption, penetration, fusion, slag control, length and sound of arc.

Other forms of electric arc welding include the argon arc process. Argon arc welding enables non-ferrous metals such as aluminium, magnesium, copper and ferrous metals such as stainless steel to be welded without using a flux.

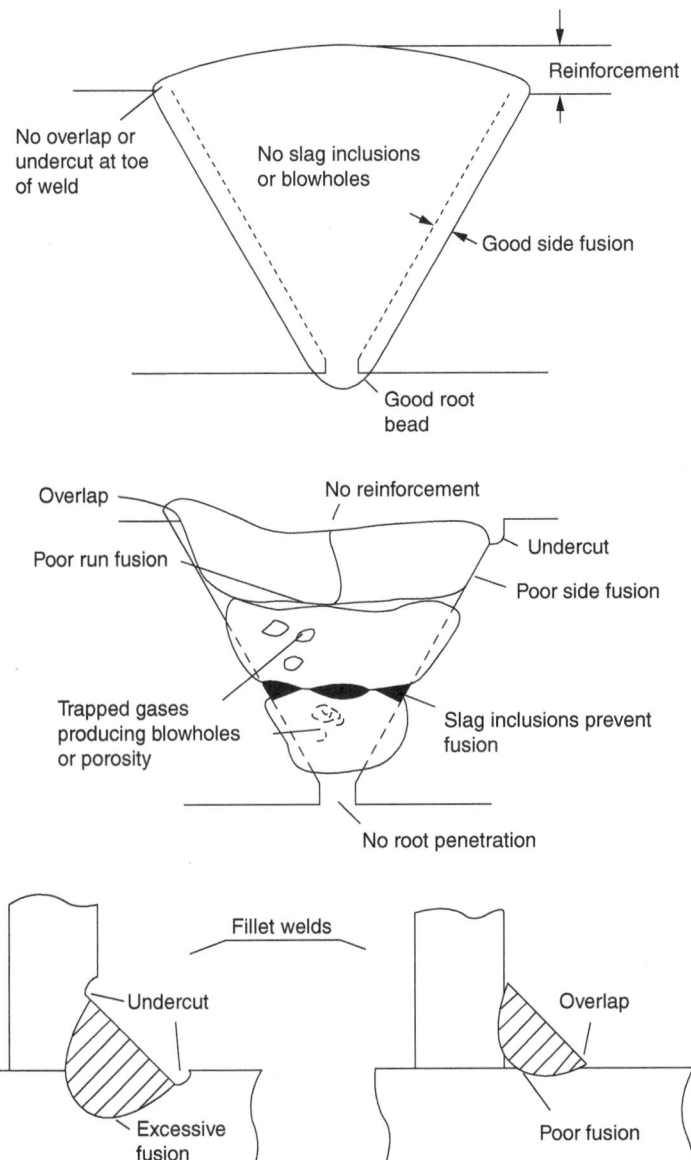

▲ **Figure 1.17** *Sound weld and some weld defects*

In this welding process (called the TIG process, that is, tungsten inert gas) the arc is struck between a non-consumable tungsten electrode and the parent metal. The arc and molten metal are completely surrounded by argon gas, which is supplied to the torch under pressure. Argon is one of the rarer inert atmospheric gases obtained from the atmosphere by liquefaction.

By completely excluding the atmosphere during welding the argon gas prevents oxidation products and nitrides being formed, thus enabling welding to take place without using a flux.

For oxy-acetylene welding, oxygen and acetylene stored in solid drawn-steel bottles under pressure is supplied to a torch. Modern torches are often of the combined type, and can be used for either welding or cutting processes; various types and sizes of nozzles are supplied for this purpose (Figure 1.18).

For welding, a neutral flame is normally required – that is a flame that neither oxidises nor reduces – and a filler metal and flux are used. Oxy-acetylene welding can be used for welding ferrous and non-ferrous metals, for example stainless steels, cast irons, aluminium, copper, etc. It is also a process that can be used for hard surfacing of materials such as stelliting.

Downhand welding

A preferable terminology is 'flat position welding', that is, welding from the upper side of the joint where the face of the weld is approximately horizontal.

Heat-affected zone

In welding or brazing it is that part of the base metal that has had its microstructure and mechanical properties altered but has not been melted.

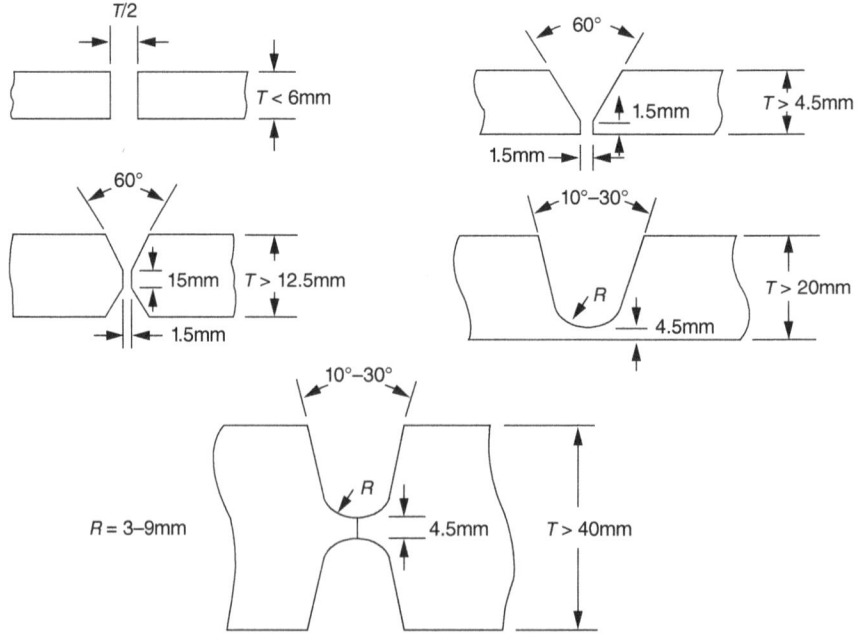

▲ **Figure 1.18** *Typical V and U butt preparations for welding*

Difference between welding, brazing and soldering

Welding: filler metal used has a melting point at or slightly below that of the base metal.

Brazing: filler used has a melting point above 500°C (approx.) but below that of the base metal.

Soldering: filler metal used has a melting point below 500°C (approx.).

Brazing and soldering processes are similar in that the filler metal must (1) wet the parent metal (2) be drawn into the joint by capillary action. Brazing filler metals are alloys of copper, nickel, silver and aluminium. Soldering filler metals are lead and tin or aluminium and zinc alloys.

A flux is used to dissolve or remove oxides. In the case of brazing, borax is used. For soldering resin in petroleum spirit is used.

Gas Cutting

The cutting of irons and steels by means of oxy-acetylene equipment is a very common cutting process that most engineers will have encountered at some point in their lives. Flame cutting or burning as it is sometimes called is convenient, rapid and relatively efficient and inexpensive.

A flame cutting torch differs from a welding torch in that it has a separate, valve-controlled supply of cutting oxygen in addition to the normal oxygen and acetylene supplies (Figure 1.19).

When cutting, for example, steel plate, the plate is first preheated by means of the heating flame until it reaches its ignition temperature, which is usually distinguishable by colour (bright red to white). Then the cutting oxygen is supplied and immediately burning commences. The cutting oxygen oxidises the iron to a magnetic oxide of iron (Fe_3O_4), which has a low melting point. This oxide easily melts and is rapidly blown away by the stream of cutting oxygen.

Once the ignition point is reached the cutting process is rapid, since the heat is supplied, in addition to that given by the heating flame, by the oxidation of the iron. It should be noted that the iron or steel is itself not melted but is oxidised or burnt.

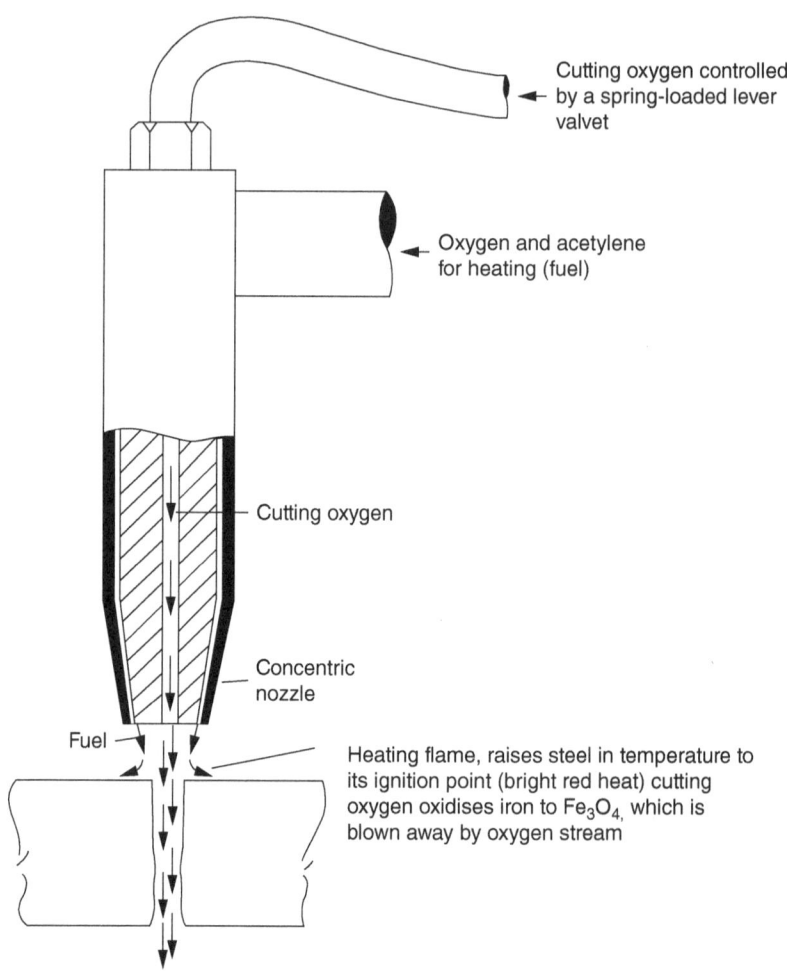

Cutting oxygen controlled by a spring-loaded lever valvet

Oxygen and acetylene for heating (fuel)

Cutting oxygen

Concentric nozzle

Fuel

Heating flame, raises steel in temperature to its ignition point (bright red heat) cutting oxygen oxidises iron to Fe_3O_4, which is blown away by oxygen stream

▲ **Figure 1.19** *oxy-acetylene gas cutting (Others gases – oxy-hydrogen – oxy-propane or oxy-coal gas)*

Due to the rapid cooling of the plate edge that takes place once the torch has passed, local hardness generally occurs. Hence dressing of the plate edges by machining or grinding to remove the hardened material is normally desirable, otherwise surface cracking may develop.

Tungsten inert gas (TIG) and metal inert gas (MIG) welding

It has been found that with some metals, such as aluminium alloys, coated electrodes do not work very well. The coatings cause the aluminium to corrode and, being heavier than the aluminium, slag remains trapped in the weld. It is nevertheless necessary to protect the arc and in such cases an inert gas such as argon may be used for this purpose. Modern equipment uses tungsten rods and argon arc or metal wire welding. In each case argon is passed through a tube, down the centre of which is the tungsten electrode. An arc is formed between the work piece and the electrode while the argon forms a shield around the arc. A separate filler wire of suitable material is used to form the joint. The tungsten electrode may be water cooled. This system of welding may be used for most metals and alloys, although care must be taken when welding aluminium, as an AC machine may be required. TIG welding is used for welding metals such as aluminium/brass (Yorcalbro), stainless steels and acid-resistant steels.

Techniques: to master the art of electric welding a person must practice, practice, practice. However, it might be useful to think about some preparation beforehand.

Striking an arc: it may seem obvious but once the welder puts the welding mask in front of their face they cannot see the work area until the arc is alight. Therefore, there is very much an element of hand/eye co-ordination coupled with judgement and knowledge, which come together to form a mental picture that bridges the gap between being able to see the work with no mask in the way and seeing the weld progress while using the mask.

It is easiest to start with a flat work piece and looking down on the weld. The welder will learn the basic welding process before trying to weld a vertical or overhead weld, which are much more difficult.

As the welding progresses the tip of the welding rod must be held to keep a constant gap between it and the work piece. However, the rod is melting away; therefore, with MMA and TIG welding, the welder must continually adjust the position by moving the holder closer to the work as the length of the welding stick reduces.

With MIG or wire-feed welding the filler material is pushed through the holder and the welder does not have to adjust their position relative to the work. In both cases a neat, consistent weld is produced only if the speed at which the weld is progressing is correct.

The angle at which the rod is held relative to the work is also important. If the angle is not correct then the heat going to the work piece will be uneven and may cut into the parent metal in a way that is not correct for the current weld to maintain strength.

2

FUEL TECHNOLOGY

Liquid Fuels

Crude petroleum can be broadly classified into three types.

1. Paraffin base in which the residue after distillation contains more than 5% paraffin wax.

2. Asphalt base in which the residue after distillation contains less than 2% paraffin wax and is mainly composed of asphalt (bitumen).

3. Mixed base in which the residue after distillation contains between 2% and 5% paraffin wax mixed intimately with asphalt.

The type obtained depends on the source and also determines the type of refining necessary and nature of the end products produced.

The raw petroleum at the well head is often associated with natural gas, which has a high methane content; this gas can be directly utilised and is piped off for domestic use. Primary separation, by heating and cooling, will allow a yield of well head motor spirit (straight run gasoline). The bulk of the crude is taken to the refinery for processing into a wide range of products depending on the type of crude. Asphalt is mainly found in residual oils and is an indefinite substance, both hard and soft, being mainly combustible although hard asphalt can cause considerable gum deposits in IC engines.

Composition of petroleum

Petroleum in all its forms consists of hydrocarbons, with small amounts (up to 5%) of nitrogen, oxygen, sulphur, metallic salts, etc., together with water emulsified in the oil and associated with natural gas.

Hydrocarbons

The exact proportions and composition decide the character of the petroleum and hence the refining and processing required. Types of hydrocarbons are made up of at least nine recognisable series, a series being a range of products with the same molecular structure pattern, from $C_n H_{2n+2}$ to $C_n H_{2n-12}$. The four main series are as follows:

Paraffins: $C_n H_{2n+2}$

    For example, methane (CH_4), butane (C_4H_{10})

Naphthenes: $C_n H_{2n}$

    For example, cyclo–butane (C_5H_{10})

Aromatics: $C_n H_{2n-6}$

    For example, benzene (C_6H_6)

Olefines: $C_n H_{2n}$

    For example, ethylene (C_2H_4)

The first two are usually classified as *saturated* and the latter two as *unsaturated*. Unsaturated series are rarely found in crude petroleum but tend to be found by molecular bonding alteration during later processing. Although olefines and naphthenes have the same C:H ratio they are distinguished by an important difference in molecular structure.

The lowest members of any series are gases, graduating to liquids as the molecular structure becomes more complex, thence to semi-solids and to solids. Consider, for example, the paraffin hydrocarbon series: methane (CH_4) to butane (C_4H_{10}) are gases, pentane (C_5H_{12}) to nonane (C_9H_{20}), which are all liquids of decreasing volatility. By octadecane ($C_{18}H_{38}$) there is a mineral jelly and further up the series lies paraffin wax solid ($C_{21}H_{44}$). With slight deviations from the molecular grouping system, millions of different combinations called 'isomers' are possible. Composition and characteristics then tend to become chemically complex, which particularly applies to high-grade gasoline for aviation and motor vehicle fuels.

Crude oil is first treated for water and dirt removal, natural gas and straight run gasolines being commonly tapped off, and the bulk of the crude is passed to the refinery for distillation. Any refinery must be fairly flexible to cope with reasonable variations of crude type and variation in market demands for the output of distillates.

The distillation process

Modern refinery processes can be separated into two basic parts the atmospheric and vacuum distillation phase and the further catalytic and thermal cracking process. From the 1980s, the complex refineries have replaced the straight run-only refineries. As a first stage the refinery distils the crude petroleum, in the straight run process, into basic products. This depends on the lightness of the various fractions and their distillation temperatures. This process produces stable products of which a typical primary distillation would be approximately as shown in Figure 2.1.

The temperatures for distillation into the various fractions would be approximately as shown in Figure 2.2.

The actual stages as shown may not be so clearly defined, that is, the cut point often has some degree of temperature overlap.

Straight run refinery layout

As seen in Figure 2.3, the crude is fractioned into the various distillates by heating in fractioning towers, the distillates being tapped off at the necessary points.

The actual layout is slightly more complex due to recirculation for stripping, reflux for enriching, provision of condensers for gas cooling, etc., all with the object of improving the quality of the distillate. The provision of the vacuum stage is to reduce the required temperatures of distillation for the heavier fractions and to avoid uncontrolled oil cracking.

Complex processing with catalytic cracking and visbreaking

To increase the quantity of the 'lighter' products, the residual fuel from the straight run process is subjected to further processing by thermal cracking. This has the effect of reducing the viscosity of the residual oil and producing a small amount of lighter products. The extant of the thermal process is determined by the need to keep the residual oil stable.

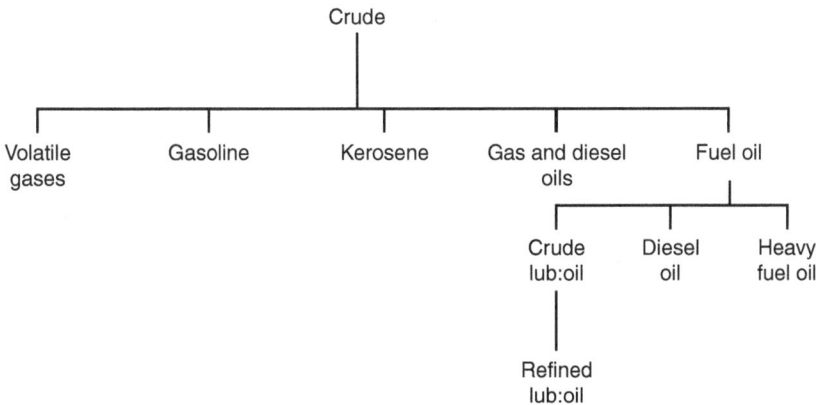

▲ **Figure 2.1** *Primary distillation of fuel*

The heavy oil output from the catalytic cracker is blended with the output from the visbreaker to further enhance the stability of the heavy fuel oil. It is the result of this part of the process that is important to the marine engineer. The catalytic cracking process uses aluminium silicate as the active ingredient, which is supposed to be removed as part of the refining process, although the removal is not 100% efficient and some

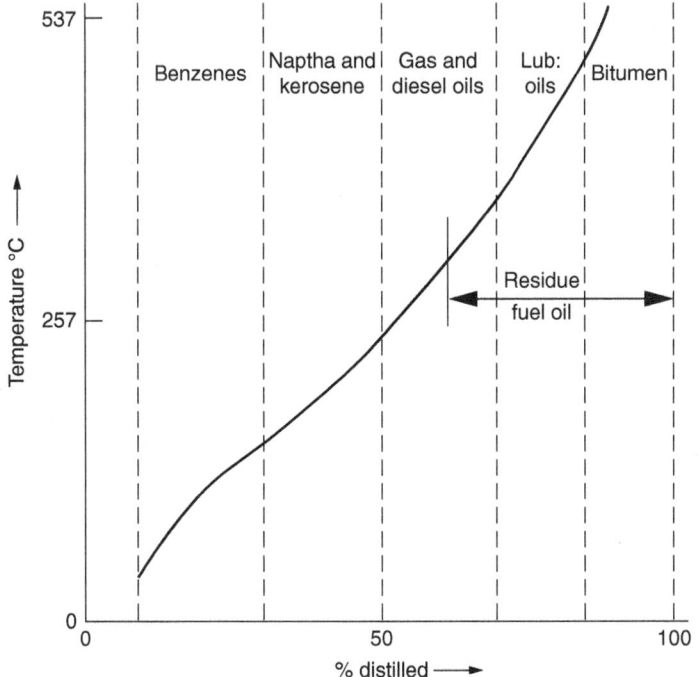

▲ **Figure 2.2** *Temperature dependent distillation markers*

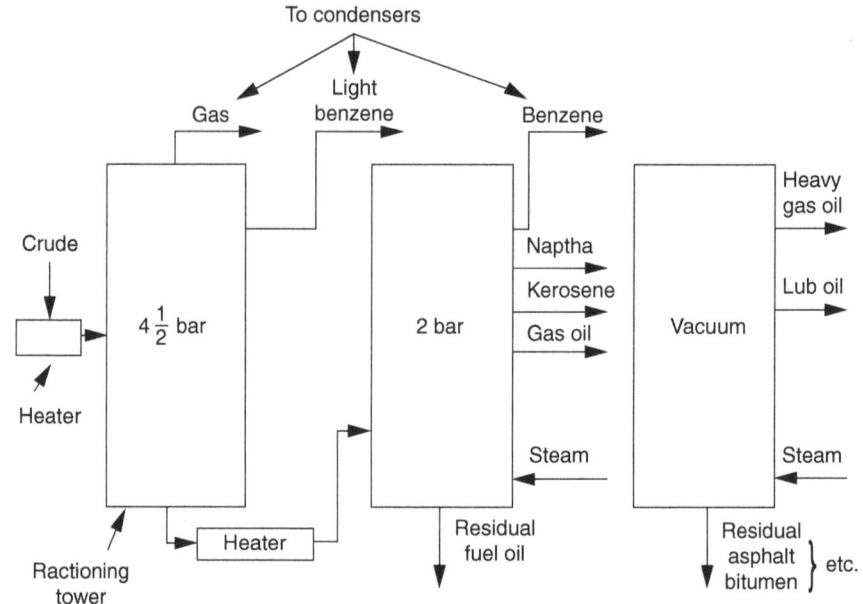

▲ **Figure 2.3** *Simple refinery layout*

'catalytic fines' (cat fines) are left over. These are harmful to fuel pumps, especially to modern high-precision fuel injection systems (Figure 2.4).

The standard for marine fuels is set out in ISO 8217, and from the fourth edition to the latest edition the standard specifies a maximum level of aluminium and silicon fines; however, marine engineers on board must be aware that the bunkers supplied to the vessel might not be to the required specification, so samples must be taken and sent for analysis. Flag state examinations will be expecting engineers, presenting themselves for examination, to know this procedure due to the problems that can occur if seriously 'off spec' bunkers are used for any length of time.

Testing of Liquid Fuel and Oils

(1) Density (ρ)

Storage of liquids is often based on volume and some correlation such as density for the mass-to-volume relationship is required. This is important for bunker capacities, choice of heating arrangements, injectors, purifiers, etc.

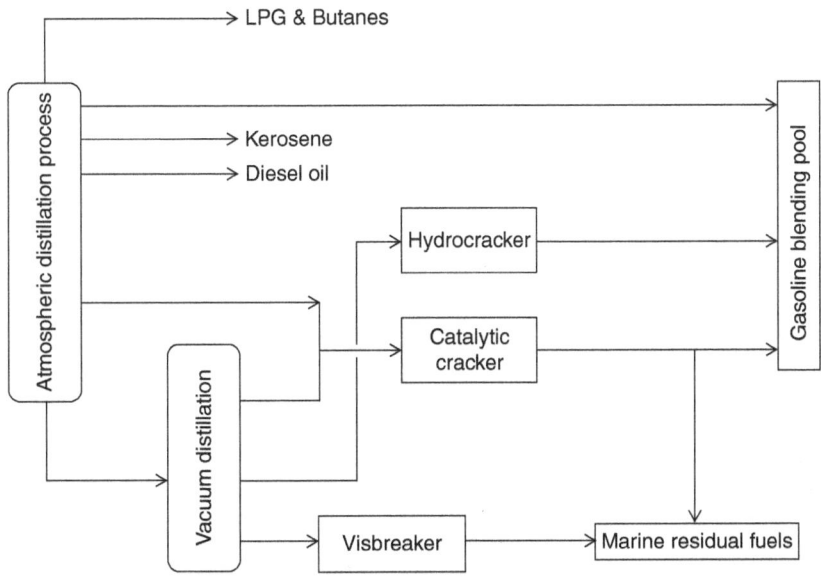

▲ **Figure 2.4** *Complex refining process*

$$\text{Density } (\rho) = \frac{\text{Mass } (m)}{\text{Volume } (v)}$$

The units are usually expressed in kg/m³ with fresh water being 1,000 kg/m³. Table 2.2a (see page 77) shows that the specification for fuel is to measure the density at 15°C. If the temperature cannot be fixed at 15°C (e.g. with high-viscosity oils) then a correction factor per degree C above 15°C is added to the observed density, or if measured below 15°C is subtracted from the observed density. Density is taken by hydrometer. The datum temperature is 15°C, water (fresh) has its maximum density of 1,000 kg/m³ at 4°C and the reciprocal of density is specific volume (m³/kg). Density is a very important consideration for the marine engineer because when the fuel is ordered by weight and delivered by volume there is always a chance of the density not being correct and the density is proportional to the energy that is extracted. Therefore there will be a reduction in performance if the density is lower than ordered. Marine fuel density meters are available for use on board by the marine engineer. On-board test equipment can also measure the Calculated Carbon Aromaticity Index (CCAI) for residual fuel oil, the maximum value of which is given in Table 2.2b (see page 78). Mass flow meters for the measurement of the incoming bunker fuel is another advancement in the measurement of the bunker fuel coming on board.

(2) Viscosity

Viscosity may be defined as the resistance of fluids to a change of *shape*, this being due to the internal molecular friction of molecule against molecule of the fluid, producing the frictional drag effect.

Absolute (dynamic) viscosity as used in calculation is difficult to determine, being numerically equal to that force to shear a plane fluid surface of area 1 m², over another plane surface at the rate of 1 m/s, when the distance between the two surfaces is 1 m. Kinematic viscosity is the ratio of the absolute viscosity to the density at the temperature of viscosity measurement.

$$F = \eta A \frac{dv}{dy}$$

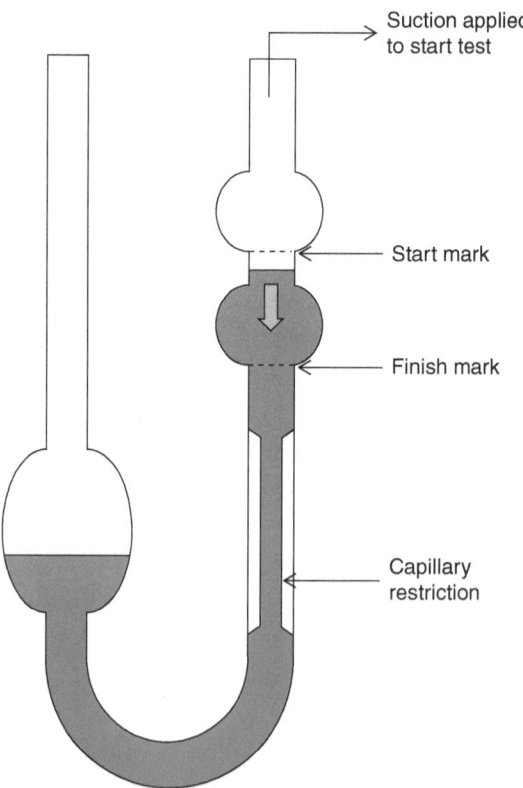

▲ **Figure 2.5** *Capillary U-tube viscometer*

$$\text{Dynamic } (\eta) = \frac{Fdy}{Adv} \left[\frac{ML}{T^2} \cdot \frac{1}{L^2} \cdot \frac{LT}{L} \right] = \left[\frac{ML}{T^2} \cdot \frac{T}{L^2} \right] \text{Ns/m}^2$$

$$\text{Kinematic } (v) = \frac{\eta}{\$} \left[\frac{ML^3}{TLM} \right] = \left[\frac{L^2}{T} \right] \text{m}^2/\text{s}$$

The kinematic method is the one chosen to determine values in the ISO fuel specification (8217). The traditional value was in centistokes at 50°C (1 mm²/s = 1 cSt) is the measurement (sometimes 40°C or 80°C) and the SI unit is mm²/s. Kinematic viscosity is measured by capillary flow of a set liquid volume from a fixed head (Poiseuille) (Figure 2.5), a similar method (Ostwald) is frequently used by the oil industry; a technique using a steel ball falling through the liquid (Stokes) can also be applied.

For practical purposes viscosity is still often measured on a time basis. It is expressed as the number of seconds for the outflow of a fixed quantity of fluid through a specifically calibrated instrument at a specified temperature. The time for 50 ml outflow is taken by stopwatch and temperature accuracy is vital, a variation of ± 0.1°C is a maximum for temperatures up to 60°C. Again on-board testing equipment can be purchased or the samples can be sent to a laboratory for testing.

Viscosity scales

In British practice the Redwood viscometer was used. Redwood No. 1, the outflow time in seconds of 50 ml of fluid, used up to 2,000 s. Redwood No. 2, for oils with outflow times exceeding 2,000 s (usually, but not always), designed to give ten times the flow rate of the Redwood No. 1 orifice.

In American practice the Saybolt Universal and Saybolt Furol were used in a similar manner to the above, employing a different orifice size as in the Redwood.

In European practice the Engler viscometer was used, which compares the outflow times of oil and water, with results being quoted in Engler degrees.

International standardisation has become fixed on the kinematic method. ISO 8217 uses the standard of mm²/s at 40°C. Figure 2.5 shows the U-tube test equipment. The oil is raised to the start position by a suction applied at the top of the right-hand section of the tube. As this position is reached the suction is released and the oil is left to fall through the capillary under its own weight. The time is measured from the start point until the oil level reaches the end point.

Temperature

Increase in temperature has a marked effect on reducing fluid viscosity. Temperature and viscosity are closely related in the choice of oil for a particular duty. For atomisation of fuels it is necessary to heat high-viscosity oils so that the viscosity is about 20 cSt at the injector and preferably near 13 cSt for internal combustion common rail engines (the viscosity of diesel oil being about 7 cSt at 38°C).

It is essential to specify the temperature at which the viscosity is quoted otherwise the value becomes meaningless for correlation.

The scale readings between viscometers can be related to each other by graphs or the use of constants. It is not possible to calculate viscosities at different temperatures without the use of viscosity–temperature curves. Each oil and blend type differs with the effect of temperature change so a curve is required to be plotted for each type. Four typical viscosity–temperature curves are shown in Figure 2.6. These represent some of the most commonly used fuels. However, as can be seen by the timeline shown in Figure 2.6, the sulphur content is falling dramatically due to the pressure on the industry to reduce atmospheric pollution.

Factors influenced by viscosity may be summarised as: frictional drag effects, pipe flow losses, flow through small orifices (atomisation), load capacity between surfaces, fouling factor, spread factor, etc.

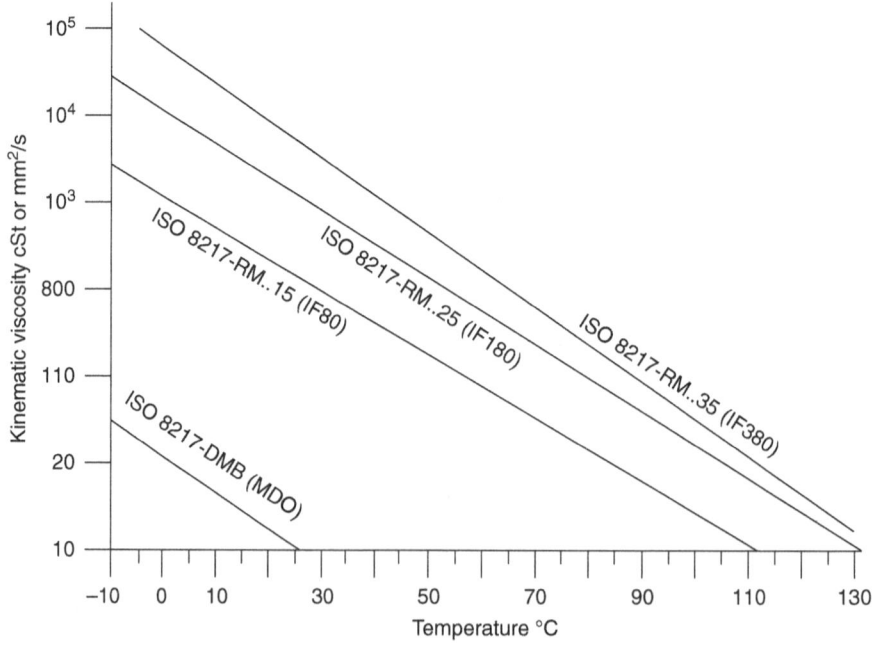

▲ **Figure 2.6** *Changing viscosity of oil with temperature*

Viscosity index is a numerical value that measures the ability of an oil to resist viscosity change when the temperature changes. The index is mostly used in the automotive industry to describe a 'multigrade oil'. A high viscosity index would refer to an oil capable of maintaining a fairly constant viscocity value in spite of wide variations in the temperature. The value of viscosity index is usually determined from a chart based on a knowledge of viscosity values at different temperatures. On board ship this measurement would be important for deck equipment, emergency generators or lifeboat engines, which may be required to operate at a large range of temperatures during the winter months.

(3) Flash point

This is the minimum temperature at which an oil gives off flammable vapour, which on the application of a flame in a specified apparatus would cause momentary ignition.

The test may be *open* or *closed* depending on whether the apparatus is sealed or not. The closed flash point is always *lower* because the lid seal allows accumulation of the volatiles above the liquid surface. The test applied for oils above 45°C, which is the usual marine range, is the Pensky Marten Closed Flash Point test. For oils below 45°C the Abel apparatus would be used.

As seen in Figure 2.7, when the operating handle is depressed the shutter uncovers the ports (down movement of the handle opens a shutter just below the ports by means of a ratchet; further movement and ratchet travel gives a flame insertion, this detail is omitted on the diagram for simplicity). The flame element is depressed through one port *above* the oil surface. Starting at a temperature of 17°C below the judged flash point, the flame is depressed, left and quickly raised in a period of under 2 s, at 1°C temperature intervals.

Just before the flash point is reached a blue halo occurs around the flame, the flash is observed just after, through the two observation ports, stirring being discontinued during flame depression. A fresh sample must be used for every test and care must be taken that no trace of cleaning solvents are present in the oil cup.

Some aspects relating to oils and the use of closed flash point may be considered as follows:

- Oils with flash point below 22°C are classified as *dangerous – highly flammable*, such oils are gasolines, benzenes, etc.
- Flash points in the range 22–66°C would relate to kerosenes and vaporising oils.

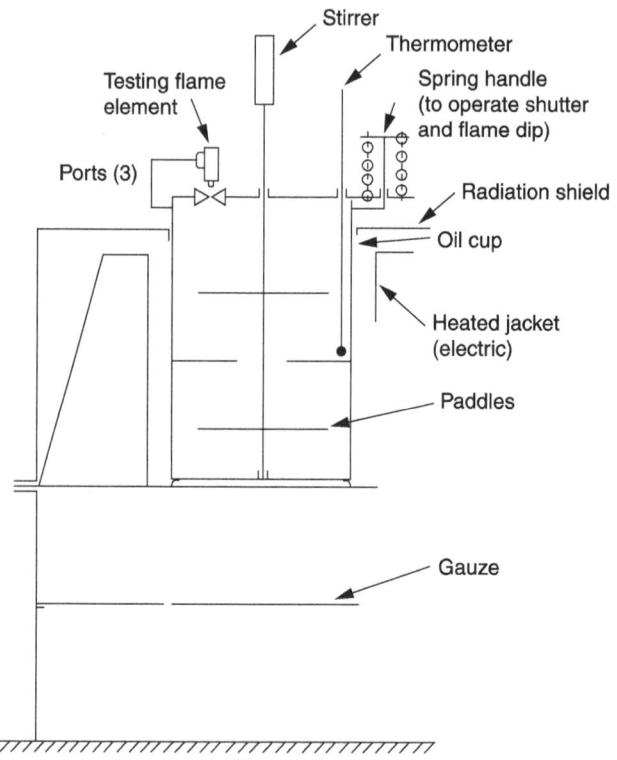

▲ **Figure 2.7** *The Pensky Marten (closed) test*

- Flash points above 60°C are classified as *safe*, (for marine purposes) and include gas, diesel and fuel oils.
- Approximate closed flash point values for different oils are given in the following table.

These values are only averages; the grade and type cause wide variations, for example the term petrol could relate to values between –60°C and 25°C.

As flash point is indicative of fire and explosion risk during storage and transport, it is an important property of the oil. The Safety Of Life At Sea (SOLAS) convention states that fuel on board ship should have a minimum of 60°C. SOLAS II-2 regulation 4 sets out the basic requirement for main engine and auxiliary engine fuels. It does, however, state that fuel with a minimum flash point of 43°C is permitted for use in the emergency generator so long as the reserves are not stored in the machinery space. It is also prudent best practice that the oil in storage should not be heated above 52°C. The fuel should also not be stored between 15°C and 40°C as this is the temperature range that promotes bacterial growth and subsequent degradation of the fuel.

Pentane	−49°C	Petrol	−17°C
Carbon disulphide	−30°C	Paraffin	25°C
Acetone	−18°C	Diesel oil	95°C
Benzene	−11°C	Heavy fuel oil	100°C
Methanol	10°C	Lubricating oil	230°C

In special cases when high-viscosity oils are used and high degrees of heating are required to produce atomisation, etc., it is allowable to heat the oil to within 20°C of the closed flash point. Great care should always be taken regarding the control of heat to heaters situated on the *suction* side of the fuels pumps so as not to cause oil vaporisation and the possibility of explosive vapour formation; any vapourisation will also cause the pumps to lose suction and stop pumping.

(4) Calorific value

This is the heating value from the complete combustion of unit mass of fuel, that is, MJ/kg, kJ/kg, etc. The approximate heat energy values of fuels are given below:

The value quoted is the *higher* calorific value, in every case. This value includes the heat in water vapour formed from water as the products of combustion are cooled, vapours condensed and hence latent heat becomes re-available for heat utilisation.

The *lower* calorific value is more realistic, from the boiler engineer's viewpoint, being the actual heat available for boiler water evaporation, but this does not detract from the fact that this is a fault of utilisation, and the higher calorific value is the actual heat available and is therefore the *preferred* value for quotation. Fuels always exhibit a fall in calorific value to some extent during storage.

Coal	34 MJ/kg
Fuel oil	42 MJ/kg
Diesel oil	45 MJ/kg
Pure hydrocarbon	50 MJ/kg (85% C, 15% H_2)

There are numerous makes of bomb calorimeters but the differences are only slight. The test as conducted is very closely detailed and only a brief synopsis is outlined here. For further close details, if desired, the reader is referred to the relevant BS specifications.

Consider Figure 2.8, the oxygen supply is to give an internal pressure of 26 bar and should not be less than $2\frac{1}{2}$ times the theoretical oxygen required. The interior of the bomb must be resistant to condensed acidic vapours from combustion. The thermometer used can be read by means of a lens to 0.002°C, and the temperature of the enclosing water, of amount 15–20 l, should be steadily maintained during the test.

A small specimen is fired by electric charge under conditions of pressurised oxygen and the temperature rise of apparatus and coolant is noted. Distilled water (0.01 kg) is in the bomb to absorb sulphuric and nitric acid vapours (from sulphur trioxide and nitrogen). Mass of fuel × hcv = WE of apparatus complete × its temperature rise. The above calculation, using masses in kg and temperatures in °C gives the hcv of the fuel (MJ/kg or kJ/kg). The water equivalent (WE) of the apparatus is determined by a test using benzoic acid. This is the calorific value reference fuel, hcv 26.5 MJ/kg, showing relatively no deterioration of calorific value during storage. Correction factors are now applied for acids formed under bomb conditions only, radiation cooling effect, etc. The temperature of test is based on 15°C approx. It should be noted that under the bomb's combustion conditions (high excess air and pressure), sulphuric and nitric acids are

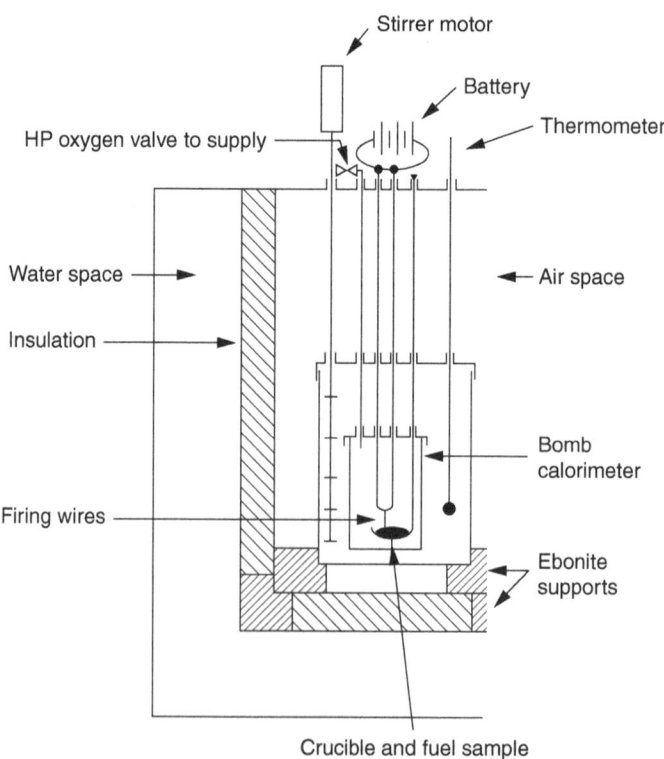

▲ **Figure 2.8** *The bomb calorimeter*

formed. Whereas under furnace combustion conditions, sulphur is burned mainly to sulphur dioxide with no acid formation, thereby no trioxide, and nitrogen would pass off in the free state.

(5) Pour point

This is a determination of the lowest temperature value at which oil will pour or flow under the prescribed test conditions. This value is an important consideration for lubricating oils working under low temperature conditions, for example, refrigeration machine lubricants, telemotors, etc.

As seen in Figure 2.9, various mixtures are used in the bath. For very low temperatures, solid carbon dioxide and acetone are used. At 11°C above the expected pour point the test begins. At temperature intervals of 3°C the test jar is removed, checked for surface oil tilt and replaced in a time interval of 3 s maximum.

When the surface of oil will not tilt, for a time interval of 5s, note temperature, add 3°C and this is the pour point. The oil is heated to 46°C before the test and is cooled in progressive stages of about 17°C in different cooling agent baths; in each case the jar must be transferred to another bath when the oil reaches a temperature of 28°C above the bath temperature. The pour point is also important for fuel oil as this is the lowest temperature at which the fuel can be stored and used without the formation of wax. If a fuel is below the pour point, the wax can build up on tank bottoms and heating coils. If the fuel is used in this condition the wax may also block the fuel filters, making it impossible to transfer from bunker tank to settling of service tank; depending upon where in the systems the wax forms it may even stop a running engine. When the fuel is heated again, the wax may not re-dissolve because of its insulating nature, therefore necessitating manual cleaning of tanks or filters.

To avoid these operational difficulties it will be necessary to keep the fuel temperature at least 10°C above the pour point. Referring to the 8217:2010 specifications, the pour point for residual fuel oils is a maximum of 30°C. This means that the fuel should be kept at a minimum of 40°C. As noted on page 4139 this is in line with industry best practice of between 45°C and 50°C.

Fuel transfer pumps are usually designed to operate at a maximum viscosity of 800–1,000 mm^2/s. Again, if we look at Table 2.2, the fuel should be at between 45°C and 50°C so that it is at the correct viscosity for efficient transfer. For less viscous fuels, that is, RMA 10 and RMB 30, the pour point will be an important parameter for handling purposes especially if tank heating is not fitted.

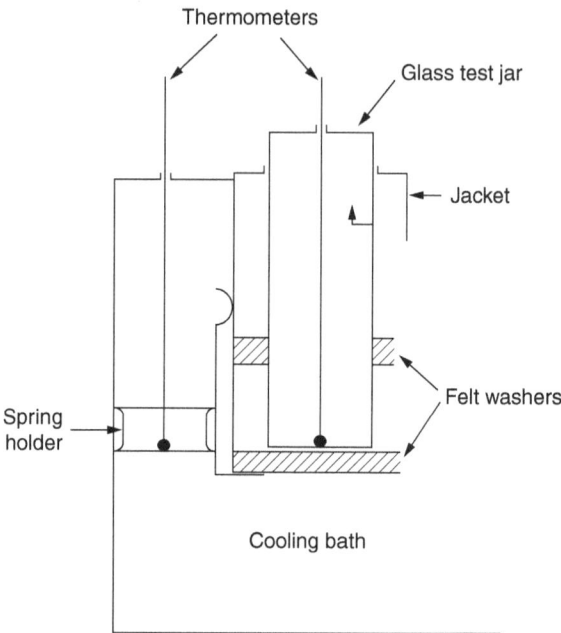

Thermometers

Glass test jar

Jacket

Felt washers

Spring holder

Cooling bath

▲ **Figure 2.9** *Pour point apparatus*

(6) Carbon residue (Conradson method)

This test indicates the relative carbon-forming propensity of an oil. The test is a means of determining the residual carbon, etc., when an oil is burned under specified conditions. This test has been used much more in recent times in line with the use of high-viscosity fuels in IC engines.

The mass of the sample placed in the silica crucible must not exceed 0.01 kg.

Initial heating period 10 minutes $\pm$ $1\frac{1}{2}$, vapour burn-off period 13 minutes $\pm$ 1, further heating for exactly 7 minutes, total heating period 30 minutes $\pm$ 2. The covers must be a loose fit to allow vapours to escape.

The heating and test method are closely controlled. After removal of sample and weighing, the result is expressed as 'Carbon Residue (Conradson)' as a percentage of the original sample mass. The test is usually repeated a number of times to obtain a uniformity of results (see Figure 2.10).

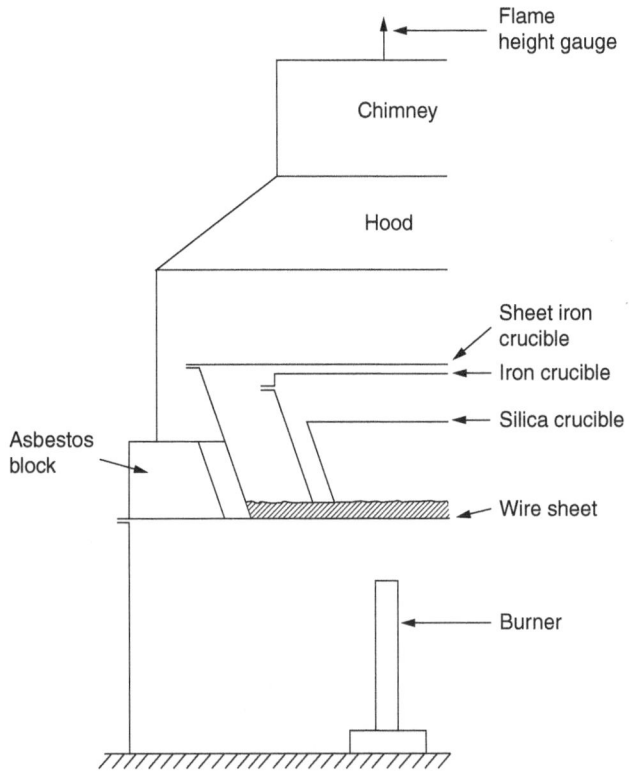

▲ **Figure 2.10** *Conradson carbon test apparatus*

(7) Water in oil

A quick test for the presence of water in a substance is to add a sample to white copper sulphate ($CuSO_4$), which turns to blue copper sulphate ($CuSO_4 \cdot 5H_2O$) in the presence of water.

The following test, as seen in Figure 2.11, is more suitable for oil.

The test usually conducted is the LP (Laboratory Procedure) standard method, in which100 ml of sample is mixed completely with 100 ml of special high-grade gasoline having standard properties. Steady heat is applied for about 1 hour. Water vapours are carried over with the distilled gasoline and are condensed in the condenser and measured in the lower part of the receiver. The result being expressed as say 1% Water, IP Method. Note that this diagram is very much simplified. The actual apparatus must be constructed to specific and exact BS dimensions, which are highly detailed. The test must also be carried out under closely controlled conditions. ISO 8217:2010 set

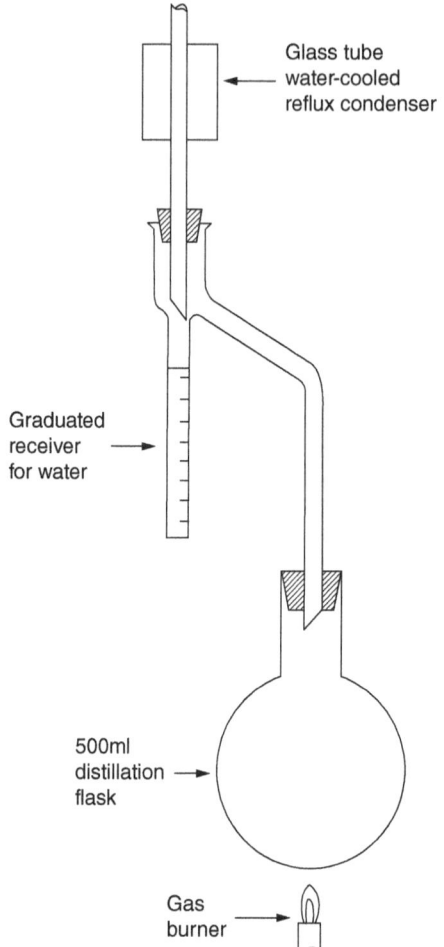

Glass tube
water-cooled
reflux condenser

Graduated
receiver
for water

500ml
distillation
flask

Gas
burner

▲ **Figure 2.11** *Water test IP method*

the maximum water content in the fuel at 0.5% by volume for residual fuel and 0.3% for diesel oil. The water content is usually much lower but it is very important that the ship board engineers check very carefully to ensure that water has not entered with the fuel. The correct on-board handling of the fuel will help if water is present. If the bunker tanks are kept at the correct temperature the water will start to separate out but it is in the settling tanks that the water can be drained with the purifiers taking out the last remaining drops.

If water does get though to the fuel that is entering the engine it might not have been introduced at the bunkering stage and the engineers must look for other sources, such as the purifiers malfunctioning or a broken steam heating coil.

(8) Calculated Carbon Aromaticity Index

The combustion of a residual fuel oil used in marine engines is a complicated multi-stage process. It is a combination of the efficiency and effectiveness of the injection system on the IC engines or the burner on oil-fired boilers and the ignition quality of the fuel. The modern refining process, described at the start of this chapter, uses a combination of techniques to extract the lighter products from the basic crude oil. This process has led to the possibility of fuels being produced that are unstable. One of the problems that resulted from the new processes was an increase in the aromatic compounds found in the residual fuel. These compounds have a high resistance to auto-ignition and during the 1990s some vessels were having difficulties because the level of aromatic compounds in the bunker fuel had reached the point where the fuel would not auto ignite at the temperature of the compressed air in the diesel engine; in one case the auxiliary engines, operating on heavy fuel, stopped running. The problematic fuel was some of the low-viscosity, high-density fuels and following research studies commissioned by Shell an equation involving density and viscosity, known as the Calculated Carbon Aromaticity Index (CCAI), was devised. Further studies on test engines demonstrated the correlation between the CCAI and ignition delay and hence the calculation was included in the 2010 revision of ISO 8217. The specification sets the CCAI for residual fuel at 870, which is an important indicator for the marine engineer to know.

A second reason for the importance of this indicator is for the marine engineer to have the knowledge that the fuel will not cause major harm to the engine. Combustion is the rapid oxidation of fuel that results in the release of heat but the ratios of air/fuel and the ignition temperature all have to be correct and as the fuel is introduced to the combustion space there is a time lag until the conditions are correct for the start of the oxidation process. The longer ignition delay means that a larger amount of unburnt air/fuel mixture will have built up before the initial combustion takes place. As the combustion process moves on, the next stage is determined by how rapidly the oxygen and remaining vaporised fuel can be mixed as the initial supply of oxygen close to the fuel droplet's boundary has been used in the first stage of the combustion. However, if this all happens too fast then very rapid rates of pressure rise in the cylinder can result in shock waves being transmitted, which could cause broken piston rings and overheating of metal surfaces.

(9) Acidity (or alkalinity)

This is indicated by the neutralisation (or saponification) number. This number is the mass, in milligrammes, of an alkali, which is often potassium hydroxide, needed to neutralise the acid in 1 g of sample. The oil is often alkaline. In this case the acid to neutralise it is in turn neutralised by the alkali and the result is then expressed as a base neutralisation number. Alternatively the quantities can be expressed in p.p.m. for 1 ml of oil sample (usually dissolved in industrial methylated spirits). Phenolphthalein can be used as the indicator. Total base number (TBN) is often used for alkalinity indication for lubricating oils.

(10) Ash, sodium and vanadium

A sample of oil (250 ml minimum) is cautiously and slowly evaporated to dryness and ignition continued until all traces of carbon have disappeared. The ash is then expressed as a mass percentage of the original sample. Ash usually consists of hard, abrasive mineral particles such as quartz, silicates, iron and aluminium oxides, sand, etc. A residue test (% by volume after heating to 350°C) is sometimes used; see Table 2.2 for the maximum ISO values. Sodium can enter the fuel due to it being in a marine environment but the ISO standard sets limits for bunker fuel. Vanadium is present in all crude oil but the amount varies depending upon the source of the crude oil. Vanadium is very hard and can have an abrasive and corrosive effect on the engine.

(11) Other tests

These are numerous. Fire point, for example, is the temperature at which the volatile vapours given off from a heated oil sample are ignitable by flame application and will burn continuously. The fire point temperature can be anything up to about 40°C higher than the closed flash point temperature for most fuel oils.

Other examples are tests for asphaltenes, sediment, suspended solids, oxidation, emulsion number, cloud point, setting point, precipitation number, etc. These are more complex laboratory tests whose description is difficult to simplify and therefore are not discussed further.

Three other tests, however, not mentioned previously, are regarded as being of extreme importance in IC engine practice. Therefore, these tests – octane number, cetane number and crankcase oil dilution – will now be considered.

(12) Octane number

This is indicative of the knock rating. Knocking or pinking are characteristic of some IC engine fuels, particularly in spark-ignition engines. It can cause pre-ignition, overheat and damage.

Normally on spark initiation the flame front proceeds through the mixture at a speed of about 18 m/s. If, due to engine conditions or type of fuel used, the mixture in front of the flame front has its temperature and pressure raised above the spontaneous ignition point then auto-ignition occurs. This means that by the time the last gas charge is reached the flame front speeds can reach 2.2 km/s and detonation, temperature rise and heavy shock waves occur. Knocking tendency is dependent on many variables such as rev/s, compression ratio, turbulence, mixture strength and if a 'hot spot' has developed due to a build-up of carbon.

Test

Iso-octane (C_8H_{16}) has very good anti-knock properties and is taken as upper limit 100. Normal heptane (C_7H_{16}) has very poor anti-knock properties and is taken as lower limit zero. Therefore octane number is the percentage by volume of iso-octane in a mixture of iso-octane and normal heptane, which has the same knock characteristics as the chosen fuel. The test is conducted under fixed conditions on a standard engine, which usually has electronic detonation detection. Modern fuels, for aviation, etc., have octane numbers over 100. The octane rating does not equate to the energy content of the fuel, it is only a measure of its tendency to burn in a controlled way at the lower end or explode uncontrollably at the higher end. In the past tetraethyl lead (TEL) was used in specified proportions to lower the iso-octane rating but this practice has been phased out since the 1970s.

(13) Cetane number

Is an indication of the ignition quality of diesel fuel with the CCAI being the measure applied to residual fuels (see Table 2.2). In a compression ignition engine, cold starting is often required; here the time interval between fuel injection and firing, called ignition delay, must not be too long otherwise collected fuel will generate high pressures when it does ignite and diesel knock results. Paraffin hydrocarbons have the best ignition quality and are thus most suitable. Speed and cetane number can be correlated: for high-speed engines (above 13.3 rev/s) a cetane number of 48 may be regarded as a minimum, ISO 8217:2010 sets 35 as a minimum for diesel oil.

A diesel fuel used in a hot-petrol engine would cause detonation, that is, it has a low octane number.

Test

Cetane is a paraffin hydrocarbon, hexadecane ($C_{16}H_{34}$) being its correct designation, of high ignition quality and is taken as the upper limit of 100. Alpha-methyl-napthalene is of low ignition quality and is taken as the lower limit of zero. Thus cetane number is numerically the percentage by volume of cetane in a mixture of cetane and alpha-methyl-napthalene that matches the chosen fuel in ignition quality.

There are a number of tests. One is by measurement of the delay period when running, by use of a cathode ray tube on a standard engine. Another, which is probably the best, is to use a standard engine running under fixed conditions with a variable compression ratio to give a standard delay, and using the compression ratio as an indication of cetane number.

An alternative method called diesel index can be used but it is not as reliable as cetane number. Density is often indicative of cetane number especially in the middle ranges, that is, density 850 kg/m³, cetane number about 61, density 950 kg/m³, cetane number about 37. Some success has been achieved by the use of additives such as acetone peroxide.

(14) Crankcase oil dilution

This is the percentage of fuel oil contamination of lubricating oil occurring in IC engines, which can seriously reduce the lubricity of the oil. The lubricating oil sample is mixed with water and heated, and fuel volatiles are carried over with the steam vapour formed. By condensation of these vapours and separation, the fuel content can be measured and can be expressed as a percentage of the original lubricating oil sample by mass.

It is also important to check the lubricating oil for water contamination, and for this purpose a similar separation test by heating is satisfactory. Severe corrosion of crankshafts has been caused by sulphur products from fuel oil mixing with any water in the lubricating oil to form sulphuric acids, which are carried round the lubricating oil system.

(14) Sulphur in the fuel

Restricting the sulphur in marine fuel oil is seen as key to the industry meeting its obligations for restricting air pollution. See Figure 2.12 for the timeline for these changes. This action is driving considerable technical change throughout the industry. The problem is that the very low-sulphur option for use in the emission control areas (ECA) is only available in diesel oil, which is much more expensive than residual fuel oil and in short supply. Another problem is that the specification for marine diesel is virtually the same as for automotive diesel but the flash point is different at 60°C for marine and 55°C for automotive. There would need to be a change in SOLAS for the maritime industry to have access to the wider availability of the automotive diesel. The values and timeline for the implementation of MARPOL Annex. VI is shown in Figure 2.12. The limits on the sulphur content of marine fuel is to reduce the SO_x and particulate matter from the emissions of power plants on ships. Annex. VI regulation 4 allows flag administrations to approve alternative means of compliance with the exhaust emission limits that are at least compatible with the limits for engines using the low-sulphur fuel. The after engine exhaust gas scrubbing alternatives are described in Chapter 11 of Volume 12 in the *Reeds* series.

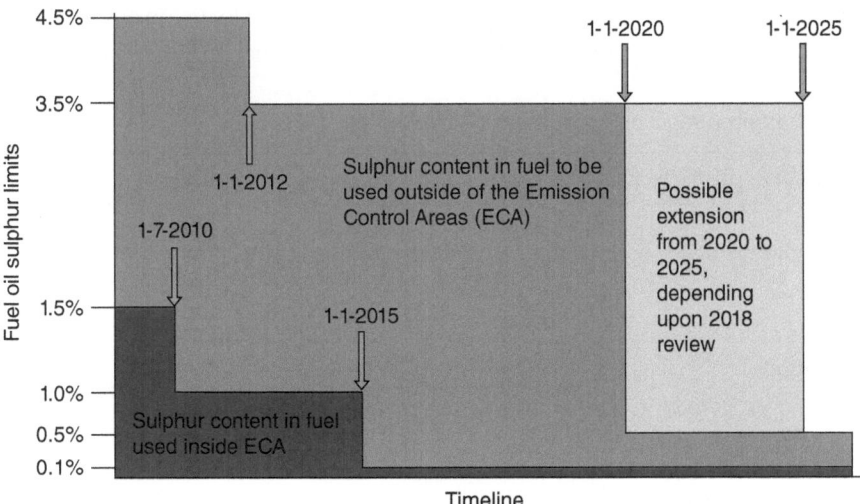

▲ Figure 2.12 *Timeline for exhaust gas emission restrictions*

Analysis of fuel oils (typical)

It is not practice to assume that a trend with one variable will apply to another. As a generalisation the 'heavier' the oil the higher the viscosity and flash point and the lower the calorific value. This would indicate systems for extra heating, purification, etc.; reduced storage volume for a given bunkering mass; increased fuel demand for a given power (Table 2.1).

ISO Specification for marine fuels

Table 2.2 shows details of the basic ISO 8217 Fuel Standard. The standards have been updated between 2005 and 2017. Bunkers should be ordered to the most up-to-date specification and checks should be made to ensure that the 'as delivered' bunkers matched the specifications. When reading this specification please note that the designation used for the different fuel products carries the ISO category F labels. DMX and DMA would normally be considered representative of marine gas oil (MGO) while DMB and DMC would normally be considered as marine diesel oil (MDO). The residual fuel products are as follows: RMA and RMB are fuels for use at low ambient temperature, they have a low pour point and would not necessarily require heat in the storage tanks; RMD, RME and RMG are fuels that would require a standard treatment plant on board; RMK is the very heavy residual fuel oil that would require a special treatment plant to handle such fuels.

DMX, as can be seen from Table 2.2, is the low-sulphur product for use in emission control areas (ECAs). (See Section 14 in this chapter for the inclusion of sulphur in marine fuels.)

Table 2.1 *Components and properties of different marine fuels*

Constituent or property	Petrol	Kerosene	Diesel	Residual
Carbon, %	85.5	86.3	86.3	86.1
Hydrogen, %	14.4	13.6	12.8	11.8
Sulphur, %	0.1	0.1	0.9	2.1
Density, kg/m^3	733	739	870	950
Higher calorific value, MJ/kg	47.0	46.7	46.0	44.0
Lower calorific value, MJ/kg	43.7	43.6	43.3	41.4
Viscosity, cSt at 50°C	1.5	1.6	5	350
Closed flash point,°C	0	50	85	90

Combustion of Fuel

The combustible elements in a fuel are carbon (C), hydrogen (H_2) and sulphur (S). These combustibles when supplied with oxygen (O_2) from atmospheric air combust and liberate heat. *Exothermic* reactions are those involving heat evolution, as are most combustion reactions, but some are *endothermic* and require heat supplied externally.

Table 2.2a *Marine Distillate Fuels ISO 8217 Basic Fuel Standard*

	Unit	Limit	DMX	DMA	DMB	DMC
Viscosity at 40°C	mm²/s	Max	5.500	6.000	6.000	11.00
Viscosity at 40°C	mm²/s	Min	1.400	2.000	3.000	2.000
Micro carbon residue at 10 residue	m/m	Max	0.30	0.30	0.30	
Density at 15°C	kg/m³	Max		890.0	890.0	900.0
Micro carbon residue	m/m	Max				0.30
Sulphur	m/m	Max	1.00	1.50	1.50	2.00
Water	V/V	Max				0.30
Total sediment by hot filtration	m/m	Max				0.10
Ash	m/m	Max	0.010	0.010	0.010	0.010
Flash point	ODC	Min	43.0	60.0	60.0	60.0
Pour point, summer	ODC	Max		0	0	6
Pour point, winter	DC	Max		−6	−6	0
Cloud point	DC	Max	−16			
Calculated cetane index		Min	45	40	40	35
Acid number	mgKOH/g	Max	0.5	0.5	0.5	0.5
Oxidation stability	g/m³	Max	25	25	25	25
Lubricity, corrected wear scar diameter (wsd 1.4 at 60°C)	um	Max	520	520	520	520
Hydrogen sulphide	mg/kg	Max	2.00	2.00	2.00	2.00

Table 2.2b Marine Residual Fuels Basic Standard ISO 8217

Parameter	Unit	Limit	RMA	RMB	RMD	RME	RMG				RMK		
Kinematic Viscosity			10	30	80	180	180	380	500	700	380	500	700
Density at 15°C	kg/m³	Max	920	960	975	991					1010		
Micro carbon residue	m/m	Max	2.5	10	14	15	18				20		
Aluminum + silicon	mg/kg	Max	25	40	-	50	60						
Sodium	mg/kg	Max	50	100	-	50	100						
Ash	%m/m	Max	0.04		0.07		0.1				0.15		
Vanadium	mg/kg	Max	50		150		350				450		
CCAI	-	Max	850		860		870						
Water	% v/v	Max	0.3	0.5									
Pour point (upper), summer	°C	Max	6		30								
Pour point (upper), winter	°C	Max	0		30								
Flash point	°C	Min				60							
Sulphur	% m/m	Max				Statuatory requirements							
Total sediment, aged	m/m	Max					0.10						
Acid number	mgKOH/g	Max					2.5						
Used lubricating oils (ULO): calcium and zinc; or calcium and phosphorus	mg/kg		The fuel shall be free from ULO, and shall be considered to contain ULO when either one of the following conditions are met: calcium > 30 and zinc > 15; or calcium > 30 and phosphorus > 15										
Hydrogen sulphide	mg/kg	Max				2.00							

Combustion of carbon

$$C + O_2 \quad \rightarrow \quad CO_2$$
$$\text{Relative masses} \quad 12 + (16 \times 2) \quad \rightarrow \quad 44$$
$$1 + 2\tfrac{2}{3} \quad \rightarrow \quad 3\tfrac{2}{3}$$

Thus 1 kg of carbon requires $2\tfrac{2}{3}$ kg of oxygen and forms $3\tfrac{2}{3}$ kg of carbon dioxide (CO_2). This chemical process liberates about 33.7 MJ/kg of carbon burned. If the carbon is incompletely burned to form carbon monoxide (CO) and that gas leaves the system unburned, the following results.

$$2C + O_2 \quad \rightarrow \quad 2CO$$
$$\text{Relative masses} \quad (2 \times 12) + (16 \times 2) \rightarrow (2 \times 28)$$
$$1 + 1\tfrac{1}{3} \quad \rightarrow \quad 2\tfrac{1}{3}$$

Thus 1 kg of carbon requires $1\tfrac{1}{3}$ kg of oxygen and forms $2\tfrac{1}{3}$ kg of carbon monoxide. This chemical process liberates about 10.25 MJ/kg of carbon burned, which represents a 70% heat loss due to incomplete combustion.

Combustion of hydrogen

$$2H_2 + O_2 \quad \rightarrow \quad 2H_2O$$
$$\text{Relative masses} \quad (2 \times 1 \times 2) + (16 \times 2) \rightarrow (2 \times 18)$$
$$1 + 8 \quad \rightarrow \quad 9$$

Thus 1 kg of hydrogen requires 8 kg of oxygen and forms 9 kg of water vapour (steam) (H_2O). This chemical process liberates about 144.4 MJ/kg of hydrogen burned, assuming the products of combustion are cooled and the latent heat is extracted from the water vapour. If the steam escapes from the plant uncondensed, which is usual, it takes the latent heat with it then the net (or lower) heat liberation/kg of hydrogen burned is 144.4 – (9 ×2.465) =122.2 MJ. The 2.465 figure is the specific enthalpy of vaporisation of steam at 15.5°C (the standard temperature). The higher and lower calorific values of hydrogen are:

hcv hydrogen 144.4 MJ/kg
lcv hydrogen 122.2 MJ/kg

Combustion of sulphur

$$S + O_2 \quad \rightarrow \quad SO_2$$
$$\text{Relative masses} \quad 32 + (16 \times 2) \quad \rightarrow \quad 64$$
$$1 + 1 \quad \rightarrow \quad 2$$

Thus 1 kg of sulphur requires 1 kg of oxygen and forms 2 kg of sulphur dioxide (SO_2). This chemical process liberates about 9.32 MJ/kg of sulphur burned. At high temperatures the very small percentage of the sulphur present as sulphates in the fuel is partly expelled as sulphur trioxide (SO_3). The percentage of SO_3 to SO_2 during normal combustion is low, usually of the order of 1%, and the chemical reaction is:

$$2S + 3O_2 \rightarrow 2SO_3$$

This reaction is of negligible importance from a heating viewpoint but is very important from the viewpoint of corrosion. The residual oils in regular use today contain appreciable amounts of sulphur (up to 6%) and so increase the amount of SO_3 in the flue gases. The presence of SO_3 has been shown to raise the acid dewpoint of the gases. This results in the condensation of acid vapours on colder metal surfaces, and the formation of sulphurous and sulphuric acids from the sulphur with resulting corrosion. Mild steels are most severely attacked by the dilute acids at temperatures of about 28°C below the dewpoint, for cast irons 50°C.

Low sulphur fuel oil

Virtually all the sulphur dioxide found in the atmosphere has been placed there by human activity. The reaction that occurs in Internal Combustion (IC) engines makes a significant contribution to this total.

Sulphur dioxide is detrimental to human health and is formed when the sulphur in the fuel combines with the oxygen during the heat of the combustion process. The illustration in the previous paragraphs show that the sulphur in the fuel contributes very little to the total of the heat energy produced by burning the fuel.

Therefore, why is the sulphur there at all?

The sulphur is in most crude oils that are taken from the earth, and it has not usually been removed during the refining process. However, the sulphur is now being removed to make available the low-sulphur fuel oil that is required to give the low-sulphur emissions in the diesel exhaust gas.

The acids formed from burning the sulphur in the fuel also cause a very relevant problem in air heaters and economisers of boiler plants. In IC engines the problem can affect cylinder liners and exhaust valves, especially where the exhaust valve has a water-cooled cage. The dew point can also be reached when the exhaust flows through economisers and exhaust gas boilers causing corrosion of the heating surfaces of these items. It is very important for the on-board engineers to ensure that the scavenge air temperature is kept at the correct value because if the engine is run with a low scavenge air temperature then the corresponding exhaust temperature is lower. This will mean that the dew point is reached and acids form at an earlier stage than was the intention of the designers, and components will be subjected to these acids that are not designed to withstand them for long.

Chemical reaction:

$SO_3+H_2O \rightarrow H_2SO_4$ (sulphuric) or $SO_2+H_2O \rightarrow H_2SO_3$ (sulphurous) then
$2H_2SO_4+O_2 \rightarrow H_2SO_4$

Certain catalysts are known to accelerate the formation of sulphur trioxide, such as vanadium pentoxide, which can be present in the fuel, also metallic catalysts such as platinum.

Sulphurous acid is relatively harmless and unstable, and in the presence of free oxygen, which invariably exists in flue gases, turns to the more harmful sulphuric acid.

Dilute acids attack the metal continuously but strong acids cause an initial attack, which is then stifled by a passive gas layer or skin developed on the metal surface.

Calorific value

It can be assessed from the approximate empirical formula:

$$hcv = 33.7C + 144.4\left(H_2 - \frac{O_2}{8}\right) + 9.32S$$

$$lcv = hcv - 2.465(kg\ H_2O)$$

Value varies from 32.5 MJ/kg for coal to 44.5 MJ/kg for fuel oils, as discussed previously.

Air for combustion

Atmospheric air is composed of nitrogen (N_2) and oxygen (O_2) in the ratio 77% to 23% by mass, neglecting proportions of other gases as small in comparison.

Nitrogen takes no part in the combustion process, merely representing a large but unavoidable heat loss: being inert it is heated and passes up the uptakes to waste. However it is a major problem when considering the composition of the exhaust gases due to the potentially harmful chemical compounds produced as a result of the combustion process. The oxygen required for combustion on a theoretical basis can be determined from the combustion equations.

$$\text{Theoretical oxygen} \times \frac{100}{23} = \text{Theoretical air}$$

In practice, the traditional method to approach complete combustion was achieved by supplying excess air. In a boiler plant about 30% excess air (much less in modern practice) is the minimum normally required, that is, about 12 kg of air/kg of fuel. In the IC engine of the compression ignition type, about 100% plus excess air is normally required, that is, about 40 kg of air/kg of fuel, as complete and rapid combustion is essential. This becomes apparent in that the CO_2 value (% by volume) should be about 18% theoretically whereas 12% and 7%, in the case of the boiler plant and IC engine, respectively, are typical values due to the air dilution of the gases. All the combustion equations given are based on perfect gases measured at the same temperature and pressure. Standard temperature and pressure (stp) is usually adopted as a basis in combustion work, that is, 760 mm of mercury (1.013 bar) and 15.5°C although normal temperature and pressure (ntp) is sometimes used, that is, 760 mm of mercury (1.013 bar) and 0°C. In modern engines and boilers the combustion process and the composition of exhaust gases can be monitored to modify the control of the fuel/air/egr systems thus achieving a better combustion, close to the theoretical maxims.

The terms stp and ntp are not rigid in definition and the assumptions made should be clearly stated: % CO_2 should always be quoted by volume, % CO_2 by mass is higher. As stated earlier, the nitrogen contributes no benefit to the combustion process. However, it has now become the subject of great attention due to the way that it moves through the combustion process and changes to result in the production of Nitrogen dioxide (NO_2) and nitrous oxide (NO). (Note that the two together are often referred to as NO_x although NO_x is not the chemical term for nitrous oxide). The very high peak temperatures of combustion cause the nitrogen to oxidise and nitrous oxides (NO_x) that are produced, as a consequence, are released into the engine's exhaust system and then into the atmosphere if they are not stopped in some way.

The NO_x emissions are of concern because they are associated with the increased acidity of particles, cloud water and precipitation (acid rain), causing damage to plant and marine life and are also an irritant to the human respiratory system. NO_x emissions are an important part of the formation of photochemical smog and atmospheric oxidants. However, reducing the NO_x from IC engines is another very technically challenging problem. Replacing some of the excess air with recirculated exhaust gas reduces the overall content of nitrogen and excess Oxygen available in any one combustion cycle and also has the effect of reducing the heat loss to the fresh nitrogen drawn in with the air. Engines were required to meet the so-called IMO Tier III exhaust emission standard for new marine IC engines by 1 January 2016, which requires a significant reduction in NO_x. To achieve this, engine development is concerned with reducing the peak temperatures that are producing the NO_x. The use of Exhaust Gas Recirculation (EGR) and the Miller cycle linked with two-stage turbocharging (see Volume 12 of the Reeds series, Chapter 1) are two of the main strategies but exhaust gas scrubbing technology is also to be employed. Sulphur in the exhaust gases of ships is also a problem as it combines with oxygen and hydrogen to form sulphur dioxide (SO_2) and sulphur trioxide (SO_3) and can also combine with water droplets to form acid. Any sulphur in the exhaust gas is directly related to the sulphur content in the fuel – hence the move to low-sulphur fuels.

Combustion of hydrocarbons

Liquid natural gas (LNG) is under serious consideration to be used as a fuel for powering ocean-going ships. The most significant obstacle holding up the wider use is the development of the supply infrastructure. One of the major reasons for the use of LNG is that it is a cost-effective answer to the poor emission quality of the exhaust when burning the more traditional fuels.

The composition of LNG is 85% to 90% methane, 6% to 12% ethane and around 2% propane. There are other traces of butane and pentane but the vast majority of the natural gas is methane. With methane having up to 25 times the greenhouse warming effect of CO_2 it is reasonable to ask how its use can be better than using other fossil fuels. If we consider the combustion of methane (CH_4) we can see the results:

$$CH_4 + 2O_2 \quad \rightarrow \quad CO_2 + 2H_2O$$
$$16 + (2 \times 16 \times 2) \quad \rightarrow \quad 44 + (2 \times 18)$$
$$1 + 4 \quad \rightarrow \quad 2\tfrac{3}{4} + 2\tfrac{1}{4}$$

Thus 1 kg of methane requires 4 kg of oxygen for complete combustion and forms $2\tfrac{3}{4}$ kg of carbon dioxide and $2\tfrac{1}{4}$ kg of water vapour. Therefore as long as methane is not allowed to leak out before being burnt and complete combustion is achieved, then the LNG will fare better in the exhaust emissions tests.

Other hydrocarbons, for example, acetylene (C_2H_2), hexane (C_6H_{14}), etc., are often considered in the combustion equations. Methane is also a gas released in bunker spaces, acetylene is a burning-welding gas and hexane is close to motor spirit gasoline. A series of common gases and their properties in combustion are given in Table 2.3.

Hydrocarbons usually have flammability limits between 1% and 10% but for volatile gases the range is much wider.

All the figures given in Table 2.3 relate to atmospheric pressure and temperature and are average values that are not to be applied too rigidly.

Temperature increase tends to increase the range of flammability, for example, CH_4 at 15.5°C in the range 6.3–12.9% and at 405°C range 4.8–16%. Pressure increase is similar but more marked with upper limit increasing and lower limit falling, up to 52 bar. Gas concentration and mixture will also have pronounced effects. The values of spontaneous ignition temperature are lower usually in oxygen, being affected by pressure, concentration, mixture and temperature. In the case of other fuels, the

Table 2.3 *Ignition properties of different combustible products*

Spontaneous ignition temperature		
Gases	In air (°C)	In oxygen (°C)
Hydrogen	588	588
Carbon monoxide	649	649
Methane	700	577
Benzene	742	731
Acetylene	422	416
Cylinder oil	420	–
Kerosene	294	–
Limits of flammability in air		
Gases	Lower (% vol.)	Upper (% vol.)
Hydrogen	4.0	75.0
Carbon monoxide	12.5	74.0
Methane	5.0	15.0
Benzene	1.4	7.4
Petrol	1.4	6.0
Coal gas	5.3	31.3

spontaneous ignition temperatures in common cases are cetane 236°C, diesel oil 246°C, gas oil 355°C, gasoline 399°C, steam coal 469°C, coke 555°C (average), etc.

Liquid natural gas as a marine fuel

Liquid natural gas (LNG) was first used as a fuel to power the ships that were transporting the LNG across the ocean. The gas was cooled to -160°C so that it could be pumped and transported in liquid form. This meant that the cargo holds had to be very well insulated to both contain its cryogenic state and protect the ship and crew from the low temperatures. Despite the insulated tanks it is difficult to stop the cargo from attracting some heat. This means that the pressure becomes too high and a small proportion of the gas has to be released. Originally this 'boil off' gas was vented into the atmosphere.

Early in the development of LNG carriers the boiled-off gas was used as a power source for the ship. The original ships were steam turbines and the LNG was used as a fuel for the high-pressure superheated steam boilers. This was an economical option because the power plant was not using ordinary bunker fuel. Very soon the diesel engine manufacturers developed engines that could also run by using LNG as a fuel and they started to win orders when the new LNG vessels were built. All the engine manufacturers have built up experience with LNG as a fuel and the research findings are that the engines will conform to the stringent emission regulations when LNG is used as a fuel. The use of this combination for a new build is starting to gather pace but the lack of bunkering infrastructure is holding back widespread development. There have also been reports that engines using LNG as a fuel are more prone to a phenomenon known as methane slip.

Methane (CH_4) slip is the phenomenon where some of the methane from the fuel moves through the engine and out of the exhaust without being burnt. Some manufacturers are keen to point out that this occurs more on the engines that operate on the Otto cycle rather than the diesel cycle. However, the industry is confident that as the mechanics become better understood so changes in combustion design will reduce the problem. Some suggestions for how methane can bypass the combustion process include being injected early or late in the combustion cycle and the gas is therefore caught in the scavenge port and gets sucked through during the overlap period. Another possibility is that the air/gas mix in the Otto cycle can be caught just above the piston ring where it remains unburnt and escapes with the exhaust.

It then follows that older, fuel oil, combustion space designs could be more prone to these imperfections than would new engines that are designed with methane slip in mind. It also follows that any reduction in fuel injection performance could make the situation worse.

As the engine design improves so will the combustion efficiency and therefore less unburnt fuel will pass through the process, making the modern, purpose-built 'gas' engine less and less prone to methane slip.

The wider industry view is that methane slip is a real issue. However, it is only part of the issue. CIMAC (The International Council on Combustion Engines) discussions focused on reducing all engine emissions and not looking at any one part in isolation.

Using gas as a fuel reduces the CO_2 in the exhaust gas considerably, cuts the NO_x by 90% and reduces the sulphur oxide emissions to practically zero according to the in-service experience of Rolls-Royce, which now has in excess of 30,000 hours, operational experience from which to draw upon. In the face of so much saving of emissions a temporary small amount of methane is a good transient solution.

Flame temperature

This varies mainly with the fuel type. Typical figures for gaseous fuels would be methane 1,872°C, hydrogen 2,037°C, carbon monoxide 1,957°C.

The values quoted are theoretical calculated values as the actual values are very difficult to measure; variables, such as gas mixtures, radiation and dissociation, etc., are allowed for by the use of charts, in every case these effects serve to reduce temperature. Theoretical values are based on cold gas with theoretical cold air supplied, excess air decreases the flame temperature, and air preheat together with rapid combustion increases the value. Hydrogen would give 6,297°C by itself with theoretical oxygen, 2,307°C with theoretical air and 1,397°C with 100% excess air. An average furnace gas temperature may be taken as 1,200°C. Flame temperature, radiation, luminosity, particle size, etc. are complex considerations still being subject to much research. Some success is starting to happen by looking at the UV radiation projected by the elements as they burn.

Oil fuel additives

Fused slag products, which give corrosive attack, are formed from the vanadium and sulphur content of the fuel oil. A particular type of additive gives the following reactions:

For sulphur dioxide:

$$SO_3 + CaCO_3 = CaSO_4 + CO_2$$

For vanadium pentoxide:

$$V_2O_5 = CaCO_3 = Ca(VO_3)_2 + CO_2$$

There is an increase in CO_2 content due to more complete combustion of the carbon. It is also usual to find an increase in flue gas temperature indicating better furnace or cylinder combustion.

Similar reactions to the above can be obtained using barium carbonate ($BaCO_3$) in place of calcium carbonate ($CaCO_3$) above. The sulphate compounds formed have a higher ash melting point and there is little risk of fused deposits. Removal of vanadium pentoxide removes one of the catalysts that assists the formation of sulphur trioxide.

Additives are usually added directly to the empty oil fuel bunker tank before bunkering in quantity about 0.03% of the fuel (for 100 p.p.m. of V or Sr compounds). The additives are usually claimed to give less carbon and soot deposits in furnaces and less gum and general deposits in IC engines.

It is unwise to ballast oil fuel tanks with salt water as this accelerates the deposit of V_2O_5 Na_2SO_4 and Fe_2SO_4 as the salt water compounds act as a flux as well as providing slag products directly. For really elevated boiler conditions, calcium hydroxide or zinc oxide additives can be used:

$$V_2O_5 + Ca(OH)_2 = Ca(VO_3)_2 + H_2O$$
$$V_2O_5 + ZnO = Zn(VO_3)_2$$

Oxides of barium, calcium, zinc, etc. have been used as fuel additives for some years to try to combat sulphur acid attack to air heaters.

With the changes in the refining of crude oil and the use of catalytic cracking processes comes a reduction in the quality and stability of residual fuel oil. This is turn has led to a rise in the use of fuel additives. There are soot-reduction additives and combustion improvement additives as well as additives to reduce the build-up of lacquer.

Analysis of Flue Gases

The analysis of the flue gases in real time is so important on modern ships. Already automotive technology relies on the composition of the exhaust gas to give feedback to the combustion controller, which in turn will modify the settings that control the start of the process to keep the engine as efficient as possible. Marine engines are following this lead but emissions from the ship's funnel have to be correct for the area that the vessel is sailing within. If the composition of the flue gases is not correct the ship could be due for a heavy fine.

Technology has come such a long way during the late 20th century and the early 21st. The development of digital techniques, lasers, chemiluminescent and the UV fluorescence principle have led to the development of real-time exhaust gas analysers for ships.

Real-time monitoring will depend upon equipment using one or more of the following scientific principles:

1. Spectroscopic absorption
2. Luminescence
3. Electroanalysis
4. Paramagnetism
5. Laser technology

MARPOL

The International Maritime Organisation (IMO) is the department of the United Nations (UN) that is responsible for matters relating to international shipping. To complete the work of devising, updating and revising relevant conventions, the IMO is divided into a number of committees and sub-committees. More information about this formal structure can be found in Chapter 12, which discusses the management of ships.

The main committee responsible for discussing all matters relating to marine pollution is the Marine Environment Protection Committee (MEPC). This committee is helped in its work by the nine sub-committees. MEPC meets each year but may also have additional meetings if the workload is high. The committee discusses issues such as the recycling of ships, air pollution, anti-fouling coatings, harmful aquatic organisms in ballast water and of course MARPOL.

MARPOL is the IMO convention that covers the pollution from ships by oil, noxious liquid substances carried in bulk, harmful substances carried by sea in packaged form, sewage, garbage, and most recently, it also covers the prevention of air pollution from

Pollutant	Cause	Possible corrective action
Sulphur oxides	The oxidation of sulphur during the combustion process	Use low-sulphur fuel or remove the sulphur from the fuel or from the exhaust gas using scrubbing technology
Carbon dioxide	Carbon in the fuel is combined with oxygen to liberate energy. The by-product is carbon dioxide. See 'Combustion of fuel' in this chapter	Burn less fuel for a given output or use an alternative fuel source other that hydrocarbons
Carbon monoxide	Incomplete combustion of hydrocarbon fuel	More efficient design of fuel-injection system or combustion space or other efficiency gains
Unburnt hydrocarbons	Incomplete combustion	More efficient design of fuel-injection system or combustion space
Visible particulates	Incomplete combustion at full or part load	Improvements in the combustion control technology
Nitrogen oxides	Nitrogen in the air (induction) oxidises at high combustion temperatures	Reduce peak combustion temperatures by use of the Miller cycle and exhaust gas recirculation (EGR)

ships with new additions to Annex. IV; the revised version of MARPOL Annex. IV entered into force on 1 July 2010.

According to IMO's study during 2007, greenhouse gas emissions from ships amounted to 2.8% of the world's total, which at the time was seen as being the same as the total contribution of Germany, sixth in the country league table of GHG emissions. However, it was also recognised that shipping moved about 90% of the world's trade and was the most efficient form of transport in terms of emissions per ton-kilometer. By 2012 the percentage had dropped to 2.2%.

International shipping has developed its business model around using the cheapest fuel source as possible for as long as possible. This mode of operation has driven engine manufacturers to produce engines that will burn the residual fuels that are left over from the crude oil refining process. When ships' engines first moved to burn heavy fuel oil they were designed to stop and start on marine diesel oil (MDO) and change over to heavy oil when the ship was travelling out at sea. However, as developments progressed the requirement to use the more expensive MDO became less and less. The problem with burning heavy fuel oil is that it is not a high-quality fuel and it also contains chemicals such as sulphur that, when put through the combustion process and combined with other chemicals, has a detrimental effect on the atmosphere when emitted from the exhaust of marine diesel engines.

IMO requires that the industry use a number of strategies to reduce harmful emissions from the ship's engines exhaust. Energy-saving strategies such as the implementation of the Ship's Energy Efficiency Management Plan (SEEMP) are covered in Chapter 12 of this volume.

The revised Annex. IV set out the road map for the reduction in the direct emissions from the exhaust of ships. A summary of the main pollutants is shown on the preceding page.

The reduction of harmful exhaust gas emissions from ships is a combination of effort from all parts of the industry. Equipment manufacturers are making more efficient machinery such as variable-speed centrifugal pumps and high-efficiency turbochargers. Fuel suppliers are ensuring that low-sulphur fuel is produced and engine manufacturers are improving engine technology. See page 75 for the timeline that is in place for the introduction of low-sulphur fuel to be used by international shipping.

Dissociation

Most combustion reactions are reversible. At high temperatures, the molecule bonds that have formed during combustion tend to disrupt and re-form molecules of the original elements/compounds, absorbing heat in the process. The chemical nomenclature is:

$$2H_2O \rightleftarrows 2H_2 + O_2$$
$$2CO_2 \rightleftarrows 2CO + O_2$$

There is an increase in volume, which is resisted by high pressures so as pressure rises dissociation reduces.

CO_2 1 bar 0.1% dissociation at 1,760 K, 6% at 2,260 K, 55% at 3,250 K

CO_2 102 bar 0.01% dissociation at 1,760 K, 0.1% at 2,260 K, 17% at 3,250 K

H_2O 1 bar 0.04% dissociation at 1,760 K, 2% at 2,260 K, 28% at 3,250 K

H_2O 102 bar 0.004% dissociation at 1,760 K, 0.3% at 2,260 K, 3% at 3,250 K

These figures only relate to the gas or vapour by itself. Gas mixtures and rich oxygen contents tend to reduce dissociation considerably.

Once the temperature falls the molecules reform (re-combustion) and heat is again evolved. Thermal decomposition is non-reversible split-up under heat, whereas thermal dissociation is reversible split-up under heat.

In an IC engine, dissociation causes reduction of maximum combustion temperatures and heat re-appearance during expansion occurs, which tends to raise the curve above the adiabatic.

In a fire there is a danger that the use of superheated steam as an extinguishing agent (say sootblowers on an air heater fire) could in fact *feed* the fire and accelerate the growth. For example, the *displacement* that occurs at about 707°C:

$$\text{Heat} + \text{Hot} \, 3\,Fe + 12H_2O \rightleftarrows 3FeO_3 + 12H_2 \uparrow$$

goes to completion giving liberation of volatile hydrogen and makes the combustion more rapid. Such fires are sometimes called 'rusting' fires. Although the total displacement reaction is reversible the main factor is *decomposition* of steam vapour whereas dissociation of steam vapour plays a relatively minor part; however, as dissociation is regarded chemically as reversible decomposition then the process is often regarded by engineers as dissociation.

The boiler combustion heat balance

A typical analysis of the heat utilisation in an oil-fired boiler would be as in Table 2.4.

It can be seen that dry combustion gases, excess air and wet vapours are losses. Minimum excess air and lowest practical flue gas temperature (bearing in mind complete combustion and corrosion in uptakes, etc.) reduce these losses together with close attention to CO_2 content, that is, to reduce unburnt combustion gas loss.

The % flue gas (FG) loss can be seen to increase with flue gas temperature increase, increase with excess air increase and increase with fall in gas CO_2 content (see Figure 2.13).

The condition of the gases leaving the funnel is often the best indication of combustion conditions. Black smoke due to insufficient air (among other things), white smoke due to too much air, blue smoke due to burning of lubricating oils (in IC engines), yellow smoke indicative of high-sulphur-bearing fuels, etc.

Table 2.4 *Boiler combustion heat balance*

Item	Heat %
In fuel (hcv)	100
In steam	80
In excess air	$3\frac{1}{2}$
In dry gases	8
In wet vapours	3
In unburnt gases	$5\frac{1}{2}$
In radiation	5
	100

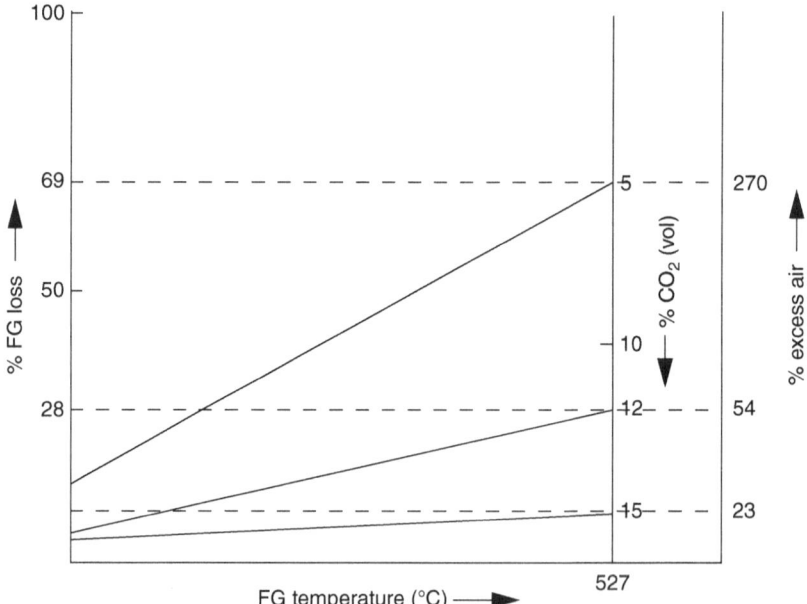

▲ **Figure 2.13** *Energy loss in flue gas relating to temperature*

However, CO_2 content is often required to give the efficiency of combustion for a particular plant. Each plant, however, will have its own optimum figure and this may vary for boilers between 10% and 14% depending on many variables.

Boiler Combustion Equipment

Good combustion is essential for the efficient running of the boiler: it gives the best possible heat release and the minimum amount of deposits upon the heating surfaces. To ascertain if the combustion is good, measure the % CO_2 content (and in some installations the % O_2 content) and observe the appearance of the gases.

If the % CO_2 content is correct (or the % O_2 content low) and the gases are in a non-smoky condition then the combustion of the fuel is correct. With correct % CO_2 content the % excess air required for combustion will be low and this results in improved boiler efficiency since less heat is taken from the burning fuel by the small amount of excess air. If the excess air supply is increased then the % CO_2 content of the gases will fall.

Condition of burners, oil condition, pressure and temperature, condition of air registers, air supply pressure and temperature are all factors that can influence combustion.

Burners. If these are dirty or the sprayer plates are damaged then atomisation of the fuel will be affected. Types include pressure jet, in various forms, rotary cup, steam jet and ultrasonic.

Oil. If the oil is dirty it can foul up the burners. (Filters are provided in the oil supply lines to remove most of the dirt particles but filters can get damaged. Ideally the mesh in the last filter should be smaller than the holes in the burner sprayer plate.)

Water. If there is water in the oil it can affect combustion by possibly leading to the burners becoming extinguished and a dangerous situation arising. It could also produce panting, which can result in structural defects.

If the oil temperature is too low oil does not readily atomise since its viscosity will be high. This could cause flame impingement, overheating, tube and refractory failure. If the oil temperature is too high the burner tip becomes too hot and excessive carbon deposits can then be formed on the tip causing spray defects. These could again lead to flame impingement on adjacent refractory and damage could also occur to the air swirlers.

Oil pressure is also important since it affects atomisation and lengths of spray jets.

Air registers. Good mixing of the fuel particles with the air is essential, hence the condition of the air registers and their swirling devices are important. If they are damaged mechanically or by corrosion then air flow will be affected. Pressure drops over the Venturi of 25 mm water gauge give air speeds of about 20 m/s. Modern swirlertype stabiliser designs give more efficient mixing with pressure drops up to 300 mm water gauge and air speeds up to 70 m/s.

Air. Excess air supply is governed mainly by air pressure and if this is incorrect combustion will be incorrect.

Figure 2.14 shows a simple pressure-jet burner arrangement for a boiler (Wallsend-Howden). Preheated, pressurised fuel is supplied to the burner tip, which produces a cone of finely divided fuel particles that mix with the air supplied around the steel burner body into the furnace. A safety point of some importance is the oil fuel valve arrangement. It is *impossible* to remove the burner from the supporting tube unless the oil fuel is shut off, which greatly reduces the risk of oil spillage in the region of the boiler front.

Figure 2.15 shows the boiler front air register (top diagram) and tip (middle diagram) for the Y-jet steam atomising oil burner, which is finding increased favour for use with water tube boilers, for the following reasons:

1. Deposits are greatly reduced, hence soot blowing and water washing of the gas surfaces need not be carried out as frequently as before (18 months or more between cleaning is possible).

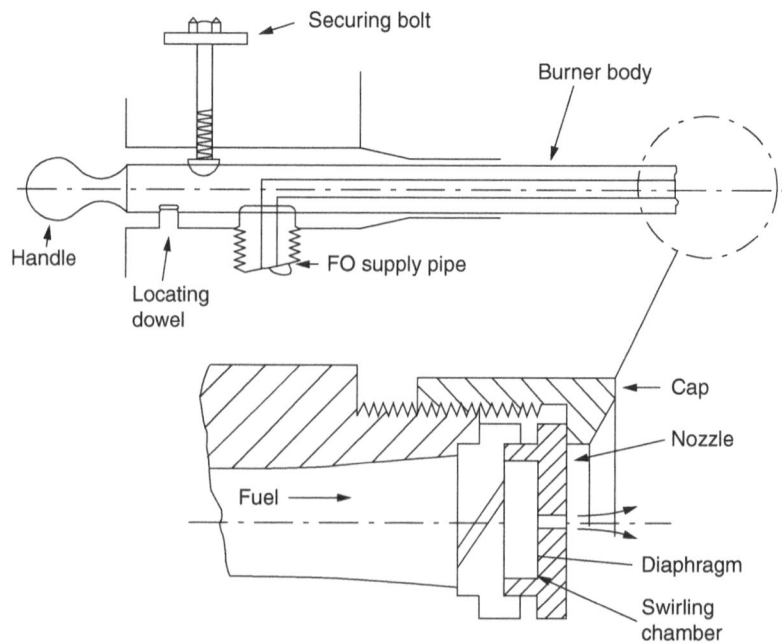

▲ **Figure 2.14** *OF burner (Wallsend-Howden)*

2.   Atomisation and combustion are greatly improved, and lower dew point gives reduced acid formation.

3.   % CO_2 reading is increased (% O_2 reading has been lowered to 1% or below) hence the boiler efficiency is greatly improved (excess air lowered from 15% to under 5%).

4.   Atomisation is excellent over a wide range of loads and the turndown ratio is as high as 0:1 (turndown ratio compares maximum and minimum flow rates).

5.   With improved combustion, and turndown ratio, refractory problems are reduced.

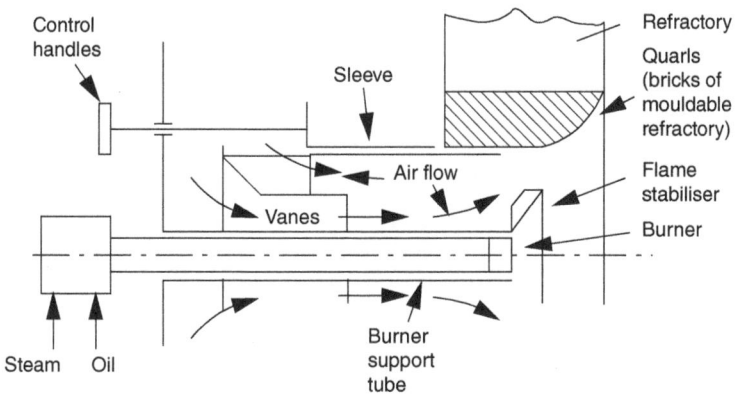

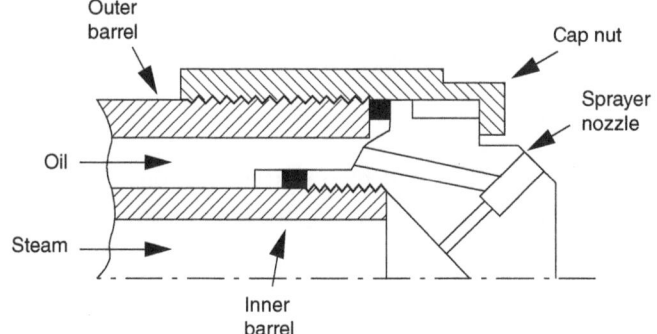

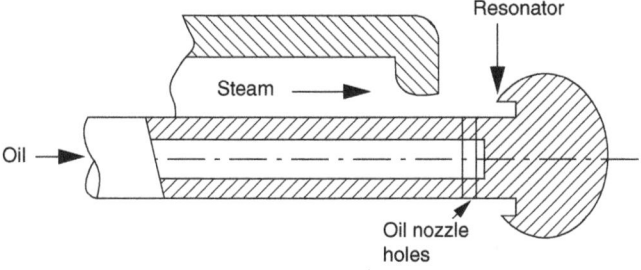

▲ **Figure 2.15** *Steam atomising equipment*

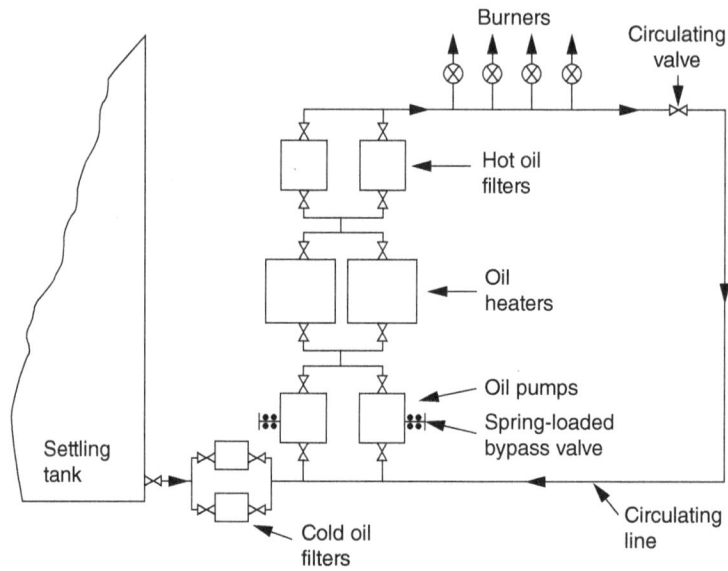

▲ **Figure 2.16** *Boiler OF supply system*

The major disadvantage of this type of burner is that it uses steam – which means water and fuel. But the steam consumption in the latest type of steam atomiser is extremely small, less than 1% of the oil consumption at peak loads. Maximum oil pressure is 22 bar and steam is supplied at 12 bar, a steam-control valve may be fitted to reduce the steam pressure at low loads.

Figure 2.16 illustrates a simple boiler OF supply system.

The ultrasonic atomiser (bottom diagram of Figure 2.15) consists of an annular steam nozzle, a resonator and a nozzle with holes. It gives high turndown ratio and low excess air. Ultrasonic energy wave vibration, 5–10 kHz, produced by high-speed steam or air flow against the resonator edge ahead of the oil holes gives atomisation with very minute oil droplets.

Viscosity control

Consider the viscometer (Figure 2.17). A gear pump is driven by a constant-speed electric motor through a reduction gear. A small, constant quantity of oil (fuel in this case) is passed through the fine capillary tube. As the flow is arranged to be streamlined (laminar), as distinct from turbulent, the differential pressure across the capillary tube is directly proportional to the viscosity of the oil. This differential pressure in two tappings, shown as + and –, is led from the viscometer to a differential pressure transmitter (DP cell).

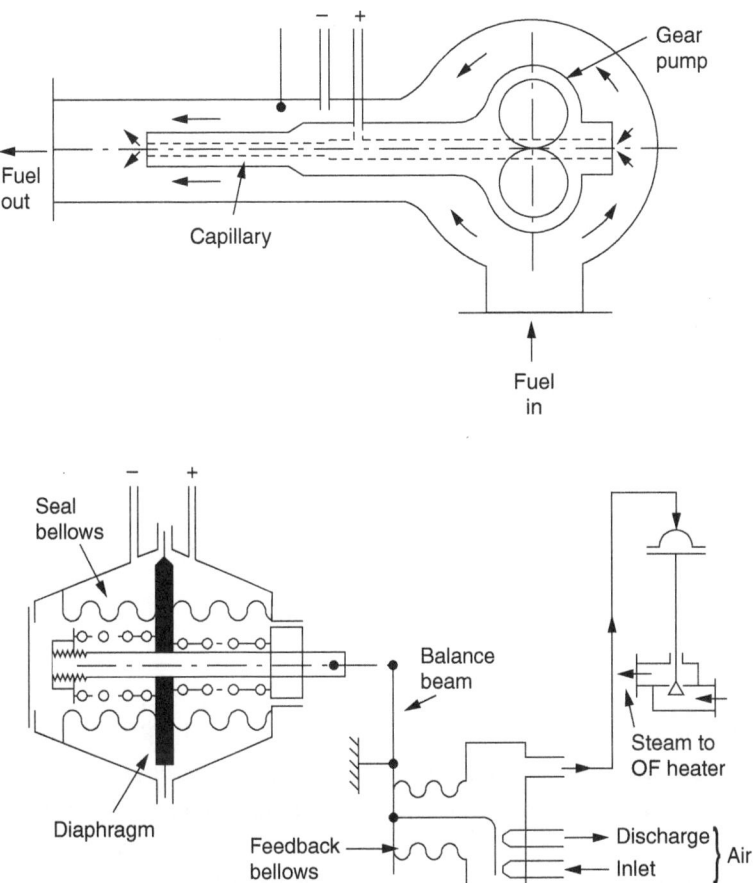

▲ **Figure 2.17** *Viscometer and DP cell transmitter*

Differential pressure from the viscometer is applied across the diaphragm of the transmitter. Increased differential pressure, caused by increased viscosity, causes the diaphragm and balance beam to move to the left. The air inlet nozzle is closed in and air pressure builds up in the feedback bellows due to a relay (not shown in Figure 2.17), supplying increased air pressure to the feedback bellows. Equilibrium occurs when the feedback force equals the originating force; under these conditions air escape is minimal. The feedback bellows pressure is the control output signal. Transmitter diaphragm chambers are filled with glycerine or silicone as oil would clog the parts.

The output signal is fed to a controller (and recorder) to control the steam flow to the oil fuel heater, which will cause viscosity adjustment. The actuator has a piston and valve positioner. The controller has desired value setting and incorporates a reset (integral) action. This detail has been simplified in Figure 2.17, lower diagram, so that the output pressure increase from the feedback bellows (due to viscosity increase) *directly* increases air pressure on a diaphragm valve to open up steam to the oil heater, to reduce fuel viscosity.

It is generally not good practice to control one variable by means of another (this induces time delays and can cause appreciable offset) but it is sometimes unavoidable.

Note: The Viscotherm unit works on a similar principle but there is no relay. The free end of the flapper is spring loaded, tending to push the flapper on to the nozzles. Movement of the balance beam left is arranged to close discharge and open supply (pressure increases). Movement right closes supply and opens discharge (pressure decreases). At equilibrium both nozzles are almost closed, which minimises air wastage.

Gaseous Fuels

There has been a steady increase in the number of 'parcel' tankers, which carry a wide range of chemicals. Such vessels are expected to comply with MCA recommendations before an IMO 'Certificate of Fitness for the Carriage of Dangerous Chemicals in Bulk' is issued under SOLAS Chapter VII. The IMO conventions require chemical tankers built after 1 July 1986 to comply with the International Bulk Chemical Code (IBC Code), which gives international standards for the safe transport by sea of dangerous liquid chemicals in bulk.

Reactions with air, with water, between incompatible chemicals and with self-reactive chemicals can arise and the United States Coast Guard has published the 'Bulk Liquid Chemicals; Guide to the Compatibility of Chemicals'. These vessels have special requirements relating to construction, materials, pumping systems, tank coatings, safety, etc. It is proposed here to outline only fuel technology aspects. LNG (liquefied natural gas) is a cryogenic, clear, colourless liquid with methane as its main constituent (about 87% by volume), ethane (about 9%), propane (about 3%) and traces of butanes and pentanes. Boiling point is about −162°C and heating value of the gas is about 10 MJ/m^3. Heat flow through insultation causes gradual evaporation, which maintains pressure above atmospheric and prevents ingress of air.

Liquid petroleum gas (LPG) is a mixture of hydrocarbon gases and includes propane, butane and ammonia. Boiling points are lower than for LNG.

Combustion equipment

Vaporisation of the liquid (boil-off about 0.2% per day) due to heat entry can be utilised as boiler or engine fuel, re-liquefied by suitable vapour pumps and compressors, or the adoption of both is possible. Utilisation as a dual fuel requires sophisticated instrumentation on the gas side and safety interlocks between the two fuel systems. The supply gas can vary appreciably in composition during the voyage, and close monitoring of composition, dryness, etc. are necessary with facility for pressure variation, variable heat input, etc. Complexities also arise due to differing air requirements, flame speeds, etc. but a suitable plant is readily available and the gaseous fuel gives rapid, efficient and clean combustion.

Explosive and toxic vapour concentration

Vessels should be equipped with duplicate combination instruments, one at least of which should be portable, and if instruments for one detection function only are provided then duplicates of each are necessary. Oxygen analysers, to indicate alarm when oxygen content falls below 18% (by volume) in a space, should also be provided.

Toxic vapours

A number of chemicals have toxic limits well below their combustible gas/air concentration ratio and it is unsafe to enter spaces even if the gas concentration is below the lower explosive limit (LEL). The threshold limit value (TLV) gives the concentration of a substance in air in p.p.m. which must not be exceeded if daily 8-hour exposure over extended time periods is intended. Typical TLV values are anilene (5), carbon tetrachloride (10), benzene (25), methanol (200) but the value can be very low, under 0.02, for certain chemicals. Detection apparatus utilises input via a bellows for a given volume, and colour reaction on selected chemicals is compared colourmetrically.

Explosive vapours

With low-flash-point products the danger may exist that the atmosphere lies between the LEL and the HEL therefore creating the risk of explosion. Typical flash points are pentane –49°C, hexane –23°C, heptane –4°C (aliphatic hydrocarbons), benzene –11°C (aromatic hydrocarbon), acetone –18°C (ketone), methanol 10°C (alcohol), carbon disulphide –30°C. Vapour pressure is that constant pressure during isothermal isobaric evaporation (or condensation), when liquid and vapour are in equilibrium (quality defined by dryness fraction). LEL and HEL are affected by variations in vapour pressure (see also Chapter 8).

Gas explosion–detector meter

As seen in Figure 2.18, the instrument is first charged with fresh air from the atmosphere using the rubber aspirator bulb (A). On-off switch (S2) is closed together with check switch (S1) and the compensatory filament (C) and detector filament (D) are allowed to reach steady state working temperature. The zero adjustment rheostat (F) can now be adjusted so that galvanometer (G) reads zero. Voltage is adjustable from battery (B) by the rheostat (E). Switch S_2 is now opened.

The instrument is now charged from the suspect gas space and while operating the bulb, the switch S_2 is again closed. If a flammable or explosive gas is present it will cause the detector filament to increase in temperature. This disturbs the bridge balance and a current flows. The galvanometer (G) can be calibrated so that the scale is marked to read '% of Lower Limit of Explosive Concentration of Gas'.

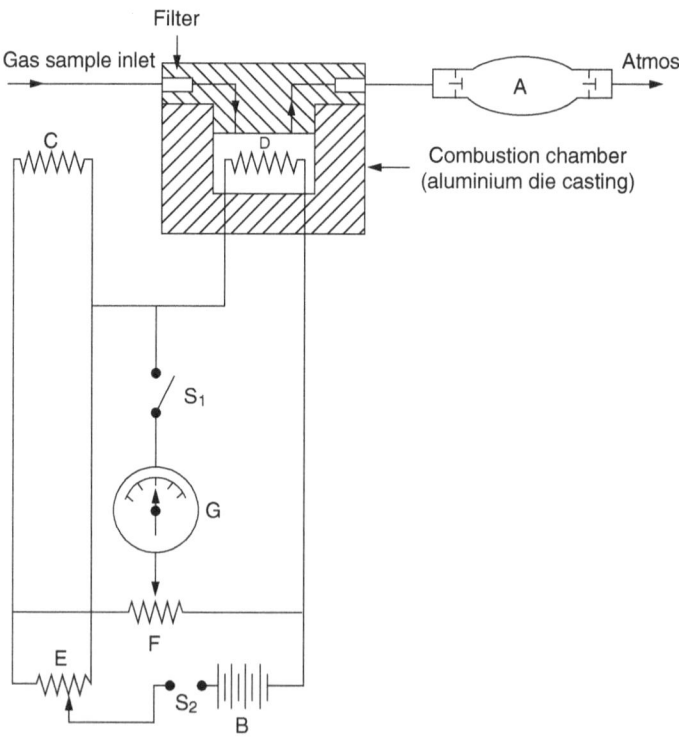

▲ **Figure 2.18** *Gas explosion-detector meter*

Alternative fuels

Alternate fuels to meet the strict pollution limits due in 2015 and 2020 are being developed. Figure 2.12 m page 75 in this chapter gives an outline of the limits for NO_x and SO_x on flue gases from the main propulsion machinery due in 2015 and 2020.

Engine manufacturers are working on ways to achieve these limits. The current thinking is that the answer lies with using a combination of technology both in the engine and after the engine.

It can also be seen from the composition of fuel that the correct choice of fuel can have a dramatic effect on the composition of the exhaust gas. However, the use, for example, of low-sulphur diesel oil will be very costly at today's prices and would seriously take up the world's supply.

LNG as a bunker fuel, for ships other than gas carriers, on the other hand is a fuel that is much more environmentally friendly and is a good cost-effective answer to the problems of the composition of flue gases.

Methane slip has been associated with LNG as it has been found that the current generation of diesel engines – that have been designed to run on fuel oil – allow some methane through the engine during combustion. However, the advantage of using LNG outweighs the disadvantage and the next generation of engines will be designed with this problem in mind.

The big issue with this fuel is supply. It is not possible at the moment to bunker LNG in all parts of the world. Also, special cryogenic systems are needed on the container ships or tankers that would use the LNG as a fuel.

Recently, MAN Diesel & Turbo ran their experimental two-stroke engine on LPG. This was successful and the thinking is that LPG can be bunkered more easily than LNG.

Hybrid systems

The general arrangement of the equipment can be found in Chapter 6 of this volume and Chapter 9 of Volume 12 of the Reeds series. The discussion here is centered upon the alternative 'energy' sources that might be used in marine applications.

The categorisation of marine systems from energy sources to power delivery can be seen in the same way as the current automotive power delivery systems. These are:

• Conventional systems
• Fully electric systems

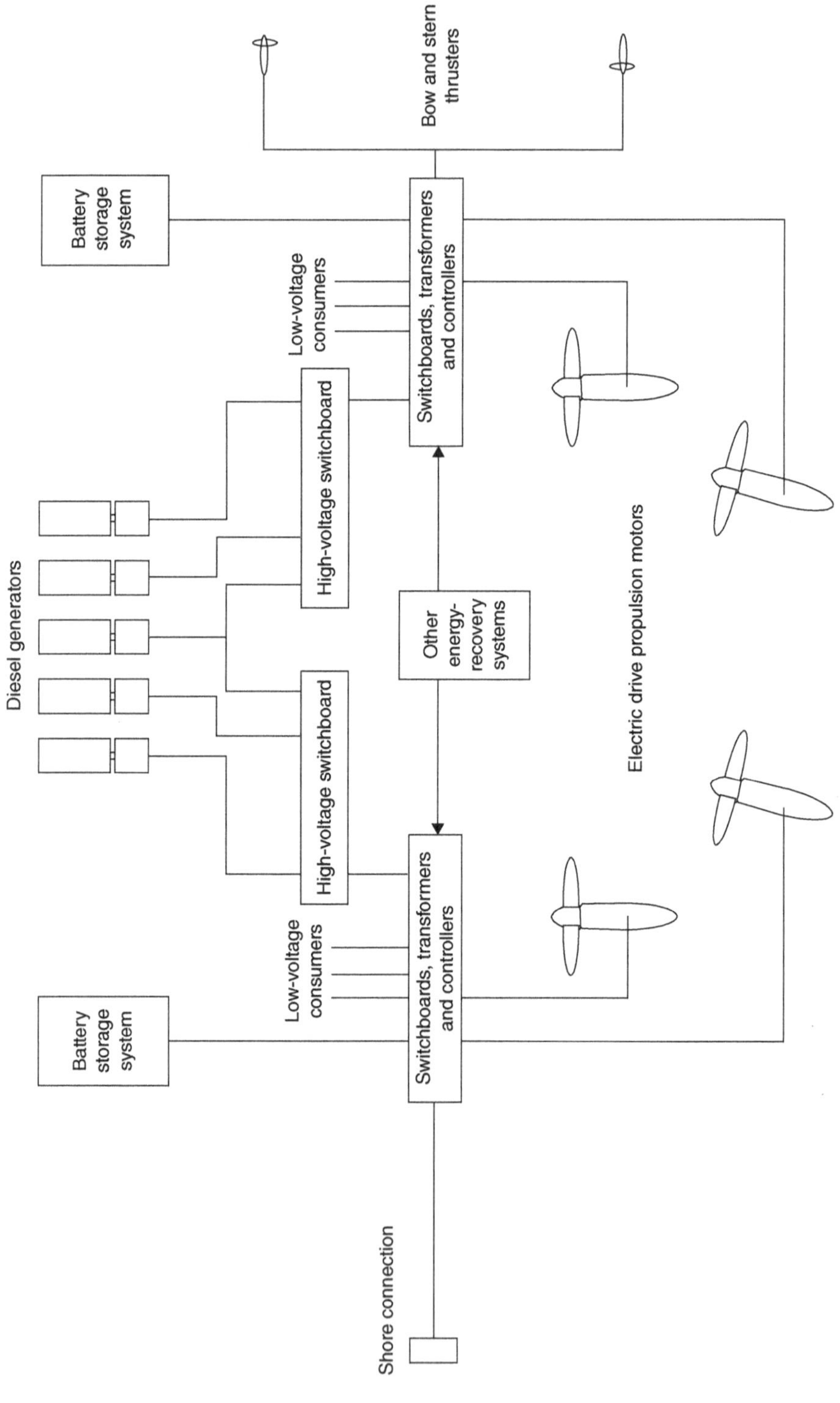

▲ Figure 2.19 *Hybrid supply system*

- Plug-in hybrid systems
- Hybrid systems

Battery technology is advancing at a rapid pace, although it is not without its difficulties. Some designs are prone to rapid overheating. The choice of material for manufacturing the cathode seems to have helped, as has the design of more effective cooling systems.

In marine applications for all fuels there will be a number of considerations for the potential users of the different technologies. The most significant of these will be:

- Overall cost of the system to run the energy source selected
- Energy density – affecting the space required within a ship for the different technologies
- Safety in handling and storage
- Efficiency in converting the energy into useful work
- Environmental impact
- Availability of supply chain
- Speed of replenishing used stock
- Useful life of the equipment
- Ease of or frequency of maintenance

The challenge, for the industry, in moving away from conventional fossil fuels is considerable. The long history of their use means that the fuel specifications and supply chains have been optimised to suit modern shipping. The massive driver for change is, of course, the adverse environmental effect globally of using the fuel in its current stage of development. Consequentially, the way forward seems to be by using a combination of different technologies that complement each other.

For example, fossil fuel has the best energy density by weight and volume apart from nuclear power. Therefore, to operate a large vessel it will be very difficult to replace fuel totally. However, the internal combustion (IC) engines used to turn the energy into power are not as efficient as they could be. Transient loads, for example, are difficult for engines to handle efficiently. Therefore if the IC engine were run only at its optimised settings and delivered its output in the form of electrical power to a battery bank or a super or hyper capacitor, the battery or capacitor could then be used to cope with the changes required when manoeuvring the vessel, thus making a more efficient plant in total. The choices could be as shown in Figure 2.19.

3

BOILERS AND ANCILLARIES

Introduction

Boilers and ancillaries are some of the most important equipment that the marine engineer will have responsibility for and flag state examiners will be very interested to ensure that engineers, presenting themselves for examination, are familiar with the equipment, especially the safety procedures.

Boilers of all shapes and sizes need particular care because they can be dangerous. Over the years the maritime industry has had its fair share of boiler explosions and some of the most lethal have been with relatively low-pressure 'fire tube' steam boilers similar to the ones currently fitted to motorships as auxiliary boilers. Another reason for the examiner being interested in the candidate's knowledge of boilers is that with the reduction in the number of ships using steam propulsion, the average marine engineer will have only have a limited experience of steam-raising equipment. However, the auxiliary boiler on a modern motorship is a very important piece of equipment because without heat to raise the temperature of the residual fuel oil the diesel engines will have to run on diesel oil, which might be in limited supply on board.

General safety rules and guidelines for the operation of boilers start with good watchkeeping practices. The burner could be prone to leaking oil and this area should be kept clean and tidy and any leaking joints or equipment should be repaired as soon as possible. Modern boilers will stop and start automatically but this does not mean they should not be checked regularly by the watchkeeper. The automatic lighting sequence will include a period of time before ignition when the boiler fan will run to remove any unburnt gases that may have been left in the furnace. The ignition will then start and finally when the fuel is introduced the boiler will light. If the watchkeeper watches this sequence carefully s/he will be able to get to know the procedure in detail and if the boiler shows signs of being reluctant to light then the engineer can have a look at the components of the burner/igniter to see if anything is becoming misaligned or is deteriorating due to wear and tear.

Modern flame-monitoring sensors use a combination of technologies to detect when the boiler is alight. The viewing heads can use ultraviolet or infrared detection or a combination of the two. Some come with self-cleaning lenses and self-checking mechanisms. One such mechanism uses a shutter that closes off the flame and the electronic circuits monitor the operation of the shutter in relation to the information coming from the flame.

Boiler combustion control is sophisticated and fairly reliable but if the flame should fail while it is supposed to be operating or if it should fail to light then the boiler will try to re-light; a second failure will lock the boiler out and the controls will give an alarm and keep the boiler off.

Re-starting a boiler following a 'lock-out' needs extra care from the engineer. S/he should be very clear about why the boiler has stopped working and the faulty condition should be rectified. When the engineer is ready to re-start having corrected the malfunction, care is still needed for safe operation. There could still be unburnt gases in the combustion space that if ignited could lead to more fuel igniting than the boiler is designed to contain. If flame inspection holes are provided the engineer must never use this while the boiler is going through the ignition sequence. Blow back of the flame through the inspection hole has in the past led to injury.

Safety valves are an essential part of any boiler and will be described over the next few pages. However, it is often asked by examiners if candidates know the difference between safety valves and relief valves. The important difference is that the safety valve must have the capacity to reduce the pressure inside a vessel despite it still working at full capacity. With respect to a boiler the safety valve must reduce the pressure inside the boiler while the boiler is still firing and the feed water pumps are operating. The European standard EN ISO 4126–1 provides the following definition for a safety valve independent of the fluid concerned:

Safety valve – A valve which automatically, without the assistance of any energy other than that of the fluid concerned, discharges a quantity of the fluid so as to prevent a predetermined safe pressure being exceeded, and which is designed to re-close and prevent further flow of fluid after normal pressure conditions of service have been restored.

Safety Valves

At least two safety valves have to be fitted to any one boiler. They may both be in the same valve chest, which must be separate from any other valve chest. The chest may be connected to the boiler with only one connecting neck. Where the boiler also has a superheater that can be shut off from the boiler, it too must be fitted with its own safety valve.

When viewing the examples shown over the next few pages, the underlying principles of operation should be noted, especially for examination purposes. With reference to Figure 3.1 it can be seen that the force generated by the steam pressure and the area of the valve exposed, $(π/4) × d^2 × P$, to that pressure is balanced by the spring force that is holding the valve shut. When the force generated by the steam pressure becomes too much for the spring, the valve will operate and as it does the steam pressure will act upon a larger area $(π/4)× D^2 × P$. This will generate a larger force against the spring, creating quick and positive opening. The pressure in the boiler will have to drop below the opening pressure before the valve will close because the valve is being helped to open by the steam acting upon the larger area.

The safety valve must be sized to release more steam than the equipment, in this case a marine boiler, has the capacity to produce. The valve is sized, positioned and set correctly so that the boiler cannot exceed its maximum allowable accumulated

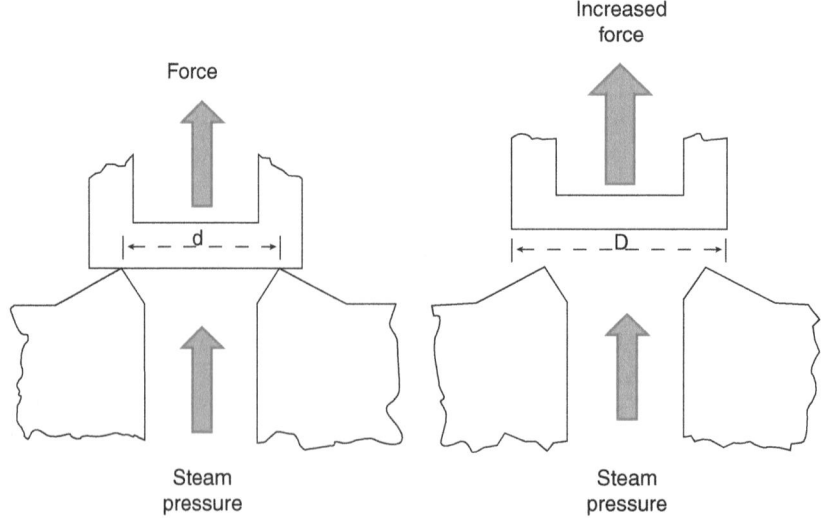

▲ **Figure 3.1** *Operating principle of the safety valve*

pressure (MAAP) even under any possible fault conditions. This means that the valves must be subjected to a testing system, required by the classification society as well as the flag state inspectors.

The Classification Society rules state that boiler and steam generators are to be fitted with two safety valves each with an internal diameter or 25 mm but steam generators with a heating surface of less than 50 m² could have one valve of not less than 50 mm. The area of the valves to give the correct discharge capacity can be calculated from the following formula:

$$C \times A \times P = 9.81 \times H \times E \tag{3.1}$$

where H = total heating surface in m², E = evaporation rate in kg of steam per m² of heating surface per hour, P = working pressure of safety valves in MN/m² absolute, and A = aggregate area through the seating of the valves in mm²,

C = a discharge coefficient whose value depends upon the type of valve,

$C = 4.8$ for ordinary spring-loaded safety valves.

$C = 7.2$ for high-lift spring-loaded safety valves,

$C = 9.6$ for improved high-lift spring-loaded safety valves,

$C = 19.2$ for full-lift safety valves,

$C = 30$ for full bore relay-operated safety valves.

If we consider a boiler operating under fixed conditions and producing steam at a constant discharge rate (i.e. $H \times E$), giving a constant pressure P then, from (3.1):

$$A \times C = \frac{9.81 \times H \times E}{P} = \text{a constant} \tag{3.2}$$

Also $\dfrac{A}{2} = \dfrac{\pi D^2}{4}$ approximately, where D is the diameter of the seating of one valve, in mm. Therefore, substituting in (3.2),

$$\therefore \frac{\pi D^2}{2} \times C = \text{a constant}$$

i.e. $\quad D^2 C = \text{a constant}$

Hence if C is increased, D must be reduced. But if D is reduced the lift of the valve must be increased in order to avoid any accumulation of pressure. This is accomplished by improving the type of valve fitted to the boiler.

Typical valve lifts are as follows:

When $C = 4.8$ lift $= D/24$ approximately

When $C = 7.2$ and 9.6   lift $= D_2/12$ approximately

When $C = 19.2$ and 30   lift $= D_3/4$ approximately

where $D_1 > D_2 > D_3$ and D_1, D_2 and D_3 are the diameters of the seating of the valves in mm.

Improved high-lift type

For low-pressure water tube boilers and fire tube boilers of the Scotch type and other varieties, the safety valve generally employed is Cockburns *improved* high-lift type. The operative parts of this valve are shown in Figure 3.2.

This valve has generally superseded the ordinary and high-lift types of safety valve. The essential differences between these three safety valves are given in Table 3.1.

Hence the *improvements* to the high lift safety valve are: (1) removal of valve wings, which improves waste steam flow and reduces risk of seizure and (2) floating ring or cylinder, which reduces risk of seizure.

The three spring loaded safety valves – ordinary, high-lift and improved high-lift – all make use of a special shaped valve seat and a lip on the valve that gives increased valve lift against the increasing downward force of the spring. The action can be seen in Figure 3.3.

For superheated steam the aggregate area through the seating of the valves is increased, and the formula is:

$$A_s = A(1 + T_s/555)$$

Table 3.1 *Types of safety valves*

Ordinary	High-lift	Improved high-lift
Winged valve	Winged valve	Wingless valve
No waste steam piston	Waste steam piston	Waste steam piston
	No floating ring	Floating ring

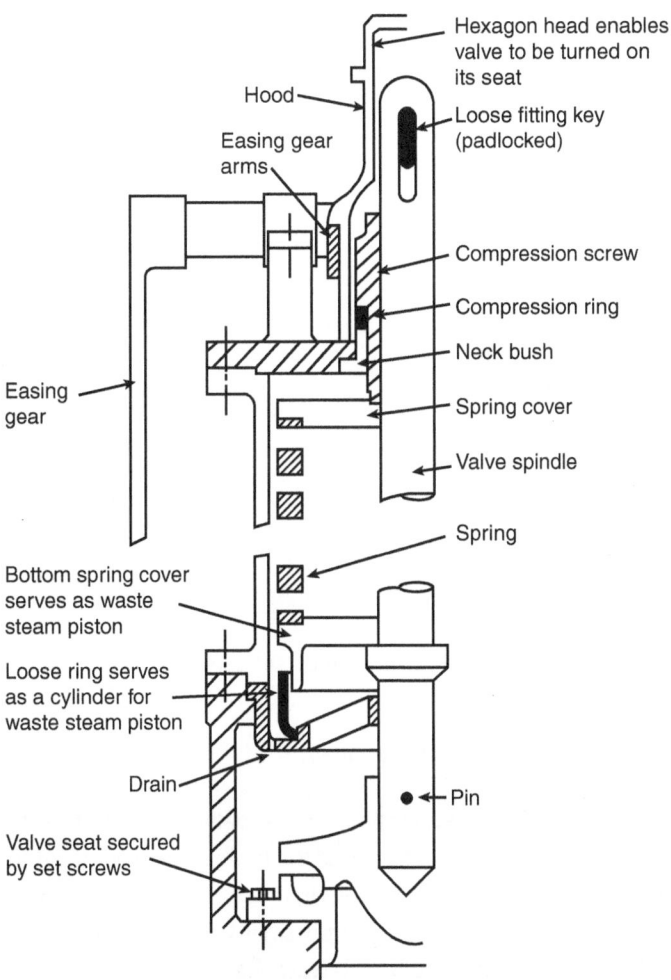

▲ **Figure 3.2** *Improved high-lift safety valve*

where A_s = aggregate area through the seating of the valves in mm² for superheated steam, A = aggregate area through the seating of the valves in mm² for superheated steam and T_s = degrees of superheat in °C.

A_s is obviously greater than A, the reason being that the specific volume of the steam has increased with the increase of temperature at constant pressure and more escape area is required to avoid accumulation of pressure (specific volume is volume per unit mass).

The area of the valve chest connecting neck to the boiler must be at least equal in cross-sectional area to one half of the aggregate area A, determined from Equation (3.1). The waste steam pipe and steam passage from the valves must have a cross-sectional area of at least:

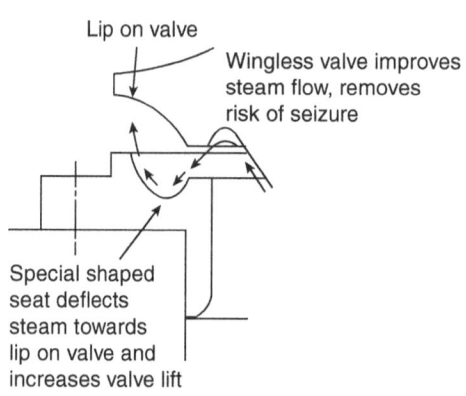

Lip on valve

Wingless valve improves steam flow, removes risk of seizure

Special shaped seat deflects steam towards lip on valve and increases valve lift

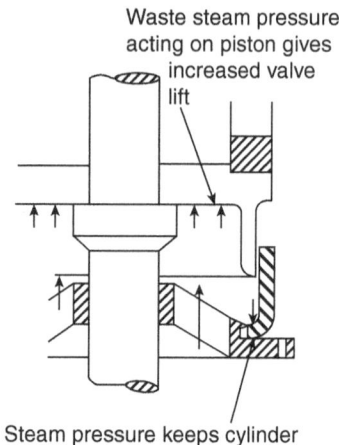

Waste steam pressure acting on piston gives increased valve lift

Steam pressure keeps cylinder in place while piston moves, also by having a floating cylinder seizure risk is reduced

▲ **Figure 3.3** *Operating detail of the safety valve*

1.1× A for ordinary, high-lift and improved high-lift safety valves,

2 × A for full-lift safety valves,

3 × A for full bore relay-operated safety valves.

A drain pipe must be fitted to the lowest part of the valve chest on the discharge side of the valves and this pipe should be led clear of the boiler. The pipe must have no valve or cock fitted throughout its length. This open drain is important and should be regularly checked, for if it became choked, there is a possibility of overloading the valves due to hydraulic head, or damage resulting due to water hammer.

Materials

Materials used for the valves, valve seats, spindles, compression screws and bushes must be non-corrodible metal, since corrosion of any of these components could result in the valve not operating correctly. Often the materials used are bronze, stainless steel or monel metal, depending upon conditions. The valve chest is normally made of cast steel. The spring is a critical element of the safety valve and must provide reliable performance within the required parameters. Standard safety valves will typically use carbon steel for moderate temperatures. Tungsten steel is used for higher temperature, non-corrosive applications, and stainless steel is used for corrosive or clean-steam duty. For sour gas and high-temperature applications, often special materials such as monel, hastelloy and 'inconel' are used.

Maintenance and adjustment

All safety valves fitted to marine boilers must have 'easing gear' included so that the valves can be checked for correct operation while the boiler is running and for surveyors. Also during inspections and surveys the maker's figures relating to lip clearances, seating widths and wing clearances, etc. must be adhered to. All working parts should be sound, in alignment and able to function correctly.

When overhauling safety valves, engineers must exercise extreme care as these are safety items of importance and they must ensure that the parts are put back in their correct order. In this modern era the use of digital cameras can be invaluable as a method of recording the sequence of dismantling. When unassembled, the individual parts can be hung by a cord and sounded by gently tapping with a hammer. If they do not ring true, examine for faults. Check drains and easing gear. The parts will all need to be surveyed and the surveyor may require additional testing with non-destructive techniques (see pages 23–26). For routine surverys, it is always helpful if replacement parts that are needed are placed next to the original part so that the surveyor can see any defects and s/he will also be able to inspect the replacement. Please note, however, that if the surveyor has been called due to a failure, the parts should not be dismantled, cleaned or replaced until the surveyor has approved the work.

Adjustment or setting of safety valves of the direct spring-loaded type: with compression rings removed, screw down compression screws, raise boiler pressure to the required blow-off pressure. Screw back compression screw until valve blows, then screw down the compression screw carefully, tapping the valve spindle downwards very lightly while doing so, until the valve returns to its seat and remains closed.

When set, split compression rings have to be fitted, then hoods, keys, padlocks and easing gear. Finally check and operate easing gear to ensure it is in good working order.

With an installation that has several boilers, raise all the boiler pressures to the required blow-off pressure, make sure the boilers are connected up, then proceed as described above, setting each valve in turn.

Accumulation of pressure test

The difference between the true 'safety' valve and a relief of pressure valve is that the safety valve will have the capacity to prevent an unsafe rise in internal pressure of the boiler even with the boiler in full operation. Classification societies require that when initially fitted to boilers safety valves must be subjected to an accumulation of pressure test to ensure the valves are of the correct discharge capacity for the boiler. To conduct such a test, all feed inlets and steam outlets to and from the boiler, respectively, must be closed, and maximum firing rate arranged. Accumulation of pressure must then not exceed 10% of the working pressure. Duration of test (water permitting) is not to exceed 15 minutes for cylindrical boilers and 7 minutes for water tube boilers. In the case of water tube boilers the test may be waived if damage to superheaters or economisers could result from the test.

Full-lift safety valve

The full-lift safety valve shown in Figure 3.4 does not incorporate a waste steam piston, instead the valve itself operating inside the guide acts as a piston in a cylinder.

Details of spring, compression nut, easing gear and valve chest etc. have been omitted for convenience since they are somewhat similar to that for the improved high-lift valve.

The operation of the valve is as follows: when the valve has lifted a small amount, the escaping steam pressure can then act upon the full area of the valve. This increases the lift until the lower edge of the valve just enters the guide. At this point the reaction pressure generated by the escaping steam with the guide causes the valve

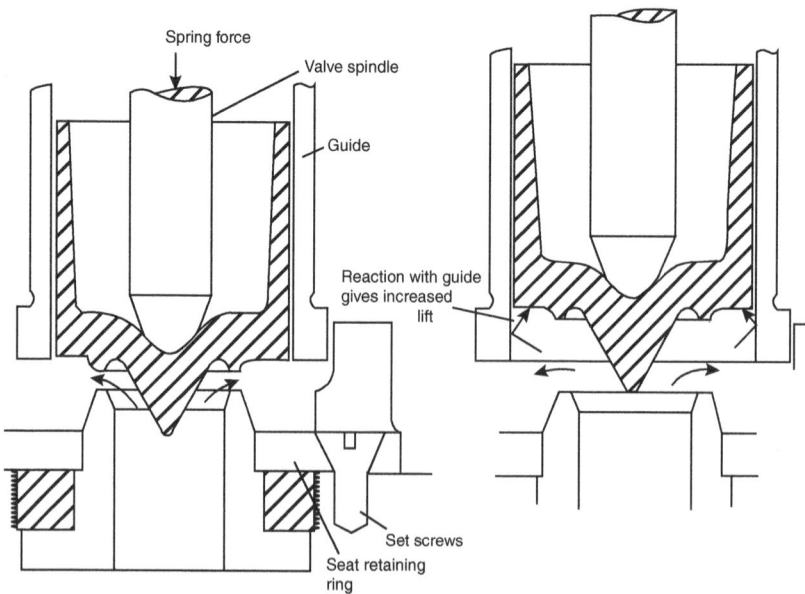

▲ **Figure 3.4** *Full-lift safety valve (for pressure up to 63 bar)*

to lift further until it is fully open. When the valve is fully open the escape area is said to be equal to the area of supply through the seating.

Full-bore safety valve

Figure 3.5 is a diagrammatic arrangement of a full-bore, relay-operated safety valve suitable for water tube boilers whose working pressure is in excess of 21 bar [2.1 MN/m^2].

The operation of the valve is as follows: when the boiler pressure reaches the desired blow-off pressure the relay valve lifts, blanking as it does so a series of ports leading to the atmosphere (see Figure 3.5a).

Steam is then admitted via the connecting pipe into the cylinder of the main valve, and since the area of. piston is about twice that of the valve, the valve opens against boiler pressure.

When the boiler pressure falls, the relay valve closes, uncovering as it does so the ports above it (see Figure 3.5b). This links the cylinder of the main valve with the atmosphere and the boiler pressure then causes the main valve to close rapidly.

This type of valve is suitable for high-pressure boilers, since the greater the boiler pressure the more rapidly the valve will close, and hence the greater the saving in steam.

The main valve spring assists closing of the valve and also ensures that the valve will be closed when the boiler is cold.

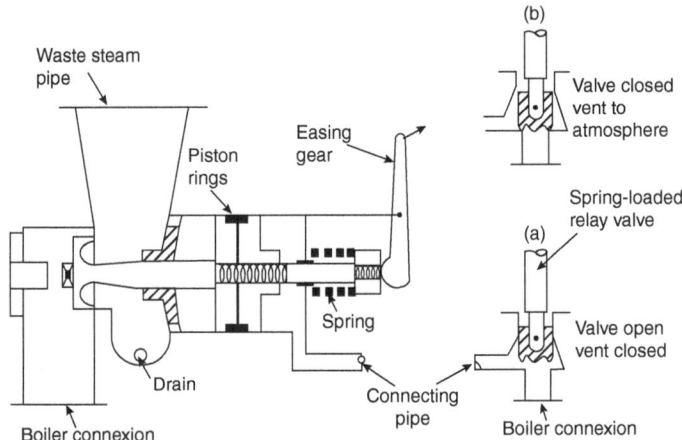

▲ **Figure 3.5** *Full-bore relay-operated safety valve*

If the valve is to be used for the saturated steam drum of a water tube boiler, the main valve and relay valve connection are sometimes made common to the drum. If, however, the valve is to be used for superheated steam then the relay valve connection is taken separately from the main valve connection to the saturated steam drum. This arrangement subjects the relay valve and main valve piston and cylinder to lower-temperature operation.

Water Level Indicators

Classification societies and other regulators require that every boiler designed to carry water at a specific level must be fitted with two independent means of measuring that level. At least one of those should be a directly connected gauge glass. Water level indicators, which may also initiate boiler shut-down functions, are very important safety items and as such the flag state inspector will be watching closely to see that the candidate being presented for examination is knowledgeable about their requirements, operation and possible faults. Examinations will concentrate on testing that watchkeepers are sure that the gauge glass is giving a true reading. Proving that they are clear relies on knowing the type of gauge fitted and the procedure for 'blowing' the glass clear. The following pages give the details of the procedures to be used with the different types of glass.

Consider Figure 3.6. This glass water gauge is fitted directly to the boiler shell and is suitable for boilers whose working pressure does not exceed 34.5 bar [3.45 MN/m²].

Blowing procedure

1. Close steam and water cocks then open the drain. Nothing should then blow out of the gauge if the steam and water cocks are not leaking.
2. Open and close the water cock to check that the water cock connection to the boiler is clear.
3. Open and close the steam cock to check that the steam cock connection to the boiler is clear.
4. Close the drain.
5. Open the water cock. Water should then gradually rise up to the top of the gauge glass.
6. Open the steam cock and the water in the glass should fall to the level of the water in the boiler.

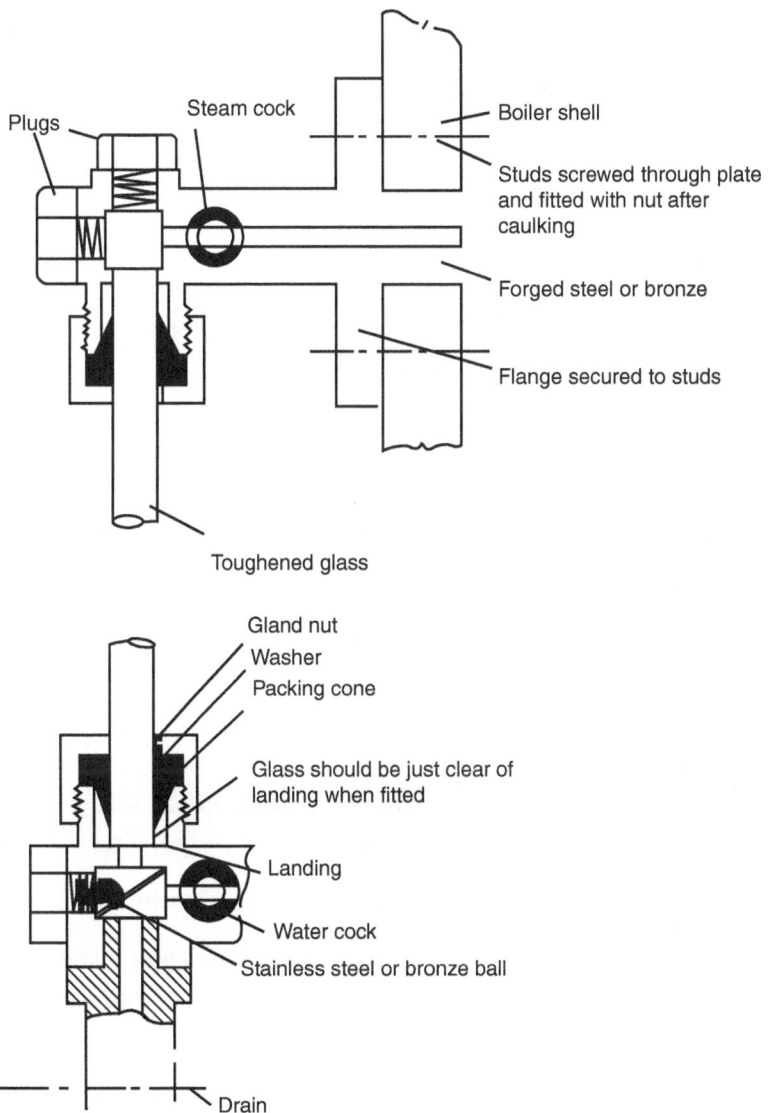

Plugs

Steam cock

Boiler shell

Studs screwed through plate and fitted with nut after caulking

Forged steel or bronze

Flange secured to studs

Toughened glass

Gland nut
Washer
Packing cone

Glass should be just clear of landing when fitted

Landing

Water cock

Stainless steel or bronze ball

Drain

▲ **Figure 3.6** *Glass water gauge*

If when (5) is reached the water cock is opened and water does not flow up the gauge glass, the water level in the boiler is below the water cock connection to the boiler and it is unsafe to put feed water into the boiler.

If when the water clock has been opened the water flows to the top of the gauge glass and then when the steam cock is opened the water flows down and out of the glass, the water level is between the water cock connection to the boiler and the bottom of the gauge glass. In this case it is safe to put feed water into the boiler.

If after (5) when the glass is full of water, the steam cock is opened and the water in the glass does not descend in the glass, the water level is above the steam cock connection to the boiler and there is a danger of priming the boiler if any additional feed is put into it.

This glass water gauge arrangement is similar to Figure 3.7a except that the gauge is connected to a large-bore pipe fitted to the boiler. The pipe has plugs fitted, two at the top and two at the bottom, which can be removed during boiler cleaning in order to clean out the pipe.

The blowing procedure for this fitting is the same as for Figure 3.7a.

These fittings 3.7c and 3.7d of the hollow and solid column types, respectively, are convenience fittings. They bring the water gauge glass clear of other boiler fittings such as gas uptakes, etc. so that the gauge glass can be easily seen by the boiler room personnel. In addition, by shutting off the terminal cocks on the boiler they should enable the water gauge steam and water cocks to be overhauled while the boiler is steaming.

To determine whether the pipes and the terminal cocks are clear, a blowing procedure sometimes referred to as cross-blowing is adopted. This procedure should be completed in the following manner and with reference to Figure 3.7c: cocks A, C and D should be open and cocks B and E should be closed; this checks that cock A, pipe X and the column are all clear.

Then when the cocks E, B and D open and cocks A and C are closed, the path of steam checks that cock E and pipe Y are clear.

Next, blow the water gauge glass as described for Figure 3.7a with A and E open.

In this case there is no direct communication between the pipes X and Y, and hence to check whether the pipes and cocks are clear the blowing procedure employed for Figure 3.7a should be used.

If either of the water cocks are choked, water will again fill the glass due to the steam condensing in the upper connections.

When a water gauge of the types 3.7c and d are blown through and all connections are clear and all cocks are in operative order, the water level in the glass will be the water level in the boiler. However, after a period of time (which depends upon conditions, e.g. ventilation arrangements, etc.) it will normally be found that the water level in the glass will have fallen. This is due to: (1) the cooling of the water in the pipe Y, thereby increasing its density; (2) the reduction of condensation of steam in the pipe X, which is caused by an accumulation of air in the upper connections due to steam condensing.

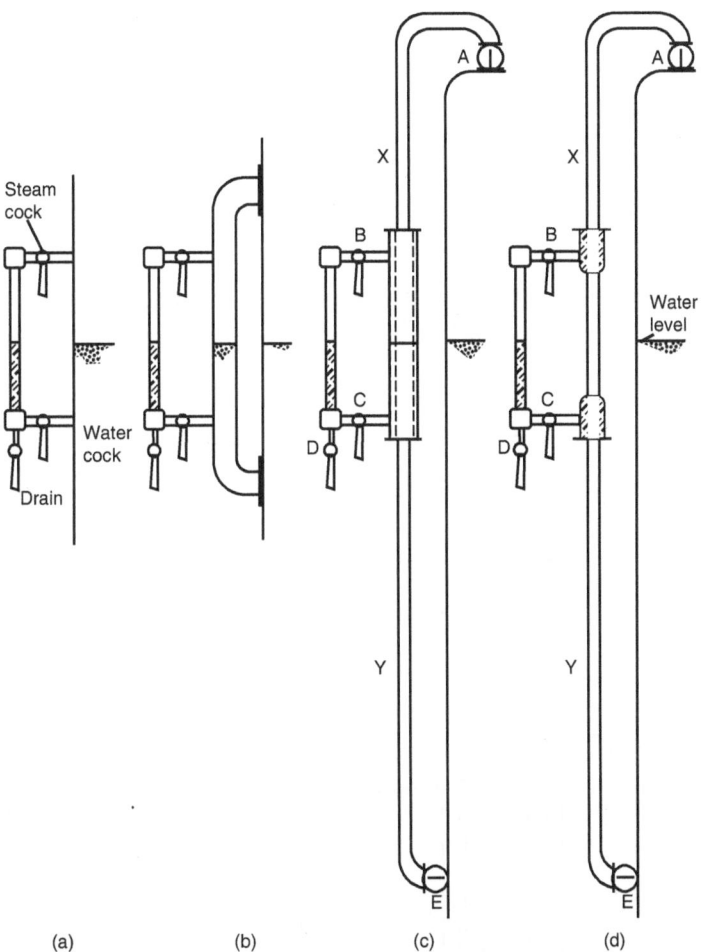

Steam cock

Water cock

Drain

Water level

(a)   (b)   (c)   (d)

▲ **Figure 3.7** *Water level indicators*

Hence when blowing through a water gauge of either of the types 3.7c and d, check the water level in the glass before blowing with the water level in the glass immediately after blowing. The difference in levels must then be taken into account whilst operating the boiler.

When any gauge glass fitting is in operation the cock handles should be vertical. If they are arranged horizontally and the gauge is in operation, vibration effects may cause the cock handle to gradually tend to take up a vertical position, thereby closing the cock in the case of steam and water cocks, and opening in the case of the drain. (The steam and water cocks for Figure 3.7d *cannot* be used as test cocks.)

The relation of handle position to correct working position is also important from another aspect, since if the handle became over-strained in relation to the plug body of the cock, the handle may be in the correct working position but the cock may be closed.

It is normal to fit extended controls for the cocks so that the gauge can be blown through from a remote and safe position.

A protective glass arrangement should be provided that partly surrounds the gauge glass to prevent injury to personnel in the event of gauge glass breakage under

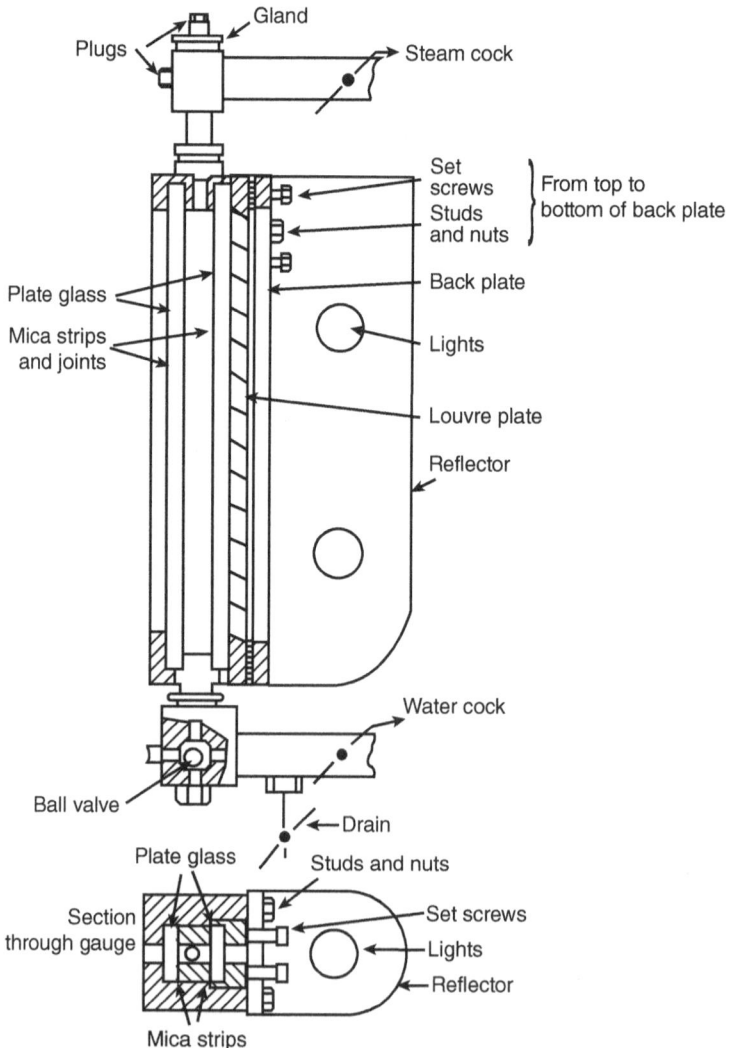

▲ **Figure 3.8** *Plate type of water gauge*

steam. A steam restrictor and water shut-off ball valve (see Figure 3.8) are sometimes incorporated with the fitting to reduce the severity of breakage.

Care must be taken when renewing a gauge glass to ensure that it is of the correct length in relation to the fittings. If it is too long, blockage of the steam connection may occur due to accumulation of deposits around the top of the glass. If the glass is too short and is not fully inserted into the packing, the packing may work its way over the open end of the gauge glass, causing a blockage.

Figure 3.8 is a plate type of gauge glass suitable for high pressures of up to 79 bar [7.9 MN/m²]. The toughened soda lime glass plate is capable of withstanding severe mechanical stress and temperature but it has to be protected from the solvent action

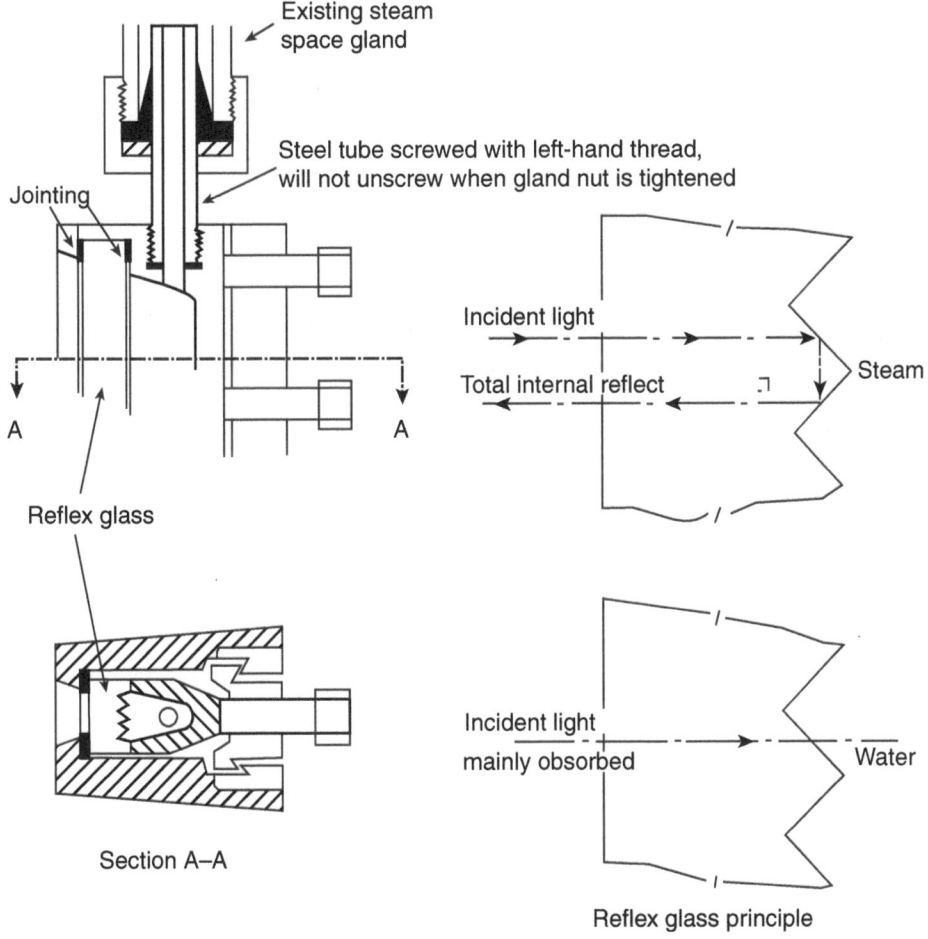

▲ **Figure 3.9** *Klinger reflex glass (for steam pressures up to 20.6 bar)*

of the boiler water. This is achieved by interposing a mica strip between the glass and steam joint so that the water does not come into contact with the glass. Light is deflected up through the louvre plate and is reflected downwards by the water meniscus, which then shows up as a bright spot.

Figure 3.9 shows the Klinger reflex glass that can be fitted new with its own glands and cocks or can be installed into existing gauge cock fittings. Steel tubes, which have spanner flats, enable the gauge to be fitted in place of a glass tube without having to dismantle the cocks. In operation, the light is reflected from the steam space and absorbed in the water space thus giving a bright and dark strip respectively whose contrast can be clearly seen at a distance. No protective glass is required but the reflex glass is only suitable for pressures up to 20.6 bar since as the pressure and temperature increases the solvent action of the water also increases.

Remote water level indicators

For safety reasons it is still a requirement that a direct visual means of assessing the boiler water level is provided mounted to the boiler. However, modern operation of machinery space dictates that remote monitoring of the boiler water is also available.

There are various types of remote water level indicators. Their purpose is to bring the water level reading to some convenient position in the engine or boiler room where it can be distinctly seen. On more modern ships this will be the machinery control room (MCR). These indicators when fitted are normally in addition to the statutory requirements for water gauge fittings for boilers.

Figure 3.10 is a diagrammatic arrangement of the 'Igema' remote water level indicator. The lower portion of the 'U' tube contains a red-coloured indicating fluid that does not mix with water and has a density greater than that of water.

The equilibrium condition for the gauge is $H = h + {}_Q x$ where ${}_Q$ is the density of the indicating fluid. H, h and x are variables.

If the water level in the boiler falls, h will be reduced, x will be increased and H must therefore be increased. The level of the water in the condenser reservoir is being maintained by condensing steam. If the water level in the boiler rises, h will be increased, x will be reduced and H must therefore be reduced. Water will therefore flow over the weir in the condenser reservoir in order to maintain a constant level.

A strip light is fitted behind the gauge, which increases the brightness of the red indicating fluid, which enables the operator to observe at a glance from a considerable distance whether the gauge is full or empty.

Figure 3.11 is another type of remote water level indicator. In this case the operating fluid is the boiler water itself. The operation of the gauge is as follows.

If we consider a falling water level in the boiler, the pressure difference across the diaphragm 'h' will increase, causing the diaphragm to deflect downwards. This motion

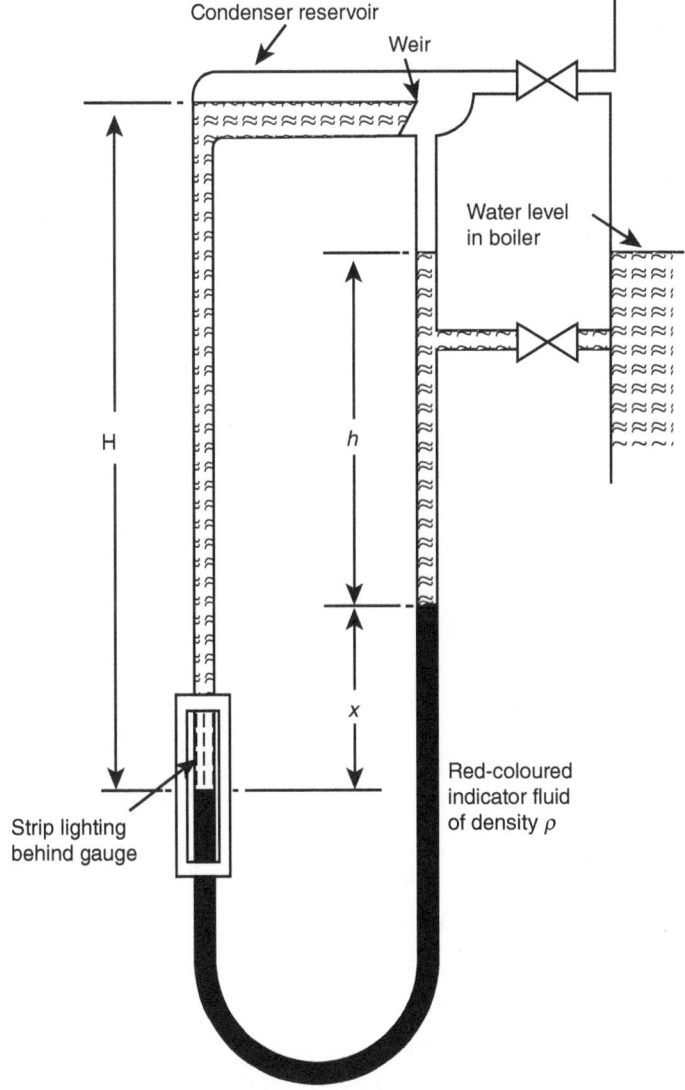

▲ **Figure 3.10** *Remote water level indicator*

of the diaphragm is transmitted by means of a linkage arrangement (inset in Figure 3.11) to the shutter, which in turn moves down, pivoting about its hinge, causing an increase in the amount of red colour and a decrease in the amount of blue colour seen at the glass gauge.

It will be clearly understood that if the water level now rises then the red will be reduced and the blue increased.

Separating the blue and red colours, which are distinctive and can clearly be seen from a considerable distance, is a loose-fitting black band that moves with the shutter, giving a distinct separation of the two colours.

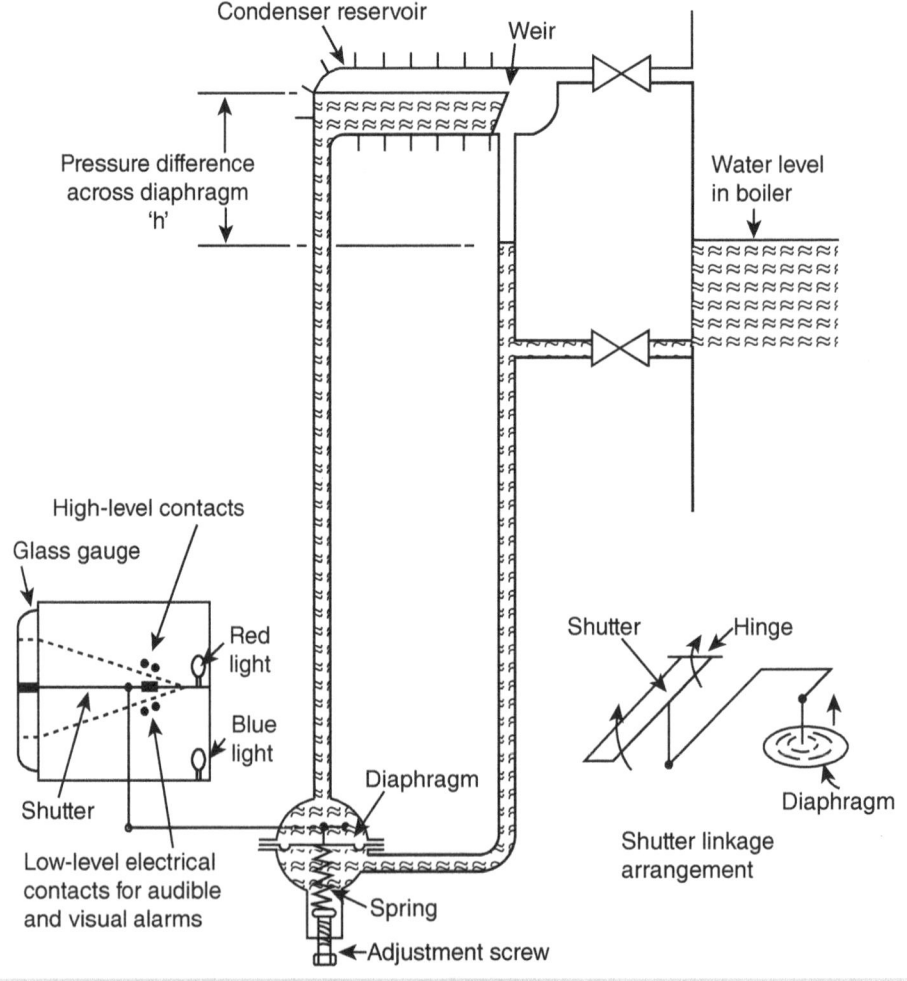

▲ Figure 3.11 *Remote water level indicator with transmitter*

An adjustment screw and spring are provided to enable the difference in diaphragm load to be adjusted. Hence correct positioning of the shutter and band in relation to the reading of a glass water gauge fitted directly to the boiler is possible.

The most modern ships have MCRs that are such a distance from the boiler that it is impracticable and unsafe to have the water at boiler system pressure being led around the engine room for remote reading.

What can be accomplished, however, is the placement of a differential pressure cell in place of the 'diaphragm' shown in Figure 3.11. The output from the cell can be led to a gauge located in the MCR.

Some of the latest techniques involve measuring the electrical conductivity of the water. Checking at what level the values change will give the precise point of the water level. This value can then be transmitted via the (CAN-bus) data transmission system to the remote monitoring point, usually the MCR or other places such as the chief engineer own monitoring station.

Boiler Controls

Modern boilers are designed to operate automatically, which has the following implications:

- Starting and stopping the combustion to suit the steam demand
- Adding water when required
- Not overfilling the boiler with water
- Ensuring that the boiler is safe at all times

The flag state examinations for General Engineering Knowledge will need to make sure that all engineers that pass out successfully understand the safe operation of oil-fired boilers. This is especially true where automatic systems are involved because the engineer in charge of the operation will need to satisfy him/herself that the boiler will not fail while left alone to operate automatically.

The boiler combustion process and equipment is covered in Chapter 2 of this volume and in much more detail in Volume 9 of the Reeds Series. However, the control of the burner will be carried out by a combination of sensors, electronics and mechanical equipment.

The initial action to start the burner will come from the need to produce more steam. This is usually detected by a drop in pressure in the steam drum due to the steam being used by the various services.

At the pre-determined pressure the burner sequence will start as long as there is sufficient water in the boiler and the low water level cut-out is not stopping the automatic sequence. The boiler's forced draught (FD) fan is started and will run for a few seconds to clear any unburnt gases from the furnace. This is known as the purge cycle.

Following the purge process the spark igniter will start followed by the opening of the oil-flow solenoid valve. Using this sequence means that as soon as the oil is introduced combustion should start and therefore no unburnt fuel will be introduced into the furnace. If the boiler does not light the sequence will be tried once more and after that the boiler will stop and go into lock-out mode.

The lock-out mechanism initiates an alarm and the engineer has to attend the boiler and rectify the fault before the boiler can be re-started. If the boiler lights successfully then a detector will signal the controls allowing the boiler to carry on running until the pressure is raised to the upper value when the burner will be stopped.

The 'flame failure' alarm will sound if the boiler has tried but failed to light. This could be for any one or combination of the following reasons:

- No fuel due to
  - a blocked filter or burner nozzle,
  - poor-quality fuel,
  - faulty fuel supply pump,
  - fuel tank empty;
- No spark due to
  - high-voltage leads becoming disconnected,
  - dirty electrical leads or connections,
  - failed transformer,
  - burnt electrodes,
  - poorly adjusted electrodes;
- Not enough air due to
  - fan stopped,
  - air register blocked,
  - drive between the motor and the fan broken,
  - fuses blown,
  - electrical motor overheating.

The most common fault will be with the ignition system, especially in slightly older boilers. The electrodes are set at a specific position in relation to the burner tip, so that the fuel is atomised just as it comes into contact with the heat source (the igniter). However, due to the forces from the air blowing past and the combustion taking place the electrodes become out of adjustment and they fail to ignite the fuel. Simple re-adjustment rectifies the problem but the engineer must be sure that all the other parts of the boiler system are safe and that the boiler goes through its purge cycle before re-lighting.

Equally as important for the safety of the boiler is the correct level of water. The metal parts close to the very hot products of combustion rely on the water level being correct in order to stay cool and not fail. In the past the loss of water has led to overheating of the furnace, which has ruptured.

A failure of this kind in a steam boiler becomes serious very quickly. This is because the water is under pressure while it is still in the boiler; however as it is released to atmospheric pressure it will very quickly turn to steam. The ratio is approximately 1:1,600 parts water to steam and even boilers that are classed as auxiliary boilers running at 7 bar will cause a lot of damage if they rupture due to overheating of the furnace.

The water that is introduced to the boiler is called 'feed water', therefore the associated system is called the feed system. There are feed pumps, feed check valves and feed water pipes, valves and tanks. Modern feed water systems are classed as 'closed' systems, that is, the water is kept away from the air as much as possible. Without the oxygen in the system corrosion is kept to a minimum.

For the feed water level to be controlled in the boiler there will have to be some means of detecting the water level. The visual water level indicator or sight glass is a very good manual method but the automatic controls will need other methods.

These could take the form of float valves, capacitance probes or conductivity probes such as in Figure 3.12.

It is important for the efficiency of the steam-raising plant that the feed water temperature is as warm as possible. Therefore, the water only has to be cooled enough to pump it back into the boiler. The feed water pump is set as low down in the system as possible so that the pressure of the water is such that it is not subject to gassing up. This is a state where steam bubbles start to appear in the water, and if the condition is extreme it will lead to a time when the pump stops working. This means that water is not entering the boiler and therefore the water level is dropping. If this continues the boiler will stop working due to the low water cut-out and the alarm is activated.

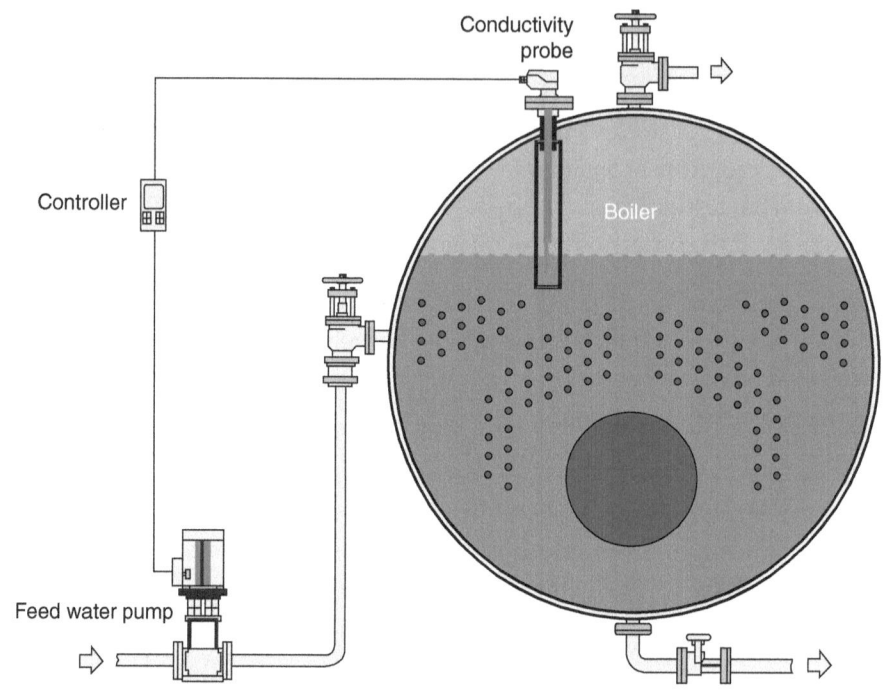

▲ **Figure 3.12** *Use of conductivity probe to detect water level (With kind permission of Spirax Sarco)*

Other Boiler Mountings

The term 'boiler mounting' is used to describe equipment that is attached to the pressure parts of the boiler necessary for its correct operation. The primary boiler mountings are described below.

Feed check valves

Feed check valves, as their name suggests, are designed to stop feed water returning back down the feed water pipe while the feed water pump is not running. For main and auxiliary boiler purposes they are normally of the double shut-off variety. This is shown diagrammatically in Figure 3.13.

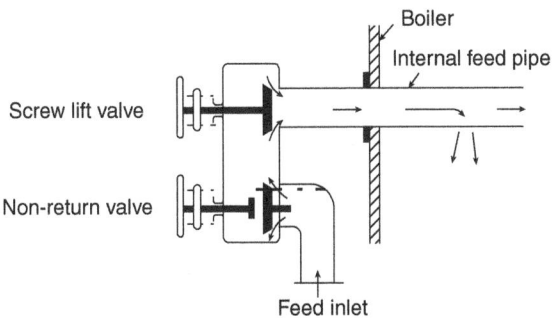

▲ **Figure 3.13** *Diagrammatic arrangement of feed check valves*

The double shut-off arrangement does enable the non-return feed check valve to be overhauled while the boiler is steaming, since if the screw-down valve is shut the non-return valve is then isolated from the boiler. But overhaul of these valves is best done when the boiler is opened up for examination.

The non-return valve is necessary since if a feed line fractures or a joint in the line blows, then the boiler contents will not be discharged out of the feed line. Also, double shut-off reduces the risk of leakage into the feed line while it is under repair and the boiler is steaming.

Main stop valve

This is the valve that allows the steam from the boiler into the steam range. The valves are not designed as control valves and will be either fully open or fully closed. The valves need to be strong to withstand full steam pressure and for this purpose modern valves are made of either cast steel or bronze. Care needs to be taken in opening the valve as a sudden rise in pressure and steam flow can cause 'water hammer' in the steam service pipes.

Blow-down valve

These valves are used to release some of the water and sediment from the bottom of the boiler so that fresh feed water can be introduced into the system, which increases the quality of the water.

Air vents and vacuum breakers

When filling the boiler following repairs the air is released via the air vent. On the other hand as the boiler is allowed to cool and the steam condenses back to water, a vacuum is created inside the boiler. The vacuum on the inside means that pressure is placed on the boiler from the outside. To ensure that no damage occurs, a vacuum breaker is fitted to the boiler shell that will allow air into the boiler to remove the vacuum.

Other Boiler Fittings

Soot blowers

Between periodic boiler cleaning, the gas surfaces of the boiler tubes should be kept as clean as practicable. To facilitate this, soot blowers, steam or air operated, are often fitted. They enable the tube surfaces to be cleaned of loose sooty deposits rapidly without shut-down of the boiler (Figure 3.14).

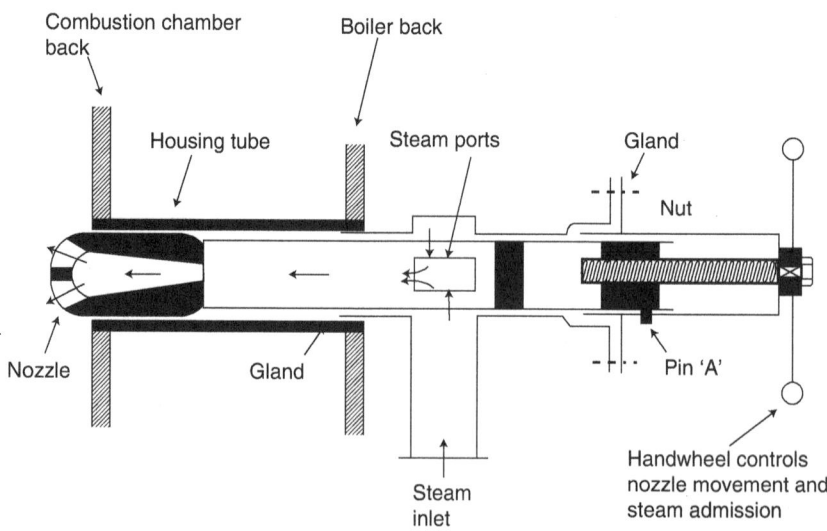

▲ **Figure 3.14** *Soot blower*

With steam supplied to the blower and the steam supply line thoroughly drained, rotation of the blower hand wheel causes the supply tube and nozzle to move towards the combustion chamber. Nozzle and tube are rotated as they move inwards by means of a scroll cut in the nut and a stationary pin A in the body assembly that runs in the scroll. Ports in the tube communicate the steam supply line with the nozzle.

The arrangement enables rotating, fine, high-pressure jets of steam to be discharged to the tube plate over a considerable area.

When not in use, the retractable nozzle of the blower is well within the housing tube and is therefore protected from overheating, which could cause burning and distortion of the nozzle.

Too frequent use of the blower should be avoided as this could cause wastage of the tube plate. It is advisable to operate the blower regularly even if the boiler tubes are clean (in this case without steam supply to the blower) to ensure the blower unit is free and in operable order.

The most up-to-date soot blowers use ultrasonic wave technology, which when applied will dislodge the soot from the tubes without the use of steam, thus saving energy. Soot blowing is important on modern ships with exhaust gas boilers, especially if the engine has been operating at a lower load than its maximum continuous rating (MCR).

Boilers

Waste heat boilers

With diesel machinery the amount of heat carried away by the exhaust gases varies between 20% and 25% of the total heat energy supplied to the engine. Recovery of some of this heat loss to the extent of 30–50% is possible by means of an exhaust gas boiler or water heater.

The amount of heat recovered from the exhaust gases depends upon various factors, some of which are steam pressure, temperature, evaporation rate required, exhaust gas inlet temperature, mass flow of exhaust gas and condition of heat-exchange surfaces.

Composite boilers are often used in conjunction with diesel machinery, since if the exhaust gas from the engine is low in temperature due to slow running of the engine and reduced power output the pressure of the steam can be maintained by means of an oil-fired furnace. Steam supply can also be maintained with this type of boiler when the engines are not in operation.

All the various types and makes of waste heat boilers are not examined in detail, since the Maritime Coastguard Agency (MCA) (E.K. General) written examination requires only a brief outline of the basic principles involved.

The Cochran boiler whose working pressure is normally of the order of 7 bar (0.7 MN/m^2) is available in various types and arrangements, including:

Single pass composite, that is, one pass of the exhaust gases and two uptakes, one for the oil-fired system and one for the exhaust system (Figure 3.15). Double-pass composite, that is, two passes for the exhaust gases and two uptakes, one for the oil-fired system and one for the exhaust system. Double-pass exhaust gas, no oil-fired furnace and a single uptake. Double-pass alternatively fired, that is, two passes from the furnace for either exhaust gases or oil-fired system with one common uptake.

The material used is good-quality, low-carbon boiler steel plate. The furnace is pressed out of a single plate and is seamless.

Connecting the bottom of the furnace to the boiler shell plating is a seamless 'Ogee' ring. This ring is pressed out of thicker plating than the furnace. The greater thickness is necessary since circulation in its vicinity is not as good as elsewhere in the boiler and deposits can accumulate between it and the boiler shell plating.

Hand hole cleaning doors are provided around the circumference of the boiler in the region of the 'Ogee' ring.

The tube plates are supported by means of the tubes and by gusset stays, the latter supporting the flat top of the tube plating.

Tubes fitted are usually of a special design (Sinuflo), being smoothly sinuous in order to increase heat transfer by baffling the gases. The wave formation of the tubes lies in a horizontal plane when the tubes are fitted, thus ensuring that no troughs are available for the collection of dirt or moisture. This wave formation does not in any way affect cleaning or fitting of the tubes.

Figure 3.16 shows the method of attachment of the furnace and 'Ogee' ring for Cochran and Clarkson welded boilers, welded to Class 1 Fusion Welding Regulations.

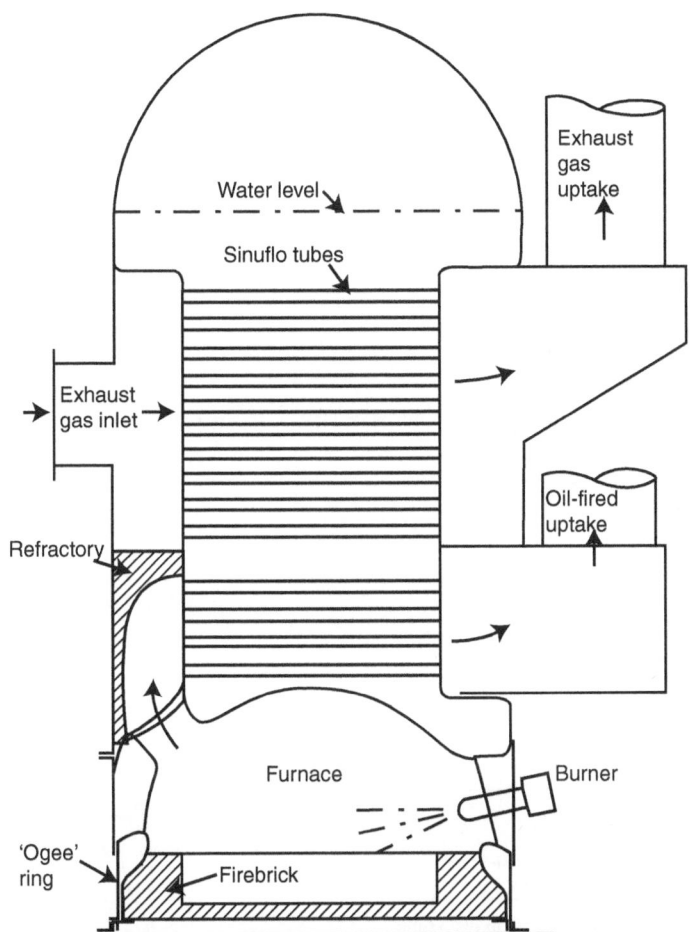

▲ **Figure 3.15** *Diagrammatic arrangement of a single pass composite Cochran boiler*

'Telltale' holes drilled at equal circumferential intervals in the boiler shell enable leakage between the 'Ogee' ring and boiler shell to be detected.

Also in Figure 3.16 are shown different methods of tube manufacture and arrangement. All, enabling gas path, gas velocity and turbulence to be increased with better heat transfer, more heat extracted and tubes maintained in a cleaner condition.

The Sinuflo tube is fitted to Cochran boilers, the Swirlyflo to Spanner boilers, and the plain tube with the twisted metal retarder is common to a wide range of auxiliary and in some cases main tank-type boilers (Figure 3.17).

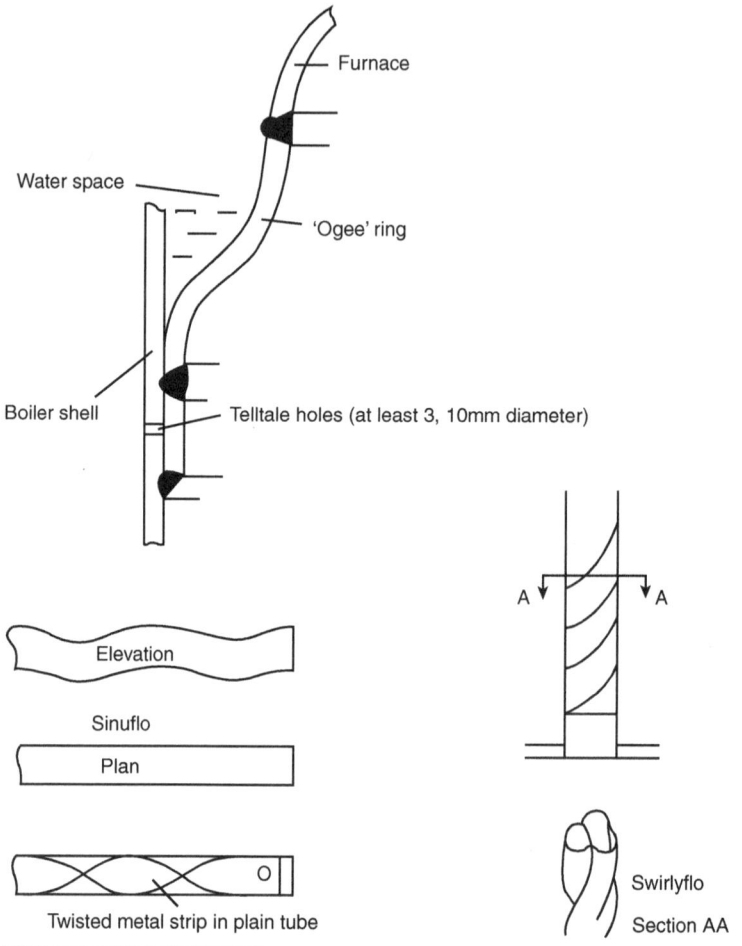

▲ Figure 3.16 *Furnace fixings*

This auxiliary boiler has an all-welded shell, a seamless spherically shaped furnace and small bore tubes. The advantages of this boiler compared to older designs are as follows:

1. Increased steam output for the same size, mass and cost of earlier designs.
2. Increased radiant heating surface.
3. Efficient and robust (∪ 80%).
4. Easy to maintain.
5. No furnace brickwork required – apart from burner quarls.
6. With small tubes, fitted with retarders, gas velocity and turbulence are increased. This gives cleaner tubes and better heat transfer.

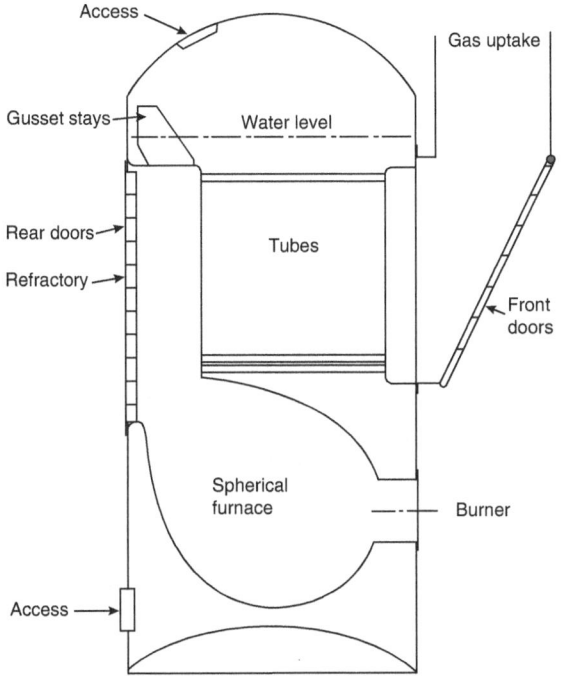

▲ Figure 3.17 *Cochran spheroid boiler*

The boiler can be supplied in various sizes ranging from:

Boiler sizes and efficiency

	Diameter (m)	Height (m²)	Heating surface (bar)	Pressure (kg/h)	Evaporation rate
1.	1.445	26	17.2	995	3.734
2.	2.591	120.8	10.3	4,550	6.325

Cochran exhaust gas boiler

This consists of all-welded tube and wrapper plates made of good-quality boiler steel. Tubes are made of electric resistance, welded mild steel and are swelled at one end and expanded into tube plate. The boiler is provided with the usual mountings: blow-down, feed checks, gauge glass, safety valves, steam outlet and feed control, etc. If it is to be run as a drowned unit (Figure 3.18) then the mountings would be modified accordingly.

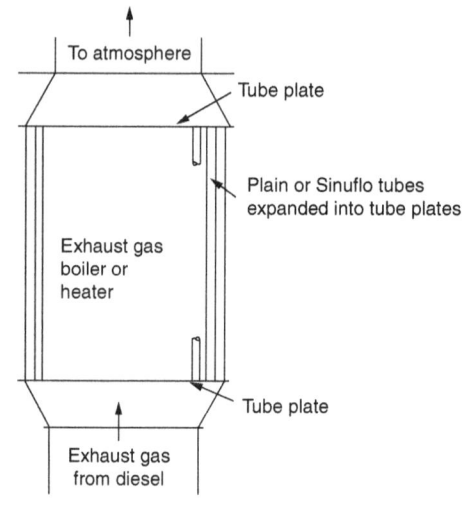

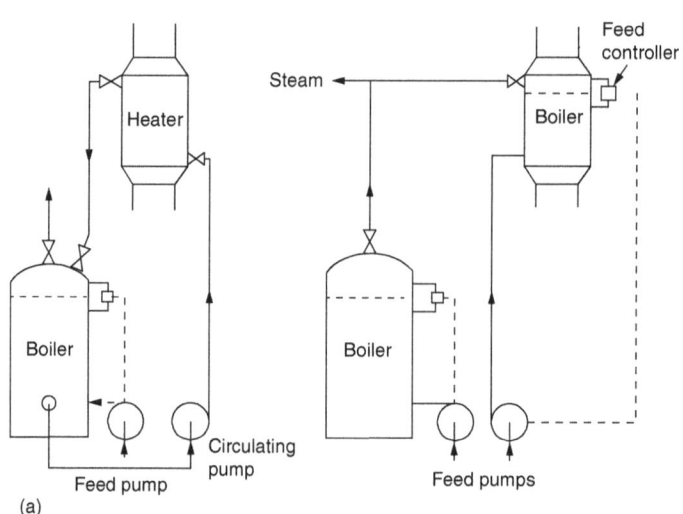

(a)

▲ **Figure 3.18** *Cochran exhaust gas boiler*

Some Defects, Causes and Repairs in Auxiliary Boilers

Furnaces

Defects that could occur in a furnace are deformation, wastage and cracking.

Deformation

With cylindrical furnaces, this can be determined by sighting along the furnace or by use of a lath swept around the furnace or by furnace gaugings.

The causes of deformation are scale, oil, sludge or poor circulation, resulting in overheating of the furnace and subsequent distortion.

Local deformations could be repaired by cutting through the bulge, heating and pressing back the material into the original shape, and then welding. Cutting through the bulge prior to heating and pressing facilitates flow of metal during pressing. Alternatively, the defective portion could be cut out completely and a patch welded in its place.

If the furnace is badly distorted then the only repair possible may be renewal.

A weakened furnace may be repaired temporarily by pressing back the deformation and welding plate stiffeners circumferentially around the furnace on the water side.

Wastage

The causes of wastage are corrosion and erosion. If it is great in extent then renewal of the furnace may be the only solution. Localised corrosion could be dealt with by cutting out the defective portions of the furnace and welding in a new piece of material.

Cracks

Cracks may be found circumferentially around the lower part of the connecting necks. These cracks are caused by mechanical straining of the furnace and the defect is generally referred to as grooving.

If the groove is shallow compared to plate thickness (depth can be ascertained by drilling or by ultrasonic detection) it is usual to cut out the groove and weld. However, if the grooving is deep the material is cut right through and welded from both sides.

Cracks, due to overheating, may be found where deformation has occurred, but these must be made good in the manner described above.

Combustion chamber

The defects that can occur in a combustion chamber are similar to those that can occur in a furnace.

Deformation

In addition to the causes of deformation listed for a furnace must be added that of water shortage. The combustion chamber top would be the first place to suffer overheating and subsequent distortion due to water shortage.

Local deformations can be repaired by cutting out the defective portion of plate, generally through the line of stays or tubes, and welding in a new piece of plate. By cutting the plating through the centre of stay or tube holes avoids a continuous weld and reduces the risk of defects that could occur due to contraction stresses.

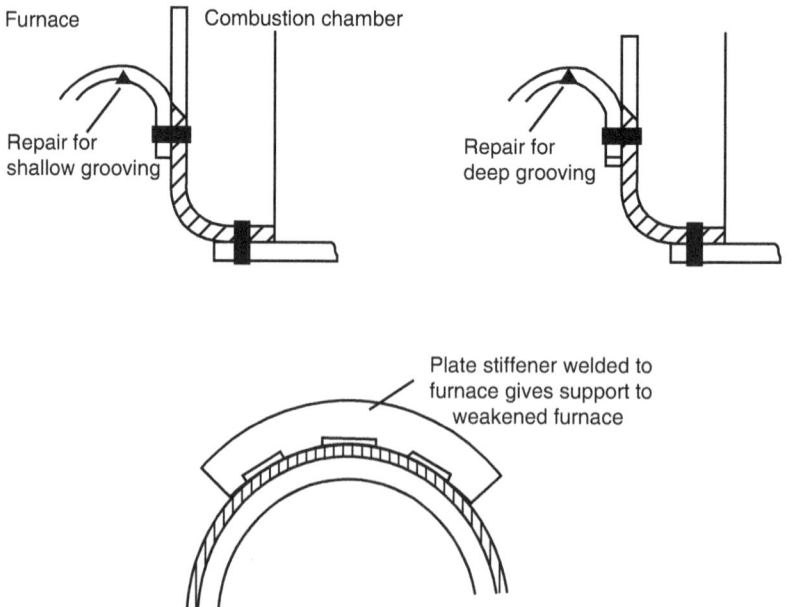

▲ **Figure 3.19** *Boiler constructional details*

Slight distortion of combustion chamber and smoke box plating could occur due to the boiler being operated in a dirty condition. This defect is common, and if there is no leakage past the stays or tubes no repair would be necessary, but it would be essential to keep the surfaces of the plating clean to prevent further distortion. Badly distorted combustion chamber plating is best renewed.

Another cause of combustion chamber plates bulging could be corrosion of stays or tubes leading to a reduction in the support for the plating. The remedy in this case would be renewal of stays or tubes.

Wastage

Leakages past tubes, stays and through riveted seams could cause wastage. If the wastage is not extensive then the defective portion of plate can be built up by welding and the tubes or stays renewed where required. If extensive, then the defective portion of plating should be cut out and a new portion welded in, and stays or tubes where required should be renewed.

Cracks

These can develop due to overheating and mechanical straining.

Likely places are the landing edges of combustion chamber seams on the fire side due to doubling of plate thickness (riveted only) and impairing of heat transfer, and around tubes and stays due to straining of the boiler and or scale build-up around the necks of the tubes or stays.

If the cracks in the seams are not extensive and they are dry they may be left. However, if they are extensive they should be cut out, filled in by welding and have their rivets renewed (Figure 3.19).

Radial grooving of the plating around stays and or tubes if not extensive can be repaired by cutting out the crack and filling in with welding. If the grooving is extensive the defective portion of plate should be cut out and a new portion welded in its place.

Shell and end plates

The principal defects to which the shell and end plating may be subjected are wastage and cracking.

Wastage

This generally occurs at places of leakages, such as riveted seams and boiler mountings.

Leakages at seams and between boiler mountings and shell in the water region of the boiler lead to salt deposition due to water flashing off to steam, leaving behind as it does so some of the salts it contained. These deposits of salts must be thoroughly cleaned away and the plating is then available for inspection for wastage and cracking.

The cracking that could occur may be due to caustic embrittlement, which is dealt with in Chapter 4.

Repairs for wastage may be built up by welding if the wastage is not excessive, or renewal of the defective portion of plate if the wastage tends to be excessive.

If seam leakage is slight and discovered early and upon examination the material is found to be sound, then the only repair necessary may be re-caulking of the seam. Care must be taken, however, to ensure overcaulking does not take place as this can lead to lifting of the plates, one from the other, and deposits could accumulate between the plates.

Cracking

In addition to the cracking that may occur due to caustic embrittlement, grooving of flanged end plating may occur, especially where the furnace front plating is flanged inwards to take the furnace.

Repairs for grooving are of the nature previously described for furnaces.

Repairs for cracks due to embrittlement generally necessitate renewal of the affected portion of plate. If caustic cracking of the main seam of a boiler is extensive then the only repair may be as drastic as boiler renewal.

Boiler Testing

Hydraulic test

When repairs have been carried out on a boiler it is customary to subject the boiler to a hydraulic test.

Before testing, the boiler must be prepared. All equipment and foreign matter must be removed from the water space of the boiler and the repairs should be carefully examined.

Any permanent welded repair must be completed by a 'coded' welder and the welds tested with suitable non-destructive testing techniques.

The boiler safety valves have to be gagged and all boiler mountings, apart from the feed check and air cock, closed. The boiler can then be filled with clean fresh water and purged of air.

Using a hydraulic pump unit connected by a small-bore pipe to the boiler directly or to the feed line, pressure can be gradually applied. The testing pressure is normally $1\frac{1}{2}$ times the working pressure, applied for at least 30 minutes.

With the boiler under pressure it can now be examined for leakages and faults. Weld repairs should again be given repeated blows with a hammer to see if they are sound.

Blowing down and opening up a boiler

If repairs or an examination of the boiler have to be carried out it will have to be emptied. It would always be better, if time is available, to allow the boiler to cool down in its own time after shut-down, then pump the water out. In this way the relatively sudden shock of cooling due to complete blow-down would be avoided (Figure 3.20).

If the boiler has to be blown-down to the sea, allow as much time as possible after shut-down before commencing. The ship-side blow-down cock must be opened first and then the blow-down valve on the boiler can be *gradually* opened up. In this way the operator has some measure of control over the situation, if, for example, the external blow-down pipe between boiler and ship's side was in a corroded condition, then if the operator opened up the boiler blow-down valve first, this could lead to rupturing of the blow-down pipe and a possible accident resulting while he is engaged in opening up the ships-side cock. Figure 3.20 shows the arrangement of a ship-side blow-down cock. When the handle is removed the cock must be in the closed position; this is a safety measure to ensure that the cock is not accidently left open.

Our senses tell us when the blow-down process is coming to a close: the noise level falls and the pressure will be observed to be low. Care must be taken to ensure that no cold sea water gets into the boiler. The boiler when empty of water would still contain steam, which could condense and cause a vacuum condition. This in turn could assist the entry of cold sea water. To help prevent sea water entry, the boiler blow-down is usually non-return (in some water tube boilers a double shut-off is provided) but even with a non-return valve it is strongly advisable to start closing the boiler blow-down valve when the pressure is low enough, and when it is down to the desired value, the valve must be closed down tightly and the ship-side cock closed.

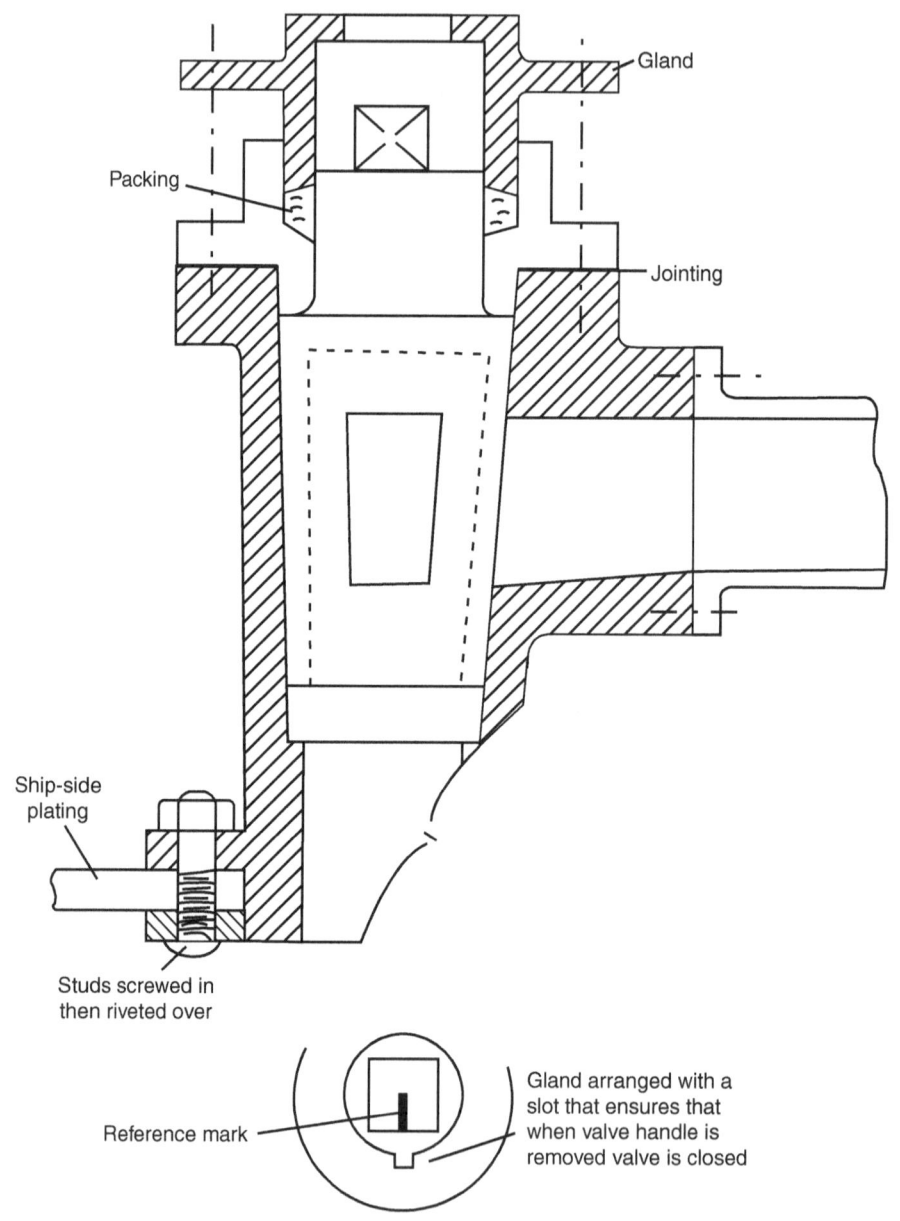

Packing

Gland

Jointing

Ship-side
plating

Studs screwed in
then riveted over

Reference mark

Gland arranged with a
slot that ensures that
when valve handle is
removed valve is closed

▲ **Figure 3.20** *Ship-side blow-down valve*

At this stage allow as much time as possible for the boiler to cool down and lose all its pressure. When the pressure is atmospheric, open up the air cock and gauge glass drains to *ensure* pressure inside boiler is atmospheric.

Either boiler door can be unbolted and the manholes 'knocked in' at this stage, top or bottom, but not both, provided sufficient care is taken. If it is the top door, secure a rope to the eyebolt normally provided, and make the other end of the rope fast. Slacken back but *do not remove* the dog retaining nuts, take a relatively long plank of wood, stand well back and knock the door down. The door is now open and the dogs can be completely removed. Do not immediately open up the bottom door since if the boiler is hot this would lead to a current of relatively cool air passing through the boiler and subsequent thermal shock.

If it is the bottom door that is to be opened first, slacken back on the dog retaining nuts by a very small amount, use a large plank of wood and break the door joint from a safe distance so that if there is any hot water remaining in the boiler no injury will occur to anyone. Again, do not immediately open up the top door of the boiler until the boiler has cooled further.

Packaged Auxiliary Boiler

Figure 3.21 shows in a simplified diagrammatic form a coiled-tube boiler of the Stone-Vapor type. It is compact, space saving, designed for u.m.s. operation and is supplied ready for connecting to the ship's services.

A power supply can be led to a small electric motor that is arranged to drive the feed pump, fuel pump (if fitted), fan and controls.

Feed water is force circulated through the generation coil wherein about 90% is evaporated. The unevaporated water travelling at high velocity carries sludge and scale into the separator, which can be blown out at intervals manually or automatically. Steam at about 99% dry is taken from the separator for shipboard use.

The boiler is completely automatic in operation. If, for example, the steam demand is increased, the pressure drop in the separator is sensed and a signal, transmitted to the feed controller, demands increased feed, which in turn increases air and fuel supply.

With such small water content, explosion due to coil failure is virtually impossible and a steam temperature limit control protects the coil against abnormally high temperatures. In addition the servo-fuel control protects the boiler in the event of failure of water supply.

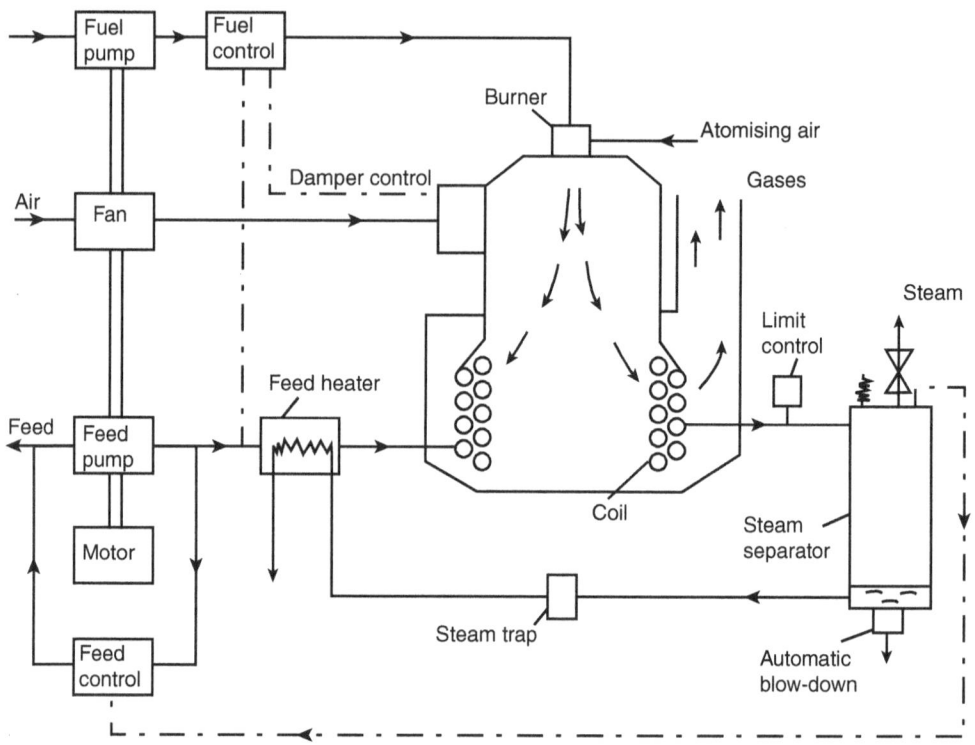

▲ Figure 3.21 *Packaged coil-type boiler*

Steam pressure	10 bar
Evaporation	3,000 kg/h
Thermal efficiency	80%
Full steam output in about 3–4 min	

Performance of a typical unit:
Note. Atomising air for the fuel may be required at a pressure of about 5 bar.

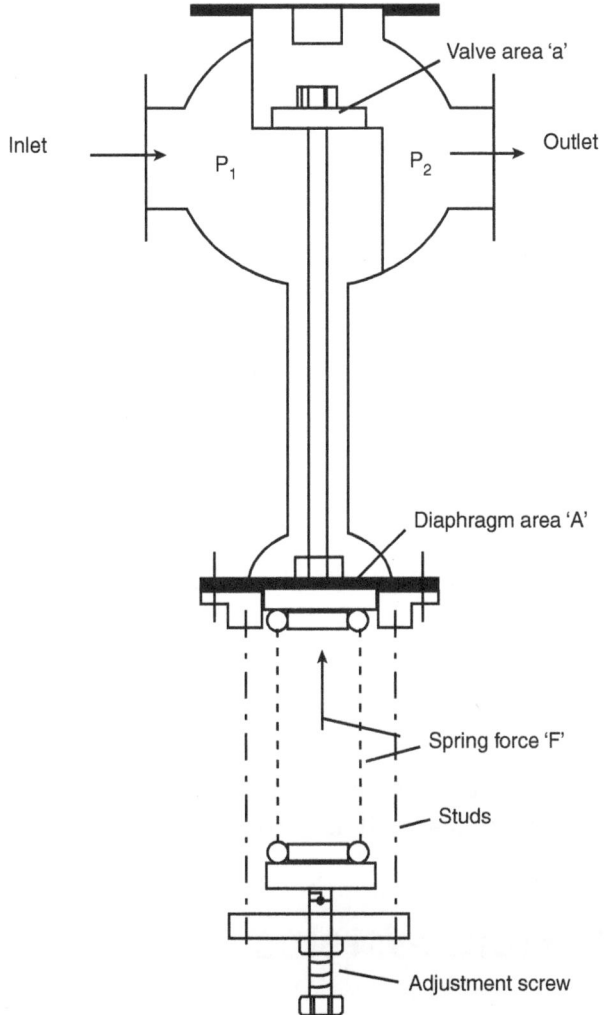

Valve area 'a'

Inlet

P_1

P_2

Outlet

Diaphragm area 'A'

Spring force 'F'

Studs

Adjustment screw

▲ **Figure 3.22** *Reducing valve*

Reducing Valve

Figure 3.22 illustrates diagrammatically a reducing valve that can be used for the reduction of steam or air pressure. As steam passes through the valve no work is done since the reduction process is one of throttling, hence the total heat before and after pressure reduction is nearly the same. When air is passed through the valve its pressure is reduced, but as no work is done by the air, or on the air, its temperature will remain nearly constant.

The reducing valve shown would have a body of cast steel or iron, a valve, valve seat and spindle of steel or bronze. Choice of materials depends upon operating conditions.

Fitted on the discharge side of the valve is a pressure gauge to record the reduced pressure and a relief valve to prevent damage to the low-pressure side of the system in the event of the reducing valve failing.

Since the valve must be in a state of equilibrium under the action of the forces that act upon it we have:

Downward forces = Upward forces

$$P_1 \times A = (P_1 - P_2) \times a + F$$

If P_1, A and a are constant we have:

$$P_2 \text{ varies directly as } F, \text{ i.e. } \quad P_2 \alpha F$$

Hence if the supply pressure is kept constant the discharge pressure can be reduced or increased at will by rotating the adjustment screw.

Thermal Fluid Systems

There are numerous requirements for the use of 'heat' at different places around a ship. The challenge is to build a system that is able to deliver a controllable range of heat to the required position on board.

The most common system for the distribution of heat is water/steam as described on the previous pages. However, the generation of steam to achieve the heating temperatures required means that it is necessary to have a pressurised system with all the associated problems and dangers that the high pressure entails.

Water also increases the potential for corrosion and if the steam condenses in pipework that is exposed to the elements then there is the added complication of freezing at low temperatures.

An alternative that is sometimes used on board is an oil-fired heater using a network of piping to circulate heated 'thermal fluid' to the different locations around the vessel.

An added advantage of the thermal fluid system is that it does not need to condense the fluid to enable it to be pumped, as must happen in the boiler/steam set-up.

Evaporators (Fresh Water Generation at Sea)

Fresh water production from sea water for domestic and boiler feed purposes has become an essential requirement aboard most vessels as this is much more cost-effective, and reliable, than purchasing water from different ports of call. The additional stowage space that would have been used for the fresh water loaded and carried can now be utilised for fuel, or extra space made available for cargo when a fresh water generation plant is placed on board, for even the simplest plant can produce about 10 tons of water for every ton of fuel used.

Various types of evaporating plant are available but the principal types used on board are the 'single-effect plants' and 'double-effect plants'.

Single effect means that evaporation takes place at one pressure only. A single-effect plant may have more than one evaporator and in this case the evaporators are arranged in parallel.

Double effect means that evaporation takes place at two different pressures (and temperatures) and the evaporators would be arranged in series.

The essential requirement of any evaporating plant is that it should produce fresh water as economically as possible and a measure of a plant's economy is its performance ratio (pr). Performance ratio is the kg of vapour produced per kg of steam supplied. For single-effect plants this may be as high as 1.1, for double-effect 1.9 and for two-stage flash evaporators the pr is about 1.5.

Although performance ratio is a good yardstick with which to compare various plants it does not give a complete picture, as the 'heat source' that is supplied to the evaporator may be either 'live heat' or 'waste heat'. This means that it may or may not have performed some useful work before arrival at the evaporator. In addition, the boiler that produces the steam may be employing heat that has been generated from the exhaust gases of a diesel engine, thus providing an extra economy by using heat that would have otherwise been 'lost' to the atmosphere. This means it would also be difficult to affix some form of performance ratio to evaporators using the waste heat in diesel engine cooling water.

The performance of any evaporating plant is adversely affected by scale formation and frequency of blowing down. If scale formation is rapid, heat transfer is reduced and the performance ratio falls. The evaporator would then be blown down and the heating element or coils subjected to cold shocking in order to facilitate scale removal. Every blow-down means heat loss of evaporator contents, hence the more infrequent these interruptions the higher will be the overall economy of the plant.

If scale could be completely eliminated, approximately 20% more water could be produced. Water treatment is available for evaporators but naturally its cost must be added to the cost of water production. Three scales that are principally found in evaporators are given below:

$$calcium\ carbonate\ (CaCO_3)$$
$$magnesium\ hydroxide\ (Mg(OH)_2)$$
$$calcium\ sulphate\ (CaSO_4)$$

Calcium carbonate and magnesium hydroxide scale formation depends principally upon the temperature of operation. Calcium sulphate scale formation is principally dependent upon the density of the evaporator contents. The reactions that take place when sea water is heated are as follows:

$$Ca(HCO_3)_2 = Ca + 2HCO_3$$
$$Then\quad 2HCO_3 = CO_3 + H_2O + CO_2$$

If heated up to approximately 80°C:

$$CO_3 + Ca = CaCO_3$$

If heated above 80°C:

$$CO_3 + H_2O = HCO_3 + OH$$
$$Then\quad Mg + 2OH = Mg(OH)_2$$

Hence if the sea water in the evaporator is heated to a temperature below 80°C calcium carbonate scale predominates. If it is heated above 80°C magnesium hydroxide scale is deposited.

If the density of the evaporator contents is in excess of 96,000 p.p.m. (3/32) calcium sulphate scale can be formed, but evaporator density is normally 80,000 p.p.m. ($2\frac{1}{2}$/32) and less, hence scale formation due to calcium sulphate should not be no problem.

Methods of Controlling and Minimising Scale (Evaporators)

1. *Use low-pressure evaporation plant:* that is, operating at a temperature below 80°C so that calcium carbonate scale predominates – that is a soft scale that is easily removed and not such a poor conductor of heat as other scales.

2. *Use magnetic treatment:* a unit consisting of permanent magnets, preceded by a filter, is installed in the evaporator feed line. The water passes through a strong magnetic field, which alters the charge on the salts so that amalgamation of the salt crystals, formed during precipitation in the evaporator, is prevented and the salt then goes out with the brine.

3. *Use flexing elements:* a heating element made of thin-gauge monel metal built like a concertina may be used. The advantage of such an element is that when pressure, and hence temperature, vary slightly the element flexes considerably thereby cracking off scale effectively and permitting longer running periods of the evaporator between shut-downs. However, when such an element is used care must be taken to ensure it is not subjected to high pressures otherwise failure can occur.

   A safeguard consisting of an air-operated trip valve in the steam line, controlled by a solenoid pilot valve actuated by a differential pressure gauge, which registers pressure differential across the element (maximum 0.9 bar pressure difference), may be fitted.

4. *Use continuous chemical treatment:*

   (a) Organic polyelectrolyte combined with anti-foam minimises scale formation and foaming, and can be used in evaporators producing water for drinking purposes (DTp). It would be continuously fed into the feed line by a metering pump (to ensure overdosing does not take place) at 1–8 p.p.m. of evaporator feed, the rate depends upon evaporator density and output. The compound is alkaline and should be treated in the same way as caustic soda; it should not be taken internally.

(b) Polyphosphate compounds with anti-foam prevent the formation of calcium carbonate scale and minimise possibility of foaming. The compound is a non-toxic, non-acidic, relatively cheap and safe-to-handle powder. The DTp allows it to be used in evaporators producing water for drinking purposes if the dosage rate is 2–4 p.p.m. of evaporator feed. It is suitable only for low-pressure plants. At temperatures around 100°C it forms a sticky grey sludge. It should only be used at temperatures below 90°C.

(c) Ferric chloride (FeCl) is a stable chemical compound supplied in sealed drums. It is a non-explosive hygroscopic, non-toxic compound and when dry is non-acidic. When it is mixed with water it becomes acidic, so for this reason protective clothing should be worn by personnel handling the chemical. It completely prevents the formation of calcium carbonate and magnesium hydroxide scales when used correctly. The concentrated solution is injected into the evaporator feed system through plastic injection equipment. This is necessary since the concentrate is intensively corrosive. When it is in the system it is very dilute and quite safe.

Note. A high-vacuum plant operating at a boiling point of about 45°C using diesel engine cooling water as the heating medium has a relatively low tube surface temperature and may not require any feed treatment. In a steam-heated plant, due to higher tube surface temperature, water treatment is usually required.

Cleaning

Heat-exchange surfaces are usually cleaned by circulating 10% hydrochloric acid solution. A pump is connected to the feed inlet to the evaporator and solution return is by gravity via the brine discharge into an open acid tank from which the pump draws the solution.

The single-effect vertical evaporator shown in Figure 3.23 is still in common use. It operates with a vapour pressure between 1.34 and 1.48 bar, and steam for the heating coils is supplied directly from the boiler. The initial cost for such an evaporator is relatively low, it is also compact and thereby space saving. The shell and dome of the evaporator is made of good-quality close-grained cast iron and the heating coils are made of solid drawn copper. Mountings provided are vapour outlet valve, steam inlet, coil drain valve, feed check valve, blow-down valve, brine ejector, safety valve, gauge glass with fittings, salinometer cock and a compound pressure gauge. In the diagram, a reducing orifice fitting is shown on the steam

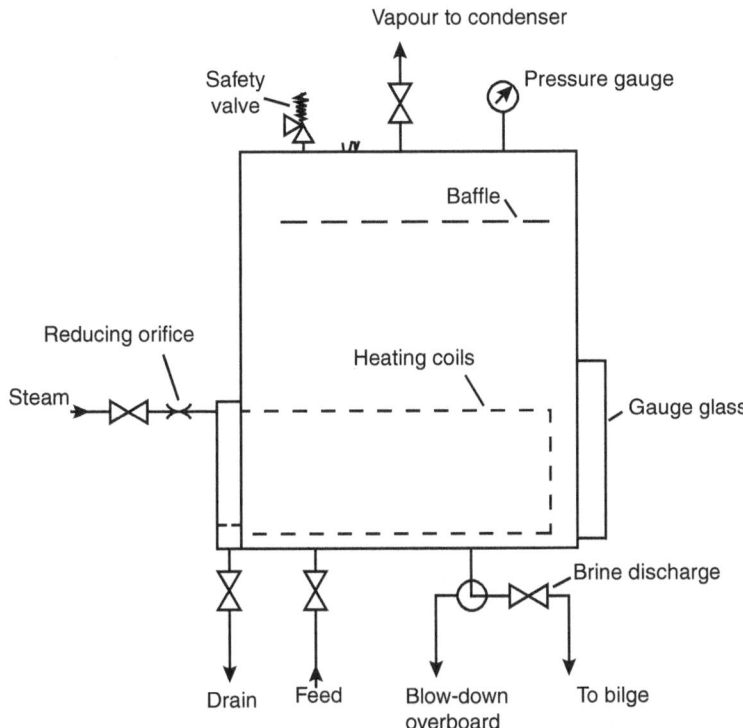

▲ **Figure 3.23** *Diagrammatic arrangement of a simple vertical evaporator*

inlet. Its purpose is to reduce the pressure of steam entering an evaporator shell in the event of failure of a heating coil.

Statutory requirements

1. If the main body is a single casting it may have a working pressure not exceeding 2 bar.
2. Cast iron should not be used above 3 bar working pressure.
3. Cast iron, bronze or gunmetal is acceptable if the temperature does not exceed 220°C.
4. Stress on studs for covers should not exceed 62 MN/m² if the studs are 22 mm diameter or over.
5. Studs should be at least 22mm diameter for covers that are to be frequently removed.

6. Where a reducing orifice is necessary it should be of non-corrodible metal and should be parallel for a length of at least 6.3 mm.

7. An accumulation of pressure test should be conducted and the accumulation of pressure should not exceed 10% of the working pressure.

Low-pressure Evaporating and Distilling Plants

This type of single-effect plant is designed to give better economy than the older single-effect vertical evaporator shown in Figure 3.23. These also have the added effect of improving the 'waste heat recovery' of the machinery plant by using the engine cooling water as a heat source, which would otherwise be lost in the engine cooler.

Low pressure (i.e. operating under vacuum conditions) evaporation plants are widely used for the following reasons:

1. Control over type of scale formed, that is, mainly calcium carbonate, which is soft and easy to remove.

2. Heating medium supplied can be at a relatively low temperature, for example, diesel cooling water or waste steam.

3. Improved heat transfer across the heating element. This is due to higher temperature differences for lower pressures than higher pressures.

Materials for low-pressure plant

The shell is usually fabricated steel (or non-ferrous metal such as the more expensive cupro-nickels) that has been shot blasted then coated with some form of protective. One type of coating is sheet rubber, which is rolled and bonded to the plate then hardened afterwards by heat treatment. The important points about protective coatings are as follows:

1. They must be inert and prevent corrosion.

2. They must resist the effects of acid cleaning and water treatment chemicals.

3. They must have a good bond with the metal.

Heat exchangers use aluminium brass in tube or plate form and with the type of fresh-water generator shown in Figure 3.24. The condenser plates are usually made of titanium, an expensive and virtually corrosion/erosion-resistant material.

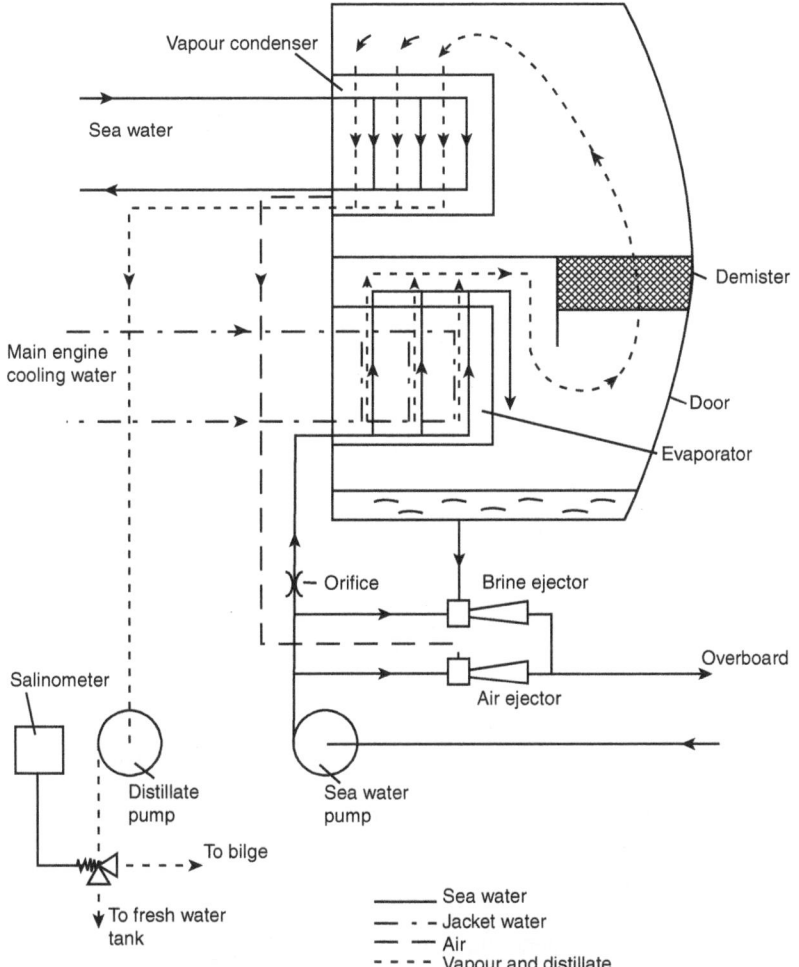

Sea water

Vapour condenser

Main engine
cooling water

Salinometer

Distillate
pump

To bilge

To fresh water
tank

Orifice

Sea water
pump

Brine ejector

Air ejector

Overboard

Demister

Door

Evaporator

———— Sea water
— · — Jacket water
— — Air
- - - - Vapour and distillate

▲ **Figure 3.24** *Fresh water generator (alfa-laval)*

A knitted monel metal wire demister that scrubs the vapour of sea water droplets is a standard internal fitting.

Heat from diesel engine cooling water is used to evaporate a small fraction of the sea water feed in the plate-type evaporator. Unevaporated water is discharged as brine and that which is evaporated passes through the demister to the plate-type vapour condenser, where, after condensation it is discharged to the fresh water storage tank by the fresh water pump.

In the event of the salinity of the fresh water density exceeding a pre-determined value (maximum usually 4 p.p.m.) the solenoid-controlled dump valve diverts the flow to the bilge, preventing contamination of the made water. Excess salinity could be caused by sea water leakage at the condenser or the evaporator priming, the former is the most likely.

Feed supply rate to the evaporator is fixed by the orifice plate and sufficient water goes to the ejectors to ensure a high-vacuum condition in the shell at all times and that all brine is easily dealt with.

This type of relatively simple, compact, space-saving unit is easily accessible for cleaning. Capacity can be adjusted by altering the number of plates in the heat exchangers and adjusting the orifice size. It can be easily arranged for u.m.s. remote operation if required.

Two-stage Flash Evaporator

A double-effect flash evaporation plant is shown in Figure 3.25.

It consists of two identical shells made of fabricated mild steel with protective internal coating, demister screens of knitted monel metal wire and vapour condensers made up of aluminium brass U tubes expanded into rolled naval brass tube plates.

Sea water is pumped through the control valve A to the second- and then the first-stage vapour condensers wherein it increases in temperature before final heating to 80°C in the steam-supplied heat exchanger.

The pressurised, heated sea water flows through an orifice into the first flash chamber whose low pressure corresponds to a saturation temperature less than that of the incoming heated sea water. Hence some of the water must be evaporated in order that its temperature can fall to around that which corresponds to the pressure in the chamber.

Unevaporated water flows through an orifice – which maintains pressure difference – into the second chamber, where more water is evaporated since the pressure is lower than in the first chamber.

A brine pump extracts low-density unevaporated water and discharges the bulk overboard. Some, however, may return to the suction side of the supply pump through the auto-valve B to maintain the feed inlet temperature at about 30°C irrespective of how low the sea water temperature may be.

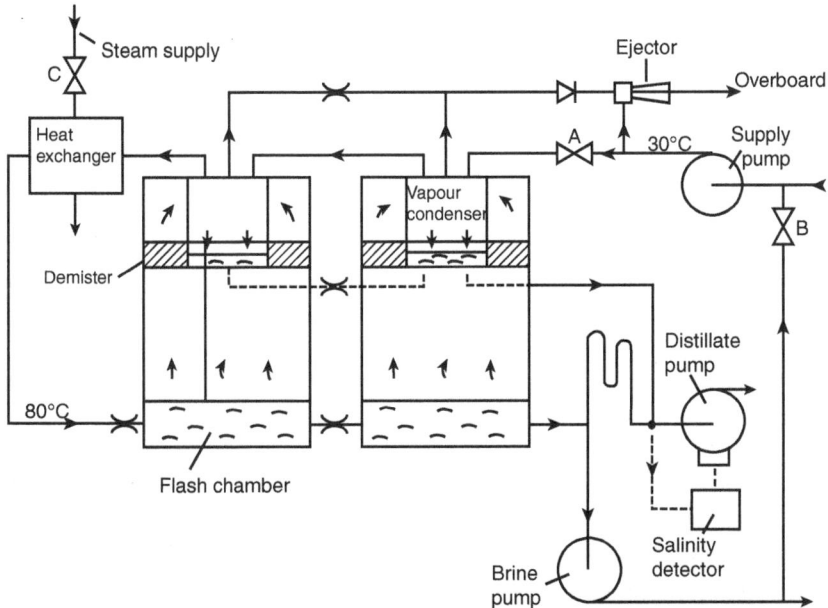

▲ **Figure 3.25** *Two-stage flash evaporation plant*

The vapour and non-condensible gases in each of the chambers pass through the demisters then over and down through the vapour condensers. Distillate flows from the first stage to the second through an orifice and then it is extracted by the distillate pump and delivered to the storage tank. A salinity detector controls the distillate pump. If the density is too high the pump stops and the distillate passes over the double-loop seal to the brine pump suction to be discharged overboard.

Non-condensible gases are extracted by the ejector, which maintains the high-vacuum condition in the chambers.

Complete automatic operation in u.m.s. vessels implies the following:

1. The steam inlet valve C would be thermostatically controlled to maintain sea water feed temperature into the first chamber at 80°C.
2. Valve A, in addition to being controlled thermostatically, would also be controlled by two high-level float switches in the first chamber.
3. The valve B and the distillate pump would be controlled as outlined above.

Output from such an evaporator could be from 13 to 250 tons/day depending upon the size.

Drinking (potable) water

It is often realised that ships are operating in an open-water environment and yet obtaining enough fresh water for consumption by the people on board is of major importance and something for the officers to think about.

MLC 2006 prompted the regulators to take a fresh look at how water suitable for drinking is managed on board.

Desalinated water made from sea water in the distillation plant on board will be safe to drink under the following conditions:

1. If it is boiled at temperatures above 75°C – most of the low-pressure plants operate at temperatures ranging from 40°C to 60°C and it is therefore recommended that these plants are fitted with an automatic disinfecting unit, usually chlorine, before discharging the water into the drinking water storage tanks.

2. Additives to diesel engine cooling water are not harmful. Those that are not allowed for health reasons are the chromates. However, sodium nitrite – even though it is considered dangerous to health – is used in some plants.

3. Inhibitors that are sometimes added to sea water systems to prevent fouling by the growth of marine organisms must not be used if sea water is used in part for supplying the evaporator.

4. Seawater inlets should be placed forward of and if possible on the opposite side of the vessel from any overboard waste water or ballast water outlet.

5. The evaporator is not used within 20 miles from the coastline. According to MSN 1845 from the UK's Maritime Coast Guard Agency (MCA), 20 miles is generally regarded as a safe distance but this may not always be the case. A risk assessment should be undertaken if an evaporator is required to make drinking water when the ship is within this range from the coast.

Water storage and distribution systems

Water suitable for use by the ship's staff and any passengers on board must be kept readily available – at least 2 days' worth is recommended under normal circumstances. In addition, thought should be given to the design and maintenances of the system including:

- Non-contaminating coatings should be used inside storage tanks
- The piping should be colour coded and clearly marked with the direction of flow
- The systems should be completely separate from non-potable water systems
- Pipelines should not be directed through tanks containing any other liquid
- Lines should not be directed through bilges
- Any fittings should be suitably compatible with potable water
- The systems should be arranged to avoid 'dead legs'

Regular inspections and maintenance are essential to avoid infections and other contamination of the fresh water systems to occur.

Fresh water tanks should be:

- Entered only by staff wearing clean overalls that have not been used elsewhere
- Opened, inspected and if necessary washed prior to refilling every 6 months
- Cleaned with a chlorine-based wash, filled with chlorinated water and flushed out on a yearly basis
- Filled with a super-concentrated freshwater solution and left in place for 12 hours during the 5-yearly docking process
- Checked for the Legionella bacteria. Showers are also an important part of the fresh water system that need special attention as this could be a potential source of the bacteria, especially if the shower has not had frequent use

As there are chemicals used in the maintenance process it is essential that all the necessary precautions are taken including the use or personal protective equipment (PPE) and following all the recommended guidelines that will come with the packaging.

Reverse osmosis desalination plant

Osmosis is the natural process where water moves from a less saline solution to a more saline solution across a semi-permeable membrane. It is the process used by plants to take water from the ground through their roots. Therefore a reverse osmosis (RO) process will involve water moving from a more saline solution (sea water) to a less saline solution across a semi-permeable membrane (Figure 3.26). The great advantage of the RO plant is that it produces a substantial amount of water that is closer to the quality of drinking water than it is to the quality of feed water. With the

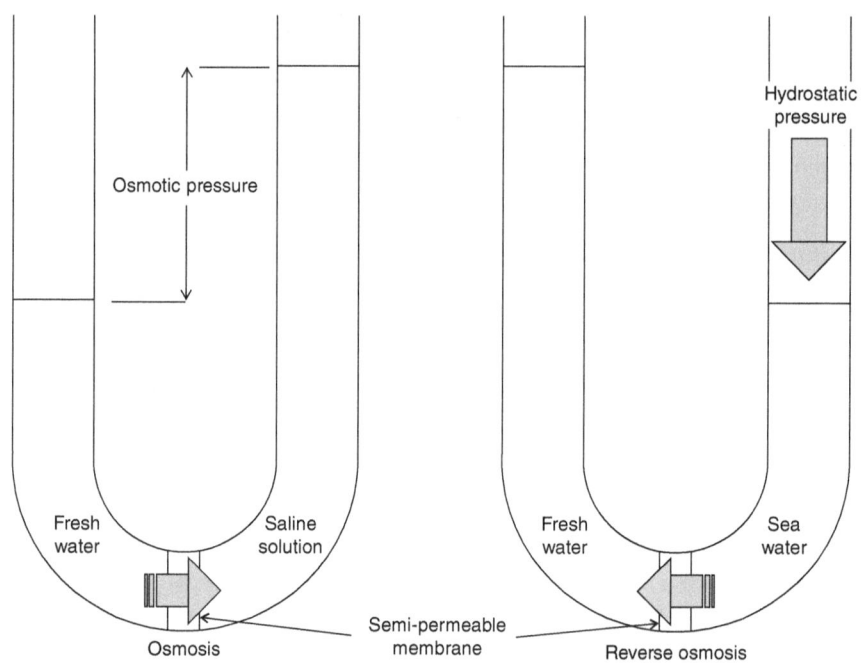

▲ **Figure 3.26** *Reverse osmosis plant*

flash evaporators the water produced is distilled water, which means that the water needs separate processing before it is suitable as drinking water. This usually means passing the water over a charcoal bed.

4

CORROSION, WATER TREATMENT AND TESTS

Corrosion

Most metals used by humans have been processed from their more natural state and the corrosion of metals may be considered as the material returning to its original form as a metal oxide. However, some metals corrode more rapidly than others in the same environment. Iron ore, for example, is an oxide of iron that is converted into the different steels and irons used in engineering. If conditions are correct for corrosion – moisture, acids, salts, etc. – the tendency is for the material to revert back to an oxide of iron by combination with oxygen. (An oxide is an element combined with oxygen, hence oxygen must be present for the transformation to take place.) Following a reaction with the oxygen in the atmosphere, some metals form a protective oxide film upon their surface that can prevent any further corrosion taking place. If this film is broken or destroyed it can in the case of certain metals be replaced very rapidly. For example, chromium, which is used in the alloy stainless steel, can form a microscopic film of chromium oxide upon the surface of the steel, which prevents further corrosion. Furthermore aluminium, which corrodes very rapidly, is quickly rendered non-corrosive owing to the passive oxide film that forms.

Corrosion is a complex subject not fully understood and research into its mechanisms will continue into the foreseeable future. Various theories have been put forward over the years, some of which have been adopted only to be discarded as further progress has been made. Formation of galvanic cells is probably the main cause of corrosion and these can be formed in near-pure boiler water, sea water or other electrolyte. Galvanic elements could be provided by the following:

- dissimilar metals, mill scale, scale and oxide film on the surface of a metal;
- differences in surface structure, inclusion or composition of the metal;
- salts, bacteria and oil degradation products in the electrolyte coming into contact with the metal surface.

Corrosion of Metals in Sea Water

Due to their molecular structure, if two dissimilar metals are placed next to each other in the presence of an electrolyte an electrical interaction will take place. This process is termed galvanic action. The electrical interaction will mean that electrons and ions will move between the two dissimilar metals and therefore some of the structure of one metal will move towards the other. Sea water is circulated, heated and stored within a vessel and it is a strongly corrosive medium because it is a good electrolyte.

With dissimilar metals in sea water, galvanic action results but the strength and direction of the action depends upon where the metals are in the galvanic series table (see Table 4.1). If the metals are close together then there is very little interaction between the two metals. If they are far apart in the table then there will be much more interaction happening. In this process it is the more anodic metal that corrodes. Table 4.1 is an extract from the galvanic series of materials in sea water. Any material in the table is anodic to those above it.

Sacrificial anodes are sometimes used deliberately to give cathodic protection to more expensive material; a well-known example is the zinc anodes that are used as cathodic protection for propellers. The anodes are placed on the hull of the vessel during dry dock and they corrode away instead of the propeller. The anodes are then replaced at the next dry dock.

Cathodic protection can also be supplied by an impressed current system. This applies a potential difference between the two dissimilar metals that is the reverse polarity of the natural galvanic action. Hence the corrosion effect is controlled.

Table 4.1 *Materials in galvanic order*

Titanium	Noble end of table (cathodic)
Graphite	
Monel metal	
Stainless steel (with oxide film)	
Inconel	
Nickel	
170/30 cupro-nickel	
Gunmetal	
Aluminium bronze	
Copper	
Admiralty brass	
Manganese steel (without oxide film)	
Cast iron	
Mild steel	
Zinc	
Aluminium	
Magnesium	Base end of table (anodic)

Listed below are the practical points and methods regarding the minimising of galvanic effects:

1. Choose materials close to each other in the series.

2. Make the key component from a more noble metal, that is, the metal to be protected.

3. Provide a large area of the less noble metal; although its corrosion is increased, it is spread over a larger area.

4. Do not use graphite grease in the presence of sea water as severe corrosion of the bronzes and steels in contact with it may result.

Graphitisation of cast iron

Cast iron contains up to 3.5% carbon, which is mainly in the form of graphite flakes (or spheroids) embedded in a metal matrix. In sea water the metal matrix corrodes – graphite being the more noble material – and the graphite is exposed. This graphitisation of cast iron may stimulate corrosion of metals; however one would expect cast iron to protect, for example, bronzes and brasses etc. since the graphite is higher in the series.

Velocity of sea water

If the velocity of the sea water relative to the material increases (due to the ship moving through the water) the corrosion rate increases – probably up to some limiting value. The reason for this is twofold: (1) increased supply of oxygen; (2) erosion of protective oxide films formed by corrosion.

Stress corrosion

A metal consists of crystals, or grains, whose atomic arrangement is regular, together with amorphous (structureless) metal surrounding them. Corrosion of the weaker amorphous metal, due to galvanic action in sea water, can take place. If stresses are 'locked up' within the metal they can be partly relieved by the corrosion, thereby exposing more amorphous metal to corrosive attack and progression of the process until it leads to possible failure.

Stress corrosion is most commonly found in brasses, but it has occurred in aluminium alloys and stainless steels. Caustic embrittlement – to be discussed later – is another form of stress corrosion.

De-zincification

Brass is an alloy of copper and zinc. In sea water, the zinc is anodic to the copper and it corrodes leaving a porous, spongy mass of copper, hence de-zincification. This should not occur to brasses in which arsenic has been added and whose zinc content is less than 37%.

A similar attack, called de-aluminification, can occur to aluminium bronzes. About 4% to 5% nickel added to the bronze can avoid this problem but trouble may still occur place at welds.

Other Corrosion Topics

Fretting corrosion

Fretting corrosion can occur where two surfaces in contact with each other undergo slight oscillatory motion, of a microscopic nature, relative to one another. Components to which this may occur are those that have been shrunk, hydraulically pressed or mechanically tightened one to the other.

The small relative motion causes removal of metal and metal oxide films. The removed metal may combine with oxygen to form a metal oxide powder that will, in the case of ferrous metal, be harder than the metal itself, thus increasing the wear. Removed metal oxide film would be repeatedly replaced, increasing the damage. Factors affecting the damage caused by fretting corrosion are as follows:

1. Damage increases with amplitude and frequency of movement.
2. Damage increases with load carried by the surfaces.
3. Damage is reduced if oxygen level is low and moisture is present.
4. Hardness of the metal affects the attack. With ferrous metals, the damage decreases as the hardness of the metal increases.

Pitting corrosion

Corrosion may be over a large area, that is, plate type of corrosion or it may be localised, that is, pitting corrosion.

Pitting corrosion is caused when there is a relatively large cathodic area and a small anodic area. Hence the intensity of attack at the anode is high. Large area differences could be caused by mill scale; oxide films; acid pockets of water; scale from salts, pores or crevices; oils, gases and ingress of metals into the boiler. It is a very dangerous form of corrosion. Its rate is generally accelerated with temperature increase, hence where metal surfaces are hottest failure may take place earlier. It should be prevented.

Corrosion fatigue

If a metal is in a corrosive environment and is also subjected to cyclic stress, it will fail at a much lower stress concentration than that normally required for fatigue failure. This is probably due to the progressive weakening effect of amorphous metal corrosion and stress relief. In boilers, microscopic examination would probably reveal the cracks to be trans-crystalline rather than inter-crystalline, which occurs with caustic cracking. The cyclic stresses may be due to the tubes vibrating or fluctuations in thermal conditions, that is, thermal pulsing.

Corrosion in Boilers

To help the reader understand, in part, the mechanism of electro-chemical corrosion it is necessary to first understand some ionic and atomic theory and to appreciate the meaning of the pH value which is an indication of acidity, neutrality or alkalinity.

Atoms and ions

An atom is composed of a nucleus with an electron or electrons in orbit around it. The nucleus is basically composed of protons and neutrons. Protons have a positive electrical charge and the neutrons are uncharged, as their name would indicate. Electrons however are negatively charged. For the atom, the number of protons present is equal to the number of electrons. Hence the resultant electrical charge will be zero, in other words the atom will be electrically balanced. If, however, an atom gains or loses an electron or electrons there will be an excess of either a positive or negative electrical charge. It is then referred to as negative or positive ion (Greek wanderer).

Water is basically composed of hydrogen and oxygen atoms but it does contain ions also.

Hydrogen ion

A hydrogen ion is an atom of hydrogen which has lost its electron. It would normally be written H^+ indicating an excess of positive electrical charge, or $H - \varepsilon$ indicating the loss of the electron ε.

Hydroxyl ion

A hydroxyl ion is a compound of oxygen and hydrogen that has gained an electron. It would normally be written OH^- indicating an excess of negative electrical charge, or $OH^+ \varepsilon$ indicating a gain of an electron.

pH values

Water contains the previously defined hydrogen and hydroxyl ions. The relative concentration of these ions is important. The product of the hydrogen and hydroxyl ion concentration in water at approximately 25°C must always equal 10^{-14} gm ion/l of solution. If the hydrogen ion concentration exceeds the hydroxyl concentration, the water is acidic, whereas if the concentrations are equal the water is neutral. When the hydroxyl ion concentration is greater than the hydrogen, the water is alkaline.

$$
\begin{aligned}
\text{For example,} \quad & 10^{-5}(H^+) \times 10^{-9}(OH^-) && \text{Solution} && \text{Acid} \\
& 10^{-7}(H^+) \times 10^{-7}(OH^-) && \text{Solution} && \text{Neutral} \\
& 10^{-9}(H^+) \times 10^{-5}(OH^-) && \text{Solution} && \text{Alkaline}
\end{aligned}
$$

Note the product of the concentrations is always 10^{-14} and the powers 5, 7 and 9 for the hydrogen ion concentration serve to indicate the degree of acidity or alkalinity of the solution. Hence the pH values now becomes apparent, p (for power) and H (for hydrogen ion). Therefore the pH value is *the logarithm of the reciprocal of the hydrogen ion concentration.*

For example, $10^{-5} = \dfrac{1}{10^5}$, reciprocal $= 10^5$, logarithm $= 5$

pH values range from 0 to 14, that is, from very acidic to very alkaline.

If the water temperature is increased, the hydrogen ion concentration increases and hence there is an increase in acidity or a decrease in alkalinity. Chemicals when added to water alter the hydrogen ion concentration and hence the pH value. Acids lower the pH value, alkalis increase the pH value.

In electro-chemical corrosion of metals the pH value is very important for it governs the degree of corrosion.

Electro-chemical corrosion

When iron is in contact with water that contains hydrogen ions, corrosion may result. The hydrogen ions in contact with the metal surface become hydrogen atoms by taking an electron from the metal. The resultant metal ion (caused through loss of electrons) combines with the hydroxyl ions in contact with the metal surface and so form a metallic hydroxide, which is soluble in the water depending upon the pH value, hence the metal is corroded. This action is similar to a battery action wherein current is caused to flow from anodic to cathodic regions. The migrating ions in the electrolyte (water) and the electrons in the metal form the circuit (Figure 4.1).

Hydrogen, which has formed on the surface of the metal due to the combination of the hydrogen ion and metal electron, may form a polarising layer upon the metal's surface. This will prevent further corrosion. If, however, dissolved oxygen is present in the water, it will combine with the hydrogen to form water and no polarisation will occur and corrosion will continue. Also, if the water is acidic enough, the hydrogen can leave the surface of the metal in the form of hydrogen gas, again preventing polarisation and

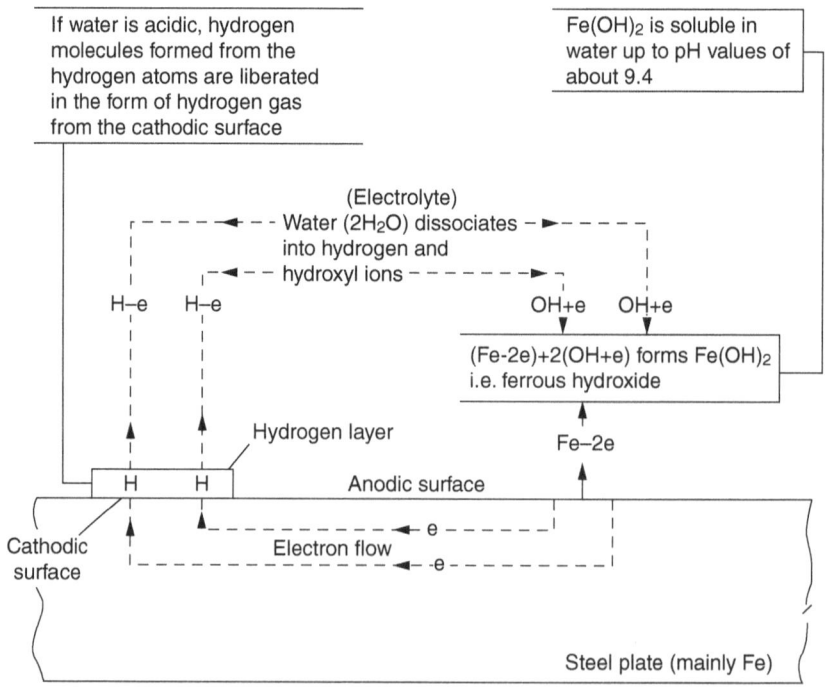

▲ **Figure 4.1** *Electro-chemical corrosion*

continuing corrosion. Hence it is vital that the boiler water should be alkaline with little or no dissolved oxygen content.

Some causes of boiler corrosion

Oils

Lubricating oils may contaminate the feed system and find their way into the boiler. This could be caused due to over lubrication of machinery and inefficient filtering of the feed water. Oils such as animal and vegetable oils can decompose in the boiler liberating their fatty acids and it is these acids that can cause corrosion. It is therefore advisable to use pure mineral oil for lubrication of machine parts where contamination of feed can result. However, oil of any description should never be allowed to enter the boiler as it can adhere to the heating surfaces causing overheating. It can also cause priming due to excessive ebullition.

Mechanical straining

This is not a corrosive agent in itself but is due to the breakdown of the surface of the metal. Pitting type corrosion could result due to differential aeration. (Differential aeration: if a portion of metal becomes partially inaccessible to oxygen it becomes anodic and corrosion may result.) Mechanical straining of boiler parts may be due to mal-operation of the boiler, raising steam too rapidly from cold, missing or poorly connected internal feed pipes, fluctuating feed temperature and steaming conditions. Grooving is caused through mechanical straining of boiler plates, and where a groove is present there is always the danger of corrosion resulting in the groove.

Galvanic action

When two dissimilar metals are present in a saline solution galvanic action may ensue, resulting in the corrosion of the more base metal. Zinc, for example, would serve as an anode to iron and iron would serve as an anode to copper. Sacrificial anodes are frequently used to give cathodic protection. In Scotch boilers, zinc plates are sometimes secured to furnaces and suspended between tube nests. These act as sacrificial anodes giving cathodic protection to the steel plating, etc., of the boiler.

Corrosion of non-ferrous metals in steam and condensate systems may result in deposits of copper on boiler tube surfaces (known as 'copper pick up'), which due to galvanic action can lead to boiler corrosion.

Caustic embrittlement

The phenomena of caustic embrittlement (or inter-crystalline fracture) is by high concentrations of sodium carbonate, which then undergoes hydrolysis to form sodium hydroxide (NaOH), which dissolves the iron in the boiler components. The added stress corrosion cracks follow the grain or crystal boundaries of the material and failure of the affected part could result. Concentrations of sodium hydroxide required for embrittlement to occur vary with operating conditions; roughly about 1,300 grains/litre at 300°C is a guide to the amount of concentration required. Normally such concentrations would not be found in a boiler, but any slight leakage would cause the water to flash off into steam, leaving behind enough solids locally that would cause the high concentrations required.

The sodium hydroxide depresses the solubility of sodium sulphate, although sodium sulphate can be made to precipitate and provide protection against caustic embrittlement for the material. Ratios of sodium sulphate to caustic soda for given pressures will provide a safeguard for the boiler parts. It is recommended that the ratio of sodium sulphate to caustic soda should not fall below 2:5 at all times. Other substances that have been used as inhibitors against caustic embrittlement are quebracho tannin and sodium nitrate.

Caustic corrosion in *high-pressure boilers* is usually indicated by gouging of the tubes and is caused by excess sodium hydroxide and a concentrating mechanism. This phenomenon results in the destruction of the protective magnetic oxide of iron film (Fe_3O_4) and the base metal is then attacked by the concentrated sodium hydroxide.

Effects of salts and gases in feed water

Feed water employed for marine boilers is usually unevaporated fresh, evaporated fresh, or evaporated salt water. The first and last of these three were only normally employed as feed for low-pressure boilers, such as the Scotch boiler. Evaporated fresh water is principally employed, along with evaporated salt water for water tube boilers. All of these waters can contain salts that could be harmful to the boiler from the point of view of scale formation and corrosion. Obviously the evaporated fresh and salt waters should be low in solids content and therefore less harmful. However, feed systems can become contaminated with salt water from, for example, a leaking condenser or an evaporator priming.

Salt water

Average sea water contains approximately 32,000 p.p.m. of total dissolved solids. These solids are made up as follows.

Analysis of Sea Water

Each of the salts in Table 4.2 will now be considered in detail, with regard to their effect under boiler conditions.

Sodium chloride (NaCl)

This is common salt. Heavy concentrations of this salt can cause foaming and priming. Under boiler conditions, the density at which sodium chloride will come out of solution increases as the pressure and temperature increases. In other words, its solubility is a variable. Each salt present in the boiler water will, in general, have varying solubility with temperature variation, and solubility curves for individual salts alone in water can also be affected by the presence of other salts or compounds. In the case of chlorides, their solubility is very high and therefore they should normally not come out of solution under normal boiler conditions. Sodium chloride could in conjunction with magnesium sulphate form sodium sulphate and magnesium chloride.

$$\begin{array}{ccccccc} & \text{Sodium} & + & \text{Magnesium} & \rightleftharpoons & \text{Magnesium} & + & \text{Sodium} \\ \text{That is,} & \text{chloride} & & \text{sulphate} & & \text{chloride} & & \text{sulphate} \\ & 2NaCl & + & MgSO_4 & \rightleftharpoons & MgCl_2 & + & Na_2SO_4 \end{array}$$

Table 4.2 *Analysis of sea water*

Salt	Chemical symbol	Approximate % of total dissolved solids	p.p.m.
Sodium chloride	NaCl	79	25,000
Magnesium chloride	$MgCl_2$	10	3,000
Magnesium sulphate	$MgSO_4$	6	2,000
Calcium sulphate	$CaSO_4$	4	1,200
Calcium bicarbonate	$Ca(HCO_3)_2$	<1	200

In a chemical equation of the foregoing nature, the number of atoms on either side of the equation must be the same.

Magnesium chloride ($MgCl_2$)

Magnesium chloride is soluble under normal boiler conditions but it can to some extent be broken down forming hydrochloric acid and magnesium hydroxides.

$$\text{Magnesium chloride} + \text{Water} \rightleftharpoons \text{Magnesium hydroxide} + \text{Hydrochloric acid}$$

$$MgCl_2 + 2H_2O \rightleftharpoons Mg(OH)_2 + 2HCl$$

Magnesium hydroxide has a low solubility and it is also the most common magnesium compound found in a boiler. Due to its low solubility it can deposit and form scale but with suitable treatment it can be precipitated into the form of a non-adherent sludge that can be blown out of the boiler. Hydrochloric acid can cause corrosion according to the following reaction.

$$\text{Hydrochloric acid} + \text{Ferrite} \rightarrow \text{Ferrous chloride} + \text{Hydrogen}$$
$$2HCl + Fe \rightarrow FeCl_2 + H_2$$

$$\text{then Ferrous chloride} + \text{Water} \rightarrow \text{Iron hydroxide} + \text{Hydrochloric acid}$$
$$FeCl_2 + 2H_2O \rightarrow Fe(OH)_2 + 2HCl$$

From the reaction it can clearly be seen that the result of the attack of the acid upon the iron is to produce a chloride of iron, which breaks down to form an iron hydroxide with regeneration of the hydrochloric acid, hence the corrosive cycle can continue. With suitable treatment this corrosion can be prevented.

Magnesium sulphate ($MgSO_4$)

Magnesium sulphate is soluble under normal boiler conditions, but if too high a density is carried it may deposit and form scale. This salt could to some extent combine with sodium chloride, forming magnesium chloride and sodium sulphate (see NaCl).

Calcium sulphate ($CaSO_4$)

This salt is possibly the most dangerous scale former present in the boiler water. It can deposit and form a hard tenacious scale that greatly affects the rate of heat transfer possibly causing overheating and subsequent failure of the heating surface. The mechanism of scale formation is complex. In its simplest form it could be described as follows. When a steam bubble is formed upon a heating surface, the plate, under the bubble becomes overheated as it is insulated momentarily from the water. The water, containing salts in solution, in contact with the plate around the periphery of the bubble, also becomes overheated. If the salts are those whose solubility decreases with increase of temperature (calcium sulphate is one), they will be deposited in the form of a crystal ring. This is because the water has become supersaturated locally with these salts. Further, when the bubble bursts, the water coming into contact with the overheated plate will again be overheated locally, causing more salt deposition. It would follow therefore that a general statement could be made regarding salts and scale formation. Salts whose solubility decreases with increase in temperature are those that form scale upon heating surfaces and sludge upon cooling surfaces. Salts whose solubility increases with increase in temperature do not normally form scale upon heating surfaces but a sludge may be formed if their saturation point is reached.

Calcium bicarbonate ($Ca[HCO_3]_2$)

This salt is decomposed when heated, liberating carbon dioxide and permitting the precipitation of calcium carbonate.

$$\text{Calcium bicarbonate} \rightarrow \text{Calcium carbonate} + \text{Carbon dioxide} + \text{Water}$$
$$Ca(HCO_3)_2 \rightarrow CaCO_3 + CO_2 + H_2O$$

Calcium carbonate has a low solubility and this solubility decreases with increase in temperature so can therefore form scale. The scale so formed is soft and porous in nature and is not such a poor conductor of heat as calcium sulphate scale.

Fresh water

Unevaporated fresh water should not be used often as make-up feed for boilers. It can contain some or all of the salts present in salt water and other salts besides, but usually in small proportions. Whether a water is classified as salt or fresh basically depends upon whether it is potable or not. An average sample of fresh water is a practical impossibility, only samples of fresh water can be given.

Analysis of a Fresh Water Sample

Table 4.3 *Analysis of a fresh water sample*

Salt	Symbol	p.p.m.
Sodium chloride	$NaCl$	50
Sodium nitrate	$NaNO_3$	35
Magnesium sulphate	$MgSO_4$	30
Calcium sulphate	$CaSO_4$	90
Calcium carbonate	$CaCO_3$	200

Hardness salts

Alkaline hardness salts are the hydroxides, carbonates and bicarbonates of calcium and magnesium. The bicarbonates of calcium and magnesium are called temporary hardness salts since they will be decomposed by heating or boiling the water, liberating carbon dioxide and leaving carbonates.

Non-alkaline or permanent hardness salts are the chlorides, sulphates, nitrates and silicates of calcium and magnesium. Hardness due to these salts is not removed by boiling or heating the water but they can be removed with chemical treatment.

Total hardness, therefore, is the sum of the alkaline and non-alkaline hardness salts present in the water. Since these are the scale-producing solids, knowledge of the feed water's total hardness is essential.

Silicates

Silica is found in most waters and is also present in the plant, especially when new from casting sand used for pipe bending and welds.

In low-pressure boilers, silica combines with calcium and magnesium to form calcium and magnesium silicates, which can precipitate and form a hard scale.

In high-pressure boilers silica may combine with other elements to form complex silica scales, which are glassy, extremely hard and difficult to remove. If the silica content of the boiler water is in excess of about 20 p.p.m. (amount decreases as boiler pressure increases) it is likely that it will volatilise and deposit on turbine blades.

Carbon dioxide

If the water contains dissolved carbon dioxide, carbonic acid may be formed, which can cause corrosion.

The carbon dioxide may have been absorbed into the feed water due to contact with the atmosphere. It can also be formed due to breakdown of bicarbonates and carbonates present in the feed.

Carbonic acid partially dissociates into hydrogen ions and bicarbonate ions, hence the hydrogen ion content of the water is increased. The bicarbonate ions can combine with the ferrous metal to form ferrous bicarbonate, which dissociates into ferrous carbonate and carbonic acid, which is re-dissolved into the water. If there is a supply of dissolved oxygen in the water the ferrous carbonate is converted into ferric hydroxide with regeneration of the carbon dioxide. Thus the process may be a continuous one providing there is a continuous supply of dissolved oxygen in the water. This reaction due to carbon dioxide is represented below in a simplified form.

$$\text{Carbon dioxide} + \text{Water} \rightleftharpoons \text{Carbonic acid}$$

$$CO_2 + H_2O \rightleftharpoons H_2CO_3$$

then

$$\text{Iron} + \text{Carbonic acid} \rightarrow \text{Iron carbonate} + \text{Hydrogen}$$

$$Fe + H_2CO_3 \rightarrow FeCO_3 + H_2$$

then

$$\text{Iron carbonate} + \text{Oxygen} + \text{Water} \rightarrow \text{Iron hydroxide} + \text{Carbon dioxide}$$
$$4FeCO_3 + O_2 + 6H_2O \rightarrow 4Fe(OH)_3 + 4CO_2$$

Iron hydroxide (ferric hydroxide $Fe(OH)_3$) may break down further to form ferric oxide with loss of water

$$\text{hence} \quad 4Fe(OH)_3 \rightarrow 2Fe_2O_3 + 6H_2O$$

Hydrogen

When acid corrosion is rapid, for example when the acid is concentrated under a deposit, damage due to newly formed (nascent) hydrogen molecules at the cathode can result. These hydrogen molecules penetrate the boiler tube metal and react with carbon $C + 4H \rightarrow CH_4$ to produce methane. This carbon loss weakens the metal and the methane gas exerts a pressure that separates the grains of steel. Hydrogen damage can also occur when hydrogen is released by caustic corrosion.

External corrosion

It must not be forgotten that corrosion of a boiler can occur externally. Causes of corrosion in this case could be sooty deposits in the uptakes in the presence of moisture that could form sulphuric acid, which can corrode a standing boiler (which is not under steam) with damp lagging and acidulated bilge vapours.

Boiler Water Treatment

The principal objects of boiler feed water treatment are given below:

1. Prevention of scale formation in the boiler and feed system by (a) using distilled water or (b) precipitating all scale-forming salts into the form of a non-adherent sludge.
2. Prevention of corrosion in the boiler and feed system by maintaining the boiler water in an alkaline condition and free from dissolved gases.
3. Control of the sludge formation and prevention of carryover with the steam.
4. Prevention of entry into the boiler of foreign matter such as oil, waste, mill scale, iron oxides, copper particles, sand, weld spatter, etc. is by careful use of oil heating arrangements (close watch on steam drains), effective pre-commission cleaning and by maintaining the steam and condensate systems in a non-corrosive condition.

Lime and soda treatment (low-pressure boilers)

Lime (calcium hydroxide, $Ca(OH)_2$) and soda ash (sodium carbonate, Na_2CO_3) are used to deal with the calcium and magnesium compounds in the boiler water.

Lime and Soda Treatment

Calcium hydroxide (lime, $Ca(OH)_2$) reacts with temporary hardness salts and magnesium compounds.

Sodium carbonate (soda ash, Na_2CO_3) reacts with the calcium compounds originally in the water and those found through using calcium hydroxide as shown in Table 4.5.

Calcium hydroxide is used to react with magnesium compounds and alkaline hardness salts. Sodium carbonate is used to react with the calcium compounds in the boiler feed including those formed through employing calcium hydroxide. This combination of lime and soda gives zero hardness and alkaline feed water.

Unevaporated fresh water used as make-up feed would contain alkaline hardness salts, which would precipitate and form a soft sludge or scale when the water is heated in the feed heater, economiser or boiler. Hence the water should be treated with lime and soda prior to its entry into the system. Tables 4.4 and 4.5 indicate the reactions that occur when lime and soda are used.

Table 4.4 *Lime's reaction with temporary hardness salts and magnesium*

$Ca(HCO_3)_2$	+	$Ca(OH)_2$	$\rightarrow$	$2CaCO_3$	+	$2H_2O$		
Calcium bicarbonate	+	Calcium hydroxide	$\rightarrow$	Calcium carbonate	+	Water		
$Mg(HCO_3)$	+	$2Ca(OH)_2$	$\rightarrow$	$Mg(OH)_2$	+	$2CaCO_3$	+	$2H_2O$
Magnesium bicarbonate	+	Calcium hydroxide	$\rightarrow$	Magnesium hydroxide	+	Calcium carbonate	+	Water
$MgSO_4$	+	$Ca(OH)_2$	$\rightarrow$	$Mg(OH)_2$	+	$CaSO_4$		
Magnesium sulphate	+	Calcium hydroxide	$\rightarrow$	Magnesium hydroxide	+	Calcium sulphate		
$Mg(NO_3)_2$	+	$Ca(OH)_2$	$\rightarrow$	$Mg(OH)_2$	+	$Ca(NO_3)_2$		
Magnesium nitrate	+	Calcium hydroxide	$\rightarrow$	Magnesium hydroxide	+	Calcium nitrate		
$MgCl_2$	+	$Ca(OH)_2$	$\rightarrow$	$Mg(OH)_2$	+	$CaCl_2$		
Magnesium chloride	+	Calcium hydroxide	$\rightarrow$	Magnesium hydroxide	+	Calcium chloride		

Table 4.5 *Calcium hydroxide's reaction with temporary hardness salts and magnesium*

$CaSO_4$	+ Na_2CO_3	→	$CaCO_3$	+	Na_2SO_4
Calcium sulphate	+ Sodium carbonate	→	Calcium carbonate	+	Sodium sulphate
$CaCl_2$	+ Na_2CO_3	→	$CaCO_3$	+	$2NaCl$
Calcium chloride	+ Sodium carbonate	→	Calcium carbonate	+	Sodium chloride
$Ca(NO_3)_2$	+ Na_2CO_3	→	$CaCO_3$	+	$2Na(NO_3)$
Calcium nitrate	+ Sodium carbonate	→	Calcium carbonate	+	Sodium nitrate

Caustic soda treatment

This could be used in place of the soda and lime treatment. Caustic soda (sodium hydroxide, NaOH) reacts with the alkaline and non-alkaline magnesium compounds, the alkaline calcium compounds, and it also forms sodium carbonate, which can react with the non-alkaline calcium compounds. Table 4.6 indicates the reactions that occur when sodium hydroxide is used.

Table 4.6 *Caustic soda (sodium hydroxide, NaOH) treatment*

$Ca(HCO_3)_2$	+ $2NaOH$	→	$CaCO_3$	+ Na_2CO_3	+	$2H_2O$
Calcium bicarbonate	+ Sodium hydroxide	→	Calcium carbonate	+ Sodium carbonate	+	Water
$Mg(HCO_3)_2$	+ $4NaOH$	→	$Mg(OH)_2$	+ $2Na_2CO_3$	+	$2H_2O$
Magnesium bicarbonate	+ Sodium hydroxide	→	Magnesium hydroxide	Sodium carbonate	+	Water
$MgCl_2$	+ $2NaOH$	→	$Mg(OH)_2$	+ $2NaCl$		
Magnesium chloride	+ Sodium hydroxide	→	Magnesium hydroxide	Sodium chloride		
$MgSO_4$	+ $2NaOH$	→	$Mg(OH)2$	+ Na_2SO_4		
Magnesium sulphate	+ Sodium hydroxide	→	Magnesium hydroxide	Sodium sulphate		
$Mg(NO_3)_2$	+ $2NaOH$	→	$Mg(OH)_2$	+ $2NaNO_3$		
Magnesium nitrate	+ Sodium hydroxide	→	Magnesium hydroxide	Sodium nitrate		

The sodium carbonate that is formed by employing sodium hydroxide should be in sufficient quantity to deal effectively with the non-alkaline calcium compounds. If however, this is not the case, sodium carbonate will have to be used in conjunction with sodium hydroxide.

Caustic Soda (sodium hydroxide, NaOH) Treatment

Care must be exercised when handling caustic soda as heavy concentrations can cause skin burns.

The foregoing treatments – lime and soda, caustic soda – have declined considerably in use. They have been retained for completeness, interest and instruction as they could prove useful in emergency conditions.

Table 4.7 lists the water treatment recommendations for boilers. The action of the chemicals listed will now be examined.

For the precipitation of scale-forming salts into sludge and to give alkalinity, phosphates are used. Phosphates will combine with the calcium in the boiler water

Table 4.7 *Water treatment recommendation (BS 1170 1983)*

Purpose	Chemical	Type of boiler
To prevent scale	Sodium phosphates	All, up to 84 bar wp
To give alkalinity and minimise corrosion	Sodium hydroxide or sodium carbonate	All, up to 84 bar wp All, up to 60 bar wp
To condition sludge	Polyelectrolytes or starch or tannins or sodium aluminate	All, up to 84 bar wp All, up to 84 bar wp All, up to 84 bar wp All, up to 31.5 bar wp
To remove traces of oxygen	Sodium sulphite or hydrazine	All, up to 42 bar from, 31.5 to 84 bar
To reduce risk of caustic cracking	Sodium sulphite or sodium nitrate	All, up to 31.5 bar All, up to 31.5 bar
To reduce risk of carry over of foam	Anti-foams	All, up to 84 bar
To protect feed and condensate system from corrosion	Filming amines or neutralising amines	All, up to 60 bar 17.5 to 84 bar

forming tricalcium phosphate ($Ca_3[PO_4]_2$), which will precipitate, as its solubility is low, to form sludge or porous scale (scale prevention can be achieved by using coagulants). Phosphates will also combine with the magnesium compounds forming magnesium phosphate ($Mg_3[PO_4]_2$) which also precipitates into the form of sludge.

Using phosphates instead of sodium carbonate for conditioning high concentrations of caustic soda are avoided since at high temperatures sodium carbonate can break down as follows:

$$\text{Sodium carbonate} + \text{Water} \rightleftharpoons \text{Sodium hydroxide} + \text{Carbon dioxide}$$

$$Na_2CO_3 + H_2O \rightleftharpoons 2NaOH + CO_2$$

Phosphate Treatment

It will be noted that the reaction in Table 4.8 is reversible, but in the case of high pressures and temperatures there is a greater tendency to the right than to the left of the equation. Therefore the caustic soda content of the boiler water would increase.

From the reactions shown in Table 4.8, it can be seen that through using trisodium phosphate, sodium carbonate is formed. The sodium hydroxide eventually formed due to the breakdown of the sodium carbonate should not normally be excessive.

Table 4.8 *Phosphate treatment*

$3CaCO_3$	+	$2Na_3O_4$	→	$Ca_3(PO_4)_2$	+	Na_2CO_3
Calcium carbonate	+	Sodium phosphate	→	Calcium phosphate	+	Sodium carbonate
$3CaSO_4$	+	$2Na_3O_4$	→	$Ca_3(PO_4)_2$	+	$3Na_2SO_4$
Calcium sulphate	+	Sodium phosphate	→	Calcium phosphate	+	Sodium sulphate
$3CaCl_2$	+	$2Na_3PO_4$	→	$Ca_3(PO_4)_2$	+	$6NaCl$
Calcium chloride	+	Sodium phospate	→	Calcium phosphate	+	Sodium chloride
$MgSO_4$	+	$2Na_3PO_4$	→	$Mg(PO_4)_2$	+	$3Na_2SO_4$
Magnesium sulphate	+	Sodium phosphate	→	Magnesium phosphate	+	Sodium sulphate

It should give the requisite hydroxyl ions (OH⁻) necessary to maintain moderate alkalinity.

Phosphates normally used are sodium hexametaphosphate, sodium metaphosphate, disodium phosphate and trisodium phosphate. The metaphosphates are normally put into the feed system as they are slower to react and therefore should not produce scale or sludge in the feed system (feed heaters, etc.). Disodium and trisodium phosphate are usually pumped directly into the boiler since they are quicker to react and could possibly form sludge or scale in the feed system.

In the presence of sodium hydroxide in the boiler water, metaphosphate, monosodium and disodium phosphate are converted into trisodium phosphate. Hence depending upon the requisite alkalinity, we can select the necessary phosphate. Due to calcium removal by the phosphates, the tendency for the silicates present in the water to form scale is greatly reduced. They tend instead to remain in solution in the boiler water.

Coagulants

The use of coagulants in the boiler water is to condition the precipitates, rendering them into the form of a sludge that is non-adherent and can be easily blown out of the boiler. Calcium phosphate, magnesium hydroxide and calcium carbonate can form scale but by using coagulants they can be rendered relatively harmless into a non-adherent sludge. Coagulants used for this purpose are: polyelectrolytes (these are synthetic organic polymers of high molecular weight, e.g., sodium polycrylate, which may be present in boiler chemical mixtures), sodium aluminate, starch, tannin, gels and casein, etc. Sodium aluminate can also be used with the lime and soda treatment. It can break down and form aluminium hydroxide, which combines with the magnesium hydroxide in a flocculent form, which turns into a scum usually on the top of the water. Other precipitates can combine with the freely flowing floc (scum) and thus be blown out of the boiler via the scum valve. The floc can also combine with any traces of oil that may be present, rendering them harmless.

Note: Coagulants form colloidal suspensions in the boiler water. Colloids generally consist of sub-microscopic particles (clusters of atoms or molecules) with like electrical charge, and therefore repel each other and prevent the formation of larger particles. They combine with precipitates of opposite electrical charge to produce a floc or scum.

De-aeration

It has been stated that for corrosion to take place oxygen must be present to accomplish the formation of metal oxides. Therefore if the air is removed from the feed water then the oxygen is also removed and hence de-aeration of the feed water is essential.

De-aeration can be accomplished either mechanically or chemically, or a combination of both. It is usual to carry a reserve of chemicals in the boiler water in order to deal with any ingress of dissolved oxygen that may result due to mal-operation of the de-aerating equipment, or some other circumstances. The oxygen scavenging chemicals used for de-aerating the water are usually sodium sulphite or hydrazine. Sodium sulphite reacts as follows:

$$\text{Sodium sulphite} + \text{Oxygen} \rightarrow \text{Sodium sulphate}$$

$$2Na_2SO_3 + O_2 \rightarrow 2Na_2SO_4$$

Sodium sulphate, which is formed through using sodium sulphite to de-aerate, remains in solution in the boiler water under normal conditions.

Hydrazine solution (60% hydrazine, 40% water approximately) is finding increasing popularity for oxygen scavenging. It reacts under boiler conditions with the oxygen to form water:

$$\text{Hydrazine} + \text{Oxygen} \rightarrow \text{Water} + \text{Nitrogen}$$

$$N_2H_4 + O_2 \rightarrow 2H_2O + N_2$$

thus having the advantage of not increasing the boiler water density.

Initially it was thought that excessive dosage of hydrazine could lead to steam and condensate line corrosion due to ammonia being produced as the excess hydrazine decomposed:

$$\text{Hydrazine} \rightarrow \text{Ammonia} + \text{Nitrogen}$$

$$3N_2H_4 \rightarrow 4NH_3 + N_2$$

However, a *controlled* excess is beneficial to the steam and condensate system as it counteracts the effect of carbon dioxide corrosion.

There may be a delay in the build-up of a reserve of hydrazine in the boiler water since it reacts with any metal oxides (apart from Fe_3SO_4) that may be present.

Hydrazine should be stored in a cool, well-ventilated place since it is a fire hazard. When handling, protective clothing should be worn – treat in the same way as caustic soda. Hydrazine should be injected into de-aerated feed.

Sodium sulphite may still be used as an oxygen scavenger. If that is the case then the following points regarding it are important: (a) pH value is important to reaction rate with the oxygen. At about 7 pH it is a maximum hence the sodium sulphite should be injected into the system before any alkaline ingredients. (b) In high-pressure boilers the sulphite can break down to give hydrogen sulphide (H_2S) and possibly sulphur dioxide (SO_2), which can attack steel, brass and copper. (c) It increases dissolved solids content.

Condensate line treatment

Where the steam is wet, and also in the condensate system, corrosion can occur due to the presence of carbon dioxide carried over with steam. To ensure alkalinity in this section of the system a volatile alkaliser may be injected into the steam line. These alkalisers are generally ammonia or cyclo-hexylamine. They combine with the steam as it condenses to form carbonates and bicarbonates, which decompose in the boiler back into the CO_2 and the alkaliser, some of which then returns to the steam system.

If the pH value of the condensate is maintained at about 9 this should ensure no corrosion in the low-temperature steam and condensate sections of the plant.

Filming amines, the most common of which is octadecylamine, are insoluble in water at room temperature but volatile in steam. Filming amines prevent corrosion by forming a protective adsorbed layer on metal surfaces. Neutralising amines are colourless, volatile liquids that can burn and whose fumes are toxic. Monocyclohexylamines or morpholine in solution is supplied in sealed containers and should be stored in a cool place.

Antifoams are complex organic compounds of high molecular weight. They are used to control the foam in the boiler drum and thus prevent carry-over. They will generally be included in boiler chemical mixtures.

Prevention of caustic embrittlement

Sodium sulphate is used for the prevention of caustic embrittlement and the ratio of sodium sulphate to caustic soda should be kept at or above the recommended value of 2:5.

Alternatively sodium nitrate may be used; the ratio of sodium nitrate to caustic soda should not fall below 0.4:1 at all times.

Boilers Not in Service

When boilers are taken out of service for short or long periods of time they must be protected from corrosion.

In the case of water tube boilers out of service for a short period of time (e.g. 2 days) the boiler can be fired at intervals to keep the boiler pressure above about 3.5 bar and the boiler water must be maintained in composition as required for the boiler when under normal steaming conditions. Alternatively the boiler could be filled while hot, with hot de-aerated alkaline feed water and about 0.5kg of anhydrous sodium sulphite added for each ton of water in the boiler. In this latter case, the boiler must be topped up periodically and any air in the system must be got rid of.

With fire tube boilers out of service for short periods the only action that needs to be taken is to ensure that the alkalinity to phenolphthalein is not less than the recommended value, or completely fill the boiler with alkaline water.

If the boiler is to be taken out of service for long periods it should be drained completely then dried out by means of heater units. Next, trays of quicklime should be placed internally in suitable positions throughout the boiler before it is sealed up. Blanks should be fitted to the pipe connections in the event of steam being maintained in other boilers and the blow-down should be blanked in any case. The lime should be renewed at least once every 2 months. This is becoming increasingly important as more ships are laid up for extended periods of time.

Cleaning of New Boilers

The purpose of pre-commission chemical cleaning is mainly to remove surface rust and mill scale that occur during boiler erection and manufacture and also dirt and traces of oil.

A comprehensive example of the treatment that would be carried out by a firm of specialists in boiler treatment could be, in order, the following:

1. Boil out the boiler at atmospheric pressure with an alkaline solution to remove traces of oil and dirt.

2. Wash out the boiler with a heated acid solution to remove rust and mill scale.

3. Rinse the boiler with a weak acid solution.

4. Flush the boiler out repeatedly to remove debris.

5. Subject the boiler to a passivation process. This would be carried out under pressure with hydrazine.

The feed system would be subjected to a similar process but the alkaline boil out would obviously be omitted and the passivation would be done at atmospheric pressure with hydrazine.

Boiler Water Tests

Boiler water should be regularly tested, and the treatment of the boiler water should be conducted according to the results obtained from the tests.

For low-pressure fire tube vertical-style or horizontal 'packaged' boilers the salinometer and litmus papers are now very rarely used as part of the testing procedure. However, it is just as important to keep the feed water for these boilers in good condition as it is for the larger 'water tube' boilers.

A considerable part of the cause of corrosion in boilers lies with the dissolved oxygen that enters the feed/boiler water. The closed-feed system limits the chances of the water coming into contact with oxygen in the air and therefore reduces the problem. Therefore, keeping this system in the correct working order is important for minimising any corrosion to the rest of the steam raising plant.

Maintaining the temperature of the feed water higher than 85°C further reduces the water's ability to mix with and carry oxygen around the system. There is, of course, the danger of allowing the feed water temperature to creep too high, which could lead to the feed pump being affected due to cavitation, or by steam starting to form and giving rise to the effect known as 'gassing up' of the pump.

These good 'watchkeeping' actions will mean that any further 'dissolved oxygen' in the boiler water can be treated with the minimum amount of chemicals. The best of the chemicals will be the 'oxygen scavengers' such as sodium sulphite and bisulphite.

Table 4.9 *Typical test values*

Frequency of test		Daily	Daily	Weekly	Daily	Daily	Daily	Daily	Daily
Boiler water test		Alkalinity to phenolphthalein	Chlorides (max)	Caustic alkalinity	Dissolved solids conductivity at 25°C (max)	EDTA hardens (max)	Sulphite excess	Phosphate reserve	Hydrazine reserve
Result expressed		p.p.m. as $CaCO_3$	p.p.m. as $NaCl$	p.p.m. as $CaCO_3$	$\mu S/cm$	p.p.m. as $CaCO_3$	p.p.m. as Na_2SO_3	p.p.m. as PO_4	p.p.m. as N_2H_4
Pressure (bar)	Boiler Type								
Up to 17.5	Package, General fire tube boiler or steam to steam generator	150–300	350	75–250	3,000	5	50–100	30–70	–
Up to 17.5	Water tube	150–300	350	75–200	2,250	5	50–100	30–70	–
17.5–31	Water tube	150–300	150	100–250	1,500	5	50–100	30–70	–
31–42	Water tube	100–150	100	50–100	750	1	20–50	30–50	0.1–1.0
42–60	Water tube	50–100	50	40–60	600	1	–	30–50	0.1–1.0
60–80	Water tube	50–80	30	40–60	450	1	–	20–30	0.1–1.0

It will be important to check that all the joints in the chemical dosing equipment are secure and are not drawing air into the main system.

Another important thing to stop is the formation of scale. This precipitates out of solution and deposits onto the heating surfaces within the boiler. As the scale is about 100 times worse at conducting heat than the steel surface is, the ability of the boiler is considerably reduced.

Salinometer

The range of the scale is normally from 0 to $\frac{4}{32}$ and when the salinometer is floating in pure water at 93°C, which has a relative density at that temperature of unity, the salinometer reading is zero. When the salinometer is floating in solutions of common salt at 93°C the salinometer reading is $\frac{1}{32}$ (approx. 32,000 p.p.m.) when the relative density of the solution is 1.025. (The relative density of salt water at 93°C is approximately 1.025 or $\frac{1}{32}$ on the salinometer.)

If sea water is used as make-up feed for low-pressure boilers it is recommended that the boiler density should be maintained as close as possible to $\frac{4}{32}$ (approx. 125,000 p.p.m.). This would be attained by resorting to blow-down. The use of sea water as make up feed for boilers should be avoided as far as possible, but if it has to be used a certain amount of protection for the boiler can be provided by using soda ash.

Litmus papers

These are used as a very rough approximation to the degree of acidity or alkalinity of the water. A litmus paper when inserted into a sample of boiler water may change colour, turning blue if the water is alkaline or red if the water is acidic. The degree of colouration is a very rough indication of the pH value of the boiler water.

For accurate testing of the boiler water, the foregoing salinometer and litmus paper methods are inadequate. Table 4.9 gives recommended values for low- and high-pressure boilers. To ensure that these values are being maintained, more refined testing methods must be used.

Alkalinity

Tests for alkalinity are as follows:

(1) Alkalinity to phenolphthalein

Take 100 ml sample of boiler water,

add 1 ml (10 drops) of phenolphthalein,

add N/50 sulphuric acid to clear the sample.

Calculation: ml of N/50 acid used $\times$ 10 = p.p.m. $CaCO_3$

Phenolphthalein is less alkaline than hydroxides or carbonates, and when it is added to a sample containing hydroxides and/or carbonates it will turn pink in colour. The acid used after this colouration will first neutralise the hydroxides, forming salts. It will then react with the carbonate molecules present, forming bicarbonate molecules. Bicarbonate molecules are less alkaline than phenolphthalein, hence the pink colouration will disappear once all the hydroxides and carbonates have been dealt with by the acid. One bicarbonate molecule is formed from two carbonate molecules, hence in the test the quantity of acid used is a measure of the alkalinity due to the hydroxides (caustic) present and half the carbonates.

(2) Total alkalinity

Take alkalinity to phenolphthalein sample,

add 10 drops of methyl-orange, result yellow colouration,

add N/50 sulphuric acid until pink.

Calculation: ml of N/50 acid used for both tests $\times$ 10 = p.p.m. $CaCO_3$

Methyl-orange indicator is less alkaline than phenolphthalein and bicarbonates. It can be used initially in place of phenolphthalein or, as is more usual, as a continuation after the alkalinity to phenolphthalein test. If no yellow colouration results when the methyl-orange is added to the alkalinity to the phenolphthalein sample no bicarbonates are present. Therefore, the alkalinity as determined in the alkalinity to phenolphthalein test has been due to hydroxides alone. *Note*: Hydroxides and carbonates can co-exist in a solution but hydroxides and bicarbonates cannot.

(3) **Caustic alkalinity**

Take 100 ml sample of boiler water,

add 10 ml of barium chloride,

add 10 drops of phenolphthalein, result pink colouration,

add N/50 sulphuric acid to clear the sample.

Calculation: ml of N/50 acid used × 10 = p.p.m. $CaCO_3$

In this test, barium chloride is first added to the boiler water sample in order to precipitate all the carbonates that are present. The test is then carried out as for the alkalinity to phenolphthalein test but in this case only the hydroxides (caustic) will be measured.

Chloride test

Take alkalinity to phenolphthalein sample,

add 2 ml of sulphuric acid,

add 20 drops of potassium chromate indicator,

add N/35.5 silver nitrate solution until a brown colouration results.

Calculation: ml of N/35.5 solution used × 10 = p.p.m. Cl or ml of N/50 silver nitrate solution used × 10 = p.p.m. $CaCO_3$

Chlorides may be present in the boiler water sample and it is essential that they be measured as they would be an indication of salt water leakage into the feed system, through some means such as either a leaky condenser or a primed evaporator, etc. The alkalinity to phenolphthalein sample taken has had the hydroxides and carbonates dealt with and they will play no further part in the test now conducted for chlorides. The sample is made definitely acidic by the addition of a further small quantity of acid. This is to speed up the chemical reactions that next take place. Silver nitrate has an affinity for potassium chromate and chlorides, although its principal preference is for the chlorides. When it has neutralised the chlorides present in the sample, it is then free to react with the potassium chromate. In doing so, it produces a reddish-brown colouration. It is therefore apparent that the amount of silver nitrate solution used is a direct measure of the chloride content of the boiler water sample. *Note*: As the drops of silver nitrate strike the sample, a reddish-brown local colouration results, which quickly disappears if chlorides are present. This should be ignored.

Sulphite test

Take 100 ml of boiler water sample,

add 2 ml of sulphuric acid, add 1 ml of starch solution

add potassium iodide–iodate solution until sample is blue in colour.

Calculation: ml of iodide–iodate solution used $\times$ 12.5 = p.p.m. Na_2SO_3

The boiler water sample is made slightly acidic to speed up the chemical reactions that are to take place. Potassium iodide–iodate produces a blue colouration through reaction with starch, but it has a preferential chemical reaction with sulphite if it is present in the sample. Hence when the potassium iodide–iodate solution has dealt with all the sulphite present, it is then free to react with the starch present in the sample, producing a blue colouration. It is therefore apparent that the amount of potassium iodide–iodate solution used is a direct measure of the sulphite content of the boiler water sample. As far as is possible, the atmosphere should be excluded in this test otherwise an incorrect result may occur. If the test indicates that an adequate reserve of sodium sulphite is present in the boiler water there is no need to conduct a test for dissolved oxygen.

Phosphate test

Take 25 ml of filtered boiler water sample,

add 25 ml vanadomolybdate reagent,

fill comparator tube with this solution and place in right-hand compartment of comparator,

in the left-hand compartment place a blank prepared by mixing equal volumes of vanadomolybdate reagent and de-ionised water. Allow the colour to develop for at least 3 minutes and then compare with the disc.

Calculation: phosphate reserve in p.p.m. (mg/1) from the disc reading.

Hardness test

Take 100 ml of filtered boiler water sample,

add 2 ml (20 drops) of ammonia buffer solution,

add 0.2 g of mordant black 11 indicator and stir until dissolved.

If hardness salts are present the solution turns wine-red. The test is to titrate with EDTA solution until colour changes to purple and then blue (with some waters a greyish-coloured end point is reached).

Calculation: ml of EDTA solution used $\times 10 = $ p.p.m. $CaCO_3$

pH value

A boiler water's pH value can be obtained by three basic methods.

1. Litmus papers.
2. Colourimetrically (see Figure 4.2).
3. Electrolytically.

The litmus paper method has already been described but the test does not give a very accurate pH result, indicating merely if the water is acidic or alkaline. Tests (2) and (3), however, give a reasonably accurate pH value.

Colourimetric method

Take sample of boiler water,

place one thymol blue tablet in a 50 ml Nessler cylinder,

add 50 ml of sample to Nessler cylinder and ensure tablet is dissolved, (1)

put 50 ml of sample into the other Nessler cylinder, (2)

place (1) in right-hand compartment of the Nessleriser,

place (2) in left-hand compartment of the Nessleriser.

Place appropriate disc in Nessleriser and match the colours, then read the pH value from the right-hand window.

Electrolytic method

An electric cell, using the boiler water as an electrolyte and two special electrodes both made of glass, is used. The potential difference between the electrodes is directly dependent upon the hydrogen ion content of the electrolyte (boiler water). This potential difference is measured by a sensitive voltmeter connected into the external circuit of the cell and calibrated to read pH values.

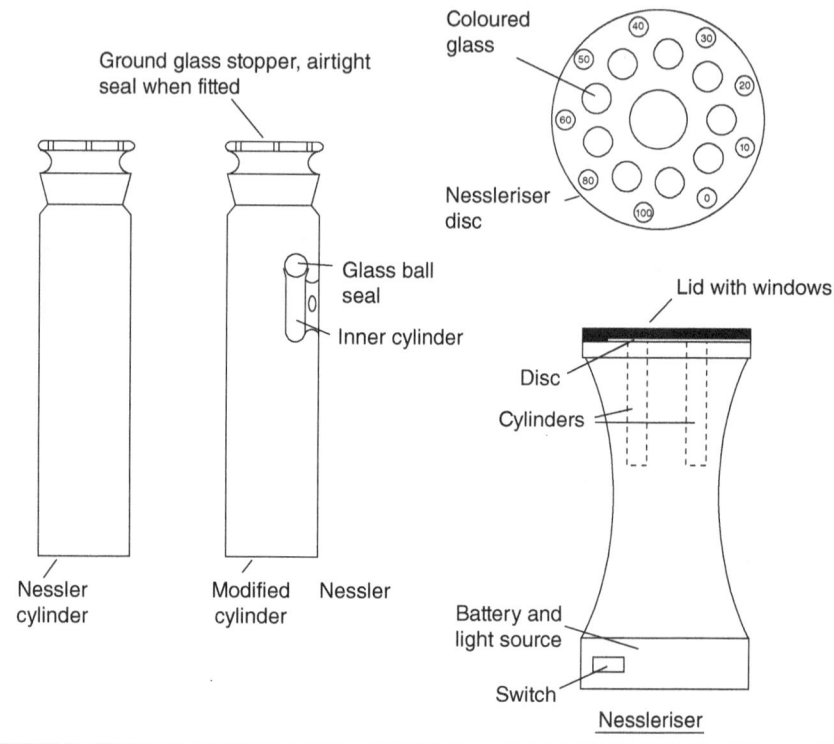

▲ **Figure 4.2** *Nessler cylinder*

Dissolved oxygen test

Take 500 ml of boiler water sample,

add 0.3 ml of manganese chloride,

add 0.3 ml of potassium hydroxide,

add 1 ml of hydrochloric acid,

add 2 ml of ortho-tolidine.

In this test it is essential that the atmosphere be excluded from the sample being tested. To arrange for this a specially designed sampling flask is used. After the addition of the various chemicals to the boiler water sample, the resulting solution is compared colourimetrically with a colour chart or a series of indicator solutions whose dissolved oxygen content is known. Where colours of sample and indicator coincide, the dissolved oxygen content of the boiler water sample is read from the indicator.

Alternative test

To 8 ml of prepared stock reagent, made up of indigo carmine glucose tablet dissolved in glycerol, 2 ml of potassium hydroxide solution is added to make Leuco reagent.

Some of the Leuco reagent (which must be used within 15 hours) is used to fill the inner tube of a modified Nessler cylinder.

A glass ball is then placed on to the inner cylinder sealing in the reagent – no air must be trapped below the ball. The boiler water sample is run into the Nessler cylinder for at least 2 minutes to obtain a good sample and exclude all air. Next, the cylinder stopper is quickly fitted and then by inverting the cylinder the glass ball will fall from the inner cylinder and the reagent will mix with the sample. When the two are thoroughly mixed, the cylinder is placed into one compartment of a Nessleriser and a cylinder containing boiler water sample only is placed in the opposite compartment. An appropriate Nessleriser disc is then fitted and rotated until the colours match. When this occurs a reading in ml/l is obtained in the small window.

If the dissolved oxygen content is high (above 0.2 ml/l) it is recommended that no attempt should be made to reduce it by means of sodium sulphite (Na_2SO_3), otherwise considerable quantities of this chemical would have to be used increasing the total dissolved solids content of the boiler water.

When dissolved oxygen content is high, it could be due to a leakage into that part of the system that operates at a pressure below atmospheric, faulty de-aeration equipment, air ejectors, etc. The matter should be corrected as early as possible to reduce risk of corrosion.

Total dissolved solids

These are ascertained by use of a hydrometer or electrical conductivity meter.

Hydrometer: Usually graduated in grains per imperial gallon (to convert grains per imperial gallon to p.p.m. multiply by 14.3). Care must be taken when using the hydrometer to account for the water meniscus and to ascertain accurately the temperature of the sample. Temperature correction tables for the hydrometer are usually supplied with it.

Conductivity meter: A portable, battery-operated, electrical conductivity meter is used in this test. The removable conductivity cell is washed out and filled with a treated boiler water sample (treatment consists of cooling to 15–20°C, adding phenolphthalein and removing pink colouration with acid). The filled cell is plugged into the meter, its temperature checked and the temperature control set to correspond. A range switch is set to approximate range of reading expected then a central control is operated until 'null' balance of the electrical bridge circuit (the cell forms one resistance) is achieved. Position of the central control indicates the total dissolved solids in the water usually in p.p.m. but it may be conductivity in micromhos (to convert: p.p.m. total dissolved solids = conductivity in micromhos $\times$ 0.7).

Hydrazine test

Take 250 ml of boiler water sample, exclude air and cool to 16–25°C,

add 15 ml of 0.5 N hydrochloric acid to each of two Nessler cylinders,

add 25 ml of boiler water sample and 10 ml of 4-dimethylaminobenzaldehyde to one cylinder, (1)

Add 35 ml of boiler water sample to other cylinder. (2)

Place (1) in right-hand compartment of the Nessleriser. Place (2) in left-hand compartment of the Nessleriser. Match samples against disc colours.

$$\text{Calculations}: \frac{\text{Disc reading}}{25} = \text{p.p.m. hydrazine}$$

The hydrazine reserve in the boiler water should be between 0.1 and 1 p.p.m.

Ballast water has been used since the introduction of steel hulls as a method of stabilising a vessel especially when the vessel has little or no cargo. Unfortunately, along with the water came everything that was living within it. The species travelled with the ship and were then discharged at the port of call where they were released into an environment completely different to their origins. Some will not have survived but others have taken over to the detriment of the indigenous species. In some cases, the invasive species have had a devastating effect on the local aquatic wildlife. This has led to the development of the Ballast Water Management Convention, which requires every ship to manage their ballast water and sediments to a predetermined standard according to their ship-specific ballast water management plan. Further all ships will have to carry a ballast water record book.

Diesel Engine Cooling Water

It is just as important to look after the diesel engine cooling water as it is the feed water going into a boiler. The correct cooling effect of the water depends upon an efficient transfer of heat through the metal wall of the surfaces surrounding the combustion space to the water circulating on the other side of the metal divide. If scale or other contaminants are allowed to precipitate out of the water solution and settle on the metal surface, an 'insulating' layer forms and the ability of the water to remove the heat is reduced. In extreme cases the build-up of scale, debris or metallic particles can lead to cracking in engine cooling channels (jackets). The debris circulating in the water can also affect the condition of jointing components such as 'O' rings, reducing their operational lifetime.

There are other ways in which the cooling water can have a detrimental effect on the engine. These are if:

- The cooling water is allowed to become acidic (any sulphur or nitrogen, possibly from the products of combustion, combining with the water could form acids strong enough to cause corrosion).
- Oxygen is able to build up in the system, as this could combine with the water to corrode the metal in the cooling system.
- Erosion occurs due to non-laminar flow of the water through the system. This could be due to design faults, vibration from the machinery or a build-up of debris causing obstacles for the water to overcome and causing turbulence.

The diesel engine fresh water cooling system will cause premature corrosion to the engine if it is not treated and monitored to stop any of these adverse processes happening. It is also normal to complete regular tests of the cooling water to ensure that the following are all within the engine and/or the specialist chemical supplier's limits:

- the pH value is stable (between 7 and 10 at 20°C, as a rough guide);
- the corrosion inhibitor's concentration level;
- the total dissolved solids (chloride concentration) – a rise in this level could indicate sea water contamination of the fresh water system (50 p.p.m. as a rough guide).

It is also important to ensure that any bacteria are prevented from contaminating the system. This can be monitored by checking the nitrite level. If this drops without loss of water then further investigation should be carried out.

Usually the suppliers of the chemicals for treating the cooling water will also supply the testing kits necessary to complete the condition monitoring process of the water. This will then determine the chemical dosing required for a given engine running under different conditions. The testing of the cooling water on board should take place at least once a week and it is recommended that samples are also sent ashore for further testing every 3–4 months.

On engines with aluminium components it is important to complete the protection process and also maintain a low pH value. The modern products also form an oxide film on metal surfaces thus preventing electrolytic corrosion. It is vital that manufacturer's specifications are followed very closely.

With nitrite-based protective systems the concentration of nitrate is important. The level must be high enough to build up and maintain a thin protective layer over the metal components. A low level will lead to the water increasing in conductivity and enhancing corrosion as a consequence.

The introduction of the chemical to the cooling system should be carefully controlled. If the chemical is in a powder form it should be mixed thoroughly before being applied to the engine cooling system. If the cooling water header tank is used then the engineers must ensure that the inhibitor is circulated around the entire system.

During the handling of all chemicals, engineers must ensure that the correct personal protective equipment (PPE), according to the chemical specification sheet, is used at all times.

It may become necessary to drain the system flush through and refill with a fresh charge. This may be due to the completion of extensive work on the system or it may be that the system has become so contaminated that a fresh charge is the best course of action. If this is the case it is important that the system is drained and refilled with clean water, and that this is circulated and heated to about 60°C which should dissolve any grease left over from the maintenance work.

It is important to check that the heated water circulates around the whole system and that any cross-over valves or control valves are opened to allow the flow to enter every section of the system. At this point the system can be checked for any leaks etc.

The 'flushing' water is drained from the lowest point in the system. If it has been necessary to use an acid wash (to de-scale a system) then it will be necessary to continue to flush through the system until a neutral pH value is recorded at the drainage point.

The fresh charge is added from the desalinated supply and the correct initial charge of chemical added as directed by the chemical manufacturer's/supplier's instructions.

Careful monitoring should happen, especially for the first 24 hours of machinery operation following the cleaning process.

5

MANOEUVRING AND STEERING

Steering Gears, Stabilisers, Rudders and Thrusters

This chapter discusses all the equipment associated with the manoeuvring and stabilisation of ships. The correct use of this equipment and ensuring that it is available when required is very important for the safety of the vessel. The text will include telemeter (transducer) systems, power (amplifier) units, actuator (servo) mechanisms and conclude with a short account of related principles as utilised with ship stabilisers and other equipment that is used when manoeuvring a vessel.

IMO have developed rules for the construction and use of steering gear on ships. The rules to be complied with are set out in Chapter II of the SOLAS convention. The regulations were set out when the conventional system for steering a ship was one or two rudders being turned by an electro-hydraulic steering engine such as the 'four' ram type or the rotary vane-type equipment. However, with the development of azimuthing propulsors and water jets for the propulsion and steering of large ships the rules set out within SOLAS have had to be clarified. For example, regulation 29.3.2 states that the steering gear must be capable of driving the rudder from 35° port to 30° starboard or vice versa, within 28 seconds. There were few alternatives when the regulation was written but now a modern vessel may not have a rudder and testing the steering on a ship with podded drive consists of turning each pod through 360° in either direction.

> *Redundancy in the ship's steering ability is the most important feature of the mechanisms on-board any vessel. With the modern set-up the electrical signals from the bridge to the steering gear controls should be duplicated and the cables, or data signals, for each system should follow different routes.*

With electro-hydraulic systems there should be two hydraulic pumps driven by different electric motors and any hydraulic circuit should be arranged so that two separate hydraulic systems are available as back-up in the case of failure. On very small craft a mechanical 'tiller' can be used as the emergency back-up steering mechanism. One area that will cause a problem if lost is the oil itself. Therefore, all vessels are required to carry a fresh charge of oil in case the original is all lost in some sort of failure.

Steering Systems

Regulations 3 and 29 of SOLAS give the definition/description for the constructional arrangement of both the main and auxiliary ship's steering systems. The traditional method of steering a ship is with the aid of a rudder (Figure 5.1). On small vessels this was operated by hand by using the tiller. On larger vessels with heavy, bulky rudders they must be turned with the use of power. Regulation 29.3.2 also gives some basic minimum performance requirements for the steering mechanism of a SOLAS-compliant vessel. This standard, first specified in the 1960s, requires the 'rudder'.

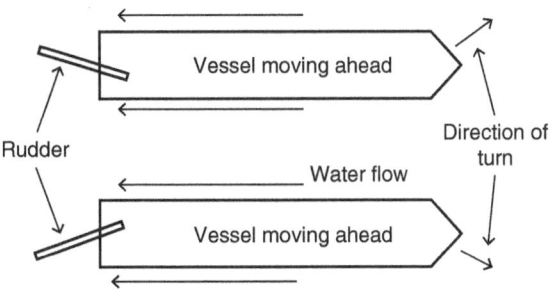

▲ **Figure 5.1a** *Conventional steering action of a ship*

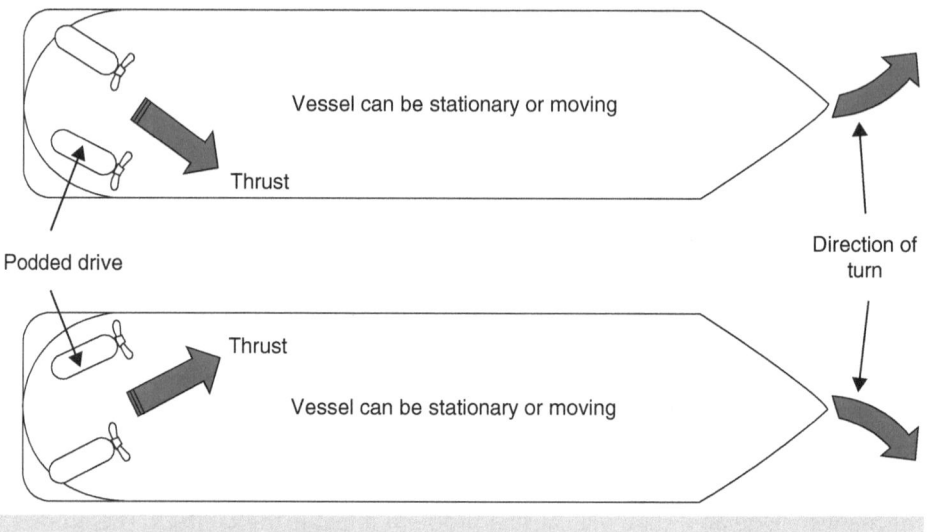

▲ **Figure 5.1b** *Steering action with azimuth thrusters*

Steering control

Chapter 2 Regulations 29 and 30 of SOLAS includes the requirement for the steering to be operated from the bridge of the ship and due to the distance between the bridge and the steering engine requires a control system. The power for the steering engine included on modern ships is either electro-hydraulic or electrical. When operated by auto-pilot a closed-loop control system will be used to keep the vessel on the chosen route. The most up-to-date systems can even manoeuvre and berth the vessel without the intervention of humans.

The very first 'powered' systems had a purely mechanical control mechanism that used to employ a 'master and slave' principle. The transmitter was situated on the bridge and the receiver at the steering gear unit. Mechanical movement was transmitted hydraulically: as the bridge steering wheel was moved to starboard the rotating pinion causes the right-hand ram to move down, pushing oil out to the receiver unit along the right-hand pipe. The left-hand ram moves up, so allowing a space for oil to come from the receiver unit. Liquids are virtually incompressible; hence any down movement of the right-hand ram produces an identical movement at the receiver unit. This in turn displaces the same quantity of fluid, which is taken up in the extra space created by the left-hand ram moving up. The fluid in the replenishing tank acts as an oil reservoir. The moving linkage of the receiver unit is connected directly to the control lever of the steering engine, which will in turn operate the steering engine therefore turning the rudder in the correct direction and to the correct angle. These arrangements will only be found on historical craft and on small boats where some outboard engines are turned using this arrangement.

Electro-hydraulic system

The electro-hydraulic arrangement replaces the long lines of hydraulic pipe with electrical cables, originally using the 'synchro' amplification system. This system is made up of two rotary, electro-mechanical position-sensing devices, which look like small three-phase electric motors. Each device has a stator, with the field coils arranged in either star or delta, and a rotor. The stator windings are connected together and the two rotors are interconnected to the same voltage. Sometimes these are described as interconnected variable transformers.

The arrangement of the system means that the two rotors will tend to align themselves. Therefore when the master unit, situated behind the wheel on the bridge, is moved the two will be out of alignment. This will produce an offset signal that can be used to activate the solenoid valves controlling the direction of flow from a small oil pump to one side or the other of the telemotor ram, which in turn is connected to the actuator control rod for the main steering pump. When the rudder comes into line with the helm order the second rotor is turned into line with the first and the offset signal is removed and the main steering pump will stop pumping (Figure 5.2).

Telemotor fluid

Good-quality mineral lubricating oil is used with the following properties: (1) low pour point, (2) non-sludge forming, (3) non-corrosive, (4) good lubricating properties, (5) high flash point, (6) low viscosity, to reduce frictional drag, but not too thin to make gland sealing difficult. Typical properties would be density 880 kg/m³ (at 15.5°C), viscosity 12 cSt (at 50°C), closed flash point 150°C, pour point 30°C.

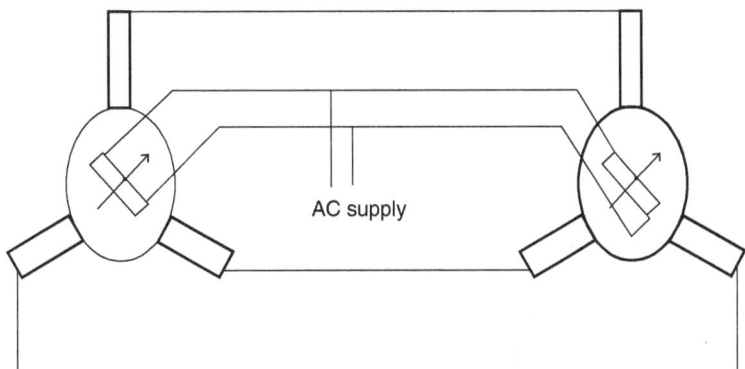

AC supply

▲ **Figure 5.2** *Synchro wiring arrangement*

Testing and Emergency operation

Regulation 30.2 in Chapter 2 of SOLAS gives details of the need for at least two independent electrical supplies so that in the event of a partial or total failure of any part of the steering system there must be provision for a back-up system or built-in 'redundancy'. In the past this will usually have been carried out by a direct gear from the aft steering wheel station to the power unit control. On the modern vessel the telemotor or control circuitry is arranged so that two independent sets of wiring follow different routes from the bridge to the steering flat. Therefore it is highly unlikely that the signal will be lost, especially with the latest Can-BUS technology where data signals are routed through wiring that may have several alternative solutions.

However, as a requirement of SOLAS, there must also be an arrangement for contact between the steering flat and the bridge and in the event of the telemotor not working the engineer can operate the controls directly on the telemotor receiver after reacting to the command from the bridge.

Both the receiver unit and hand gear unit linkage operate the receiver control unit through a sliding rod. When the telemotor pin is fitted to the *receiver linkage hole* the receiver motion is given to the control unit, the hand gear merely sliding in the sleeve. If the pin is removed and put into the *hand gear sliding linkage hole* this operates the control unit and the telemotor connection merely slides in the sleeve. Only one telemotor pin is provided to be used in the required position, in port, on emergency steering, the pin should be in the hand position.

Pin detail can be seen in Figure 5.7.

Testing the steering gear

SOLAS Chapter 5 Regulation 26 sets out the requirement to test the ship's steering machinery within 12 hours prior to departure from any port. It is logical to make this requirement standard practice for the ship's officers and incorporate a procedure in the ship's safety management system (SMS). In the interest of efficient operation and meeting statutory obligations the steering gear can be tested as part of the pre-sailing checks carried out by the Bridge Officer of the Watch and the Engineering Officer of the Watch.

With the two responsible officers working together, one operating the controls and the other checking the operation, physical verification of correct operation can be completed just prior to sailing.

The equipment to be checked should be:

- main and auxiliary steering gear;
- all the remote steering control systems (and from the different bridge positions);
- emergency power supply;
- the rudder indicators (that they match the actual position of the rudder);
- all power failure alarms;
- any automatic equipment and isolating arrangements.

If the vessel is sailing into US waters then their government's Code of Federal Regulations (CFR) Part 164 (Navigation Safety Regulations) requires that the steering gear is also tested 12 hours prior to 'Entering' US Navigable Waters. As it is most likely that the steering gear will be in constant use before entering the US it will be a matter of ensuring that all the different systems are tested.

Electrical telemotor

Traditional bridge remote control is either electric, hydraulic or gyro pilot. The system of Figure 5.9 has been reduced in size and grouped into an oil bath box as shown in Figure 5.3 in which the principle is almost identical but the input is electric in this case.

A bridge lever moves rheostat B and unbalance current flows to rotate the control torque motor and hunt rheostat A back to equilibrium when the motor will stop (see details given later of Ward Leonard electrical steering gear).

Electrical input is most common in modern practice and motor drive via a flexible coupling (or electro-magnetic clutch) rotates the screw shaft in the control box. This causes the screw block to move and, through the floating lever, causes movement of the actuator control rod. This electrical-mechanical transducer also has limit switches and may utilise synchros and gear trains.

To change to local mechanical input control the electrical control is switched off and the spring detent on the handwheel lifted while the handwheel shaft is pushed home so that the spur gear engages when the detent is released to lock the shaft (electrical remote input, mechanical local input).

Control terminology

It is useful to re-cap on our control terminology – for more information please refer to Volume 10 of the Reeds series (*Instrumentation and Control*). Manufacturers of steering

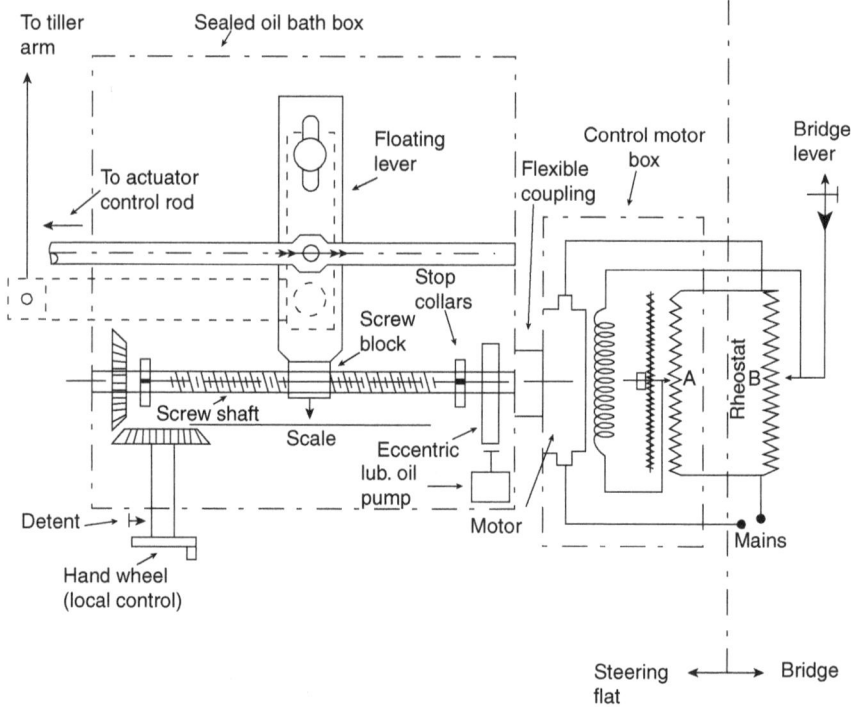

▲ **Figure 5.3** *Electrical telemotor*

control systems now use the highly reliable CAN-bus technology. These modular systems can be configured in different ways depending upon the application.

Ships may have twin rudder or podded drive systems. They may also have multiple steering positions where the input control can be changed from a central position on the bridge to one of the bridge wings. This is most common (essential) to ro-ro ships and cruise ships where control is changed to the bridge wing for berthing or leaving port.

The design of this arrangement is very important as the controls must be 'synchronised' to give a smooth transition of control from one station to another.

The CAN-bus technology will also easily integrate the rudder, or pod, angle into the system, which allows either manual or automatic steering to be adopted. These can further be described as 'follow-up' and 'non-follow-up'. The non-follow-up systems are designed to directly move the rudder following an input action from the helm. The follow-up systems allow the helm to request a specific rudder angle and the system will keep turning the rudder until that 'desired' angle is reached.

The control system can also be connected to the gyro-compass and/or magnetic compass in 'auto-pilot' mode. This keeps the ship sailing in the selected direction without input from the helm. Following on from this will be a link to the GPS system allowing the control system to link up the different equipment in controlling the vessel to a specific position (dynamic positioning).

The control (CAN-bus) network will also have outputs to the voyage data recorder, central alarm system, compass repeaters as well as interface with the engine control system.

Increasingly, steering control systems are being expanded to become part of a totally integrated 'Navigation' system. This includes electronic charting systems and weather routing ability. Eventually the aim of some manufacturers is to give a vessel total autonomous operation.

Survivability in the case of collision or fire is much easier to build into the control systems that work on the transmission of data instead of electrical switching signals. This is due to the information having several different pathways to follow when connecting any two pieces of equipment, such as the helm and the steering engine.

Power (Amplifier) Units

The function of the power unit is to amplify (and possibly transduce) the receiver output signal in the correct direction, for transmission to the final controlling actuator operating the rudder.

Electrical supplies and systems

Most 'rudder based' steering systems on modern ships have electro-hydraulic type systems. The most popular of these have 'four ram' and the 'Rotary Vane' type actuators (see pages 210–212). Reliable performance depends upon the electrical supply. Due to its importance, the supply lines must be arranged to follow two different independent routes, from source to the electric service. The destination will either be the motors supplying the power to move the steering directly or the motors to operate the hydraulic pumps.

Although each supply could come from the main switchboard, one could be routed from the 'emergency switchboard'.

Hydraulic systems

A variable delivery oil pressure pump is used and two designs will now be considered. These are the constant speed, variable delivery hydraulic pump, of which the Hele–Shaw pump is still the most common example; the alternative is the VSG or swash plate pump, which does the same job but is arranged slightly differently. These are highly reliable pieces of equipment that have proved themselves with many years of service on many different ships. They are very important and are often the subject of the flag state examiner's questions for students undertaking their Engineering Certificate of Competency.

The Hele-Shaw pump

As seen in Figure 5.4, the shaft is stationary and the cylinder body forming the cylinders rotates around the shaft, being driven at a constant speed and direction by the electric motor (or steam engine). The plungers are connected to slippers that run in annular grooves inside two circular rings on each side of the plungers. When the centre of the rings coincide with the shaft centre (O) the pump travel is at mid-position. At this position the plungers rotate at a fixed radius distance from the shaft centre (distance OC). This means there is no *relative* motion between the plungers and the shaft and no pumping action takes place.

If now the circular slipper rings are moved to the right by the operating rod, through the casing from the telemotor rod, then the centre of rotation of the slippers and plungers is at *B*, which is eccentric to the centre of the shaft, O. This means the *greatest* distance the plunger gudgeons are from O is OG and the *shortest* distance is OF. With the direction of rotation as shown, in travelling round from G to F the plungers are moving *in* relative to the fixed central shaft and ports, hence the top port T acts as a *discharge*. In completing the circular route from F back to G the plungers are moving *out* relative to the central shaft and ports and the bottom port B acts as a *suction*. The path is shown on the diagram as a dotted line, likewise the relative plunger movement at four positions.

If the circular slipper rings are moved left so the centre of plunger rotation is at A then the *shortest* distance is OD and the *greatest* distance is OE, that is, plungers are moving *out* in the top half of rotation and T is a *suction* and *in* during the bottom half of rotation and B is a *discharge*. The path and plunger movement are shown here as dashed lines.

As the stroke of the plungers depends on the movement of the slipper path *horizontally* and hence the eccentricity, so the pump is of the variable delivery type. Also direction

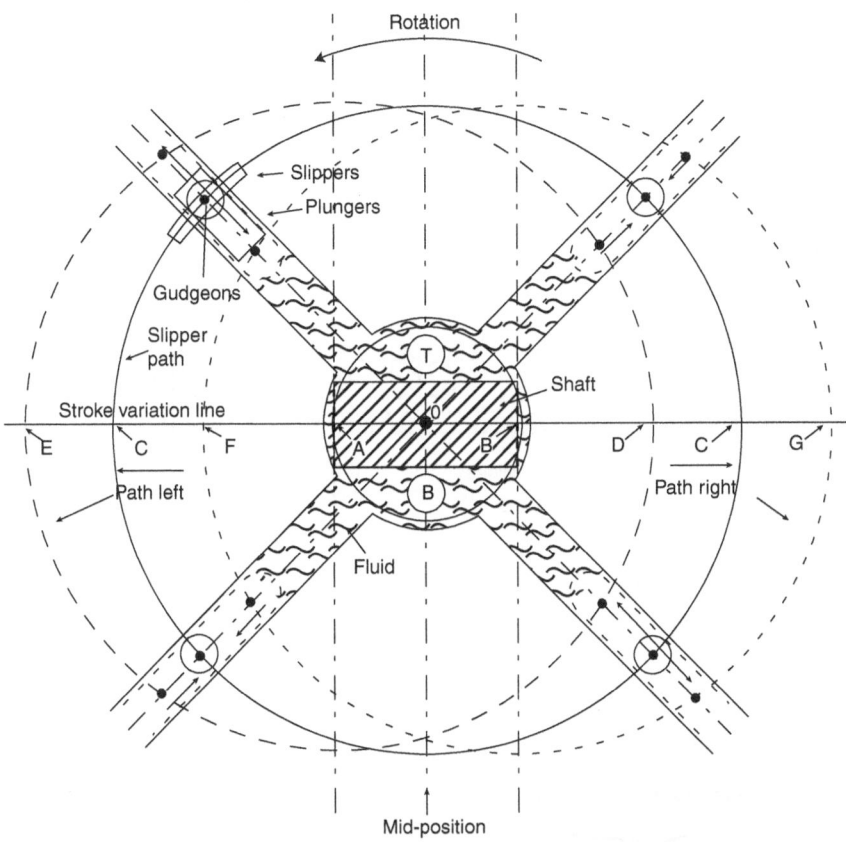

▲ **Figure 5.4** *Hele-Shaw principle*

of flow depends on movement left or right of central position so that for unidirectional rotation the direction of flow is reversible.

In the simplified construction diagram as shown in Figure 5.5, the drive for the cylinder body can be seen rotating in roller bearings outside the fixed central shaft with the oil block fastened to the casing. Plunger and gudgeon pin connection to slippers can be seen with the slipper running in the annular groove of the circular ring. The circular rings are not rigidly fixed but are free to rotate as floating rings on roller bearings, which reduces oil churning and friction losses.

The control or actuating spindle passes through the casing and moves the floating ring *horizontally* left or right by means of the floating ring guide on horizontal slides.

In practice, the pump is usually provided with an odd number of cylinders, usually seven or nine, which produces more even hydraulic flow and better pump balance.

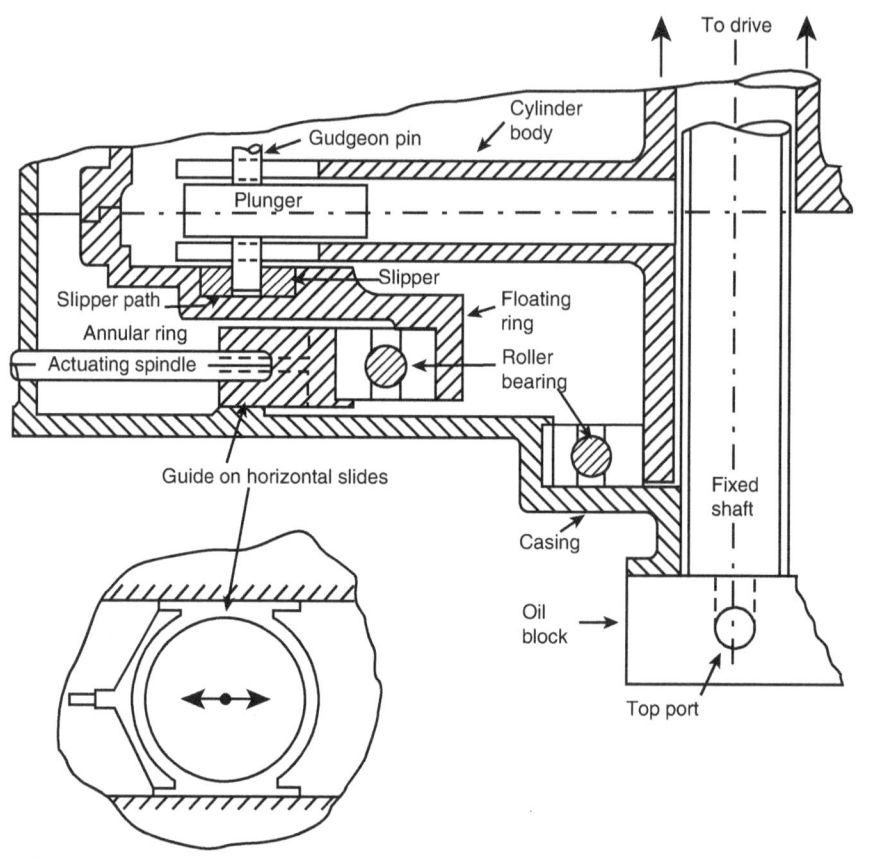

▲ **Figure 5.5** *Hele–Shaw pump detail (1/4 plan)*

Variable delivery pump – alternative design

This type of pump has been used with ship stabiliser units (see later). Consider Figure 5.6. This utilises a principle similar to the Hele–Shaw but has axial piston drive from a tilting swash plate. Similar designs utilise axial piston drive from a tilting trunnion or from cam faces (see ball piston type, Figure 5.15).

Slipper pads bear against the swash plate face and the plungers are driven in and out axially for each revolution of the rotor. For one direction of tilt ports on one side of the horizontal centre line become suction and on the other side of this centre line become discharge. For the opposite direction of tilt the direction of flow is reversed. The quantity of discharge depends on the angle of tilt. In mid-position no relative movement exists between piston and end plate and no pumping action takes place.

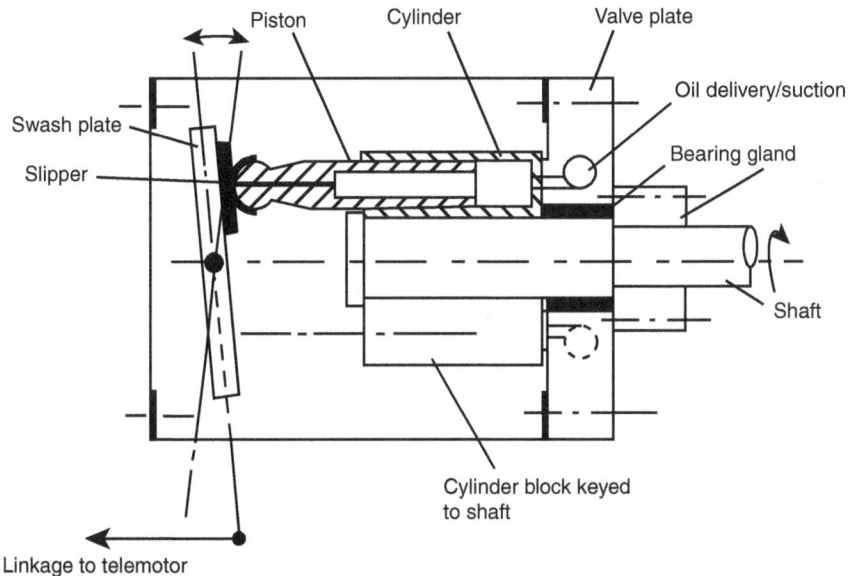

▲ **Figure 5.6** *Variable delivery pump*

Actuator (Servo) Mechanisms

There are two main types of steering 'engine' now in use, namely electro-hydraulic and all electric. Two designs are given for each:

1. Electro-hydraulic: ram, rotary vane.
2. All electric: Ward Leonard, single-motor.

Electro-hydraulic ram steering gear

The pump unit delivers to rams that are virtually directly coupled to the rudder stock forming the actuator mechanism.

Referring first to the diagrammatic plan view of the electro-hydraulic steering gear given in Figure 5.7, consider a movement of the wheel to starboard and hence ship's head to starboard; the rudder movement will be to starboard so that the rams will move starboard to port (right to left).

The steering telemotor moves from right to left (as considered previously) but is mounted on the joist bracket through 180° so that the movement on Figure 5.7 is left to right. The receiver motion is given to a lever that is fixed at the centre (fulcrum) so that the other end moves right to left. There is a hand gear control, two positions for the telemotor pin, and movement stops. The movement right to left of the lever draws out the pump stroke control lever, to which is connected the actuating lever for the stroke variation and control for the pressure pump. The pump driven by an electric shunt-motor at constant speed now *delivers* oil to the starboard ram and *draws* from the port ram. The rams therefore move right to left along the guide joists. Stops are provided, on the joist, to limit travel.

As the rams slide across they push on the ram crossheads moving the tiller arm to port, the arm sliding through the swivel bearing. The crosshead detail is shown in Figure 5.8. A wear-down rudder allowance of 19 mm is provided so as not to induce bending stresses on the ram. With the tiller arm going to port the rudder moves to starboard. The rotating stock movement is led back by a spring link to the pump control floating lever. This constitutes the hunting gear (feed-back) in that when the telemotor movement stops, the floating lever stops going to the left. The bottom of the lever is being pushed to the right and so the stroke control of the pump is almost immediately brought back to pump mid-position. This means the pump stops pumping and the unit is virtually fluid locked at the required rudder position.

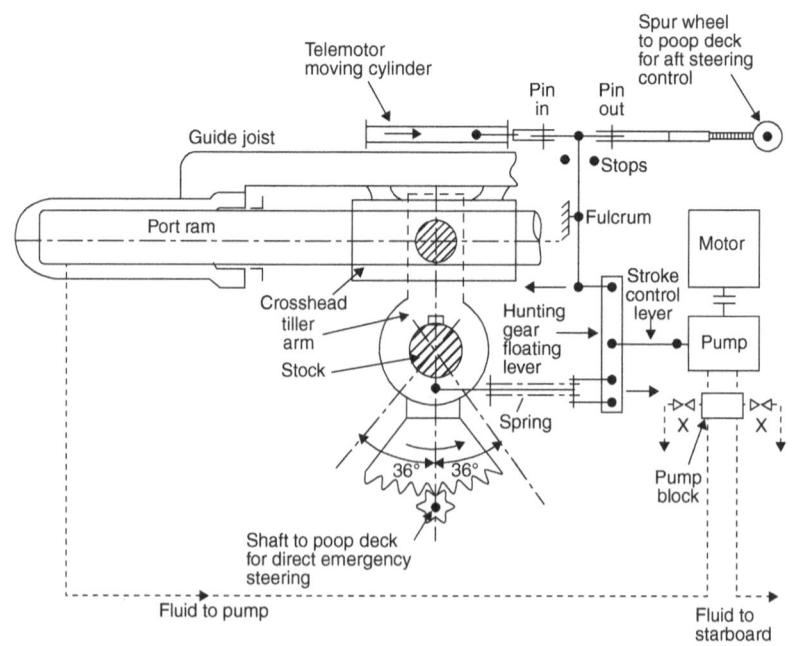

▲ **Figure 5.7** *Electro-hydraulic steering gear*

The tiller-rudderhead bearing and carrier usually have the main casting of cast steel, with a large machined base for fitting to the deck. A bronze thrust ring is on top of this casting and the tiller boss has a machined ring face to go against this thrust face. The thrust ring is in halves and dowelled against rotation and lubrication is provided. The main gland bush, in halves, is usually of gunmetal and is grease lubricated.

Fabricated assemblies are common in modern practice.

Rudder, stock and other parts have weight transmission to the tiller by means of a steel support plate and eyebolt on top of the rudder stock.

At the pump block are non-return valves and connections leading to the sump or replenishing tank to act as suction and replenishing leads (Figure 5.7). The ram pipes to and from the pump to the rams are also led into the pump block. The oil used in the system is well-filtered, pure lubricating oil. The gear is filled by coupling up to the hand steering and rotating port and starboard with the motor running, having previously filled suction sump or replenishing tank and ram cylinders, replenishing valves and bypass valves open. The bypass valves are then shut and the gear fully rotated port and starboard while the air is purged from the ram cylinders etc. at the air cocks. The bypass valves are two-fold units in the block, consisting of bypass and isolating valves. Two other valves, of the *spring-loaded* type, act as double shock relief valves. Each valve connects both sides of the system when the pressure in either ram cylinder reaches 80–190 bar (depending on the design) the valve lifts, so letting the rudder give way when subject to severe sea action. When giving way the pump actuating spindle is moved and the pump acts to return the rudder to the previous position when the loading reduces. The relief valves, when operating, are effectively providing feedback (with increased offset between set and desired values in the short term, which will be reduced as soon as normal conditions prevail when relief valves close).

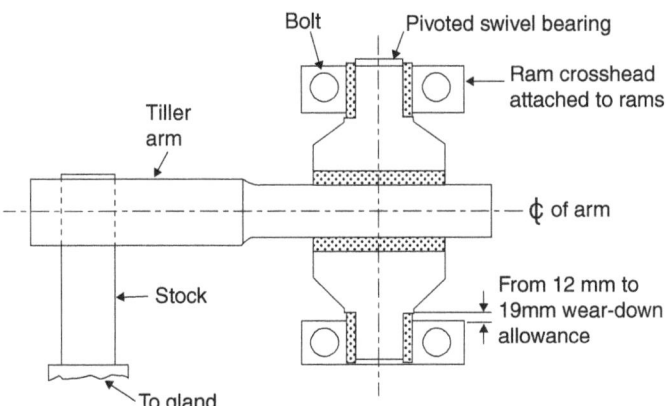

▲ **Figure 5.8** *Crosshead arrangement*

The relief valves lifting pressure setting therefore fixes the maximum loading on the rams. This in turn limits the maximum torque that is exerted on the rudder stock and the maximum torsional stress is hence limited to about 34.5 MN/m². The gear works on the well-known principle of the 'Rapson Slide' and knowing the maximum lifting pressure of the relief valves. Then the ram load is fixed, applying the leverage for distance to stock gives the torque exerted, which allows size calculations for the stock diameter, and horse power and sizes for the motor and pump. Higher-pressure systems have relief valves acting at about 190 bar. For details of the pump block, see Figure 5.9.

Emergency operation

In many installations four rams are provided, two on each side of a double tiller arm, together with twin motor and pump units. All connections are normally open with one pump unit in service, the other is a standby that can be quickly switched on if the service motor fails. See also Figures 5.9, 5.11.

During manoeuvring in dangerous waters both pump units are often used together. Such installations are usually arranged to operate from emergency essential service

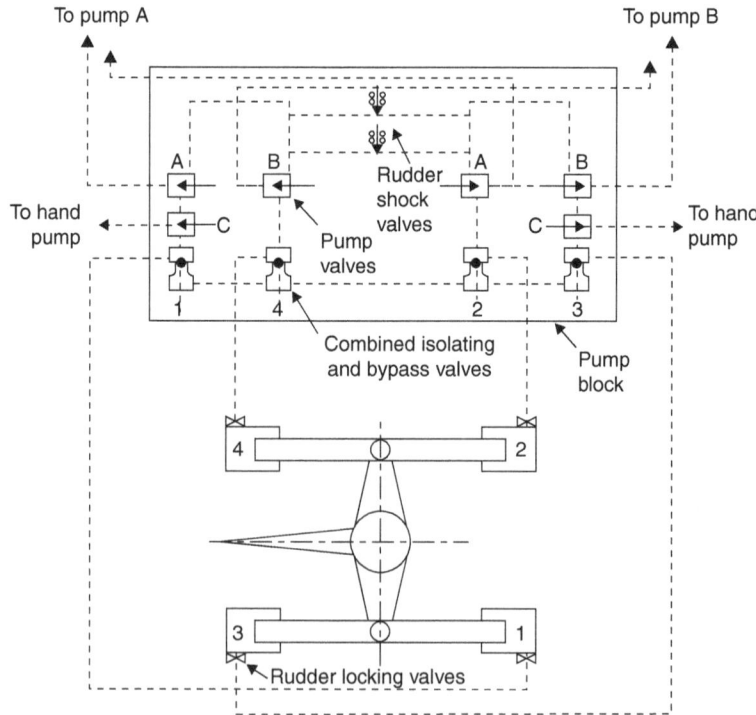

▲ **Figure 5.9** *Control valve block*

battery *and/or* generator circuits in the event of a main power failure. The above considerations satisfy the most onerous required regulations.

For *less* onerous rule requirements a hand pinion drive (as shown in Figure 5.7) or hand pumping may be acceptable. In the extreme case of an emergency a block and hawser arrangement could be rigged up.

A complete armature and field coil would normally be carried as part of the spare gear.

Control valve block

Refer to Figure 5.9, which illustrates a system that is acceptable, but by no means standard, for electro-hydraulic-type gears.

Rudder locking valves on all cylinders are open except in cases of emergency.

The main valve block contains three groups of valves: (1) two rudder shock (relief) valves, (2) four pump isolating valves (two valves for each pump A and B) and (3) four combined ram cylinder isolating and bypass valves connecting to each cylinder.

When a pump is not in use it is prevented from motoring and relieved of a starting load by a spring-loaded discharge valve. This valve is kept in a bypass position until the pump is started up and its discharge pressure is sufficient to move the valve. The bypass connection is then automatically closed and the isolating connection opened, by valve movement, thereby connecting the pump hydraulically to the steering gear. This detail is *not* shown in Figure 5.9. An alternative is to use a centrifugal coupling, between motor and pump, whose pawls open out when running but when stopped the pawls engage a ratchet so locking against rotation.

Under *normal* conditions only one of two pumps with four rams is in use and to bring in the other pump only requires the operation of the starter. The four pump valves marked ABAB are open and the two hand pump valves marked CC are closed.

For *emergency* conditions with pumps shut down the four pump valves are closed and the two hand pump valves opened. Five combinations of rams are usually possible, that is, all four, two port, two starboard, two aft, two forward. A diagonal arrangement is not usually possible. As sketched the combined valves are all open on the isolating connection and closed on the bypass connection. To operate with rams 1 and 2, valves 1 and 2 are open, and valves 3 and 4 shut. To operate with rams 2 and 4, valves 2 and 4 are open, and valves 1 and 3 are shut. Similar procedures apply for the other two combinations. A valve is classed as open if the isolating connection is shut (bypass connection is obviously open when the isolating connection is closed, and vice versa).

Fork-type tiller (including the 'Rapson Slide')

This is a more recent design in which the rams as a single forging act upon a codpiece that slides in slots that are machined into upper and lower jaw pieces of the tiller (see Figure 5.10).

This illustrates the increasing with angle mechanical advantage (frictionless) of this well-known mechanism.

Students should study figure 5.10 carefully, especially the formulae showing the rationale for the mechanical advantage (frictionless) for pinned actuators. This mechanical advantage increases with angle and is 1.53 at 36° (for rotary vane-type steering gear the mechanical advantage is unity for all angles).

Four ram steering gear

Figure 5.11 illustrates diagrammatically the well-known Hastie type of four ram gear with a valve arrangement differing from the previous example. In this case only three combinations are possible: (1) all four cylinders in operation; (2) with valves B, C and F open, A, D and E closed as cylinders 1 and 2 are operational; and (3) with valves A, D and E open, B, C and F closed as cylinders 3 and 4 are operational.

The hunting gear arrangement is similar to the two ram system and two pumps are normally employed.

Electro-hydraulic rotary vane steering gear

Details are as sketched in Figure 5.12. Rotation depends on which side of the vane is connected to the pump pressure feed, and this should be clear from the plan view as sketched. The rotary vane unit is normally designed for a maximum pressure of about 90 bar as distortion and leakage are liable to occur at higher pressures. The design is simple and effective and has proved popular in practice. In fact the apparent space and weight saving is not as great as may be imagined due to the higher pressures and integrated construction utilised in modern hydraulic ram designs. There is, however, a definite space saving but the first cost is usually higher. Absorption and transmission of torque relief is essential to avoid excess radial loading of vanes. Support has resilient shock absorber mountings, which also allows for small misalignment – where spade-type rudders are used axial and radial thrust bearings are provided. The three vane-type is used for rudder angles of 70° and for larger angles a two vane unit would be used. Vanes, of SG iron, are secured to the

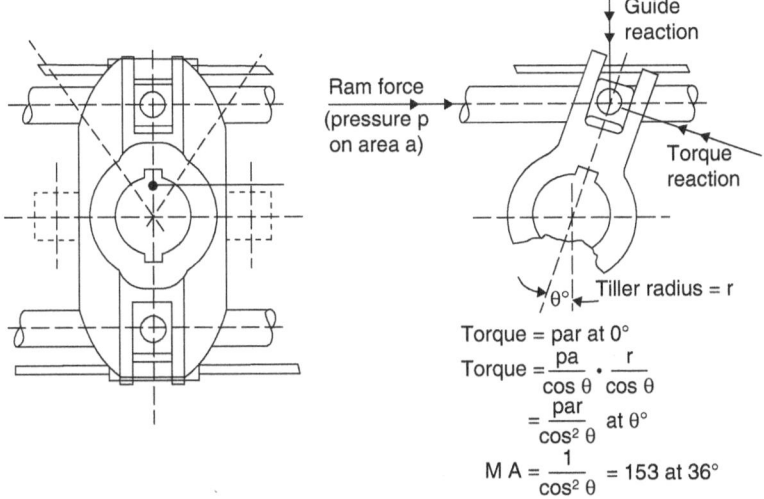

Torque = par at 0°

$$\text{Torque} = \frac{pa}{\cos\theta} \cdot \frac{r}{\cos\theta}$$

$$= \frac{par}{\cos^2\theta} \quad \text{at } \theta°$$

$$MA = \frac{1}{\cos^2\theta} = 153 \text{ at } 36°$$

▲ **Figure 5.10** *Rapson slide fork type tiller*

rotor or stator by dowels and keyed (full length). The vanes themselves act as rudder stops. Steel sealing strips backed by synthetic rubber are fitted into grooves along the working faces of rotor and stator vanes.

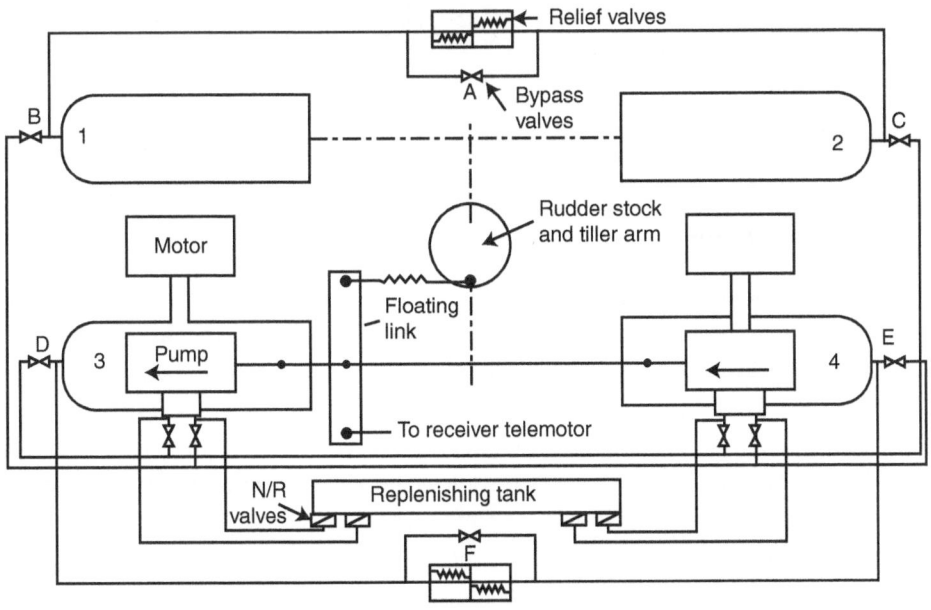

▲ **Figure 5.11** *Four ram steering gear*

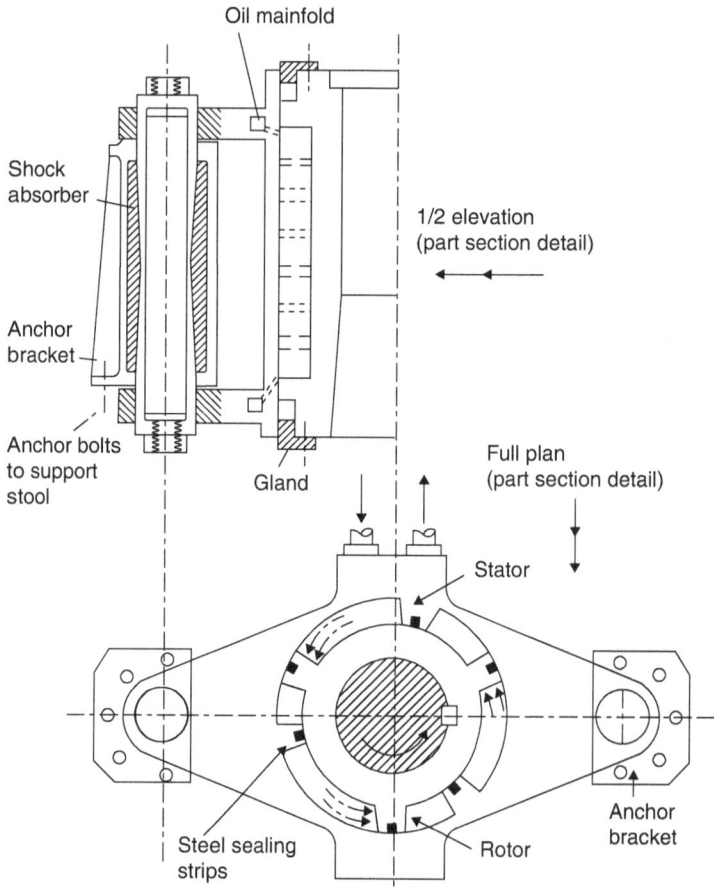

▲ **Figure 5.12** *Rotary vane unit (3 vane)*

Comparison of units

Torque is dependent (for one actuator) on pressure, area and effective leverage. The ram design is more adaptable to increase of these variables – pressure is certainly limited on the vane type to about half that on rams (due to sealing difficulties). Up to a certain torque the vane unit may well be smaller and lighter. However, integral design produces problems of construction, weight and size when the variables are increased. Provided alternative hydraulic pressure sources are available emergency operation is readily achieved with either type although the ram type is more flexible to alternative mechanical leverage. Rudder support and shock loadings require more careful consideration with vane units because of the very close and direct connection between rudder and actuator.

Risk assessment – (failure mode)

Along with the modern trend, designers and builders should complete detailed risk assessments to ensure that the system retains the correct 'minimum level of operation' following a failure of any equipment or pipework.

Automatic fail-safe system

Donkin utilises the bedplate oil tank, with control division plate up to two-thirds height, so each side is connected to its own pump. Each pump is associated with a pair of rams (optionally mounted one above the other to save deck space) so that two complete half-power steering gears are formed (giving up to 20° rudder movement at maximum ship speed or full manoeuvring at two-thirds maximum ship speed). In normal operation (at full tank) the two systems are joined by two common lines each fitted with a solenoid-operated spool valve (normally open) so allowing a free balanced flow of oil between the circuits.

If leakage of oil does occur, the first stop of the oil tank float switch closes, which isolates one of the pump motors and simultaneously closes its associated valve on the line between the two systems, putting the ram cylinders of the suspect circuit into the bypass condition. Operation is instantaneous with no interruption to steering; audible and visual alarms are fitted on the bridge.

As the choice of circuit isolated is preset, if the incorrect circuit has been chosen, the oil level in the tank will continue to fall and eventually close the second stage of the float switch. This at once changes over the pump motors and their corresponding isolating valves so leaving, by process of elimination, the correct circuit in use. After the operation of the second stage of the float switch, two-thirds of the tank contents still remain, thus ensuring an adequate supply of oil for continued working of the rudder actuator. As the defective circuit is now completely isolated, any repairs that are necessary can be carried out without interruption to steering.

Two or four ram units have been fitted to vessels up to 90,000 g.r.t.

The Ward Leonard electrical steering gear (an important but old system not usually found on modern ships)

In this case telemotor, power and actuator units are considered in one. A brief and simplified description of the electrical principles involved, sufficient only for examination purposes, is given.

There are four electrical facts that are important in the system:

1. If a direct current generator is driven at constant speed and direction then the magnitude and direction of the voltage is dependent on the magnitude and direction of the current through the field windings.
2. The magnitude and direction of the armature current to a direct current motor having constant field excitation in one direction decides the magnitude and direction of the output torque.
3. When a steady current flows in a uniform conductor there is a steady voltage drop along the length of the conductor.
4. If a voltage is applied to the ends of two uniform conductors joined in parallel then a current between zero and maximum can flow in either direction by connecting suitable points on the conductors.

Considering Figure 5.13, the control gear employs two rheostats connected as a 'Wheatstone bridge' circuit connected to the mains supply. With the two contacts in the same positions (*equal* electrical voltage), that is, correspondence of rheostat position on the conductor or rheostat of bridge and rudder, then no current flows between them.

If the wheel is moved say from amidships to starboard the contact moves on the screw towards *B* and alters its position on the bridge rheostat. This means the two contacts are at *different* voltages (Fact 3). Current therefore flows between the contacts in a fixed direction; if the wheel had been moved to port, that is, towards A, current would have flowed in the reverse direction, the magnitude of the current depending on the amount of movement of the steering wheel contact along its rheostat (which gives the necessary voltage difference between the contacts). Thus a variable magnitude *and* direction current can be made to flow in the exciter shunt field (Fact 4).

The main motor drives the main generator and exciter, the motor taking current from mains supply. With the contacts in equivalent positions no current flows in the exciter field and no current is induced in its armature even though it is rotated, similarly for the main generator, hence it produces no voltage and no current is supplied to the rudder motor armature. The rudder motor is field excited from the ship's mains but this will not produce torque without armature current so the motor is stationary.

With the bridge contact moved to starboard, current flows in *one* direction through the exciter field, and the armature now produces volts, which sends a current through the generator field. A current now flows through the rudder motor armature and the rudder motor rotates the rudder. The hunting gear now functions so that the rudder movement moves the contact on the rudder rheostat to follow the bridge

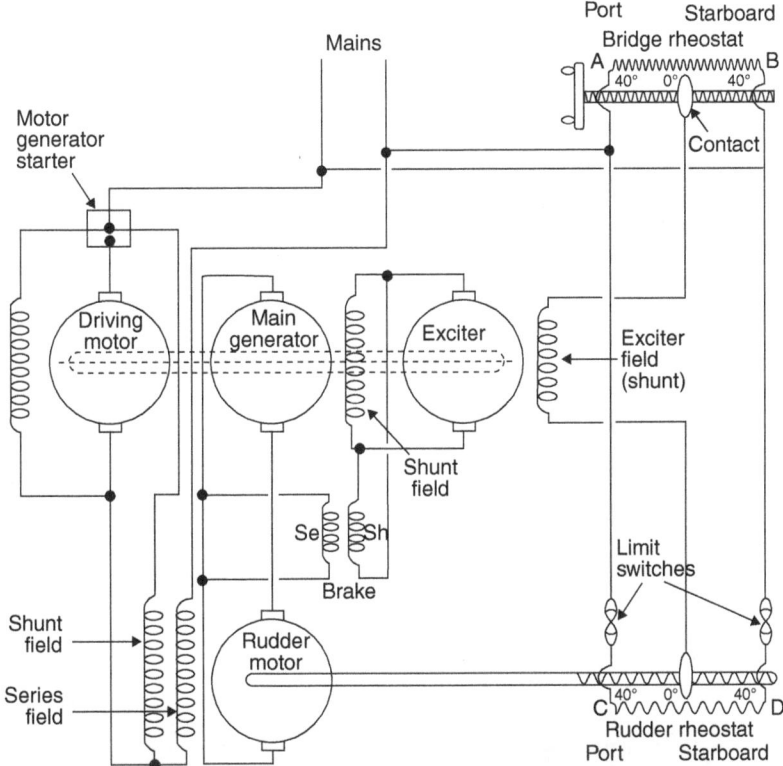

▲ **Figure 5.13** *Electrical steering gear (Ward Leonard)*

rheostat, that is, *towards* D. When the bridge wheel is stopped, the hunting gear brings the two contacts into equivalent voltage again to cause no current to flow through the exciter and subsequent circuits and so stop the rudder motor in the correct position.

If the wheel is moved to port, current flows in the reverse direction so that the generator produces *reverse* direction current (Fact 1). This current will produce *opposite* direction of rotation for the rudder motor (Fact 2) and the contact of the rudder rheostat is hunted *towards* C until equilibrium again exists. The exciter is really a current amplifier to reduce the current required at the rheostat contacts while giving sufficient current through the generator field. The series field of the rudder motor automatically gets a boost current when the driving motor comes on extra power with the generator and exciter producing current, and this boost serves to overcome the inertia of the rudder gear. Limit switches are fitted on the contacts to cut off current at about 36° position before the mechanical stops are reached. The brake is kept on while no current flows in the rudder motor armature and when current flows another resistance comes into parallel thereby *reducing* the total resistance and allowing rotation of the rudder. The brake functions to slip at a predetermined load so producing current, and this boost serves to overcome the inertia of the movement and is usually transmitted by a pinion, wheel and spur gear or by worm and wheel to a rudder quadrant.

To stop excessive hunting a damping coil in the exciter circuit is provided, which is wound in opposition to the exciter field winding.

Emergency operation

Many installations are provided with completely separate twin electrical equipments on one quadrant. A changeover switch allows independent operation and either equipment can be directly operated from a rheostat aft for hand emergency steering. The electric circuit will be on the essential services emergency circuit, being battery operated or battery and emergency generator operated. In addition, a spur gearing from poop to quadrant teeth can be provided. The above conditions would serve to satisfy the most onerous rules applicable. Suitable spare gear for all essential parts would need to be supplied.

Electric single-motor steering gear

As shown in Figure 5.14, the armature of the telemotor is fed directly from the mains and so is the potentiometer rheostat. If B is moved, say down, by the wheel then current flows due to the difference of potential between A and B. The telemotor field is now excited and the telemotor rotates so as to bring A into line again and restore equilibrium.

Through a screw nut, frame and fulcrum arrangement the reverser switch is moved up and so closed. Mains current then flows through the brake field (to release the brake), through the rudder motor series field and through the rudder motor armature. The shunt field of the rudder motor (not shown) is permanently connected to the mains but this is insufficient to cause rotation unless the series field is also excited. Rotation of the rudder motor is arranged to hunt back the rheostat contact A through a floating lever frame and screw nut arrangement as well as opening the reverser switch.

If the reverser is moved in the other direction the current direction is reversed through the rudder motor armature, but not through the field, so that rotational direction changes.

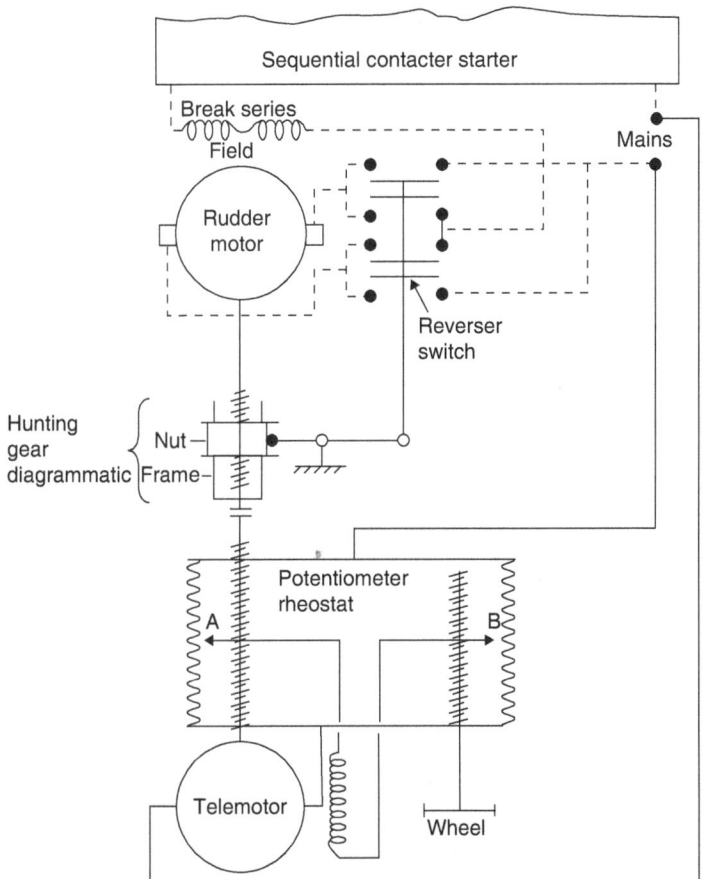

▲ **Figure 5.14** *Electrical steering gear (single-motor)*

Principal Rules Relating to Steering Gears

All the rules for the construction and use of steering gears can be found in Chapter 2 regulation 29 and 30 of SOLAS, and the Classification societies also have their rules. These do concentrate on the more conventional forms of steering ships.

The development of the 'Azipod' drive/steering systems have not resulted in a change in SOLAS, however classification societies have added the requirement for the azimuthing thruster to rotate at a minimum speed in rev/min.

1. All vessels must be provided with efficient main and auxiliary steering gear of power-operated type. An auxiliary gear is not required if the main gear is provided with duplicate power units and duplicate connections up to the rudder stock.

2. The vessel must have means provided to allow steering from a position aft.

3. Two tillers, or their equivalent, are required unless the working tiller is of special design and strength.

4. Power-operated gears must be fitted with a device to relieve shock on the rudder from outside forces such as a heavy sea.

5. Power supplies:
   - Must be arranged so that at least two separate cables are provided for each electric or electro/hydraulic arrangement (containing one or more power units).
   - Steering gears should only fail if a short circuit occurs and therefore the electrical cables and fuses must allow for 200% over current for a minimum of 60 seconds.
   - Electrical cables should be protected from fire damage and conform to the Regulations within part D of SOLAS, which relates to 'Electrical Installations' including the regulations for electrical supplies.
   - Each of the electrical supplies can be from the main switchboard or alternatively one supply can be from the emergency switchboard.
   - The electric motors for steering gears should be rated according to the (International Electrotechnical Commission) IEC60034 standard for rotation electrical machines. The 'S6-25%' rating is required for electro-hydraulic systems and 'S3-40%' for electro-mechanical systems.

6. Moving parts of steering gears should be guarded to avoid injury to personnel.

7. Hydraulic systems should employ non-freezing fluid.

8. A clear view from the steering position is required and the wheel, tell-tale indicators and rudder movement must correspond in the correct amount and in the correct direction for the ship's head.

9. During the ship's operating trials, the steering gear systems will be tested to ensure that the limits of movement, time of operation, angle of heel at speeds, etc. all meet the regulatory requirements.

   The requirements are that, for a conventional rudder system, the power system should be capable of:
   - operating the rudder from 35° of helm in one direction to 30° of helm in the other within 28 seconds;
   - moving the rudder to 35° in either direction (port and starboard);
   - operating the rudder from 35° of helm in one direction to 30° of helm in the other within 20 seconds for Ice-class vessels, tugs and supply vessels;

Each of these actions must be completed while the vessel is at its maximum draught and service speed. In cargo ships this will be with BOTH power units operating and in passenger ships the requirement should be met with any ONE power unit driving the system.

In the past serious accidents have resulted from vessels losing their steering ability and therefore the IMO regulations relating to duplication of rudder actuators have been developed to reduce the risk of such an occurrence in the future. With this is mind students should take note of the automatic 'fail-safe' systems as well as the following:

Every oil tanker, chemical tanker or gas carrier of 10,000 gross tonnage and upwards and every other ship of 70,000 gross tonnage and upward, shall comply with the following:

1. The main steering gear shall be so arranged that in the event of loss of steering capability due to a single failure in any part of one of the power actuating systems of the main steering gear, excluding the tiller, quadrant or components serving the same purpose, or seizure of the rudder actuators, steering capability shall be regained in not more than 45 seconds after the loss of one power actuating system.
2. The main steering gear shall comprise either:
   (a) two independent and separate power actuating systems, each capable of meeting the requirements; or
   (b) at least two identical power actuating systems which, acting simultaneously in normal operation, shall be capable of meeting the requirements. Where necessary to comply with this requirement, inter-connection of hydraulic power actuating systems shall be provided. Loss of hydraulic fluid from one system shall be capable of being detected and the defective system automatically isolated so that the other actuating system or systems shall remain fully operational.
3. Steering gears other than of the hydraulic type shall achieve equivalent standards.
4. Ships should carry a fixed storage tank with a fresh charge of oil for at least one power actuating system.

Watchkeepers should take extra care to complete a regular inspection of the steering gear and all the associated equipment. Early warning of any failure will be extremely beneficial to the safe operation of the vessel and relying on remote monitoring and alarm systems is not good practice. The visual and auditable checks will consist of checking that:

- electrical motors are free of excessive vibration;
- all the holding down bolts are secure and not coming loose;

- there are no leaks;
- there are no unusual smells such as overheating machinery;
- the operation is correct.

For oil tankers, chemical tankers or gas carriers of 10,000 tons gross tonnage and upwards, but of less than 100,000 tons deadweight, other solutions to an equivalent standard may be acceptable. Such regulations may well extend to all vessels.

The Ship Stabiliser

As the operating principle of this equipment is very similar to that of an electro-hydraulic steering gear, it is advisable to have a working knowledge of stabilisers.

The electric circuits and hydraulic relays involved are numerous and a detailed consideration of the equipment is not necessary. A simplified presentation is therefore given with the system considered in three parts, namely electrical control, hydraulic operation and gear, and the fin detail.

A knowledge of the operating principle from the aspect of the fin action of the ship rolling, with an understanding of the control system (using the analogy of the electro-hydraulic steering gear, avoiding the rather complex electrical, oil pump, and motor descriptions) is adequate.

Note. Development in the stabiliser field has been very rapid. Even the type of gear described may be regarded as obsolete in a few years. Modern designs are available to reduce the large space athwartships required for the design given. This is achieved by types in which the fin gear is retracted by swinging round on trunnions into the ship's hull. The fin itself rotates on the fin shaft, being rotated by an oil motor. Hydraulic systems have also been simplified.

Electrical control

A selection switch on the bridge gives settings related to hand control, normal stabilising or automatically controlled rolling, together with an output control switch for beam sea, following sea conditions etc. The electrical control is identical with the hydraulic telemotor principle but functions with electrical relays. This means that a transmitted signal produces a corresponding movement at the stabiliser station, through a hunting gear, which is converted to a mechanical movement with hydraulic amplification to operate the fin operating gear.

There are two gyroscopes. One is a *vertical keeping* gyroscope whose signal goes through two selective transmitter magslips to a follow-through magslip that is similarly operated by a rolling velocity gyroscope. The combined selected signal is transmitted to the hunter magslip of the oil motor and pump. The mechanical movement of a gyroscope alters the rotor position of the transmitter magslip and the current flow moves the rotor of the hunter magslip to a corresponding position (just as transmitter and receiver telemotors). The rotor movement of the hunter magslip operates to allow oil to be pumped from the pump to the oil motor, which rotates. The pump is driven at constant speed and direction by a small motor. The oil motor can rotate in either direction from neutral depending on the direction of movement of the hunter magslip rotor. As the motor rotates, a mechanically driven resetting transmitter magslip serves as a hunting gear and tends to fetch the hunter magslip back to the neutral position and stop rotation of the oil motor.

Hydraulic operation

This has two distinct functions: the first is to *extend* or house the fins and the second, which is under gyroscope control, is to *tilt* the fins.

Oil is supplied from the storage tank to the servo pump, which is driven at constant speed and direction by a motor. The pump is of the variable delivery tilting box (or swashplate) type but there is no reversal of suction and discharge lines.

The pump supplies oil in two pressure ranges: low pressure (29 bar maximum) to tilting control and high pressure (77 bar maximum) for fin housing, selection by changeover valve and control by control cylinder. When extending the fins the oil is supplied to the housing piston rod and flows *through* an inner tube into the fin shaft, which it pushes out to the extreme position. Oil behind the piston flows through a port and along the *outside* of the central tube back to the pump suction, surplus oil due to volume of piston rod is accommodated in the storage tank during the housing operation when the flow of oil is reversed. Control valves for this operation are located in a central control box.

Consider now the fin-tilting operation: with the fins extended the tilting operation is controlled by the gyroscope signal, which is transmitted to the hunter magslip and functions to decide the amount *and* direction of movement of the oil motor. The output shaft of the oil motor is connected by gearing to the reset magslip rotor, which serves to hunt the gear back to neutral position and stop motor rotation at the required position.

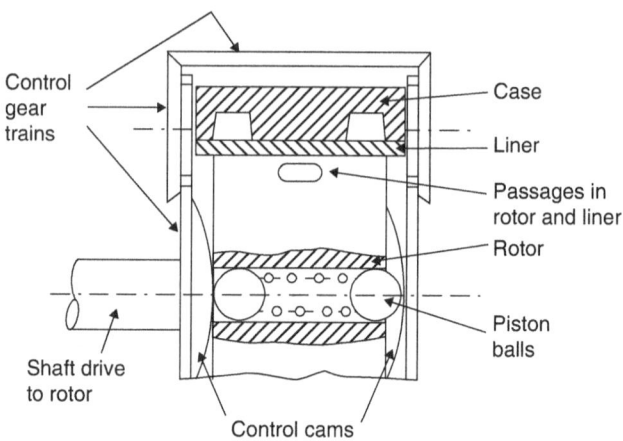

▲ **Figure 5.15** *Oil pump (part view)*

The oil pump and motor are in line. The electric motor *continuously* drives the pump shaft on which is splined the rotor, running in a liner and casing. Usually seven pump cylinders are formed by *axial* bores in the rotor and each cylinder contains two balls separated by springs, and the ball faces as they rotate run on the faces of control cams (see Figure 5.15).

In the diagram shown the balls have no *relative* velocity to each other and this is the non-pumping position. The cams are rotatable in opposite directions by use of the gear trains. In the *maximum* pumping position the two cam peaks face each other so that during rotation the balls approach and leave each other. This produces a discharge (at about 7 bar pressure) and a suction during each revolution, a port at mid-length of each cylinder leads out from the rotor. Four equidistant radial ports in the liner lead oil to wells, two suction and two discharge, in the casing, which lead oil to the suction and discharge ports that are located at opposite sides of the pump. Thus a variable delivery pump is supplying oil to the oil motor. The oil motor is similar in construction to the pump but the cams are *fixed*. Oil pressure supplied between the balls from the pump causes rotor rotation when the balls are *descending* the cams, on the return stroke oil is discharged to exhaust. By movement of the main valve, rotation is reversed in that pressure oil is now led to the port previously acting as exhaust and exhaust occurs through the port previously acting for pressure. The pump supplies oil to a pressure controller, which functions to maintain a constant pressure under all outputs. The pump discharge to the control valve is then directed to the motor, whose rotational direction is dependent on the rotation of the hunter magslip rotor. With the main control valve shut the pump merely idles and the gear is then in the neutral position.

The final servo-operated stage is now almost identical in operation to that of the electro-hydraulic steering gear. The movement from the cam operates the tilting control valve. This allows oil from the servo pump to move the tilting control cylinder in the required direction, which in turn causes the fin-tilting pumps, which are *continuously* driven by the main power unit electric motor, to deliver oil to move the rams in the tilting cylinders across.

The tilting control cylinder is virtually acting as a receiver telemotor, with similar hunting gear arrangement. The rams instead of rotating a tiller serve to tilt the ram by means of the toothed quadrant (see Figure 5.16).

The variable delivery pump for the fin-tilting and servo unit has been described previously (see Figure 5.6).

Fin detail

The principle of operation is to impose on the hull a rolling motion *equal* and *opposite* to that caused by the wave motion. This is achieved by utilising the forward velocity of the ship through the water. On the ship rolling to starboard the starboard fin is set by the gyroscope signal so that its *leading* edge is *above* the axis of tilt so causing an *upward* thrust. The port fin is set to the opposite tilt, that is, with its *leading* edge *below* the axis of tilt so giving a *downward* thrust.

Two rectangular fins, one at each side of the ship located directly opposite if space is available, are of aerofoil section. To the trailing edge of each is a hinged tail flap that is moved automatically by a simple linkage when the main fin shaft is rotated. This flap gives a very much more pronounced restoring torque action of the fin than would a plain, large fin of similar area. The fins are mounted on stainless-steel shafts, the fin being fitted on to a taper and is bolted up internally, the fin plating being welded over the built-up internal structure. The main fin can be tilted 20° each side of the neutral horizontal position. The tail flap is inclined at 30° to the main flap when this flap is at its maximum 20° position and in the *same* direction, both being in line horizontally in neutral position (see Figure 5.16).

Fins and shafts are placed athwartships, near amidships, near the turn of the bilge, and the axes of fins and shafts are inclined downwards to the outboard ends at about 15° to the horizontal plane. The 20° of tilt is suitable at about 15 knots but this maximum is reduced to about 11° at 24 knots to keep reasonable fin loading.

The fin is supported near the load application by a fin shaft crosshead, which moves on top and bottom guides during housing or fin-extending operations. Guides and shaft are inside the fin box, of fabricated construction, which is flanged at the outboard end and welded to stools at the inboard end. The flanging is bolted to trunking that

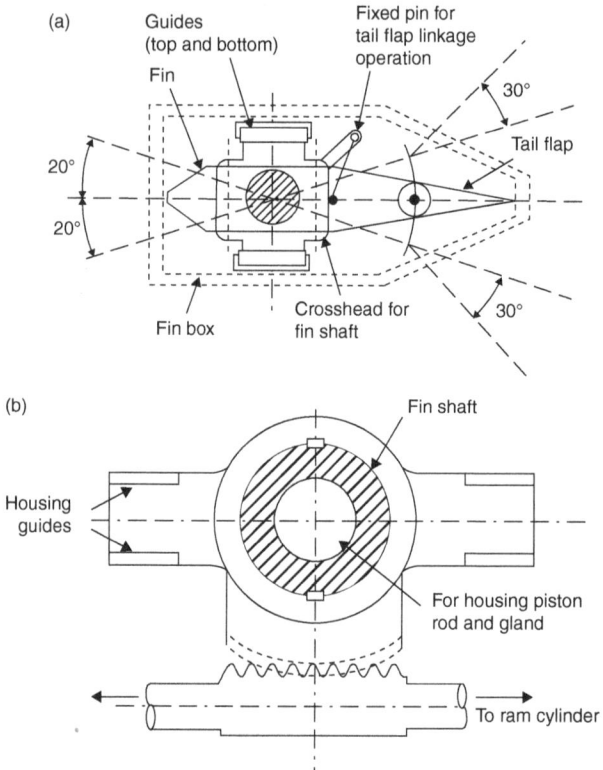

▲ **Figure 5.16** *(a) Detail of fins and flaps; (b) Detail of quadrant for tilt*

is welded to the hull adjacent to the tank top. The fin shaft passes through the fin box and trunk (which completely contain the fin in the housed position) emerging from the fin box inboard through a sea gland. When housing or extending, the fin is centralised and as it is drawn in or out by oil pressure from oil feed led in and out via the yoke and piston rod (as described previously), the guide shoes on each side of the fin shaft guide it along the extension guides (see Figure 5.16).

The crosshead guides, sea gland, etc. on the sea side are oil and grease lubricated by lubricators worked off the fin-tilting shaft ram motion.

Auto Control

The control systems on board modern ships are very sophisticated and in some cases all the propulsion, steering and navigation control and monitoring systems are integrated into a common system. Both the ship steering gear (on auto-pilot) and the ship stabiliser utilise classic control principles that are best illustrated by block diagrams. The aim of the information presented in this volume is to show the concepts only, Volume 10 of the Reeds series will cover the actual arrangements in much more depth.

Figure 5.17 shows a basic block diagram for auto steering. The microprocessor at the heart of the system (controller) will have the capacity to compare the 'Set Course' with the feedback from the compass showing any deviation between the two values. It will also have inputs for sensing a beam sea (or wind) and a delayed response to reduce the rudder movements that could be caused by small random signals or slight alterations in heading. Both rudder and ship are acted upon by external forces and, in the open sea, the vessel is quite often pushed in one direction and then pushed back in the other. Energy is required every time that the rudder is moved and therefore not correcting the vessel for every small change will make the whole process more efficient. Obviously when the vessel is manoeuvring in restricted waters the helm response needs to be as fast as possible.

Figure 5.18 shows a block diagram for stabilisation. The controller is usually two term and feedback from the measuring unit has roll angle and velocity components from gyroscopes. Every input utilises the forward velocity of the ship and the usual 'hunting action' feedback applies between amplifier (hydraulic oil pumps) and actuator (fin tilt). The ship is acted on by external forces, and selector switches allow for variation in sea conditions

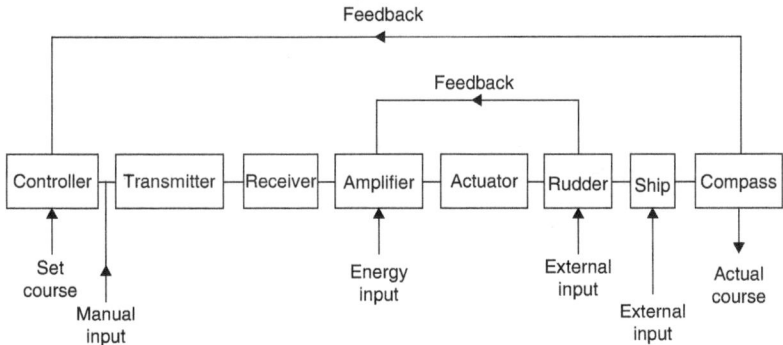

▲ **Figure 5.17** *Block diagram (auto steering)*

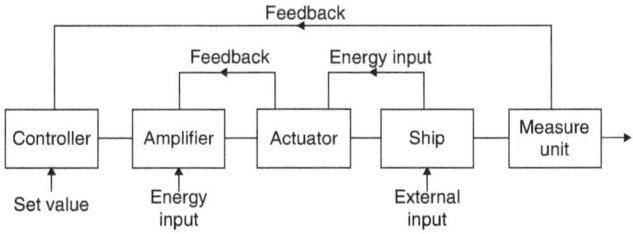

▲ **Figure 5.18** *Block diagram (stabilisers)*

6

POWER
TRANSMISSION
SYSTEMS

Introduction and Basic Design Considerations

Ships are in the commercial business of moving goods or commodities around the world in order to make a profit for their owners. This primary goal can be achieved in a number of ways. On one end of the scale the vessel could be hired or chartered to complete a specific task or project and on the other end of the scale a vessel might be purchased outright by an owner, with the intention of being operated for 25 years on the same voyage.

In between these two extremes are numerous combinations, each of which will attract a different approach to the design and operation of the vessel. For example, a potential owner might identify a trade that could be profitable and make a bid for the business based on operating the vessels at the lowest possible cost that complies with the regulations and selling them again after the contract is complete.

In another example, an already established owner might carefully calculate the income from a long-term project and use an already established support structure to spread costs and develop a business model based on offering a quality service that will attract customers in the long term.

At the beginning of both of these illustrations there will be some common questions to be asked but the ongoing decision-making and management processes might be quite different. Therefore, design of a suitable vessel will as always revolve around the ideas and aspirations of the owner.

The added complication for the shipowner is about regulation and the public perception about ships and shipping. She/he will have to think about how much longer it will be possible to operate older vessels that are environmentally more polluting than newer ones. The regulations might include clauses so that the older, more polluting vessels can be used, but will they actually win any work?

Tanker-vetting schemes mean that charters insist on minimum levels of quality that could exclude older tonnage, especially if the older tonnage is still of the single-skin design.

Therefore, potential owners might need to ask the basic questions listed here.

1. What type of cargo do I want the vessel to carry? (It is no good specifying an oil tanker to carry cars or a container ship to carry hundreds of people but there are hybrid vessels that might be suited to a combination of different duties.)

2. The second common question that operators face is what route is the vessel trading over. Is it a short sea crossing such as Dover to Calais or an inter-island ferry in the Caribbean or is the vessel expected to spend most of its life on the North Atlantic or sailing from the Middle East to Europe carrying 250,000 tons of oil? A study carried out by DNV and MAN in 2010 looked at the situation and market trends for container ships. Based on their findings, the 9,000-teu range was selected as the target case for a concept development, together with the Asia–US east coast trade route through the new Panama Canal. This project was very interesting and was of immense value to promote discussion and illustrate the concepts of vessel procurement planning.

For example, if an owner were contemplating starting a container ship trade over this area then they would have to consider the newly completed project of the widening of the Panama Canal.

The new arrangement has extended the existing two lanes with a bigger third lane and a set of increased-size lock chambers. The lock chambers are 427 m long, 55 m wide and 18.3 m deep, allowing passage of ships with a maximum breadth of 49 m, maximum passage draught of 15 m and an overall maximum ship length of 366 m. The new canal opened open in 2016, which was two years after the 100th anniversary of the existing canal.

These fundamental questions of trading pattern and operational profile must be understood before optimisation of the hull and machinery can be started. The hull form of the ship will take into account the particulars of the route, such as water depth, and a clever hull, combined with the main engines and other equipment selected, means that 49,000-ton new ships could return similar fuel consumption to the 26,000-tonne ships they replace. Other energy-saving measures include boilers that use waste heat from exhaust gas to heat water and power water purification plants.

Designing the hull and machinery for a wide range and combination of speeds and draughts is difficult. Therefore, the ideal situation is to define the route and then carefully design the ship so that it can operate as close to its optimum design characteristics as possible, for as much of the time as possible.

The required propulsion power and electrical demand is calculated for each part of the expected route. Obviously this aspect could provide quite a tight brief or it could involve several 'what ifs?' The vessel might be replacing a ship already well-established on a regular trade, in which case the operational constraints will be well known. And the hull and machinery can be optimised to give the highest efficiency over the whole route rather than only the design speed and draught. For the hull – this applies especially when it comes to the main dimensions, block coefficient, centre of flotation and bulbous bow design and for the machinery – it is the selection of main engine and auxiliary engines so that the propulsion power and electric power needed can be produced as efficiently as possible in all the different sailing legs and different operational modes.

Working according to the operational profile, for example, container ship designers have optimised the ship at the point of maximum fuel consumption, which is normally at maximum speed and maximum dwt/draught. Any savings made at this point will probably yield the maximum gain.

During the current economic conditions and as shipping is coming under increasing pressure to control emissions, there are other considerations to take into account when making the final choice of propulsion and/or electrical power together with the transmission methods chosen to be employed. It may be that designers install the traditional shaft line transmission system. The owner may have requested the design because it is simple, relatively maintenance free and well known to the engineers who are to look after the vessel. This will be the first part of this chapter.

However, there are substantial savings to be gained from using the new electric propulsion systems, especially when used on passenger ships where there is a large air conditioning and lighting load required in addition to the propulsion load.

Shaft Line Transmission

The classification society American Bureau of Shipping (ABS) defines the installation based on a vessel's shaft line transmission system.

1. Propulsion shafting is a system of revolving rods that transmit power and motion from the main drive to the propeller. The shafting is supported by an appropriate number of bearings.

2. Propulsion shaft alignment is a static condition observed at the bearings supporting the propulsion shafts. In order for the propulsion shafting alignment to be properly defined, the following minimum set of parameters (whichever may be applicable) need to be confirmed as acceptable:

   - bearing vertical offset;
   - bearing reactions;
   - misalignment angles;
   - crankshaft's web deflections;
   - gear misalignment;
   - shaft and bearings' strength;
   - coupling bolts' strength.

Alignment

Engines and gearboxes are built with very fine tolerances and need to be fixed to rigid structures, such as bedplates or strong foundations and casings, which stop any flexing that might upset the precision operation of the components. The ship, however, will experience considerable flexing as a result of the forces acting on the structure of the vessel. When the main propulsion engine is situated towards the rear of the vessel the propeller shaft will be much shorter than when the engine is situated in the middle of the ship.

Mid-ship installations with long, and thus fairly flexible, shafting require a different approach to aft-send installations with short rigid shafting. Similarly, long reciprocating engine crankshafts present different problems from geared turbine machinery and the flexibility of welded longitudinally structured vessels gives different problems from the older, more rigid, transversely structured designs.

As a main illustration the alignment for a longitudinally structured, mid-ship-engined vessel having a long reciprocating engine crankshaft fitted to a diesel engine will be described in detail after some introductory general remarks. This description, to give a complete picture, will include the shafting alignment completed in the ship, crankshaft and bedplate alignment in the machine shop and lastly the matching of the engine to the shafting as one integral unit in the ship.

Then the amendments necessary for turbine-engined vessels and aft-end installations will be described briefly.

General considerations

Hogging and sagging effects due to a combination of the state of loading, draught and the effects of the waves can quite easily be more than 1 mm per 1 m of ship length.

The effect of the waves and sea action over the main engine's rigid bedplate length of about 16 m can be ignored. However, the influence gives a realisation of how difficult it is to perfectly align a heavy engine and shafting length into the flexible beam form of a ship that is subjected to the influence of the forces of the sea. Under such forces it can be realised that a reasonable amount of flexing will result and allowances for this must always be accepted.

The problem that ship, builders have is that the flexing of the vessel, due to its immersion in water, may change the position of the transmission shaft alignment from the position that it had while in dry dock as it was being built. Therefore why not start and complete the process in the water? The reason for this is that it is difficult to establish a line through the stern tube bearing to the back end of the engine, whereas this action is easily completed in dry dock. If the hull flexing calculations can be completed with accuracy then the job becomes easier. The initial alignment is carried out in the dry dock with the final alignment in the water verified by an alignment survey.

Following on from the initial installation and in-water survey is the no-load to the full-load condition survey. These surveys will verify calculations, and stiffening in the way of tank tops and engine seating supports, together with the use of rigid bedplates, can reduce the central deflection to a maximum of about 13 mm over the engine room length and to about 2 mm maximum over the length of the bedplate.

Invariably the bedplate has a *sag* form when light ship of say 1 mm and a *hog* form when fully loaded of say 1 mm. The average ship is rarely sagged as can be shown by drawing load and buoyancy curves for average conditions. An engine crankshaft set true at light ship could when hogged 2 mm introduce static bending stresses of say 90 MN/m². Most engine builders have their own records and experiences for dealing with this problem (see Figure 6.4).

It is often suggested that the engine's main bearings in the above case should be lined down in a deflection curve, of 1 mm maximum in the *centre* of length, for the new engine. This means the shafting is true at $\frac{1}{2}$ load and maximum static bending stresses are on each side of this, that is, 45 MN/m² instead of true at light ship giving all bending on one side, that is, 90 MN/m².

The figures quoted are intended for illustration only, using high values, in many cases lining in a curve may only require a deliberate offset of about one-tenth of the quoted figures, that is, 0.1 mm. The degree of offset depends on the type of ship and engine, stiffness, variations due to loading, etc. These factors can only be established correctly by experience before the builder can decide on the best method of lining up to suit the requirements of these variables. In the following alignment description this offset will not be considered (but see Figure 6.4).

Older alignment methods used piano wires and micrometers, feelers between coupling faces, etc., whereas modern methods utilise optical telescopes and targets giving accuracies of ±2 µm per 1 m length. To fully describe the methods it will be assumed that the shafting and engine are first lined up by the older method, being checked at each stage by the modern method, although in practice such duplication may not be considered due to time factors, which is not to say that it would not be advisable. Increased power, size and cost of modern vessels and engines make it essential to ensure correct initial alignment so as to avoid continuous trouble later. Modern ships are assembled in pre-fabricated sections, the precision cutting and welding making for a very accurate finished construction. The initial design, based upon the experiences of previous vessels, will be modified using computational methods. The final design will be assembled using laser-guided systems.

Shafting alignment in the ship

Reference datum should be established and are usually the height of the shaft above the keel at the aft end, and the height of the crankshaft centre above the keel extended to the forward machinery space bulkhead (also centre athwart ships) forward. These two datum points are taken from the ship drawings. Some modern shipyards will start to mark out the engine and shafting alignment as soon as the blocks are laid.

The rough bore of the stern frame is fitted with a plate flange, which has a small hole (say 1 mm) drilled in it at the correct height above the keel. With this centre a reference circle can be drawn for setting up of the exact boring of the frame. Similarly, at the engine room forward bulkhead a small flange in the bulkhead has the small hole drilled at the correct engine height above the keel and at the mid-ship point athwart ships.

An electric light is situated behind the hole in the forward machinery space bulkhead and by looking from outside the stern frame this light can be seen through the two sight holes. Now at the aft peak bulkhead and the aft machinery space bulkhead and any water-tight bulkhead through which the shaft passes, sighting plates are used. At these points the horizontal plate is moved vertically up until the light line of sight is masked, a horizontal reference mark is now made across the bulkhead. The plate is moved vertically down until the light is masked and another horizontal reference mark made; bisection of these two lines gives the *horizontal* centre (see Figure 6.1).

The same procedure is now repeated using vertical boards, moved horizontally port to starboard; bisection of these two lines gives the *vertical* centre. Rough bores are now bridged, the centre is fixed temporarily with a tin plate and a small hole is centred. Now from aft to fore a continuous light should be visible through all bulkheads, and the reference circles can now be drawn for exact boring. The exact borings are now made and the ship is ready for the optical telescope checking (see Figure 6.2).

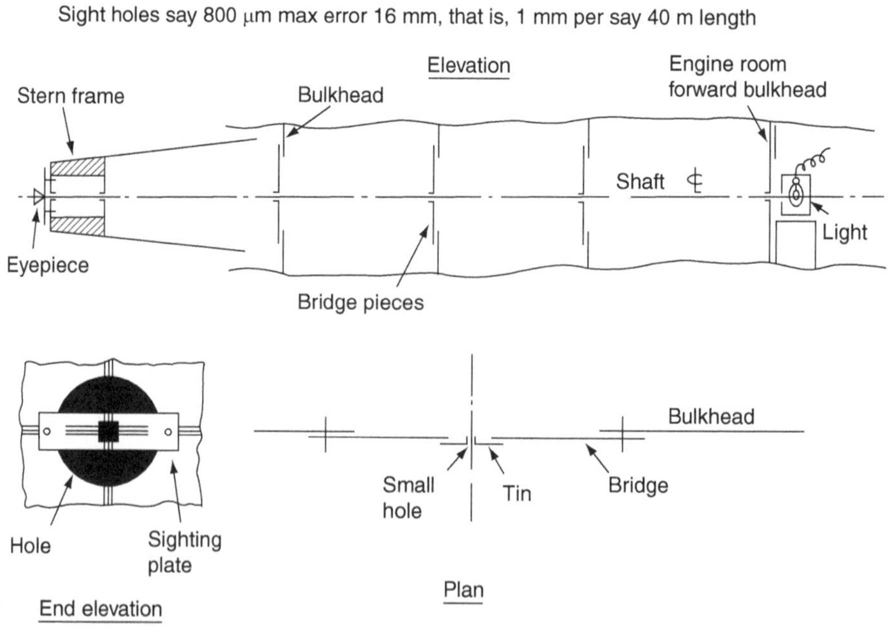

▲ **Figure 6.1** *Sighting by light*

Readings in mm, all journal diameters equal.

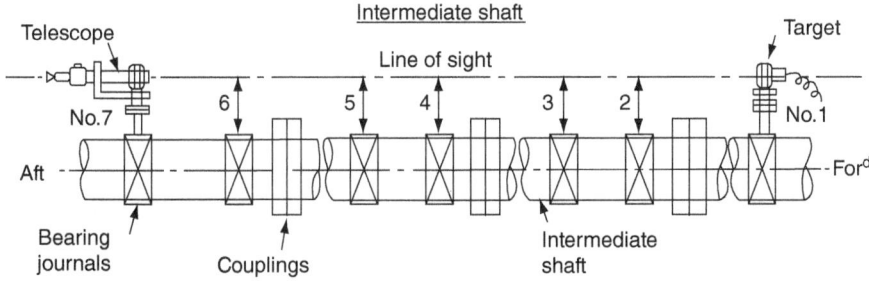

Bearings	No. 7	No. 6	No. 5	No. 4	No. 3	No. 2	No. 1
Scale	500·0	501·8	499·2	498·0	497·9	499·1	500·0
Difference	0	+1·8	−0·8	−2·0	−2·1	−0·9	0

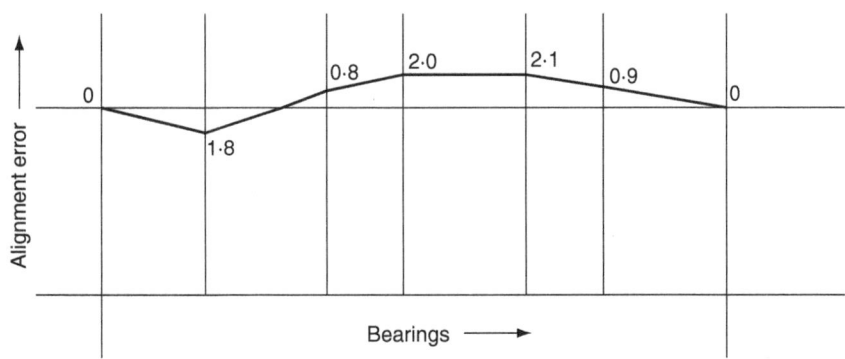

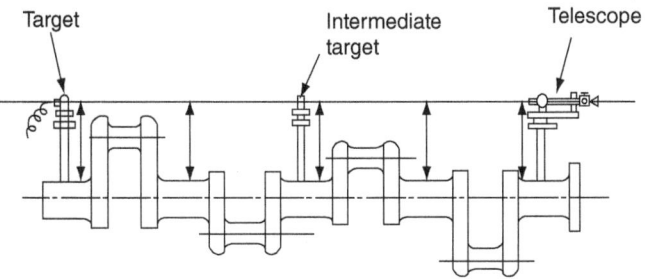

Telescope micrometers allow variation of ±1.00 mm on perpendicular crosslines but the radial markings on the telescope which are graduated upto 25 mm diameter allow much greater misalignment to be read directly from target without using micrometers at all.

▲ **Figure 6.2** *Sighting by optical telescope*

The optical telescope with eyepiece and cross lines, focus, vertical-horizontal micrometer adjustments, etc. is set up in a spherical mounting on a base with adjustment bracket (magnification about 30 times). The assembly is mounted on a spigoted plate and bolted into the aft end of the stern frame, being an exact fit. The target, which has circular, vertical and horizontal markings, is fitted into an adjustable spherical mounting with a light. This assembly is fitted to a flange bracket and set at the correct height and athwartship position on the forward machinery space bulkhead.

The telescope is now adjusted vertically and horizontally until the target is perfectly centred, and this is now the line of sight. Targets are now fitted into adaptor plates and placed with a tight fit into all intermediate bores. The telescope is focused on each in turn to give any vertical or horizontal error, and should be refocused on to the line of sight datum after each intermediate bore checking. Mirror targets can be used if it is thought desirable to check squareness. Now a check is available on the initial lining-up, which should indicate a close degree of accuracy.

At this stage the stern tube, tail end shaft and propeller would be fitted and the ship launched and taken to the fitting-out berth, or the stern frame may be blanked and these fittings made after the launching. The intermediate shaft is now fitted right up to the engine by using feelers at four points between the coupling faces and chocking the tunnel bearings with hard wood blocks on to the stools. The intermediate shaft is now ready for checking with the telescope.

The line of sight for telescope and target is fixed at the extreme ends, positioned at equal distances from the two journal diameters by cups mounted on matched stands that are in blocks strapped to the shaft. The target is usually mounted on the first bearing in the tunnel and the telescope on the last bearing before the tail end shaft, and so the line of sight is established. Readings can now be taken for all intermediate bearings by focusing on a lighted graduated scale held vertically on each bearing in turn. A graph can now be plotted of any misalignment and the chocking adjusted until a true line is effected. It is also advisable to check all journal diameters for equality and also to take a horizontal alignment to ensure correct port and starboard line. All tunnel bearings are now permanently chocked and bolted down, coupling bolts fitted and thrust block (if separate) fitted and bolted to place. The ship is now ready for the engine fitting.

Laser alignment tools are the modern method of completing the same process. The tools are light, easy to set up and relatively low cost. The laser light generation unit has a magnetic base and can be attached to any metallic surface. The precision laser is accurate to less than 0.5 mm over 15 m and intermediate calibrated targets can be set up anywhere along the length of the beam.

Crankshaft and bedplate alignment in the machine shop

The bedplate is first levelled up on the shop floor, usually being in two parts, the whole assembly having been surface planed and the main bearing gaps machined. The bedplate is now lined up and rough chocked by the use of spirit levels. Piano wires are mounted along the full length of the bedplate, one port and one starboard, passing over end pulleys and loaded. Micrometer readings between wire and the machined top edge of bedplate are taken at say 1 m intervals, standard allowance being made for the wire sag, and the bedplate chocked up until a true horizontal reading is achieved. The bedplate is now ready for optical checking.

The telescope is mounted vertically, standing on adjustable tripod feet, the reference plane being obtained by mounting a pentagonal prism under the telescope (with a micrometer adjustment) and rotating it about an axis concentric with the telescope, this prism deviates the line of sight by 90°. The flat plane, being independent of gravity, is adjusted to pass through three definite height targets on the surface and a travelling target enables all other points on the surface to be adjusted to bring the whole area into that common plane. The bedplate lining-up is completed by a set of readings through the bearing bores by piano wire or light method, being checked by optical telescope in a similar manner to the method previously described prior to fitting stern tube and tail end shaft (see Figure 6.2).

A dummy shaft is now used to bed into the lower halves of the main bearing bushes, after which the crankshaft is bedded into place. A set of readings would be taken on the shaft, as described for the intermediate shaft, by optical telescope, the engine would then be assembled, crankshaft deflections taken and the test bed trial carried out. Before the engine is dismantled a set of piano wire readings, port and starboard, would be taken from the bedplate together with a re-check of crankshaft deflections. If the telescope was used then optical readings along the bedplate, port and starboard would replace piano wire readings and in addition a set of readings on the crankshaft through the running gear would be taken, using the oil holes through the bearing caps for the scales. The engine is now dismantled to an extent whereby it can be easily transported and lifted into place on board ship.

The description given is somewhat elaborate, due mainly to the two-method description given, but it should be remembered that if the engine setting in the shops is correct and a reference is taken, then ship alignment is much simplified by merely setting back to this correct reference alignment.

Engine alignment to the propeller shaft in the ship

The aft engine coupling is lined up to the forward thrust shaft face coupling with feelers. Screw jacks to place and temporary chocking between bedplate and tank top right the engine into position. The engine is now chocked to the final test bed readings taken before dismantling and the crankshaft deflections taken. The holding-down bolts or studs are either fitted from a template set to the forward thrust block face or fitted with the bedplate in place. As a complete check on the whole integral setting of crankshaft and intermediate shaft the use of the optical telescope is most advantageous. This also allows a spot check on alignment without disturbing the shafting at any time.

Unfortunately a complete sight from tail end shaft to forward end of the crankshaft is usually impossible due to the ship construction, as the height of sight required above the crankshaft causes the line to foul the tunnel roof a few feet aft of the aft engine room bulkhead unless laser alignment is used, which projects a precision beam with no sag.

Therefore the two lines of sight (LOS), intermediate shaft and crankshaft sights respectively, have to be used. These lines of sight are extended to give as much overlap as possible. The measurement differences at No. 1 bearing and at thrust journal should be identical, indicating parallel lines. Divergence from each other indicates a gradient, that is, millimetre per metre length. To give maximum overlap a tube may be welded into the hold above the tunnel roof with a down tube to a bearing further along the intermediate shafting. Fouling of lines of sight may occur at the turning wheel, and this can be overcome by slackening the halves and reclining the wheel over while still exerting its load on the shaft.

This completes ship and engine alignment. A typical set of readings would be as plotted in Figure 6.3.

Aft-end installations

Vessels with engines situated toward the stern may not suffer the same misalignment effects due to the short, rigid shaft length. There should be no need to line the engine down for load variations, the engine being lined exactly true, at light ship, in the standard way. The main problem with these vessels is that the large tail shaft wear-down allowed (8 mm plus) in say a 5 m shaft connection to the engine. This throws a heavy load on the aft end of the crankshaft and as a consequence these vessels are more prone to tail-

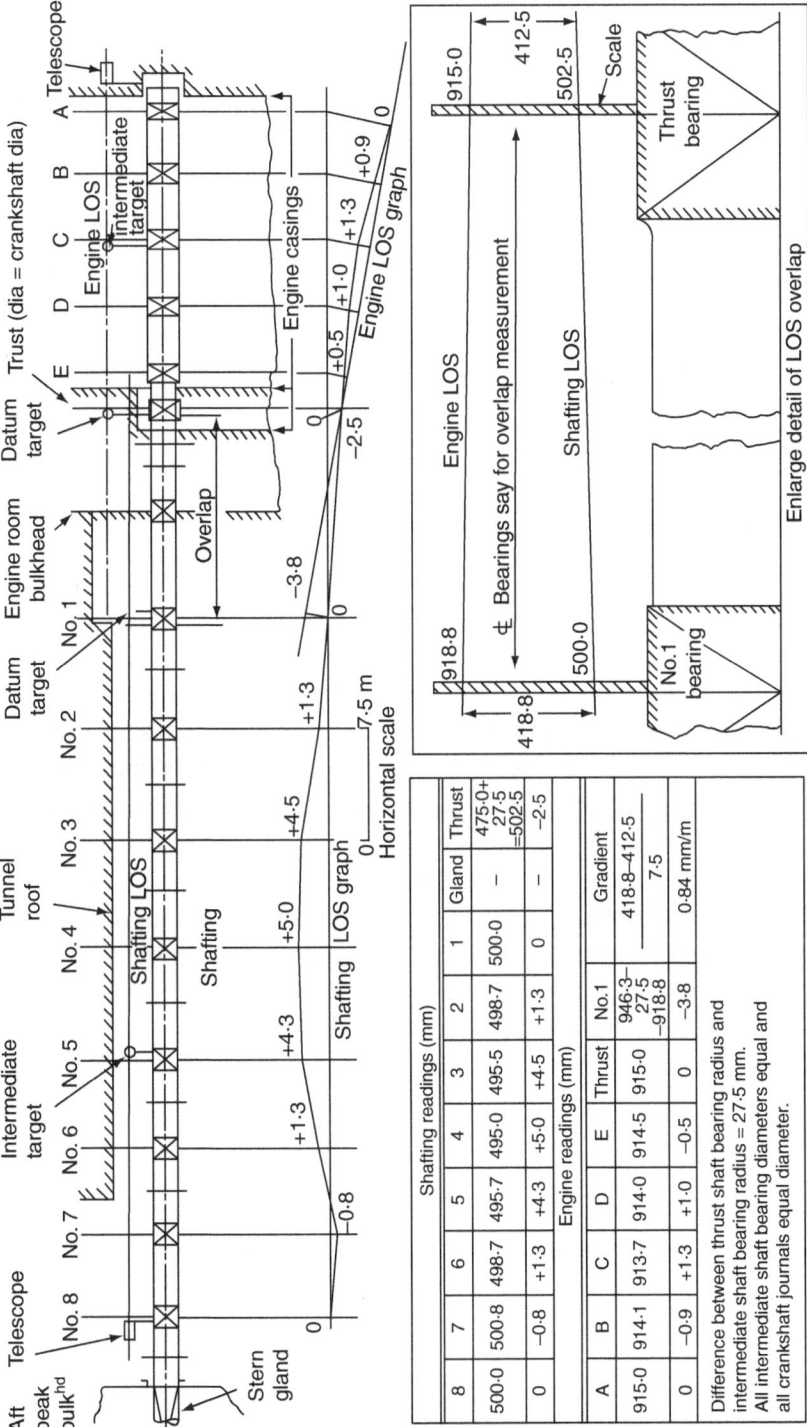

Shafting readings (mm)

	8	7	6	5	4	3	2	1	Gland	Thrust
	500·0	500·8	498·7	495·7	495·0	495·5	498·7	500·0	500·0	475·0+ 27·5 =502·5
	0	−0·8	+1·3	+4·3	+5·0	+4·5	+1·3	0	−	−2·5

Engine readings (mm)

	A	B	C	D	E	Thrust	No.1		Gradient
	915·0	914·1	913·7	914·0	914·5	915·0	946·3− 27·5 −918·8		418·8−412·5
	0	−0·9	+1·3	+1·0	−0·5	0	−3·8		7·5
									0·84 mm/m

Difference between thrust shaft bearing radius and intermediate shaft bearing radius = 27·5 mm.
All intermediate shaft bearing diameters equal and all crankshaft journals equal diameter.

▲ **Figure 6.3** *Sighting in ship*

shaft and aft-end crankshaft failure. Great care is advised when investigating torsional vibration problems. To offset the wear-down load it is advisable to fit the tunnel and thrust bearings with fitted top halves and it would be advantageous to limit wear-down very strictly. The alignment method is a more simple form than that described (due to shorter shafting) but it should be mentioned that even though the shaft length is small, alignment errors have caused serious trouble in such a short, rigid length. Slew effects as much as 6 mm aft (crankshaft) to 18 mm forward (crankshaft) with an engine hogged vertically 3 mm have occurred in the past. Optical telescopes and laser light techniques allow a continuous sight fore to aft over the extreme ends, which is most advantageous.

A lightweight carbon fibre shaft would be advantageous for this design. (See page 230 for more details.)

Turbine-engined vessels

The alignment of shafting is as described but the problem here, from the engine aspect, is virtually one of two or three turbine wheel shafts through pinions lined on to a large gear wheel thence to the thrust block. In the past, alignment errors have commonly reflected back through the gearing to cause excessive pitting, scuffing, heavy wear, etc. on the second reduction pinions. The turbine-gearing alignment was somewhat complex involving long inside micrometers, much cross-sighting to turbine stools and later to turbine wheel shafts, which with piano wires and light methods made the process highly detailed. In modern practice use of the optical telescope, of laser light together with the pentagonal prism for line of sight deviation, allows cross-sighting from a fixed standard flat reference plane (say horizontal gear case-turbine joint) to any number of points including height checks of all bearings, making the gearwheel-turbines alignment much easier. The gearwheel will have been lined to the thrust block face, having previously been checked by telescope through main wheel bearings. The following points are worthy of consideration:

1. The lift of the shafts due to the oil film should be taken into account.
2. Due to the high rotational speeds of the turbines there are processional torque effects.
3. The flexible coupling copes with a considerable degree of misalignment.
4. Turbine troubles, when they do occur, seem to persist and invariably stem from initial misalignment or vibration-characteristic errors.

Pilgrim wire method

Reference has already been made to the use of taut (piano) wires and some elaboration of this method is relevant. This technique produces fairly accurate results especially in a vibration-free situation. For crankshaft alignment with five cylinder engines and above, using telescope or wire methods, it is usually necessary to remove one connecting rod to allow the sight (or wire) to pass over the full shaft length. The alternative is to take readings with an overlap across two central main bearings, possibly at two different heights, and then adjust to a common datum. Allowance for wire sag varies with wire diameter and tension and an empirical formula is generally used. For a wire of 0.5 mm diameter, tension 200 N, an approximate expression is sag $= L^2/29.25$ where sag is in millimetre and L (half length of wire) is in metres. Interpretation of readings, and variation with ship loading conditions, are as described for telescope methods (Figure 6.2).

A light method can be used in calibration. One pole of battery is earthed and the other, in series with an indicator lamp, is connected to dial indicator touch stylus. As wire is earthed the slightest touch of stylus on wire causes the indicator lamp to light. These techniques can be used by engineers if laser technology is not available.

Crankshaft Deflections

Excessive crankshaft deflections readings in main or auxiliary reciprocating machinery will mean that there is a continuing variation in the distance between crankwebs as the shaft moves through a full engine revolution (Figure 6.4). This will cause dangerous bending stresses in the web and the fillets between crankpin and web. Therefore the measurement is an indication of actual deflection, that is, vertical hog or sag of the shaft, thus the value of these deflections are assumed to be dependent on two main factors *for a given mass per unit length* (connecting rods, pistons, etc.).

1. Distance between the supports of the shaft (i.e. main bearing inner faces), as the further apart the supports the greater the sag effect.
2. Distance from shaft centre line that the measurement is taken. This is usually close to the extreme edge of the web. For a sawcut in a shaft, the further one moves from the shaft centre the greater the gap. The web's extreme distance and size of section is usually proportional to the engine stroke.

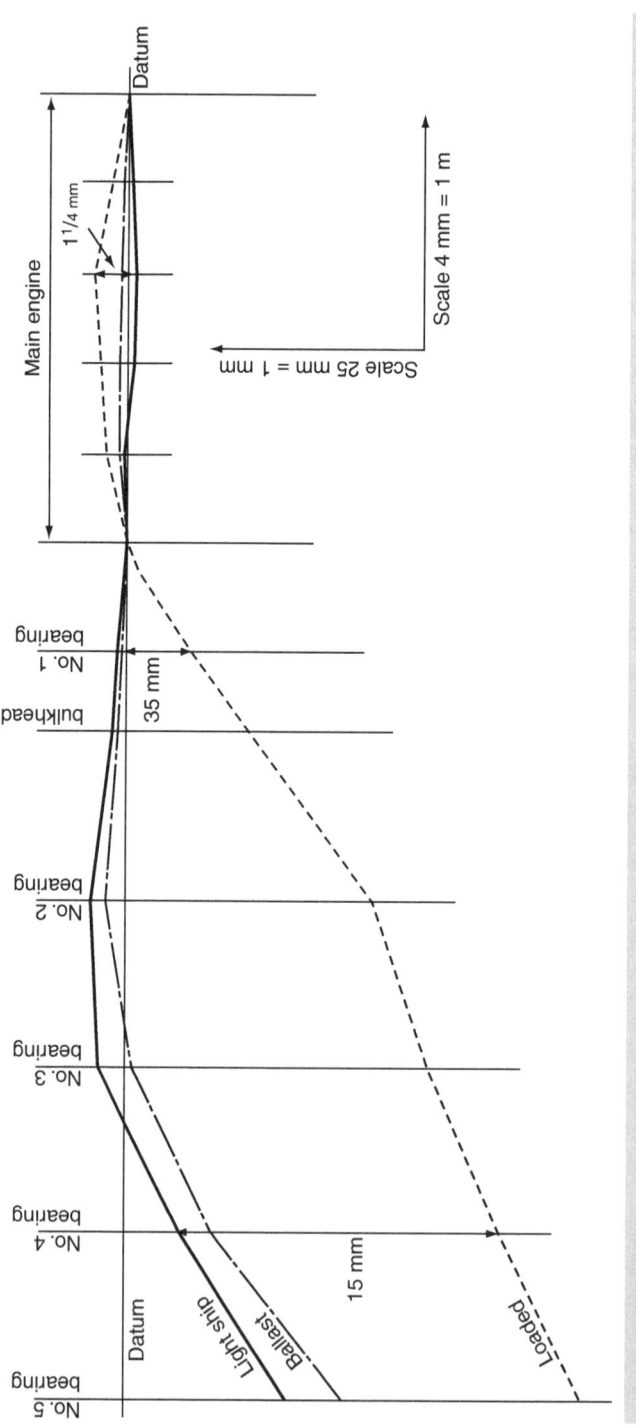

▲ **Figure 6.4** *Shafting alignment variation for different ship load conditions*

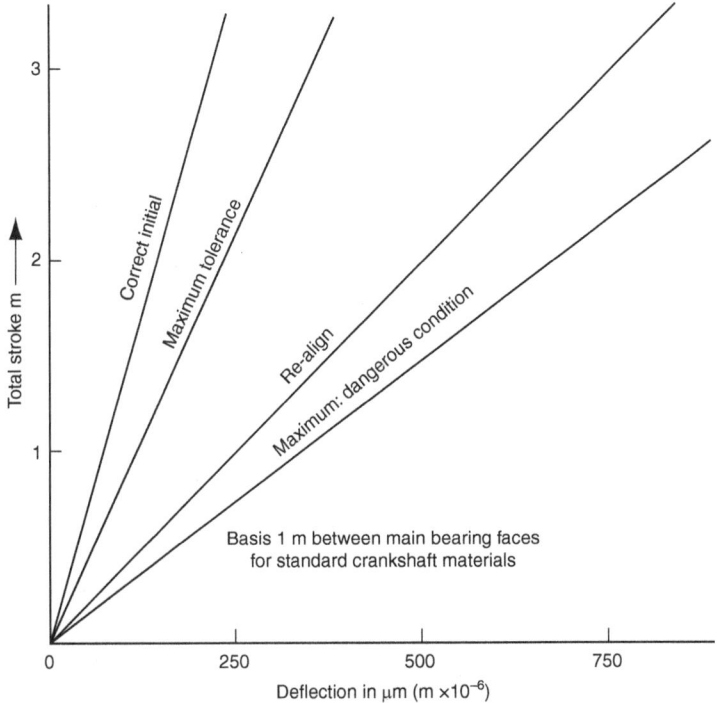

▲ **Figure 6.5** *Crankweb allowable deflections*

Note

Vertical deflection = 5 mgl^4/384 *EI* for a simply supported uniformly distributed loaded beam.

Thus as a *generalisation*, it may be said that crankshaft deflection is proportional to total engine stroke and distance between main bearing faces. Figure 6.5 has been prepared on this assumption. To illustrate its use:

Consider an engine of 1.5 m stroke and 1.5 m between main bearing faces. From Figure 6.5 allowable test bed deflections (maximum) based on 1 m between bearing faces is 167 μm, therefore for this engine (1.5 m between bearing faces) maximum initial deflection allowed is 250 μm. Based on Figure 6.5 as a rough approximation: Correct 70 μm/1 m stroke/1 m face distance (max. 118). Realign 240 μm/1 m stroke/1 m face distance (max. 330). A set of typical figures for a six-cylinder IC engine and the method of taking deflections is illustrated in Figure 6.6 and Table 6.1.

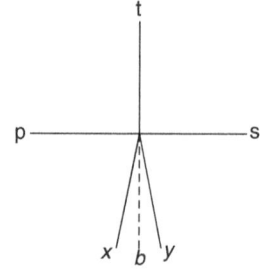

Crank positions for deflection readings

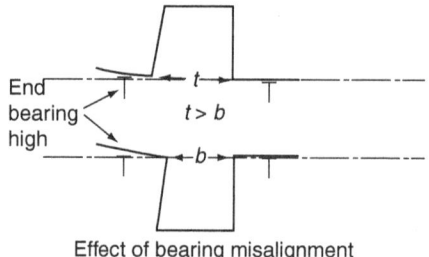

Effect of bearing misalignment

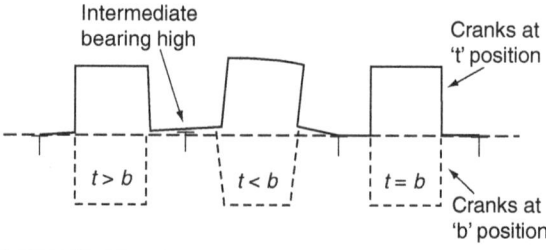

▲ **Figure 6.6** *Deflection readings and bearing heights*

Stresses caused by static deflection are difficult to assess, but as a rough guide each μm crankweb deflection (which *may* be somewhere about 1 μm central vertical deflection) *could* cause a bending stress of about 33 kN/m².

Vertical and horizontal misalignments can be checked against the permissible values supplied by the engine builder, which is often in the form of a graph. If any values exceed or equal maximum permissible values then bearings will have to be adjusted or renewed where required. Indication of incorrect bearing clearances may be given when the engine is running. In the case of four-stroke medium, or high-speed diesels, load reversal at the bearings generally occurs. Excessive bearing clearances can be accompanied by a loud knocking noise as movement takes place and then the white metal of the bearing usually gets hammered out.

Table 6.1 *Typical deflection values for a six-cylinder engine*

Crank position	Cylinder number					
	1	2	3	4	5	6
X	0	0	0	0	0	0
P	50	20	60	-80	-30	10
t	100	30	120	-140	-80	40
S	50	30	60	-80	-60	30
Y	-20	20	-20	0	0	-20
$b = (x + y)/2$	-10	10	-10	0	0	-10
Vertical plane misalignment $(t - b)$	110	20	130	-140	-80	50
Horizontal plane misalignment $(p - s)$	0	-10	0	0	30	-20

Positive deflection when crankwebs open out. Gauge readings in µm (mm/1,000).
High bearings – No. 1 (end) and between Nos. 3 and 4 cylinders.

If bearing lubrication, for a unit that has excessive bearing clearance, is from the same source as piston cooling, then a decrease in the amount of cooling oil return may be observed in the sight glass, together with an increase in its temperature.

If bearing clearances are too small, overheating and possible seizure may take place. Increased oil mist and vapour at the particular unit may be observed – together with an increase in bearing temperature, which could then lead to a crankcase explosion.

Regular checks must be made to ascertain the oxidation rate of the oil. If this is increasing then high temperatures are being encountered, and as the oil oxidises (burns) its colour blackens.

The use of wireless technology is due to take the industry by storm once systems become reliable for use in the marine environment where the transmission of information must be 100% accurate inside what is essentially a steel box. However, there are some places where the technology is already starting to take hold and one of these is with the recording of crankshaft deflections. The job traditionally was very tedious, messy and awkward. Engineers invariably had a 'crankcase' boiler suit that they would wear for jobs such as this and then discard upon completion due to the oil that would have dropped on to them inside the crankcase.

The Bluetooth-enabled measuring device can be placed in position between the marks on the crankwebs while they are close to the open crankcase door. The crankshaft can then be turned with the engineer being outside the crankcase. The receiver collects and stores the measured information from the wireless signal sent from the measuring device. The engineer can then download the information to a laptop also using a wireless connection. Dimensions are still recorded with the crankshaft in the positions indicated in Figure 6.6.

It must be pointed out that the specific figures shown must *always* be treated with caution and are only used for a general picture and guidance. Some of the most up-to-date engines have a maximum deflection difference set at 100 μm.

Interpretation of crankshaft deflections gives an indication of high and low bearings. Before any check is made it is advantageous to make sure the shaft is bedded into the lower half, that is, use of feelers. Deflections should be used in conjunction with optical telescope readings and wear-down bridge gauges.

When a bearing between two cranks is higher than those on either side of it, both sets of crankwebs will tend to open out when the cranks are on bottom dead centre and close in when cranks are on top dead centre, vice versa if there is a low bearing between two cranks. The scale ratio between vertical bearing heights compared with crankshaft deflections is often taken as 2:1, that is, 0.1 mm change in bearing height produces 0.05 mm change in crankshaft deflections).

Blended edge main bearings

The difficulty of the ship bending to accommodate the action on it by the power of the sea means that manufacturers must take this into account in the design of the machinery fitted into ships. Since the introduction of thin-shell, white-metal main bearings in large diesel engines some have failed without a clear cause being identified except for small imperfections in the bearing's geometry that could have been caused by the flexing of the crankshaft.

In response MAN Diesel & Turbo introduced the first blended edge (BE) main bearings in 2004. These were incorporated in some of the bearing positions in some of the engines and were successful in reducing the high edge loading and spreading this over a larger area. As MAN's experience with these bearings increased so they were able to introduce them to more bearing positions in more engines. The design of the outer edges of the BE bearings is a smooth radius taper, as shown in Figure 6.7, which allows the bearing journal to change its inclination without touching the outer edge and placing additional load on the oil and the BE. The improved distribution of pressure

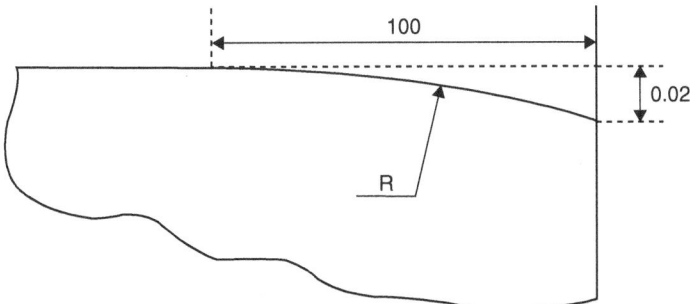

▲ **Figure 6.7** *Blended edge main bearings (all dimensions in mm)*

over a greater area often increases the minimum oil film thickness by more than 100%, which dramatically decreases the risk of main BE fatigue failure. It should be noted by the engineers on board that the new designs are marked 'BE Type'. The actual radius of the curve could be different for each bearing position but has been calculated for each bearing by computer simulations that have been enhanced by experience.

Shafting Stresses

Stress is expressed as the total load divided by the area that it is acting through. Given this fact we can consider the ratio of strengths required for solid and hollow shafting; consider a solid shaft, diameter D_1, compared to a hollow shaft, diameter outside D_2 diameter inside d_2.

$$\frac{\text{Strength solid}}{\text{Strength hollow}} = \frac{\text{Torque solid } (T_1)}{\text{Torque hollow } (T_2)} = \frac{q_1 J_1 r_2}{q_2 J_2 r_1}$$

where r is shaft radius

$$\text{as } \frac{T}{J} = \frac{q}{r} \text{ ,, } \begin{array}{l} q \text{ is working shear stress maximum} \\ J \text{ is polar second moment of area} \end{array}$$

thus for the same working stress:

Consider a solid 360 mm dia. shaft of mass 5×10^3 kg and compare this to a hollow 380 mm × 250 mm dia. shaft of the same length, this has a mass of 3×10^3 kg.

$$\frac{\text{Strength solid}}{\text{Strength hollow}} = \frac{360^3 \times 380}{(380^4 - 250^4)} = \frac{1.76}{1.66}$$

that is, approximately the same.

∴ Approx. 40% weight reduction for the *same* length. Hollow shafting is, however, more expensive. Consider next the case of a 300 mm dia. shaft with a flaw detected 25 mm deep. Now shaft power is proportional to torque:

$$\frac{T}{J} = \frac{q}{r} \text{ and therefore } T = \frac{2Jq}{D}$$

$$\therefore T = \frac{2 \times \pi D^4 \times q}{D \times 32} \qquad \therefore T = \frac{\pi D^3 q}{16}$$

Thus power ℓ torque ℓ diameter³.

Thus power reduction for 25 mm flaw:

$$\text{Reduction ratio} = \frac{250^3}{300^3} = \frac{1.56}{2.70}$$

that is, power reduction to about 58% of the original power.

Before considering the regulations appertaining to shafting sizes for the various types of shafting it would be as well to first build up a set of simple calculations for the various different shafting lengths in turn. Only the broad aspects will be considered, *much simplification* and *many assumptions* being necessary in order to clarify and present the results as a reasonably simple picture.

Intermediate shafting

This is usually involved in the first part of the calculation. The shaft is subject to torsion, based on the required horse power and taking a safe stress factor the diameter can be derived. The couplings and coupling bolt dimensions can also be calculated. The fundamental torsion equation $T/J = q/r = G\theta/l$ being the basis for most calculations.

The compressive stress induced by the end thrust, from the propeller, is small in comparison with other stresses and it acts on all the shafting. A thrust ahead of about 500×10^3 N would only induce a compressive stress of about 1.73 kN/m², this can normally be ignored except where such thrust is transmitted to the hull, that is, at the base of the thrust collar.

Bending stresses could really only arise from the ship movements and alignment variation and the effects should not be large.

As a summary, the intermediate shaft is subject to torsional shear stress, which influences the factor of safety and hence resultant working stress, a slight compensation would be allowed for due to end thrust, bending and possible variations of torque caused by the fluctuating load on the propeller as it moves deeper and then out of the water if the vessel is pitching.

Thrust shaft

Calculation is almost the same as for the intermediate shaft but virtually no misalignment bending would occur in such a short shaft length over a stiffened tank top. The thrust action on the collar would require a thicker diameter at the collar root but once clear of, say, the thrust pads the shaft could be tapered down to the intermediate shaft diameter.

Propeller shaft

The shaft is subject to torque and end thrust, as is the intermediate shaft, but torque variations due to propeller racing would be fluctuating considerably more. In addition, the shaft is subject to a bending stress due to the propeller weight, on one end, in still water. Assuming the propeller is immersed in still water and taking the loading as existing in Figure 6.8, the weight, after allowing for an up-thrust of water, is say 45 kN, and treating as a simply supported cantilever beam then the bending moment is $mgl = Wl$, that is, $45 \times \frac{3}{4} \simeq$ kN m on the shaft where it enters the hull.

This is already a considerable force but when the propeller rises out of the water due to racing in heavy seas this value increases even more, so the whole of the shafting must be assessed as a heavy, fluctuating, largely indeterminate, bending moment.

Coupled with the above it should be borne in mind that the shaft is worked in a corrosive medium and there is a *possibility* of direct contact between shaft and sea water, in spite of precautions taken to exclude this event. Some authorities consider that the fatigue strength of elements in sea water is 25% of that for the same parts in air.

Summarising for this shaft, there is fluctuating, combined bending and twisting, with a degree of uncertainty of magnitude, together with end thrust and the possibility of corrosive attack. Therefore, given these criteria, the factor of safety employed is high, usually over 12. The tail shaft may always be looked on as probably the weakest link in the system and various different types of material have been used in the past, such as nickel and chrome additions to mild steel, but the preferred material after much

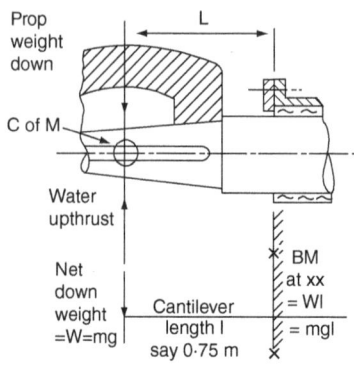

Simple analogy to point load
cantilever for prop shaft bending

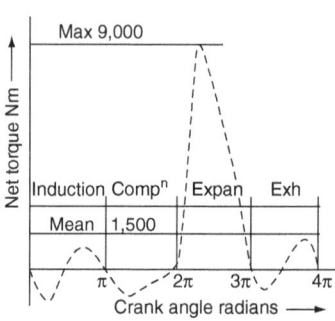

TM diagram single-cylinder SA
4-stroke IC engine

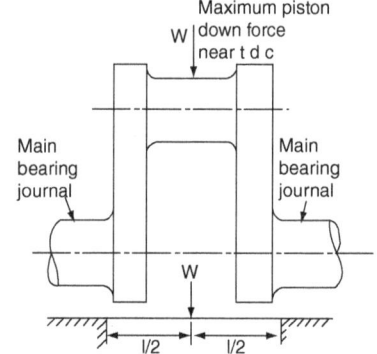

Simple analogy to point load SS beam
for crankshaft bending

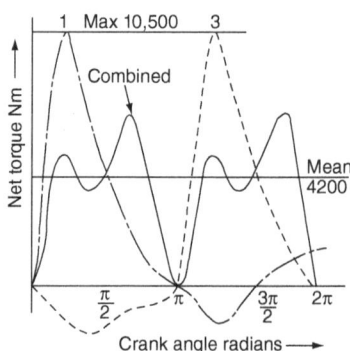

TM diagram for 4-cylinder SA
4-stroke IC engine

▲ Figure 6.8 *Shafting stresses*

experience is often still mild steel because of ductility, strength factors and fatigue resistance. Tensile strength should not normally to exceed 540 MN/m². Propeller shafts are withdrawn for examination every two years. This period is now five years for vessels fitted with continuous liners or oil-type stern tube bearings that also have special stress-reduced keyways or keyless fitting of the propellers. The same period applies to most controllable pitch propeller designs.

Some modern tonnage is being built with carbon fibre reinforced plastic drive shafts. These include smaller boats but also naval vessels, cruise ships, superyachts and fast ferries. The weight saving can be as much as 70% and there are no limits to the dimensions and the torque that can be handled by composite fibre shafts. Not only are the designers saving in the weight of the shaft but also the weight of intermediate bearings, which are not needed. (See Chapter 1 of this volume for more about composite material.) To date, up to 23,000 kW per shaft has been installed on ships but 50,000 kW projects are being designed. The shafts can be up to 12 m long and transmit torque of up to 1,000 kNm with 2,200 kNm being planned.

Crankshafts

Consider the turning moment diagrams illustrated in Figure 6.8.

The maximum to mean torque ratio for the single-cylinder IC engine is 9,000:1,500, that is, 6:1. Two facts also emerge:

1. The torsional shearing stress due to turning moments of the engine is fluctuating.
2. Variation of maximum to mean torque ratio is high with IC engines. The *mean* torque is the power transmitted to the drive per revolution but the shafting sizes must be based on the *maximum* torque.

Next, referring to the combined torque diagram it is seen that for cranks at π rad. with a firing order 1, 3, 4, 2 the ratio is 10,550:4,200 or 2.5:1, which is much improved in comparison to the single-cylinder IC engine. In this diagram only one revolution is considered and the induction, exhaust, compression effects, etc. for Nos. 2 and 4 cylinders, occurring during this 2π rad. have been omitted for simplicity, although the combined curve has taken them into account. It can be seen that the addition of cylinders gives a smoother turning moment and a reduced maximum to mean torque ratio. However, the main two facts are apparent in that there is still torque fluctuation and the torque ratio is high for IC engines. This analysis gives some idea of the torsional stresses involved and the next factor to consider is bending.

It should be noted that the net or output available torque is the summation of torque applied due to the gas load and inertia torque required to accelerate or retard the moving parts, being a numerical addition, due account taken of positive and negative torque signs at any crank angle.

The mass of the parts also need consideration.

Referring to Figure 6.8 we may consider the case for simplicity as a simply supported beam with a central point load, this load having a maximum value near t.d.c.

Consider an IC engine with 762 mm dia. cylinders and maximum firing pressure of 5.5 MN/m² then maximum firing force is:

$$W = \pi/4 \times 0.762^2 \times 5.5 = 2.5 \text{ MN}$$

Assuming the simply supported beam case:

$$\text{Bending moment} = \frac{mgl}{4} = \frac{Wl}{4} = \frac{2.5 \times 1\frac{1}{3}}{4}$$
$$= 0.833 \text{MN m} \ [833 \text{ kN m}]$$

assuming distance between main bearing faces as $1\frac{1}{3}$ m, that is, l.

The conclusions are (1) a crankshaft is subject to heavy bending loads, and (2) the loading is very rapidly applied in the IC engine, being a form of impact.

There will be some slight end thrust on the crankshaft but in comparison with bending and torsional stresses it is negligible.

After consideration of simple applications of stress to the main shaft lengths a reasonable picture of the stress factors involved should now be available and the necessary shafting regulations or rules can now be considered.

Guidelines for Transmission Shafts

Ingot steel for shafts and coupling bolts should have a UTS of 430–500 MN/m², the couplings should be forged in with the shaft or have the shaft ends hydraulically upset, separate couplings may be ingot steel or steel castings. The chemical composition, mechanical properties and heat treatment details will also need to be approved by the classification society.

The maximum stresses allowed in any one shaft will depend upon the safety factors for the application of any particular shaft. Other factors to consider will be cyclic stresses, transient operations and life expectation of the shaft.

Crankwebs for built-up crankshafts should be ingot steel or cast steel and shaft liners should be of bronze. All materials should be subject to the required tests and treatment.

Intermediate shaft

The final dimensions will need to comply with empirical formula based on theoretical principles and practical experience. There is a separate formula for each and every engine type, for example, steam turbine, turbine-electric drive, motor machinery (with all variations of number of cylinders, firing intervals, type of cycle, etc.). Computer software has been developed by several organisations and the classification society will need to know which software has been used.

Crankshafts and turbine wheel shafts

As above.

Crankwebs (built-up shafts)

Thickness parallel to shaft 0.625 × crankshaft diameter (*d*), thickness radially around crankpin 0.438 × crankshaft diameter. Webs should be securely shrunk on to pins, shrinkage allowance of about diameter/625, dowels may be fitted.

Thrust shaft

At a collar, the root diameter should be 1.15*d*. Outside of the collar the root diameter may be tapered down to the diameter *d*.

Stern tube shaft

Diameter is 1.14*d*. If any part of the shaft is in contact with sea water these sizes are to be increased $2\frac{1}{2}$ %. Note that this is the case of a shaft passing through stern tubes, which does not support the propeller weight (i.e. twin screw bracket support). (Clearance will be dependent on bearing type, lubrication method, sealing design, hammering, etc.).

Propeller shaft

This is the case of the shaft that supports the weight of the propeller.

$$\text{Diameter} = (d \times c) + P/K$$

$c = 1.14$

P = propeller diameter in mm

$K = 144$ if (a) a continuous liner is fitted, (b) the shaft is oil lubricated and sea water is excluded and (c) where the shaft material is resistant to corrosion by the water in which it will operate.

$K = 100$ for all other shafts.

At the coupling, the flange may be tapered down to 105 d.

There are strict rules for the machining of keyways as these have been a significant sources of failure in the past. Modern ships are not made with keyed propellers but there may still be some in service that the student may come across during his/her service.

Stern tube bush

The stern tube has attracted attention because the move towards oil-lubricated bearings has led to an increase in oil pollution from ships. This has led to the development of the water-lubricated stern tube bearing and also biodegradable oil that can be used so that any oil leakage will break down in the water in a short time. There are, however, rules relating to the construction of the stern tube bearing, which can be summarised in the following way.

The length of the bearing in the stern bush next to and supporting the propeller should be:

1. Water-lubricated bearings lined with an approved elastomeric bearing technology such as reinforced plastic material should have a length of not less than four times the diameter required for the propeller shaft under the liner. (This also applies to the older lignum-vitae bearings.)
2. Water-lubricated bearings lined with two or more circumferentially spaced sectors of an approved elastomeric bearing technology, where it can be shown that the sectors operate on hydrodynamic lubrication principles, should have a bearing length so that the nominal bearing pressure will not exceed 5.5 bar (5.6 kgf/cm²). However, the length of the bearing should also not be less than twice its diameter
3. Oil-lubricated bearings made of a synthetic material should have the flow of lubricant arranged so that overheating, under normal operating conditions, cannot occur. The acceptable nominal bearing pressure will be considered upon application and will be supported by the results of an agreed test programme. In

general, the length of the bearing will not be less than 2.0 times the diameter of the shaft in way of the bearing.

4. Oil-lubricated white-metal lined bearings that are provided with an approved type of oil sealing gland may have a length of approximately twice the diameter required for the screw shaft. The nominal bearing pressure will not exceed 8.0 bar (8.1 kgf/cm²) and the length of the bearing is to be not less than 1.5 times its diameter.

5. Bearings of cast iron and bronze that are oil lubricated and fitted with an approved oil sealing gland will generally have a length of not less than 4.0 times the diameter required for the propeller shaft.

6. Bearings that are grease lubricated will have a length not less than 4.0 times the diameter required for the propeller shaft.

Forced-water lubrication is to be provided for all bearings lined with rubber or plastics. The supply of water may come from a circulating pump or other pressure source. Low flow rate indicators must be provided on the water service to plastics and rubber bearings and the water grooves in the bearings should be of the correct size and shape so that the system promotes very little wear-down, particularly for bearings of the plastics type. Thordon Bearings Inc. are leaders in this field and have designed their own water quality package that takes in water through a filter before using it to lubricate the bearings. This system has the advantage of working well in areas that have a high concentration of abrasive material in the water, such as river estuaries.

Other features of all stern tube bearings are as follows:

- They must be securely fixed in the housing so that they are not able to
  (a) move along the centre line of the shaft,
  (b) rotate in the direction of the rotating shaft.
- There must be means of shutting off the water supply to the bearing where the piping is forward of the aft bulkhead or aft peak tank.
- Ships that have a worldwide brief must be designed to accommodate any differential expansion due to change in sea temperature. This is particularly true of any seal that is designed to stop any oil leakage on the oil-lubricated bearings.
- Any lubricating oil header tank within the oil-lubricated bearing system must be fitted with a low-level alarm when the tank is above the loaded water line.
- Oil-lubricated bearings should have suitable temperature monitoring and cooling systems fitted to cool the oil.
- Where in-water surveys are allowed there must be provided a way to determine the clearance in the stern bush with the vessel afloat.

Liners

Liner thickness comes from a prescribed formula and is shrunk or hydraulically pressed on, without dowels. Shaft and liner joint, at all points, must exclude entry of sea water and any cavity, that is, non-fitting strip, should be filled with a suitable composition.

Couplings and coupling bolts

The coupling bolt's diameter is from a prescribed formula. Flanges of thickness should be at least equal to the bolt diameter (propeller shaft coupling at least $0.27d$). Fillet radii on a shaft should be the diameter $\times 0.08$. *Note*: In all cases, shafting, couplings and bolts must provide resistance to astern pull.

Ultrasonic tests and final alignment verification

These are required on shaft forgings that are being prepared for shafts with a diameter of 250 mm or greater. The shaft alignment procedure is completed on a static shaft. There are software programs that will assist but the final procedure will normally be carried out in the presence of the class surveyor. The 'sighting through' and 'sag and gap' measurements are verified after the superstructure is in place, all major welding is complete and the vessel is in the water.

Extra strength for operation in ice

Consideration must be given to the extra strength of all the components if the vessel is to continually run in ice conditions.

General comment

The discussion and information given during this chapter is to ensure that the on-board marine engineer has the knowledge to understand how the machinery in his/her machinery spaces has been set up. It also allows the marine engineer to give an informed opinion if they are part of the owner's team attending refit of new builds as well as answering questions during the flag state examinations. However, during the life of the vessel the shaft support bearings are subject to wear and the engineers on board will need to be aware of the consequences of 'bearing wear-down'. The shafting will obviously come out of line if one or more of the support bearings alter the height of the shaft due to wearing down. Although the wear in any bearing is not desirable the primary concern is the stern tube bearing. The weight of the propeller already places a heavy load on the outer edge of the stern tube bearing and any wear on the inboard bearings will add to the bending of the shaft and the wear-down on the stern tube bearing itself should be measured as part of the special survey. The initial alignment of the shafting should take into account the wear-down of shaft bearings.

It should also be noted by the on-board engineers that the wear of bearing shells will not be even in the fore to aft direction along the centre line of the bearing/shaft. Therefore if the bearing has to be removed for any reason it is vital that the shells are replaced in the same way around as they came out. The bearing may run hot if the shell is reversed during an inspection procedure.

The Propeller Shaft and Stern Tube

Water-lubricated type

A design for a 6 MW shaft output is detailed in Figure 6.9, which should be self-explanatory. The maximum wear-down is best set at 8 mm with the continuous liner to be examined every three years. For twin-screw vessels it may be regarded as advantageous if the propeller shaft could be withdrawn quickly sternways; a coupling with this in mind, also shown in Figure 6.9, is more costly and adds complexity. Some patent types utilise rubber bearing surfaces but are not generally used for the larger shaft sizes. Impregnated plastic resin compounds on plastic-type bases have been used successfully in place of lignum-vitae. One such type that has been used in the past is called Tufnol.

This is a thermo-setting laminate produced from cotton fabric and phenolic resin as the main constituents. The fabric is impregnated with the resin and the layers of this impregnated material are pressurised under heat until the fabric laminations are bonded into one sheet. This material gives uniform density, hardness and swelling together with good wettability and low coefficient of friction plus an ultimate compressive

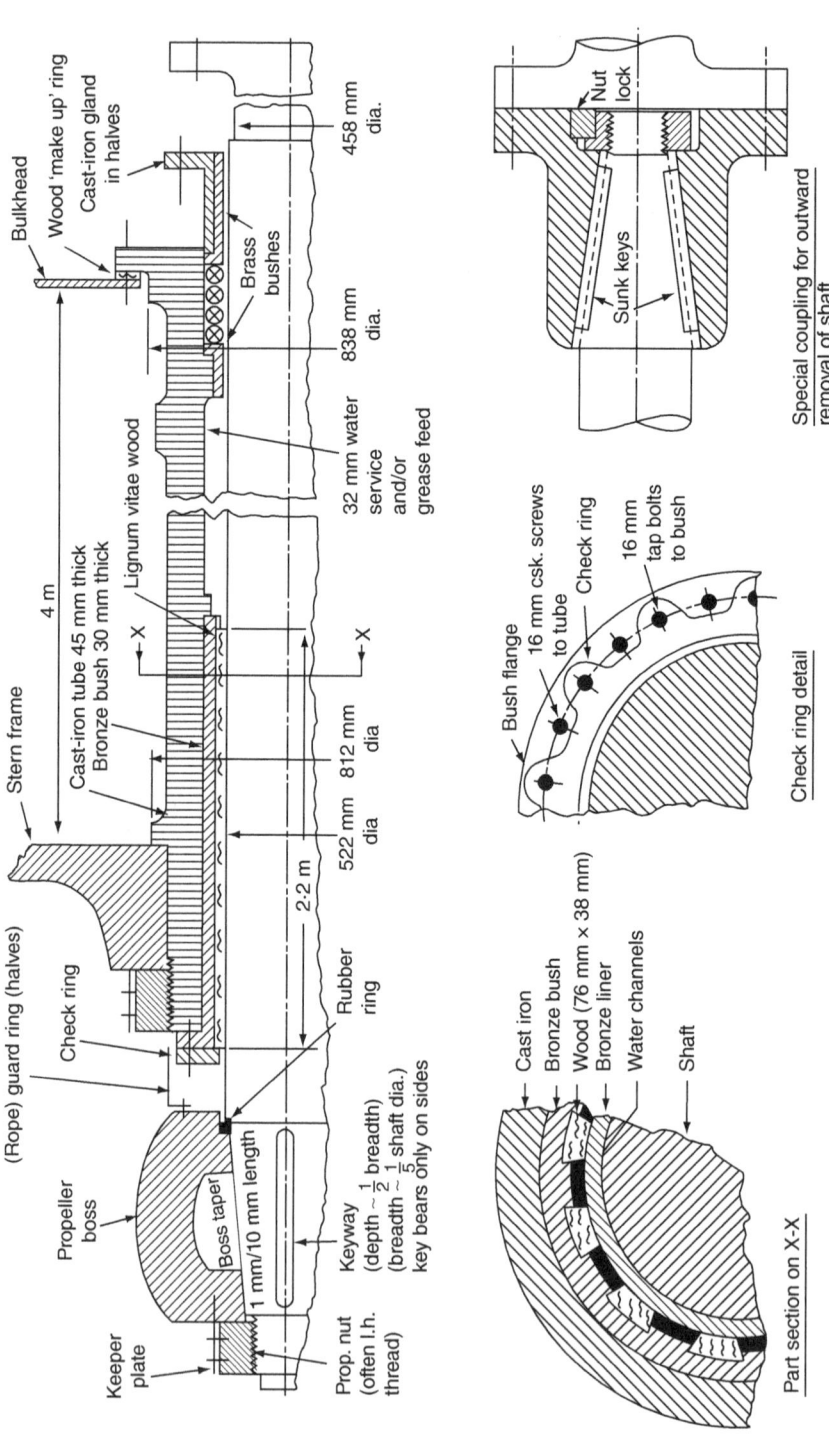

▲ Figure 6.9 *Propeller shaft and stern tube (water)*

Bulkhead

Wood 'make up' ring

Cast-iron gland in halves

458 mm dia.

Brass bushes

838 mm dia.

Lignum vitae wood

32 mm water service and/or grease feed

Stern frame

4 m

Cast-iron tube 45 mm thick

Bronze bush 30 mm thick

X

X

(Rope) guard ring (halves)

Check ring

522 mm dia

812 mm dia

2.2 m

Rubber ring

Propeller boss

Boss taper

1 mm/10 mm length

Keyway (depth ~ $\frac{1}{2}$ breadth) (breadth ~ $\frac{1}{5}$ shaft dia.) key bears only on sides

Keeper plate

Prop. nut (often l.h. thread)

Nut lock

Sunk keys

Special coupling for outward removal of shaft

Bush flange

16 mm csk. screws to tube

Check ring

16 mm tap bolts to bush

Check ring detail

Cast iron

Bronze bush

Wood (76 mm × 38 mm)

Bronze liner

Water channels

Shaft

Part section on X-X

strength (flatwise) approximately twice that of lignum-vitae. A coefficient of friction of 0.005 when water lubricated can be achieved and short length–diameter ratios with a greater continuous bearing surface could be designed at very high loadings with this material. A water supply to the inboard end of the bearing is essential and grease and oils should never be used.

The best method of fitting the staves and the most preferred water grooves (*UV* type) are as shown in Figure 6.10 (note a later type key shown in Figure 6.10). The swelling expansion due to water absorption is greatest in the direction normal to the laminate and will not normally exceed 1 mm in 40 mm thickness. Diametric clearance is about 2 mm for 500 mm shaft.

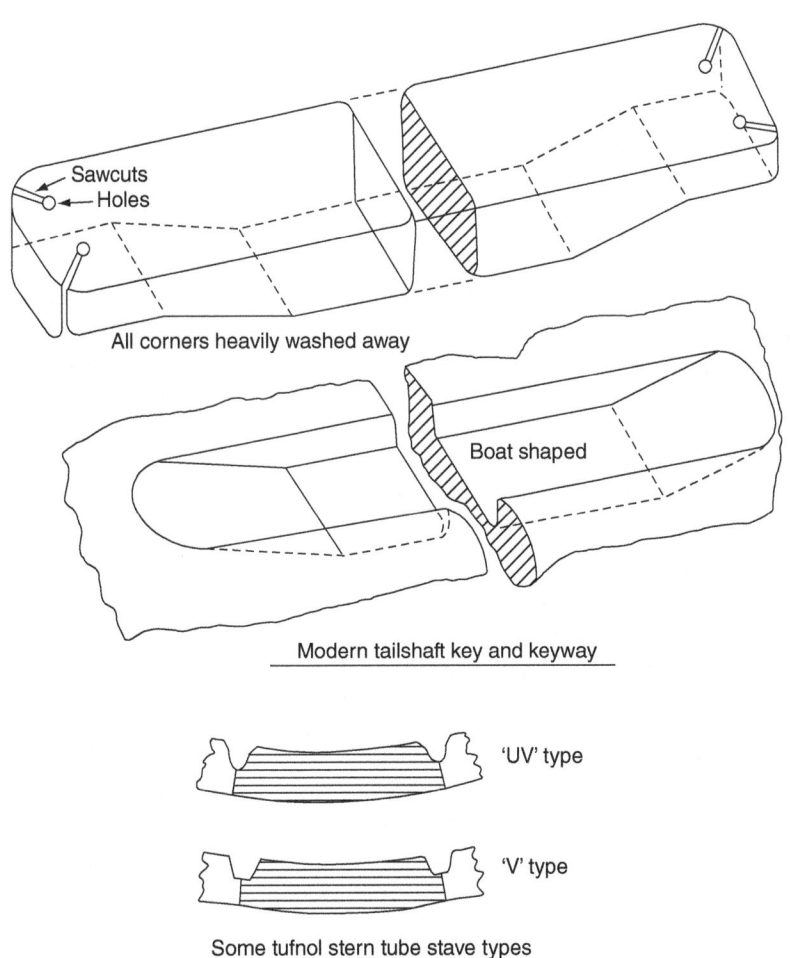

Sawcuts
Holes
All corners heavily washed away

Boat shaped

Modern tailshaft key and keyway

'UV' type

'V' type

Some tufnol stern tube stave types

▲ **Figure 6.10** *Alternative keys and staves*

This material can be used for many other duties-such as general bearings, gears, resilient mountings, flexible couplings, etc. Physical properties of a type suitable for tail shaft bearings would be as follows:

UTS	62 MN/m²
Compressive stress (ultimate, flatwise)	290 MN/m²
Shear strength (ultimate, flatwise)	100 MN/m²
Young's modulus	7 GN/m²
Impact value	1.08 mN

Due to the problems of oil leakage from the oil-lubricated stern tube bearings and the latest penalties under MARPOL, the water-lubricated stern tube bearings are returning to popularity as the system of choice. The most up-to-date bearings are made from elastomeric polymer alloy with the bottom half of the bearing made smooth to promote hydrodynamic lubrication. The top half of the bearing has grooves designed to allow the water to flow aft, back into the sea. Filtered water is led to the bearing and allowed to discharge back to the sea. This arrangement means that there is no need for an outer seal and no chance of leaking oil into the water possibly incurring a fine. Recently, the classification societies have brought the period allowed before removing the propeller shaft for examination of the lubricated shafts into line with the 'oil-lubricated' shafts.

Oil-lubricated type

This stern tube has been in use for many years. Various designs are available but the same principle is apparent – seal the ends of the tube with a gland and supply oil under pressure. Water should be regularly drained off and in port the tube should be emptied and drained. A typical oil tube is as sketched in Figure 6.11. Wear-down for the white metal should not normally exceed 2 mm to avoid hammering out and the period between inspections is about six years. A highly resilient, reinforced plastic material is often used in place of white metal. It is claimed to have superior load-carrying capacity, high resistance to fatigue and shock loading, with good lubrication properties. Stern tube seals, with oil lubrication, have also tended to use rubber rings increasingly. Fluoric rubber (Viton) with additives has been shown to be more effective than nitrile butadiene rubber for seal rings. In these designs four seal rings are usually located in the support housing aft with oil pressure supply to the middle chamber. Two similar ring seals, with oil feed between, are arranged in a floating housing at the forward end. Ceramic-coated liners can also be used.

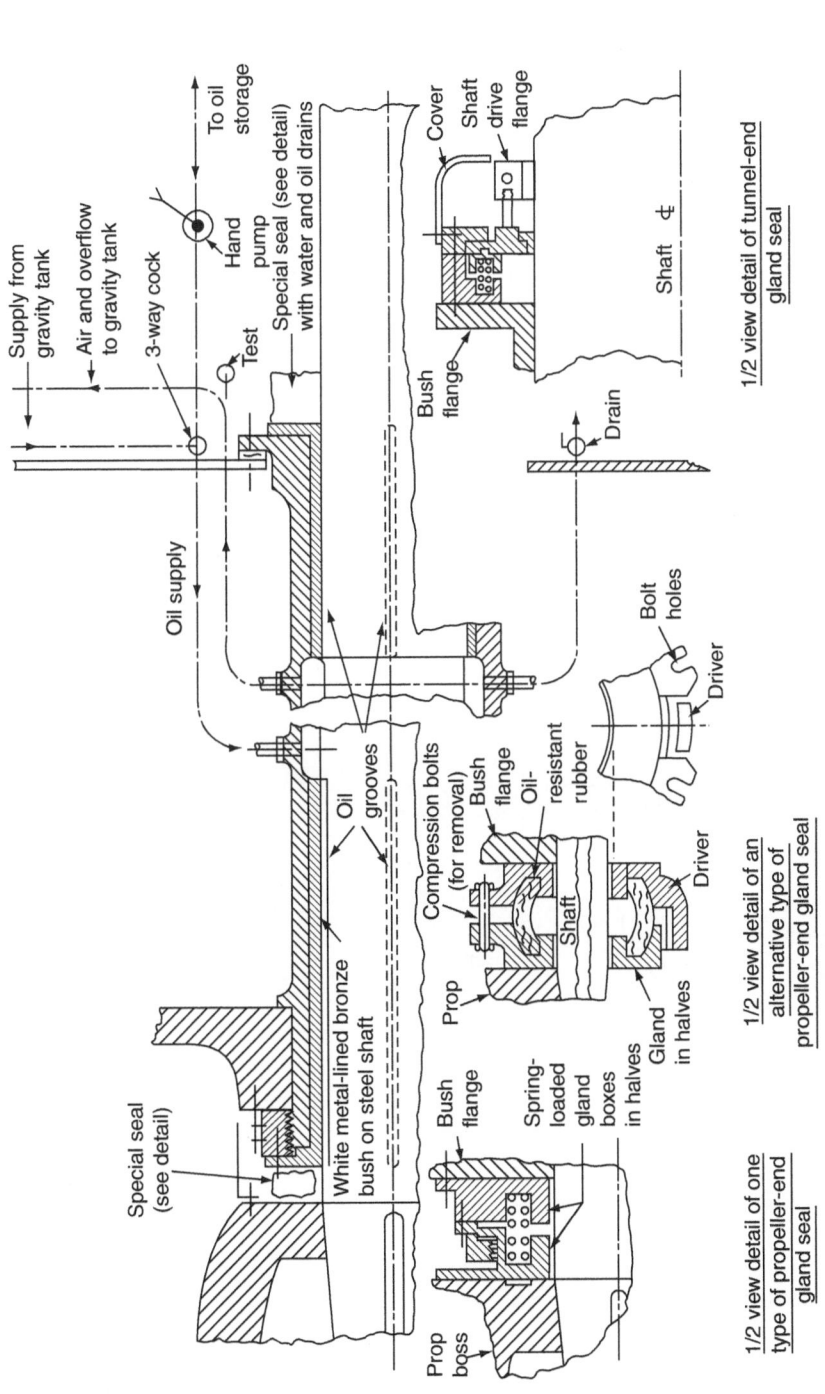

▲ Figure 6.11 *Propeller shaft and stern tube (oil)*

Withdrawable stern gear system

The advantage of this arrangement is that inspection, realignment or repair can be carried out quickly with the ship afloat and without the need to disturb the propeller or uncouple the shaft. It can be used with fixed or controllable pitch propellers, flange or cone mounted, and with most types of seal. With the bearing (split in halves) withdrawn the propeller and shafting weight is supported on a ring permanently secured to the stern frame. The overall layout is similar to Figure 6.11 but there is an integral tube-bearing, split in halves, which is fitted (or removed) from inboard (see Figure 6.12).

The whole unit, including the outer seal, can be moved along the shaft inboard for inspection. Note also the more modern practice of hydraulic floating a keyless fit (taper 1:30) with advantages of simplicity and reduced stress factors. Closure is with a hydraulically tightened nut. The unit is generally more costly than the conventional design. Short, large-diameter (1,300 mm), spherical SKF roller bearing units are available capable of inward withdrawal. Connection between intermediate and propeller shaft is by SKF oil injection coupling and standard inner and outer seals are used, clearance would be about 0.8 mm.

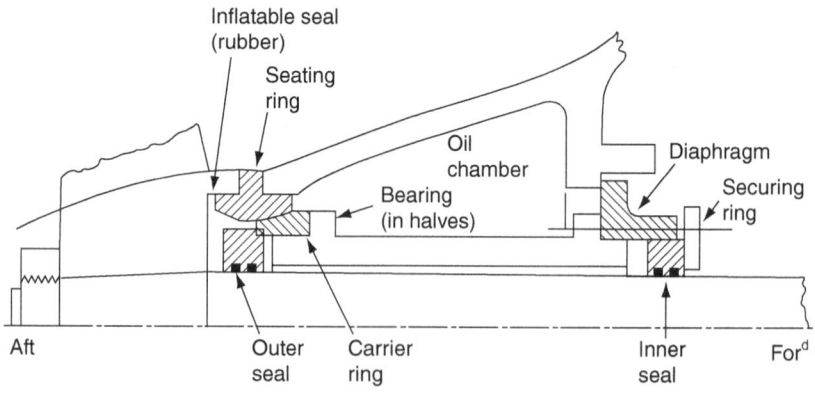

▲ **Figure 6.12** *Withdrawable stern gear*

Alternative stern gear

The tendency to fit increasing diameter and weight propellers (in excess of 70 tonnes), so as to drive very large vessels with low revolutions to give higher performance, has made increased shaft flexibility and reduced bending moments very desirable (Figure 6.13).

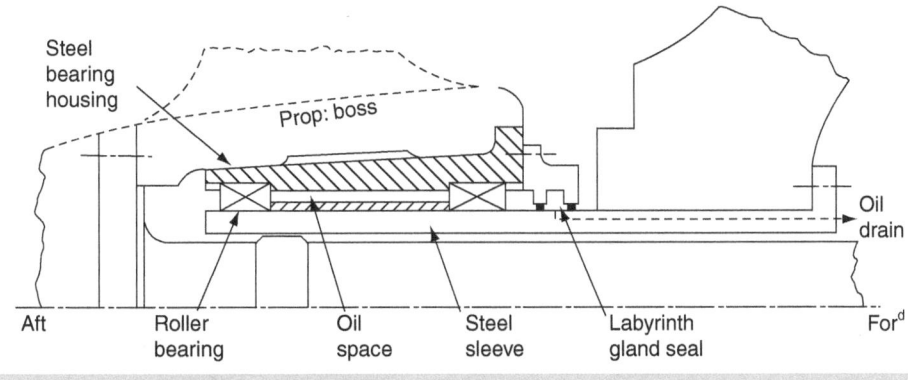

▲ Figure 6.13 *Alternative stern gear*

The propeller has its own self-contained bearing and the drive torque shaft is more flexible. Hollow helical (spiral spring) roller bearings are used giving differential radial expansion allowance and flexibility to shock loading – plain outer races allow shaft axial movement. Note the flanged connection to the propeller boss – this is simple and trouble free but it requires a special coupling at the inboard end to allow withdrawal aft. Such a coupling is shown in Figure 6.8 – astern thrust resistance is increased with an inner nut. A similar design to Figure 6.11 is available utilising a plain bearing inside the propeller boss in place of roller bearings. If shaft withdrawal inboard is essential, a cone and taper with nut arrangement can be used to secure a flanged coupling for bolting to the propeller boss (to replace the flange detail as sketched in Figure 6.11).

Propellers

The finer elements of propeller design are best studied by reference to a work on naval architecture for propeller design, pitch theory, etc. However, there are some interesting thoughts that are worth discussing at this stage.

The propeller operates by setting up a pressure difference across its surfaces that leads to an automatic acceleration of the water depending upon that pressure differential. The propeller blades are shaped in the same way that an aircraft wing is shaped. The aerofoil shape of the aircraft wing induces lift and the propeller produces thrust. The design of the propeller and the associated hull is critical to the efficient operation of the propeller.

Until the end of the 1990s it was still customary to associate excessive vibration of the hull with resonance set up between the frequency of the propeller blade and the natural frequency of the hull, coupled with the clearances between the propeller and the hull. Working from this thought process it was sometimes possible to effect an improvement in the size of the vibrations by changing the propeller for one with a different number of blades, or by increasing the propeller blade clearances by altering the position of the blades relative to the hull.

With the development of computational fluid dynamics (CFD), it has now been realised that any improvements achieved by the process of moving or re-designing the propeller were largely because the propeller blades had been placed in a more favourable wake field.

The wake field is the water that is merging back together following its disturbance by the hull that has just been pushed through it. The water leaving the hull of the vessel forms the flow of water into the propeller. The efficient performance of a vessel's propeller is determined by several factors: laminar flow of water into the propeller, cavitation produced and wake field interaction as a result of the propeller disc rotating in the inhomogeneous inflow of water after it has passed over the hull of a vessel.

However, despite careful propeller and hull design, the flow of water around the vessel and finally ending up at the propeller is not uniform. Therefore, as the conventional propeller blade revolves it travels through water of different inflow velocities, which sets up pressure pulses that are transmitted to the hull of the vessel causing vibration. This is the reason for the development of the skewed propeller, which has blade sections located at different radii so they do not enter the wake peak.

Rotating propeller blades are also travelling in different directions across the wake field at the top of their rotation than they are at the bottom of their rotation. This again causes fluctuating pressure pulses leading to inefficiency and increased vibration. Performance efficiency, on a single-screw vessel, can be improved by an asymmetrical design to the hull of the vessel, which will give a more evenly distributed inflow of water to the rotating propeller.

Research and development now concentrates on the development of ideas and equipment that will improve the flow of water to and from the propeller allowing it to carry out its function more efficiently. Some success has been achieved by using ducted propellers or ducts on the hull forward of the propeller that smooth the flow of water into the propeller. Rolls-Royce have designed a hubcap that bridges the gap between the propeller boss and the rudder (Figure 6.14). The tapered hubcap fitted to the hub of the propeller leads the water flow on to a bulb, which forms part of the spade rudder. The rudder has a twisted leading edge, optimised for the flow from the propeller, which converts some of the swirl energy that is normally lost in the slipstream,

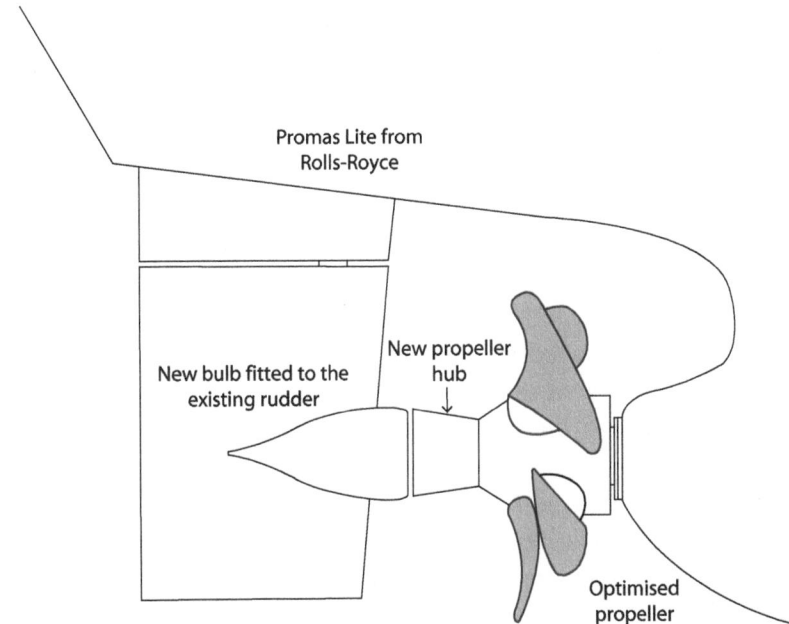

Promas Lite from
Rolls-Royce

New propeller
hub

New bulb fitted to the
existing rudder

Optimised
propeller

▲ **Figure 6.14** *Drawing of promas*

into additional forward thrust. The result is an increase in propulsive efficiency of about 6–8% depending on application.

The curves of the propeller blade designs used to describe its hydrodynamic properties are listed below:

- the radial distribution of pitch;
- chord length;
- rake;
- skew;
- maximum thickness and
- maximum camber and typically the blades are twisted to give a constant pitch along the blade from root to tip.

The objective is to generate an even force across the blade area with the thrust then being transmitted through the boss to the shaft line.

Extensive research and development of the interaction between the propeller and hull has led to changes in the propeller's design, which in turn has improved efficiencies.

A propeller blade is more efficient if it has a thin blade but a thin blade would compromise its strength and constructional techniques. Classification societies have always had strict rules for the design and manufacture of marine propellers to ensure that they are strong enough to function correctly. Traditionally, the data created to describe the design and geometry details of a propeller would have been distributed via paper drawings, offset tables and calculation sheets. More recently and especially with the introduction of highly skewed propellers, most of the design process has become software based. Digital information produced is stored in a file format called PFF (propeller file format), which is then used by the classification society to evaluate the strength of the propeller and make comparisons against their rules. Rules are being revised all the time as new research becomes available. For example, following entry into force on 1 January 2010 of the Finnish-Swedish Ice Class Rules (FSICR) Ed. 2008, the scantling criteria of propellers for use in ice was completely renewed. The strength of large propellers is achieved by the use of bronze as it is also erosion resistant. Manufacturers such as Rolls-Royce and Clements Marine all have their propellers manufactured to the ISO484 standard. ISO484/1 is for propellers above 2.5 m and ISO484/2 is for propellers of between 0.8 m and 2.5 m.

Cavitation occurs when there is fall in water pressure allowing a pocket of gas and/ or vapour to develop. If that pocket of gas then moves to an area of high pressure it will collapse. The strength of the water imploding to fill the space is sometimes strong enough to cause damage to the metal structure in the area of the collapsing bubble. The Dutch research institute Marin identify bubble, blade root, sheet, cloud, hub vortex, tip vortex and propeller hull vortex cavitation as distinct and different types of cavitation. Lloyds Register, however, also report that they have observed sheet cavitation interacting with tip vortex cavitation causing vibration in a full-size vessel. The same interaction, however, was not present in the scale-model testing, showing that there is still a need for full-scale observations.

The cavitation generated on the pressure side of the blade is more damaging than cavitation generated on the reverse or non-pressure side of the blade and therefore the aim of the designer is to reduce the harmful cavitation and live with the others.

MAN Diesel & Turbo's research is concentrating on predicting the different kinds of cavitation that propellers experience for different applications and operating conditions. Calculation methods are constantly being developed that aim to predict cavitation with a higher degree of accuracy than in the past. The fast emerging RANS (Reynolds-averaged Navier–Stokes) programs have shown to be valuable research tools.

The interesting point about fixed-pitch propellers, as far as the operational marine engineers are concerned, is how these propellers deliver consistent service to the modern requirements of shipping in the 21st century. The problem with the fixed-pitch propeller is that the design is optimised around the maximum continuous rating (MCR) of the engine. The reason for this is that this is where the vessel will spend most of its working life and this is where the most fuel-efficient setting will be. However, a vessel with a fixed-pitch propeller is not efficient at manoeuvring speeds nor is it efficient when 'slow steaming,' the latter being an increasing feature of the charterer's needs. For a given size the fixed-pitch propeller does, however, have a larger blade area than those in an equivalent size.

Controllable pitch propeller

Use of these propellers has increased with the greater use of unidirectional gas turbine and multi-diesel drives and bridge control. The engine room (or bridge) signal is fed to a torque-speed selector that fixes engine speed and propeller pitch taking – feedback from each in the resultant control output.

Consider Figure 6.15: the input fluid signal acts on the diaphragm in the valve housing and directs pressure oil via one piston valve through the tube to one side (left) of the servo piston or via the other piston valve outside the tube (in the annulus) to the other side (right) of the servo piston. Movement of the servo piston, through a crank pin ring and sliding blocks rotates blades and varies pitch.

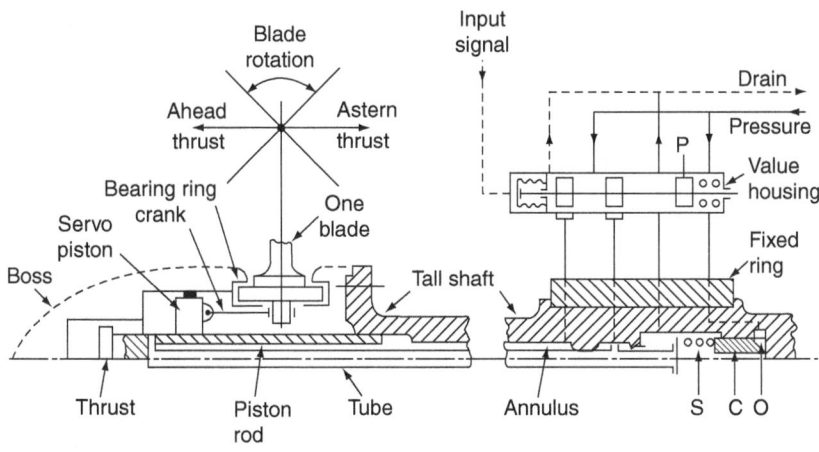

▲ **Figure 6.15** *Controllable pitch propeller*

The feedback restoring signal, to restore piston valves to the neutral position at correct pitch position, is dependent on spring(s) force (i.e. servo piston position), which acts to vary the orifice (O) by control piston (C) so fixing feedback pressure loading on the pilot valve (P) in the valve housing. The central part of the tail shaft includes a shaft coupling and pitch lock device (not shown). Emergency local pitch control, communication/alarm systems and fail-safe (navigable ahead pitch) are required.

Shafting ancillaries

In case of engine breakdowns it is usually advisable to fit a shaft friction brake at the first coupling from the propeller shaft so as to allow engine examination with safety. In case of towing, after breakdown, a trailing collar working on a tunnel bearing face is often fitted to allow shaft disconnection and reduce propeller resistance by allowing idling. Under normal conditions the collar is set at about 18 mm clear.

The torsionmeter

The first requirement is the determination of the shaft power (shaft kW or shaft MW) constant from a shaft calibration, a typical calculation would be as follows.

A shaft of 300 mm dia. and 6.5 m long is rigidly clamped at one end and the free end has a clamp and lever applied to the loads that can be added at a radius of 3 m. A load force of 222 kN produces an angle of twist of 1°.

$$
\begin{aligned}
\text{Shaft kW} &= 2\pi NT \text{ if } T \text{ in kN m} \\
&= 6.284 \times N \times 222 \times 3 \text{ in } this \text{ example,} \\
\text{Shaft kW for } 1° \text{twist} &= 6.284 \times N \times 666 \text{ in } this \text{ example,} \\
\text{Shaft kW for } \theta° \text{twist} &= 6.284 \times N \times 666 \times \theta \text{ for any case,} \\
&= 4180 \times N \times \theta \\
\text{Shaft MW} &= 4180 \times N \times \theta
\end{aligned}
$$

Therefore the meter or shaft constant = 4.18.

Thus knowing the angle of twist in degrees for the given shaft length, the shaft MW for the given rev/s can be determined. The requirement then for the torsionmeter is to measure the angle of twist in degrees between two points that are the correct datum length apart.

There are *five* types of torsionmeter:

1. Mechanical – gearing set from shaft with a differential screw reading device – this is not popular as wear immediately gives errors.
2. Acoustical – the pitch of a note from a vibrating wire varies with the torque (tension produced) – not popular at present due to difficulty of cyclic variations.
3. Optical – lag of a light flash is indicative of twist – simple but tends to give average value over a range of revolutions, no indication of cyclic variations.
4. Electrical – variation of transformer air gap due to twist – possibly most accurate and popular. The electrical type will be briefly considered (Figure 6.16).
5. Eddy-current sensors – two close-fitting, slotted sleeves are fitted inside each other and over the shaft. They are fixed to the shaft at opposite ends and therefore will be able to move relative to each other. Two electrical coils induce a high-frequency current within the sleeves and as they move relative to each other so the slots in the sleeves increase or decrease the coil inductance effect of the two high-frequency coils. Suitable electronic devices measure the changes in inductance and transmit the results which can be displayed as a measurement of the torque in the shaft.

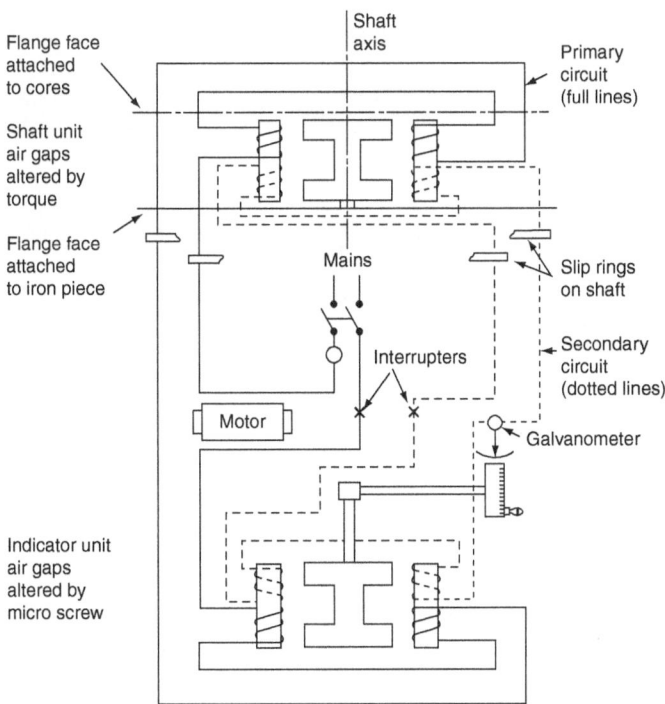

▲ **Figure 6.16** *Electrical torsionmeter*

The principle of torque measurement can be described in a practical way using the narrative below and referring to Figure 6.16. The modern systems will replace the electrical circuitry with electronic components but the principle remains the same.

Two sleeves rigidly fixed to a shaft have flanges at 180° to the shaft axis. Twist causes relative displacement between the flanges. Two cores are attached to one flange and the iron piece to the other so that relative movement between flange faces, due to shaft twist, alters the air gap of the differential transformer.

The primary circuit is wound to give the same polarity and the secondary circuits are in opposition, the provision of a motor-driven interrupter to give AC supply is required if DC mains. With no torque the air gaps are equal and the two secondary circuits are equal and opposite, but when torque is applied air gaps become unequal and a current flows in the secondary circuit, which can be read on the galvanometer. An identical unit is fitted in the indicator box and by rotation of the handle the iron piece can be moved until the air gaps in the indicating unit are identical with those of the shaft unit. This restores the electrical equilibrium in the secondary circuit, because of opposed equal currents the galvanometer reads zero, and the amount of movement at the indicating box dial is indicative of the angle of twist restoration required and hence gives the angle of twist for the length of shaft between the two flange faces in the shaft unit.

By application of the meter constant and rev/s the shaft MW is thus determined.

The dynamometer

Consider the hydraulic type of dynamometer, as sketched in Figure 6.17.

The engine under test drives the shaft to which the rotor is directly coupled. The shaft bearings are inside the casing containing the stator, which is free to swivel on trunnion supports.

Each face of the rotor has pockets or cells of semi-elliptical or oval cross-section divided from one another by oblique 45° vanes, the stator is similar. Water enters at the stator inlet channel, entering between 45° vanes and passes into rotating rotor. The water is constantly circulated around the cells in a vortex action so the torque is transmitted from rotor to stator via the water. This torque tends to turn the stator, this action being resisted by a load-measuring device so that the resisting torque will equal the applied torque and is thus being measured.

$$Bb = mr$$

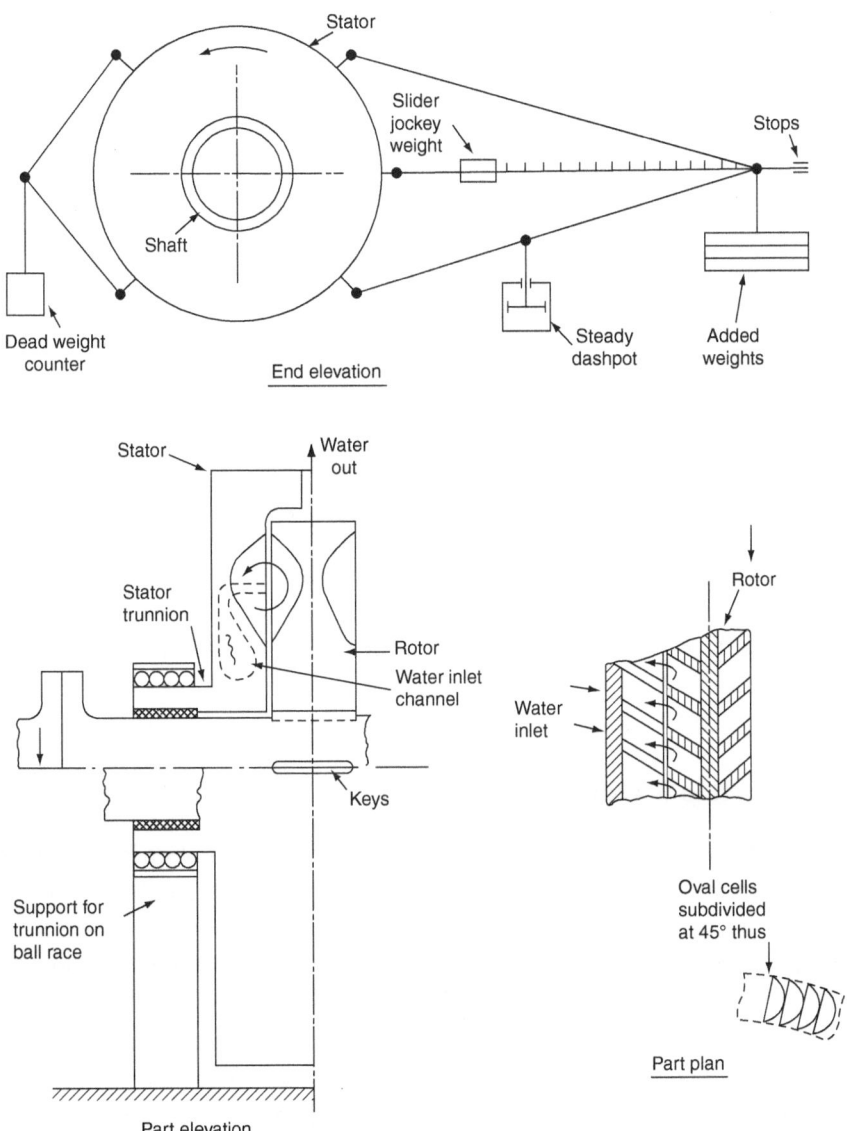

▲ **Figure 6.17** *Dynamometer principle*

For testing in both directions of rotation two rotors are provided: one used astern, the other ahead.

In modern practice the load measure devices are much simplified by the use of levers. In some designs resistance to motion is caused by a measurable electrical field coil resistance, which causes variation of eddy current resistance to rotor rotation, in place of the hydraulic resistance.

The thrust block

Modern types work on the Michell principle. The thrust of the collar is transmitted through the oil film and pads to the casing. The white-metal surface would be more likely to yield than the oil film at pressures as high as 500 bar (compressive yield of white metal say 560 bar = 56 MN/m²).

As seen in Figure 6.17, the oil scraper bears on the outer periphery of the thrust collar and delivers oil to the reservoir-stop from where it cascades on to the pads and bearings. The pads fit radially in the inverted horseshoe castings, pads being secured circumferentially by the stop. The castings back on to liners so ensuring location in fore and aft direction and fixing the clearance, which can be adjusted. The radial pivot line on the pad back varies from half to two-thirds of the pad width from the leading edge (see also Figure 6.17).

The lower-half casting acts as an oil reservoir sump, being provided with oil level gauge glass and a cooling coil. The total oil clearance is approximately 1 mm for say a 500 mm-diameter shaft. The wedges at base have a slow taper of about 20 mm/m and act to relieve the holding-down bolts of shear. The floors in the double-bottom tank below the thrust stool are closely pitched. Clearances are measured using wedges or hydraulic ram movement.

Michell thrust indicator

The standard block is modified so that a cast-steel shoe is replaced by a forged mild-steel one having a number of holes in the back making up interconnected oil cylinders. A hand pump, pressure gauge, piping and relief valve are provided. Under oil pressure from the hand pump the internally formed pistons move forward, so transferring gradually the thrust load from the liners to themselves, the thrust shoe now floats on pistons and pressure is read on the gauge. When half the axial clearance has been traversed the relief valve lifts thus preventing over-pumping. Astern thrust could be measured by a duplicate on the other side of the collar. The piston loading pressure is about 175 bar.

Ball and roller bearings

Wherever rolling friction, as with a wheel on a road, is substituted for sliding friction, as with a pin rotating in a journal, the frictional effects are much reduced. Such bearings are expensive, require minimum grease lubrication, must be sealed from dirt entry and grease escape and once overloaded are rapidly destroyed. However, they are shorter, allow more accurate shaft centring (negligible diametric clearance), have a virtually

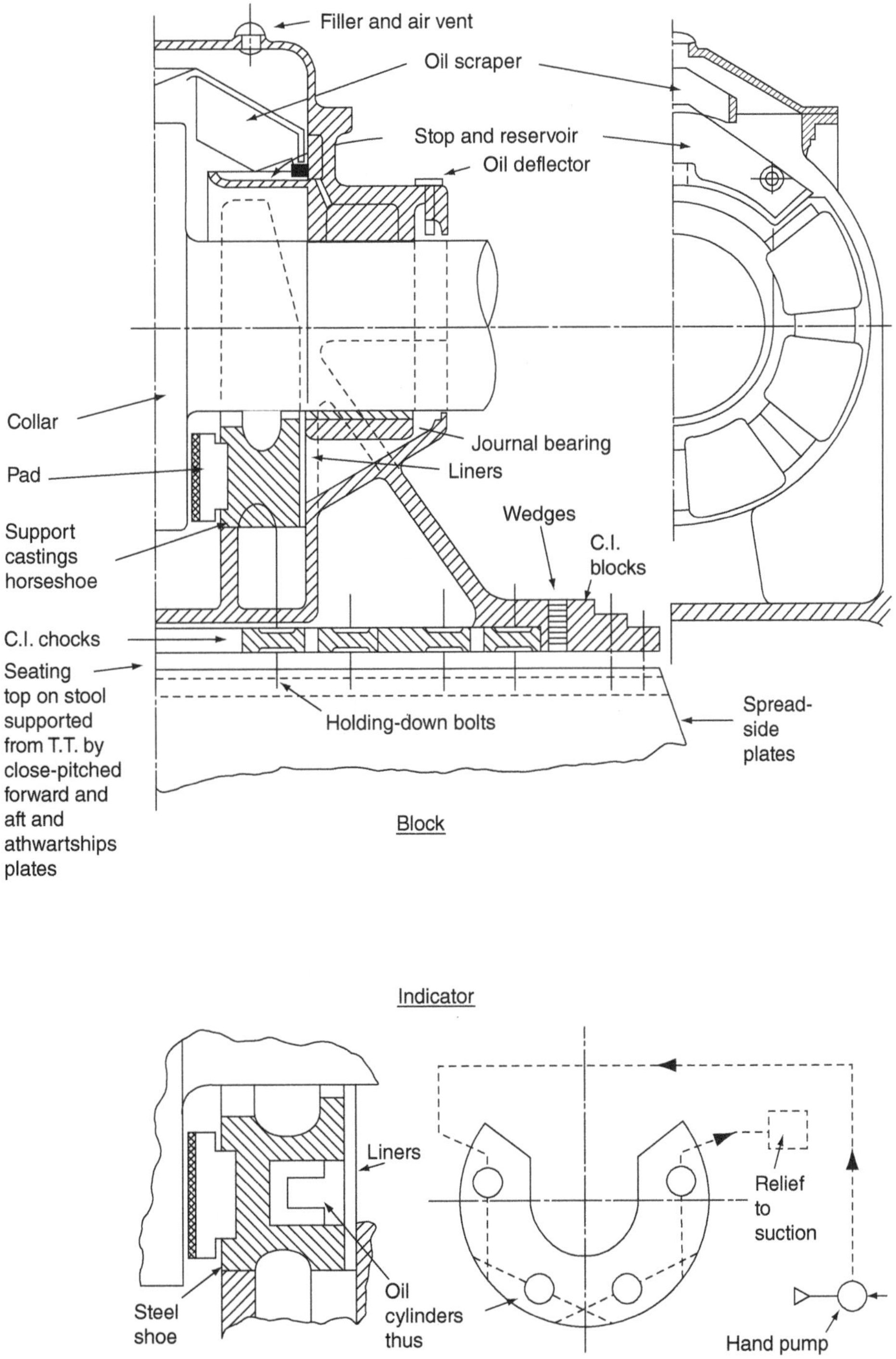

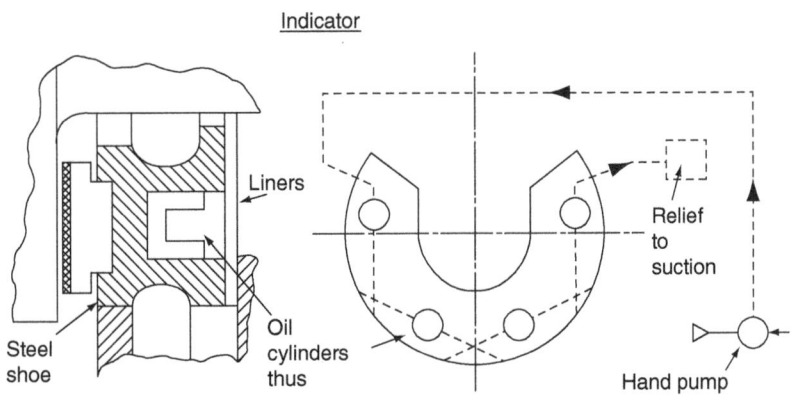

Filler and air vent

Oil scraper

Stop and reservoir

Oil deflector

Journal bearing

Liners

Wedges

C.I. blocks

Collar

Pad

Support castings horseshoe

C.I. chocks

Seating top on stool supported from T.T. by close-pitched forward and aft and athwartships plates

Holding-down bolts

Spread-side plates

Block

Indicator

Liners

Steel shoe

Oil cylinders thus

Relief to suction

Hand pump

▲ **Figure 6.18** *Thrust block*

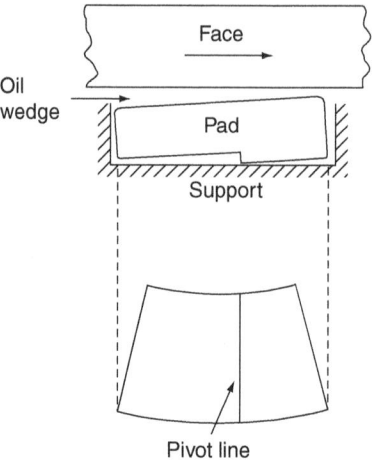

The pivot line should theoretically be nearer the outlet edge to coincide with the point of maximum pressure, in practice a central pivot is often satisfactory.

In place of a pivot line, a pivot-hardened stud is commonly used, especially on smaller types.

▲ **Figure 6.19** *Michell thrust pad design*

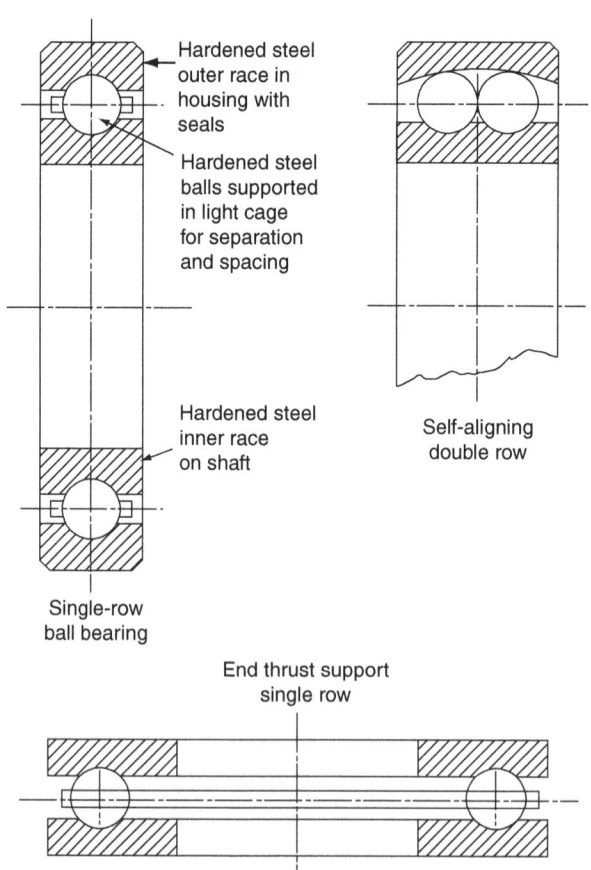

▲ **Figure 6.20** *Ball and roller bearings*

constant coefficient of friction of about 0.003 at all speeds (a journal bearing cannot match this figure until high-speed film lubrication exists) and can be self-aligning. Simple sketches of three types of ball bearing are given in Figure 6.20, and similar types of roller bearings are available, their main advantage being that greater bearing loads are possible for a given diameter.

Simple Balancing

This is a complex subject closely related to vibrations. Only some very simple fundamentals are discussed at this stage, which could be expanded a little further as a motorship problem when considering IC engines with reference to firing order, etc.

Single revolving mass

See Figure 6.21a. An unbalanced mass m at radius r, to balance introduce an opposite mass, B at radius b, in the plane of rotation such that:

$$B\omega^2 b = m\omega^2 r$$

that is, equal centrifugal force effects.

$$Bb = mr$$

Several revolving masses in one plane

See Figures 6.21b and 6.21c. Draw a mass moment (actual mass × radius) polygon and closing side giving Bb magnitude and direction.

Several revolving masses in different planes

See Figures 6.21d, e, f. A couple is a tendency to rock the shaft in its bearings in the form of an end-to-end turning moment.

Magnitude of the couple shown = Px (see Figure 6.21f)

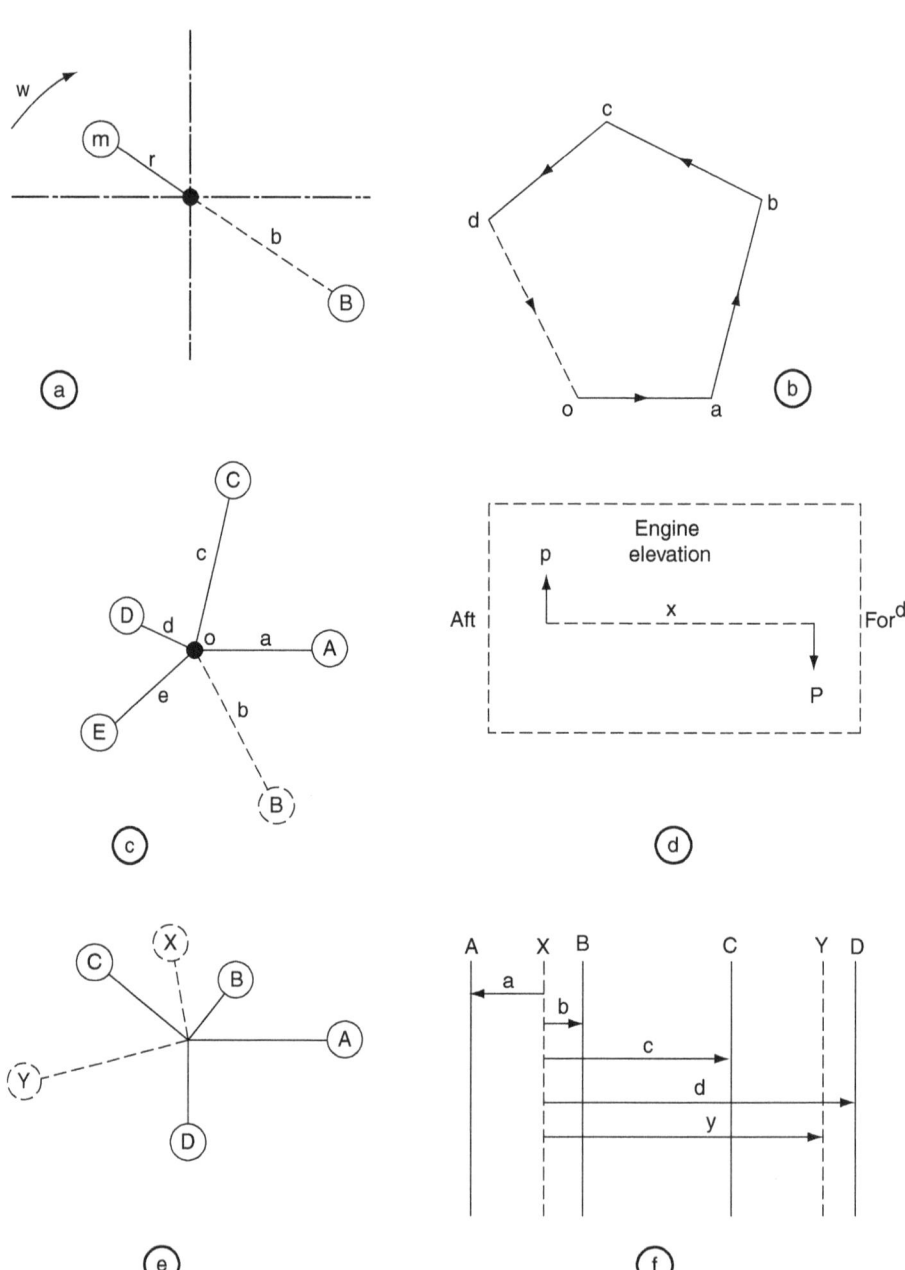

▲ **Figure 6.21** *Simple balancing*

Proceed as before and draw the mass moment polygon to find the unbalance mass moment as in Figure 6.20b. Assume it is necessary to add say two balance masses, having equivalent mass moment to that found, for convenience, at say X and Y, that is, one big mass to a particular point may not be convenient so the mass is split up (see Figures 6.21e and 6.21f). These two masses are also in different planes.

By using one of the planes, X or Y, as a fixed reference it can be fixed and then ignored, having no moment. Then a couple polygon or a tabulation is drawn for the other position. Thus the masses, radii and location of the planes for balance are determined.

Inertia force of a reciprocating mass

$$F = Fp + Fs + \cdots$$

that is, primary + secondary +... extended to further orders such as 4th, 6th, etc.

$$Fp \simeq ma\cos\theta$$
$$Fs \simeq ma\cos 2\theta/n$$

where m refers to the mass of the reciprocating parts, a acceleration of crosshead, n ratio of connecting rod to crank length, and θ the crank inclination to t.d.c. A single revolving mass cannot totally balance a reciprocating mass. However, partial primary balance can be attempted with the object of shifting the form of the unbalance to a more acceptable condition.

In-line engines

The reciprocating masses are often considered as the line of stroke component of equal mass at the crankpins. A force polygon (mass moments) on revolving forces can then be drawn and also a couple polygon, for each of primary, secondary, etc. Where possible, couple balance for some IC engines can often simply be arranged (for an even number of cylinders), by arranging one-half of the crankshaft as a mirror image of the other, firing order say 1, 3, 4, 2 (4 stroke).

Simple Vibration

General vibration

This subject is a compromise between complex mathematics and practical experience. Extensive use of computer technology is made in the actual designs and therefore only minimal details can be quoted here. There are three primary modes of vibration – transverse, longitudinal (axial) and torsional–but modern shafting is also built after consideration of both low-cycle and high-cycle fatigue.

Every vibration problem reduces to the solution of an equation of forces:

$$\text{Inertia} + \text{Damping} + \text{Spring} + \text{Exciting} = 0$$

Each force may be modified, that is, changing the size to alter inertia, providing oil type dashpots to amend damping, altering shaft stiffness to vary spring force and varying amplitude or frequency of an engine exciting force.

Consider the alternating current electrical analogy. For an R, L, C series circuit at a particular frequency (f) the inductive reactance X_L equals the capacitive reactance X_c. In this case the circuit behaves as a purely resistance circuit, that is, impedance equals resistance and current $I = V/R$ following Ohm's Law. The condition gives maximum current flow and the circuit at resonance is called an acceptor circuit (for an R, L, C parallel circuit it is rejector at resonance).

In the mechanical system the behaviour is identical, with inertia, damping and stiffness terms in place of inductance, resistance and capacitance. Resonance here gives severe vibratory forces and stresses. Resonant frequencies in electrical systems have their equivalents as critical speeds in mechanical engine systems. Variations to amend the four force terms given earlier can cause a critical speed to be moved away from a particular running speed.

Note the electrical resonance curve given in Figure 6.22.

Transverse vibration

This occurs in the athwartship direction with large reciprocating engines. It is usually due to cylinder pressure forces and inertia forces giving a resulting couple

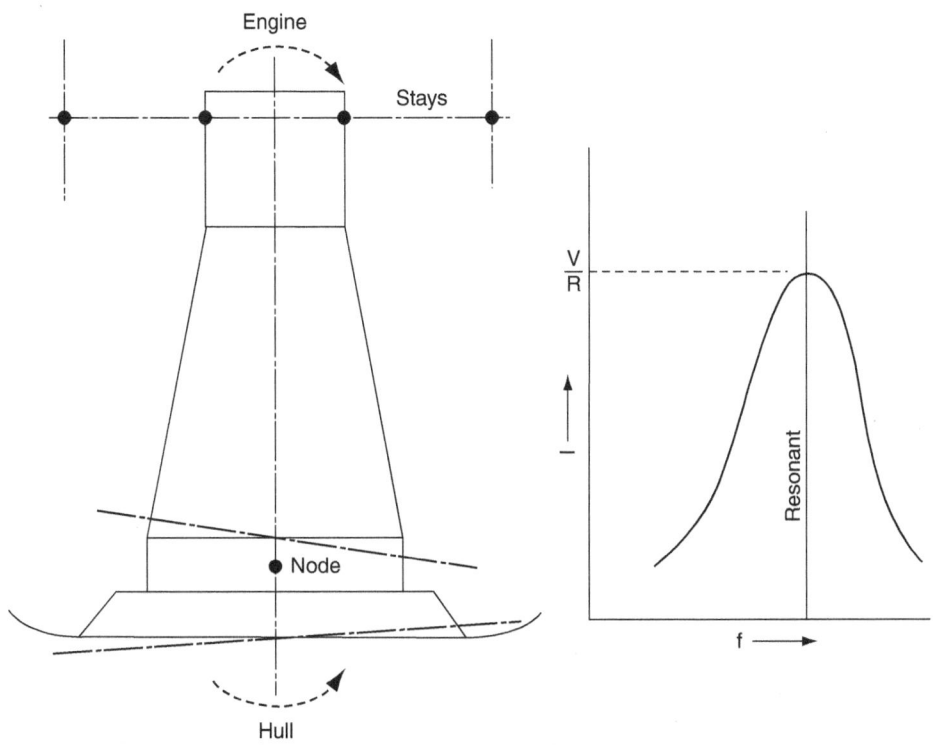

▲ **Figure 6.22** *Transverse vibration resonance curve*

about the engine crankshaft centre line and through the guide shoe. Propeller torque variations can increase or decrease this couple.

The usual solution is to stay the engine to the hull with lateral stays. Such stays must be connected to the hull by pins that would shear if the hull were distorted in collision. The hull attachment must be rigid – transverse deck beams are best. The stays must provide adequate and even stiffness to raise the resonant frequency above the service rev/s. When dealing with resonant frequencies inside the running range great care is required as minor stiffening can make the vibration worse. Doubling resonant frequency can quadruple exciting forces within the running speed.

Such vibration of the order of 1 mm can cause failure to pipes and welded joints as well as being most unpleasant in machinery and accommodation spaces. The rocking tendency can be seen from Figure 6.22.

Axial vibration

Some axial movements of amplitude ±2.5 mm have been noted, a movement of ± 1 mm can quite easily introduce crank bending stresses of 28 MN/m². Invariably these movements are propeller excited occurring as say 4th order vibrations (8 vibrations per second at 2 rev/second of shaft). In this respect a four-bladed propeller is causing the axial vibration with two blades passing the aperture every ¼ rev giving an axial pulse, the introduction of five-bladed propellers and more rigid thrust seatings have done much to reduce such amplitudes of vibration. Some experiments have been tried to utilise the principle of the Michell thrust indicator to introduce a dashpot damping effect.

Torsional vibration

A node is a point at which the shaft is undisturbed by vibration, that is, at the node the shaft can be imagined as clamped, the sections at each side vibrating opposite in phase but with the same frequency. One node gives one mode of vibration, two nodes two modes, etc., most shafting systems can be simplified to a one- or two-mode form, that is, first or second degree of vibration as at least a first assumption. This means for calculation the shaft system is considered as a three-mass system, engine in one, flywheel and propeller.

Only one serious critical occurs in the running range usually, for aft-end installations commonly above the maximum revolutions and for mid-ship installations commonly below, this being a broad generalisation. The two-node form is usually the decider in crankshafts, 9th order, two mode, at 18 vib/s (2 rev/s) while the one node is usually the decider in intermediate shafting 2nd order 4 vib/s and 4th order 8 vib/s, one mode.

Detuners and dampers

The object of the forward flexible flywheel (detuner) sometimes fitted is usually to shift the node to the crankshaft centre and reduce vibration. This flywheel causes variable stiffness with the torque variation and hence causes a change of natural frequency as load changes. Any tendency to build up torsional vibrations is reduced by frequency change and the amplitude settles down to a value below that of a rigid system. Thus this is really a flexible mass addition to the system, which is self-adjusting. Referring to Figures 6.23b and 6.23c, it is seen that the spring is supported over a long span at full loads and a short span at overloads. The spring connects the drive to the driven mass, which is thus floating.

Torsional vibratory stresses investigated for a motor ship may appear as in Figure 6.23d.

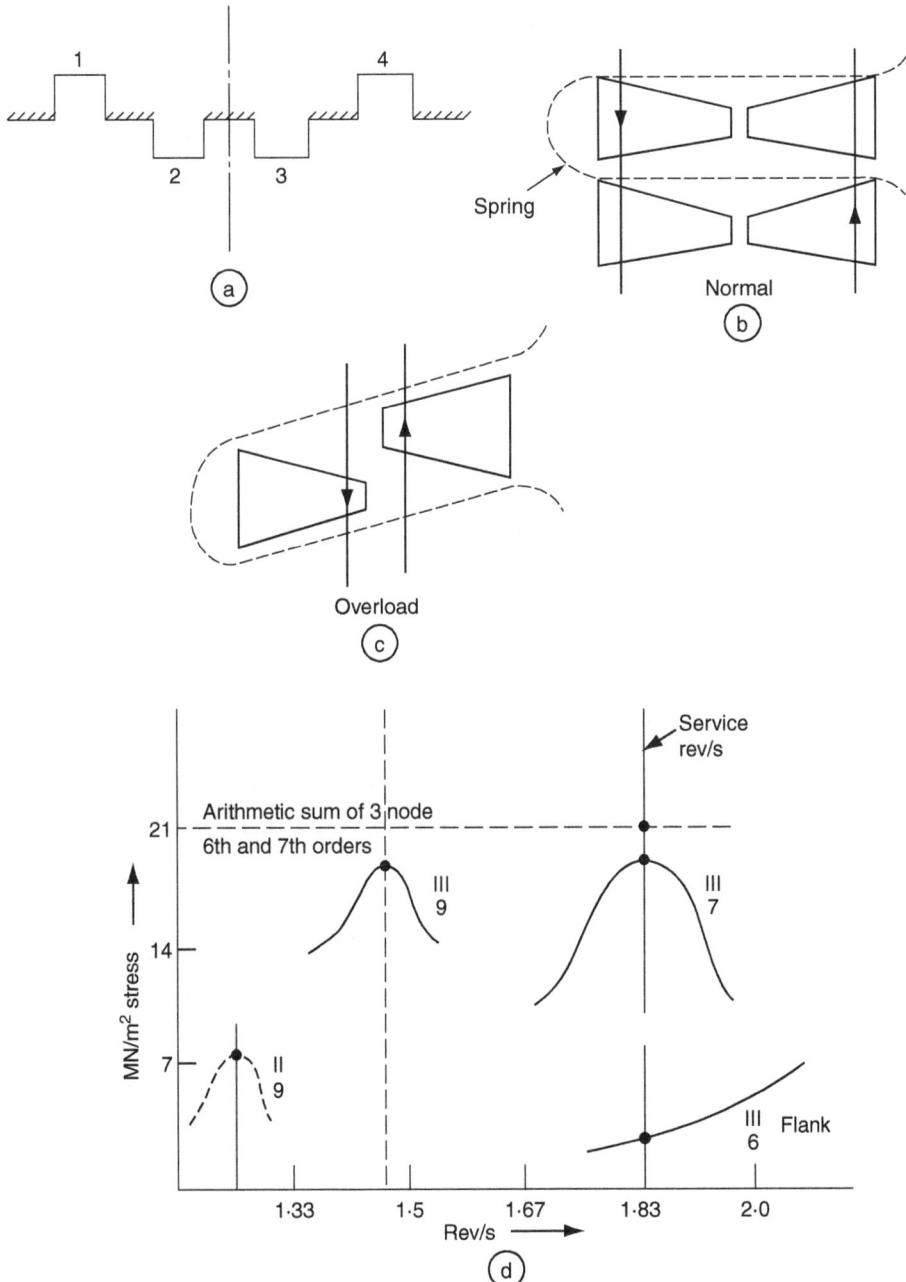

▲ **Figure 6.23** *Vibration characteristics (torsional)*

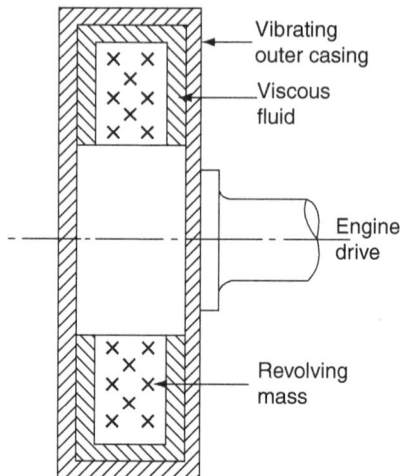

▲ Figure 6.24 *Torsional vibration damper*

Torsional vibration damper

A vibration damper will dissipate energy given to the system by exciting forces. This energy would otherwise be absorbed and appear as either strain energy at the nodes or kinetic (velocity) energy at the antinodes. Figure 6.24 is the diagrammatic principle of a 'Holset' type of damper in which the damping fluid is between the driven casing and revolving mass. Such a damper may be fitted at an antinode.

Torsional vibration eliminators often utilise damping and detuning, for example, in magnetic slip couplings, spring and centrifugal friction coupling, clutch drives, etc.

Torsional elastic curves

Figure 6.25 illustrates a four-cylinder engine flywheel and propeller system, which as a first simplification can be reduced to a three-rotor torsional system as shown. For known moments of inertia the critical speeds can be evaluated, the elastic line is drawn to represent the amplitude of vibration on a length scale. Intersection of elastic line and shaft axis gives the position of the node or nodes depending on the degree of vibration, that is, a single rotor placed at *N*, or two rotors placed at the two positions shown for *N*. Higher degrees of vibration exist.

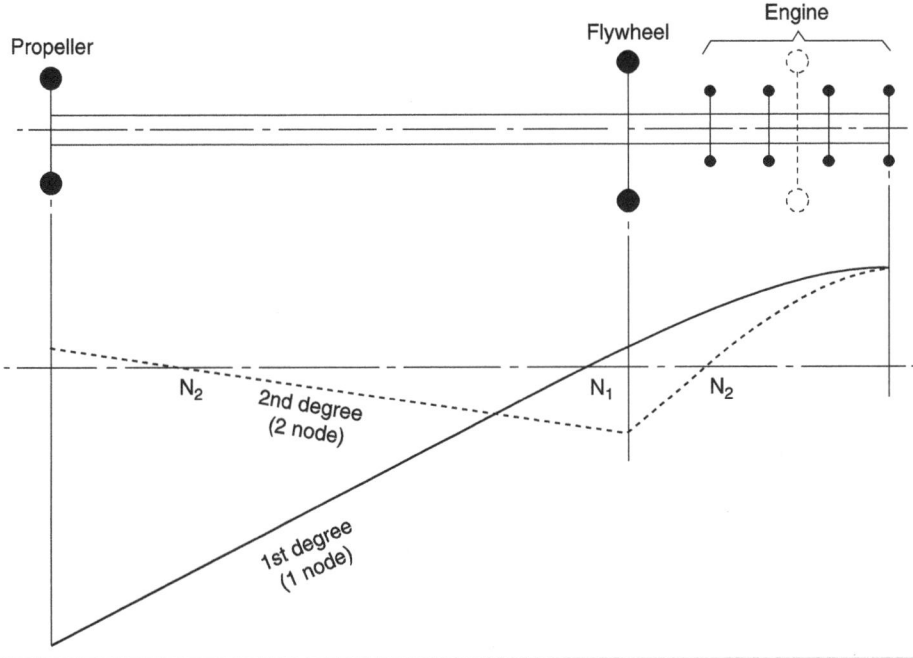

▲ **Figure 6.25** *Torsional elastic curves*

Aft-engined vessels

Modern vessels have higher powers hence giving greater magnitude exciting forces. Welded structures give less damping than riveted structures. Propeller excited forces are usually the result of insufficient hull-propeller clearances and incorrect blade form sometimes causing excess cavitation, which in turn can be the cause of the vibration. To cause hull vibration the excitation force would act on or near the antinode. The main antinode is usually at the aft end so that large, aft-engined reciprocating units are most prone to vibration effects, that is, large motor tankers.

Low- and high-cycle fatigue

Low-cycle fatigue and torque reversal will be calculated by the classification societies and the information used to establish criteria for the manufacturer of suitable propeller shafting. This criteria is applied to shafts that are subjected to load cycles of less than 10^4 during its lifetime and the causes are:

- load variations (zero to full load).
- clutching-in shock loads.

- electric motor start-up or start-delta shift
- ice shock loads
- crash stops

The design of any shaft must take into account the fact that a failure will have a major effect on the safety of the vessel by removing the power generation of the propulsive power for the vessel.

High-cycle fatigue (HCF) criteria will be applied to the construction calculations for a shaft that is subjected to more than 10^6 load cycles. The shafts will be experiencing load cycles in the region of 10^9 or 10^{10} cycles. It is based on the combined vibratory torsional and rotating bending stresses (axial stresses disregarded) relative to the respective component fatigue strengths. Typical HCF load conditions are listed below:

- Load variations due to torsional vibrations applicable for continuous operation caused mostly by engine firing pulses, and rotating bending moments due to forces from gear mesh and forces as listed in the Rules for Classification of Ships/High Speed, Light Craft & Naval Surface Craft (Pt. 4, Ch. 4, Sec. 1, F300 1).
- Load variations (torque and bending moment) due to ice impacts on the propeller for ice-classed vessels assumed to accumulate ~106 load cycles, see the Rules for Classification of Ships (Pt. 5 Ch. 1).

Classification societies have sophisticated computer modelling software to determine both the position and alignment of the propeller shaft and the size and shape of the shaft. The DNV software is called Nauticus and they have the Nauticus Machinery range, which is used to complete the shaft line design in a vessel.

Gearboxes

The most efficient propellers for any given situation are as large as possible, turning slowly with a pitch and blade profile to suit the thrust required. A main engine with enough output power to drive the propeller can be selected but depending upon other vessel design criteria the engine speed might not be correct to drive the propeller. For example, efficient propeller speeds are in the range of 100–160 r.p.m., but if a medium-speed engine is required then that will have a speed range of 450–600 r.p.m. A reduction gear of 5:1 would reduce the 550 r.p.m. of an engine down to 110 r.p.m. of a suitable propeller.

Electric drives and pods

Modern materials and design techniques, together with visionary thinking about the layout of ships' steering/propulsion systems, has led to arrangements that are moving away from the conventional designs of the past.

The electric 'podded' drive systems have become very popular on passenger ships with Rolls-Royce and ABB leading the way in the development of these units. Electric-driven pods and thrusters have also made significant gains in the propulsion and manoeuvring of vessels equipped with dynamic positioning (DP) ability and tug boats. The advancement has been in the development of variable speed AC electric motors (see Volume 12 of the Reeds series for more details). This means that the diesel propulsion motor can be operated at an economical speed for longer periods of time and without much of the 'inefficient' transient load operation that is common on vessels with more traditional arrangements.

ABB have now also introduced the DC 'transmission' grid, which means that the diesel engine can run at an economical speed/loads for even longer periods of time as the transmission system does not need a predetermined AC frequency. An added advantage is that other power sources such as batteries or solar cells can be plugged straight into the distribution system.

The podded drive arrangement – Azipods from ABB, Mermaids from Rolls-Royce and eSiPOD drive systems from Siemens/Schottel – gives the advantage of improved 'hydrodynamic hull characteristics' due to the cleaner lines of the hull causing less disturbance to the flow of the water across the hull.

A significant advantage of the podded drive is at 'manoeuvring' speeds where the turning force can be up to 200% more than with a vessel fitted with the conventional 'rudder'-based system.

The reason for this being that the turning force produced by the rudder depends upon the flow of water across the face of the rudder whereas the podded drive is generating its own thrust.

The disadvantage from this form of propulsion/steering is that the units hang below the hull of the ships and can therefore be damaged 'significantly' if they come into contact with hard objects external to the vessel.

Another issue to consider with 'pods' is that they place a significant loading on the bearings inside the casing to the drive unit. These bearings must take both the weight and the thrust from the action of the propeller and prolonged operation at low speed can mean that the 'hydrodynamic lubrication' – see page 470–472– that would normally be set up with the rotation of the shafting remains thin and results in possible damage to the bearings.

7

REFRIGERATION AND AIR CONDITIONING

J&E Hall in the UK were one of the first companies to see the potential of refrigeration on ships back in 1880 and now the industry is just at the end of another period of significant change. The changes are due to the large increase in the cruise ship market as well as the LNG carrier market, both of which are some of the biggest users of marine refrigeration and air conditioning systems.

It is also now known that chlorofluorocarbons (CFCs) that were developed during the last half of the last century, and used in refrigeration systems, are very damaging to the environment. As a consequence, the use of these refrigerant gases has now been phased out in most parts of the world. CFCs are ozone depleting substances (ODS) – see below – and the gas cannot be purchased any more as its production has stopped. However, alternatives are being used in new and existing systems and these alternatives have properties and characteristics that students should learn about.

Ozone Depleting Substances

Ozone is made up of three oxygen atoms joined together. This is slightly different to the normal two oxygen atoms that form the bulk of the free oxygen that exists in the atmosphere. The bulk of atmospheric ozone lies in the stratosphere, which is approximately 10–25 miles (15–40 km) above the surface of the earth. It carries out the important function of reflecting, back into space, some of the ultraviolet (UV) radiation that has come from the sun. Any reduction in this protective layer would mean that more harmful radiation would reach the surface of the earth.

Ozone is made and destroyed by the natural processes that are part of nature; however, this destruction had accelerated since the development, and subsequent release into the atmosphere, of CFCs. The natural production of ozone has not been able to keep up and the total amount of ozone has reduced.

CFCs are a very stable chemical that are not broken down by water and other chemicals in the lower part of the atmosphere. Therefore, CFCs remain intact until they reach the stratosphere where they are struck by intense radiation from the sun. The CFC finally decomposes under the sunlight and the chlorine released interacts with the ozone leading to its destruction. The chemicals that are most potent at reducing the ozone in the atmosphere are CFCs, carbon tetrachloride, methyl bromide, methyl chloroform and halons. Hydrochlorofluorocarbon (HCFC), which is a compound composed of hydrogen, chlorine, fluorine and carbon atoms, is being used as a temporary replacement for CFCs but even this is not allowed as a permanent replacement. The introduction of hydrogen means that the compound breaks down in the lower atmosphere before it is able to reach the stratosphere where the ozone is situated.

Basic Principles of the Refrigeration Cycle

Ice-water-steam phase changes

Most refrigerants have a vapour phase that demonstrates similar characteristics and properties to steam, except that the medium has a much lower boiling point. Therefore, we can visualise the refrigeration cycle by considering the changes that occur when 1 kg of ice at say −23°C is converted into superheated steam at say 1.013 bar, 200°C.

These temperature-heat energy changes are best illustrated graphically (Figure 7.1).

1. Heat added to raise the temperature of 1 kg of ice at −23°C to 0°C = $1 \times 2.094 \times 23$ = 48.2 kJ. (Specific enthalpy of solid ice is −48.2 kJ at −23°C, i.e. h_1.)

2. Heat added for fusion of 1 kg of ice at 0°C to 1 kg of water at 0°C= $1 \times 333 = 333$ kJ. (Specific enthalpy of fusion for ice is −333kJ, i.e. h_{if}.)

3. Heat added to raise the temperature of 1 kg of water at 0°C to saturation temperature (t_s) 419.1 kJ. At 1.013 bar, $t_s= 100°C$ and $h_f = 419.1$ kJ (see Figure 7.1). (This is liquid specific enthalpy for water.)

4. Heat added for vaporisation of 1 kg of water at 100°C to 1 kg of steam at 100°C (i.e. constant t_s) is $h_{fg} = 2,256.7$ kJ (2.257 MJ) (see Figure 7.1).

   (This is specific enthalpy of vaporisation for steam.)

5. Heat added to superheat 1 kg of steam from 100°C (t_s) to 200°C $(t) = 299.2$ kJ (see Figure 7.1).

   (This is specific enthalpy to superheat above dry saturated steam or dss.)

   Degree of superheat = (200–100) = 100°C.

6. Heat added to change water at 0°C to dss at 100°C = 2,675.8 kJ (2.676 MJ). Also see the table in Figure 7.1. (This is specific enthalpy h_g for dry saturated steam vapour.)

Note: The descriptions at number (2) and number (4) required the addition of heat energy but the temperature remains the same. This is due to the energy being required to drive the chemical phase change from ice to water.

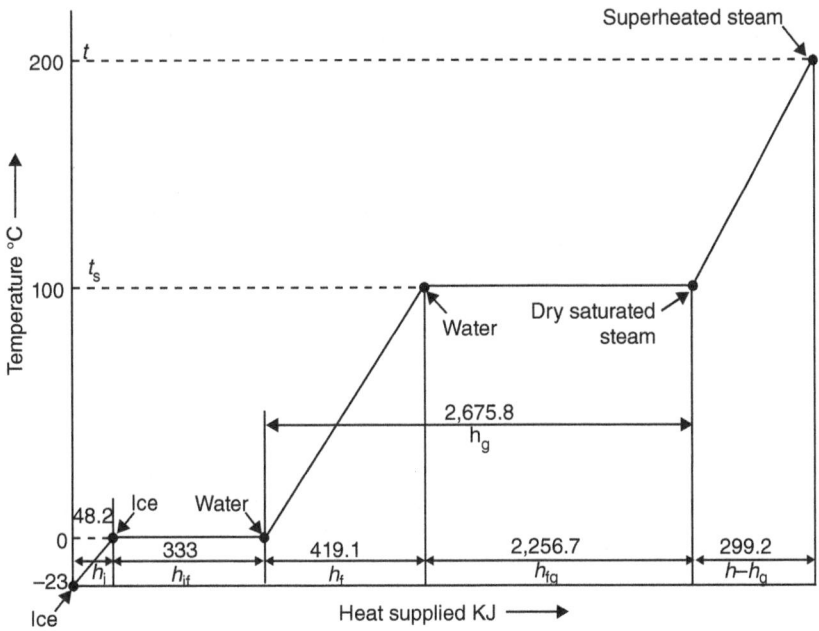

▲ **Figure 7.1** *Ice-water-steam phase changes*

The given diagram should clarify the use of terms such as specific enthalpy of vaporisation and superheat, used at a later stage when considering the properties of refrigerants. Normally, if the specific heat capacity (c_p) is constant (with temperature), then the specific enthalpy with no phase change is given by:

Heat exchange $Q = m \, c_p \, (t_2 - t_1)$, where m is the mass.

The reference datum for water is taken as 0°C (273 K), due to the phase (state) change at that point, but for refrigerants it is often taken at –40°C, this having no significance apart from the fact that it reduces the use of negative numbers (absolute zero and the Kelvin scale does the same).

The term for heat energy (liquid or vapour) is enthalpy. This is a measure of the quantity of heat that a solid, liquid or gas is holding at a given temperature. The heat per kilogram, as energy, is specific enthalpy (the specific heat capacity of ice is 2.094 kJ/kg and the specific enthalpy of fusion of ice is 333 kJ/kg).

A subcooled or undercooled liquid is liquid existing at a temperature lower than the saturation temperature for that pressure while a liquid exactly at saturation temperature is a saturated liquid, for example, water at atmospheric pressure is a subcooled liquid at 77°C and a saturated liquid at 100°C.

Wet saturated, dry saturated and superheated vapour refers to the degree of heat saturation; wet vapour has dryness fraction (or quality).

Specific volume is volume occupied by 1 kg of liquid or vapour in m³ (reciprocal of density).

Refrigerants

Desirable properties of a refrigerant

1. Low boiling point (otherwise operation at a high vacuum becomes necessary).
2. Low condensing pressure (to avoid a heavy machine, plant scantlings and to reduce the leakage risk).
3. High specific enthalpy of vaporisation (to reduce the quantity of refrigerant in circulation and lower machine speeds, sizes, etc.).

4. Low specific volume in the vapour phase, which reduces the size of the plant required and increases efficiency.

5. High critical temperature (temperature above which vapour cannot be condensed by isothermal compression).

6. Non-corrosive and non-solvent (pure or mixed).

7. Stable under working conditions.

8. Non-flammable and non-explosive (pure or mixed).

9. No action with oil (the fact that most refrigerants are miscible may be advantageous, i.e. removal of oil films, lowering pour point, etc., provided separators are fitted).

10. Easy leak detection.

11. Non-toxic (non-poisonous and non-irritating).

12. Cheap, easily stored and obtained.

Refrigerant gases

Refrigeration systems depend upon using a suitable gas that will circulate the system, be pressurised and expand and therefore transport heat from one place to another. When scientists and engineers first developed a gas most suited to this task they were totally unaware of the side effects that would occur if and when the gas was released into the atmosphere.

When scientists did start to become aware of the changes happening in the earth's atmosphere they started looking for causes and found that the substances contained with the refrigeration gases had a significant part to play in the detrimental effect on the atmosphere.

The United Nations set up the Environment Programme and following the Vienna Convention the 'Montreal Protocol' was developed in 1987, and they both gained 'Universal Ratification' on 16 September 2009.

197 nations have signed up to the protocol and are now working towards helping to protect the world's layer of ozone.

Given the situation described about the effect that some refrigerants have on the atmosphere, the United Kingdom has made reference to IMO's MARPOL Annex. VI and to the Montreal Protocol in their notice to mariners – MSN 1819 (M+F).

The M Notice explains the changeover process where Ozone Depleting Substances (ODS) gases are not prohibited from use in systems that were built before May 2005, which is the date that MARPOL Annex. VI came into force. New installations that operate on HCFC are permitted until 1 January 2020. However, already in force – from 1 January 2010 – is the EU regulation that prohibits the use of virgin HCFC in the maintenance of any system.

Therefore, only recycled gases were permitted to be used from 1 January 2015 and the use of all HCFC gases for maintenance purposes were not allowed. HCFCs could still be used in existing installations so long as there is no maintenance required for the system. Recycling means that recovered HCFCs have been placed through a filtering or moisture-removal process. The recycled gas must not then be placed on the open market; they can only be used by the recycling company for the maintenance of existing systems.

The most commonly used ODS refrigerants are given below:

- CFCs: R11, R12 and R502;
- HCFCs: R22;
- HCFC blends including: R401a, R402a, R403a and R406a.

The common trading names for refrigerant gasses are Arcton, Freane, Freon, Isceon, Solkane and Suva. Due to the restrictions in the use of CFC and HCFC as a refrigerant, one of the alternatives is ammonia (NH_3) because it is environmentally friendly. It is not, however, completely human friendly and great care must be taken when handling ammonia or using an ammonia refrigeration plant. Smaller plants also have problems finding compatible lubricating oils.

Properties of early refrigerants

No refrigerant has all the desirable properties, each one having various advantages and disadvantages. Refrigerant R12 being a CFC gas is not currently used, but its properties are included here to allow comparisons to be made with the latest replacements.

Ammonia as a gas in refrigeration plants has found favour in some new fishing vessels where the fish are processed on board and frozen very soon after catching; CO2 is also a non-ozone depleting substance and an alternative that is allowable for new plants, although very few marine plant are being fitted to ships due to the difficulties of working with a CO2 plant. Properties for these gases are as follows:

Refrigerant R12 (CCl_2F_2) (dichlorodifluoro methane)

Carbonic anhydride (CO_2) (termed carbon dioxide)

Anhydrous ammonia (NH_3) (termed ammonia)

Properties of refrigerants

The properties of the refrigerants are given in the refrigerant table (Table 7.1) and by careful analysis the advantages and disadvantages can be weighed against each other and a choice made depending on preference, experience, properties, conditions of use and now of course the global warming potential (GWP).

Various organisations produce tables giving the thermodynamic properties of the different refrigerant gases. These result in extensive amounts of data to consider and therefore in order to compare refrigerants under various conditions of working it is useful to review the differences under the same conditions.

Under test conditions these will be the pressures corresponding to saturated vapour temperatures of −15°C at compressor intake and 30°C at compressor discharge; 5°C subcooling and 5°C superheating fix the final refrigerant temperatures used at intake to expansion valve and to compressor respectively.

For the comparison here, simple conditions of −15°C and 30°C are used, discounting subcooling and superheating, and the given properties are based on the Tables in 7.1 and 7.2.

There are now so many different versions of the new gases that it is difficult to make direct comparisons from just looking at the different properties of the gases. There are additional parameters to consider such as:

- compatibility with standard lubricants;
- compatibility with other materials in the systems, such as polymers and other plastics;
- temperature glide (see page 302);
- available supplies of a 'non-contaminated' gas (some refrigerated containers have been found to contain contaminated R134a).

Further analysis for some of the principal gases is also given over the next few pages. R12 is given for comparison and R407c and R134a have been added as these are currently highly popular replacements for the older 'banned' gases.

Table 7.1 *Comparison of refrigerants*

Property	CFC (R12)***	Carbon dioxide (R744)	Ammonia
Chemical symbol	CCl_2F_2	CO_2	NH_3
Discharge pressure, bar	7.4	72	11.7
Suction pressure, bar	1.8	23	2.4
Critical pressure, (bar)	40	73.8	113.7
Critical temperature, °C	112	31	133
Specific enthalpy of liquid, kJ/kg, at −15°C	22.3	48.9	112.4
Specific enthalpy of vapour, kJ/kg, at −15°C	181	323.6	1,426.6
Specific enthalpy of liquid, kJ/kg, at 30°C	64.6	193.8	323.1
Specific enthalpy of vapour, kJ/kg, at 30°C	199.6	266.9	1,468.9
Specific volume of liquid, m^3/kg	0.0007	0.001	0.0015
Specific volume of vapour, m^3/kg at −15°C	0.093	0.017	0.51
Boiling temperature at atmospheric pressure	−30	−78	−33
Specific heat capacity of the liquid, kJ/kg °K	0.96	3.23	4.65
Corrosive	No	No	No
Toxic	No	No	No
Flammable	No	No	Yes
Explosive	No	No	Yes
Cost	Was fairly expensive	Cheap	Cheap
Leakage test	Hallide torch	Soap and water	
Compatible with mineral oils	Yes	No	Possibly
Global Warming Potential	10,050	1	100
Cost	Fairly expensive	Cheap	Cheap
Leakage test	Hallide torch	Soap and water	Wet litmus
Miscible with oil	Yes	No	Slightly

[1]NH_3 is very stable in water, highly soluble, for example, 1 m^3 H_2O absorbs 900 m^3 NH_3. Ammonia is also corrosive to brass and bronze if water is present.

[2]Liberates toxic phosgene gas from the fire.

Carbon dioxide (CO_2)

Advantages

Low boiling point, low specific volume (hence low volume flow rate), cheap, non-explosive, non-flammable, non-corrosive, non-toxic, but this medium has a very low GWP compared to R12.

Disadvantages

Very high pressures (hence heavy construction and careful joint attention is required), low specific enthalpy of vaporisation (hence high mass flow rate), low critical temperature (reduces plant efficiency at higher sea temperatures), rather low comparable efficiency, etc.

Not all the properties can be fully analysed from Table 7.1 but a sufficient number of properties are given to enable a good comparison between the refrigerants.

CO_2 is especially being used in the first stage of a cascade system or as a secondary coolant. Danfoss have been making industrial CO_2 refrigeration units for over 15 years now. Emerson climate technologies have started making 'Copeland scroll' compressors for use with low-temperature subcritical CO_2 cascade cycle application such as food storage units. These units would be ideal for use on passenger ships for cold food cabinets.

CFC refrigerants

Refrigerant R12 was the most popular CFC marine refrigerant suitable for low temperatures without negative evaporator pressures, that is, in vacuum, and R502 was used for hermetically sealed systems, that is, integral gas-tight motor and compressor. The main advantage claimed was an improved refrigerating effect for a given size of machine. It can be noted again that there is no ideal choice even in the CFC group of gases as there are advantages and disadvantages for all the different refrigerants.

HFC 407c and 134a

These are examples of the new HFC gases (replacements for the older CFC refrigerants):

R134a is the chemical 1,1,1,2,-Tetrafluoroethane. It is heavier than air and will therefore concentrate in bilges or other spaces low in the ship, if a leak occurs. It is also colourless and only has a slight odur, making it difficult to detect if released into the area within the ship. The gas is commonly used in air conditioning systems and refrigeration units attached to shipping containers.

It is not 100% compatible with the common mineral lubricants used in the past and adjustments and component changes need to be completed to a system that used to run with R12.

R407c is a blend of three different (HFC) gases, these are:

- R32 (CH_2F_2) at 23%;
- R125 (CHF_2CF_3) at 25%;
- R134a (CH_2FCF_3) at 52%;
- As an HFC gas R407c is not ozone depleting, however it does contribute to global warming and therefore should not be vented to atmosphere. If maintenance to the systems is required, then the gas should be isolated within the systems or recovered before the work is undertaken. Temperature glide must also be taken into consideration with this gas and any re-gassing must be completed to the gas manufacturer's recommendations.

Table 7.2 allows a quick comparison between the now banned CFC, the temporary replacement HCFC and the longer-term replacement HFC.

Table 7.2 *Properties of refrigerants*

Refrigerant	Mass	Formula	Boiling point	Freezing point	Critical temp	Critical pressure	Liquid density	ODP	GWP
			°C @ 1 atmos.	°C @ 1 atmos.	°C	Kpa (abs)	kg/m3		
CFC									
R-11	137.37	CCl3F	23.7	-111.1	198	4,408	1,447	1	3,800
R-12	120.91	CCl2F2	-29.75	-160	112	4,136	1,486	1	8,100
HCFC									
R-22	86.468	CHCLF2	-40.81	-160	96.1	4,990	1,413	0	1,500
HFC									
134a	102.03	CH2FCF3	-26.06	96.67	101.08	4,060.3	1,206	0	3,260
Ammonia	17.02	NH3	-33.33	77.72	132	11,333	682	0	0
CO$_2$	44.01	CO2	N/A	-56.67	31	7,377	1,032	0	1

http://encyclopedia.airliquide.com/Encyclopedia.asp for more information about the properties of gases

Therefore, after 1 January 2015 it was no longer allowable to service the HCFC systems. There are some alternatives:

- Replace the system using non-ODS refrigerant. The new refrigerant could be a suitable, chlorine-free HFC that is compliant with the regulations or it could be a 'natural' gas such as hydrocarbons, ammonia or carbon dioxide;
- Convert the system using one from a number of replacement HFC gases. The one that is chosen will need to be compatible with the lubricating oil currently in use or that will need to be changed as well;
- Depending upon the service and the components in the system some equipment may need changing or upgrading. There may also be a loss of performance as the replacement gases are a blend where the gases may have different boiling points, a feature known as 'temperature glide'. This could be as much as 7.5°K.
- The third option is to take no action but this is only an option if the system is operating on guaranteed stock of recycled gas, or if the system will run without maintenance and any failure will not cause a business-critical shut-down until the refrigeration system is replaced.

The UK M Notice MGN1819 draws attention to the SI 2002 No. 528 but the latest information can be found in the F-Gas regulations. Tables 7.1 and 7.2 give a good indication of the differences in the properties between the old and the new refrigerant gases.

The Vapour Compression System

Heat transfer

The refrigeration circuit acts as a heat pump by moving the heat from one area to another. The heart of the process is the fact that if the pressure of a fluid is lowered then the volume increases but also the temperature drops. The pressure reduction is completed by the expansion valve and carries on as it passes through the evaporator where it absorbs the heat from its surroundings.

Work is carried out on the fluid to raise its pressure again by the compressor and the condenser removes the heat at the higher pressure. The fluid moves to the expansion valve where the process starts again.

Operating cycle

Where a refrigeration system is critical for the safe operation of the vessel or protection of the cargo or crew such as with the provisions refrigeration system or the LNG cargo refrigeration system, marine practice would require complete duplication of all units. If the system was risk assessed as business critical it would also be singled out for special attention in the maintenance system.

A typical refrigeration system is shown in Figure 7.2 and would undertake the following steps during the complete cycle.

The refrigerant in its vapour phase is discharged from the compressor at 90°C and is condensed in the condenser at a condensation temperature of 25°C. For good heat energy transference rate a temperature differential of about 8°C between cooling water inlet and condensation temperatures is usual. The condenser gauge registers condensation (saturation) pressure and corresponding saturation temperature (t_s) on a dual scale. Some undercooling will occur in the condenser under standard conditions, typically this will be 5°C so the liquid will leave at 20°C. The liquid now passes through the expansion valve where it is released to the desired vaporisation pressure determined by the evaporator outlet temperature. Some flash-off of the liquid to vapour will also occur but the greater the undercooling the less will be the flash-off percentage. This flash-off represents a loss, as any vapour formed before the evaporator will not extract heat from the brine, giving a resultant loss of refrigeration effect. Ideally, the fluid should be totally wet entering the evaporator and just dry leaving the evaporator, which means full absorption of heat happens in the evaporator and the heat is extracted from the brine and not from the associated pipework leading to the evaporator. In an ideal system superheating is not advantageous, but due to the practicalities of system design the assumption is that if the vapour is superheated at entry to the compressor then it will have no liquid present to damage the compressor and there will be a higher degree of superheat upon leaving the compressor.

The vapour now leaves the evaporator under assumed standard conditions having 5°C of superheat. The evaporator gauge registers vaporisation (saturation) pressure and corresponding saturation temperature (t_s) on a dual scale. The actual superheated vapour temperature as read from the thermometer being −10°C. For good heat transference rate, a temperature differential of about 5°C between brine outlet and vaporisation temperatures is usual.

The vapour now enters the compressor to start the circuit again. It should be clearly understood that Figure 7.2 refers to *one set of definite conditions,* that is, a sea temperature

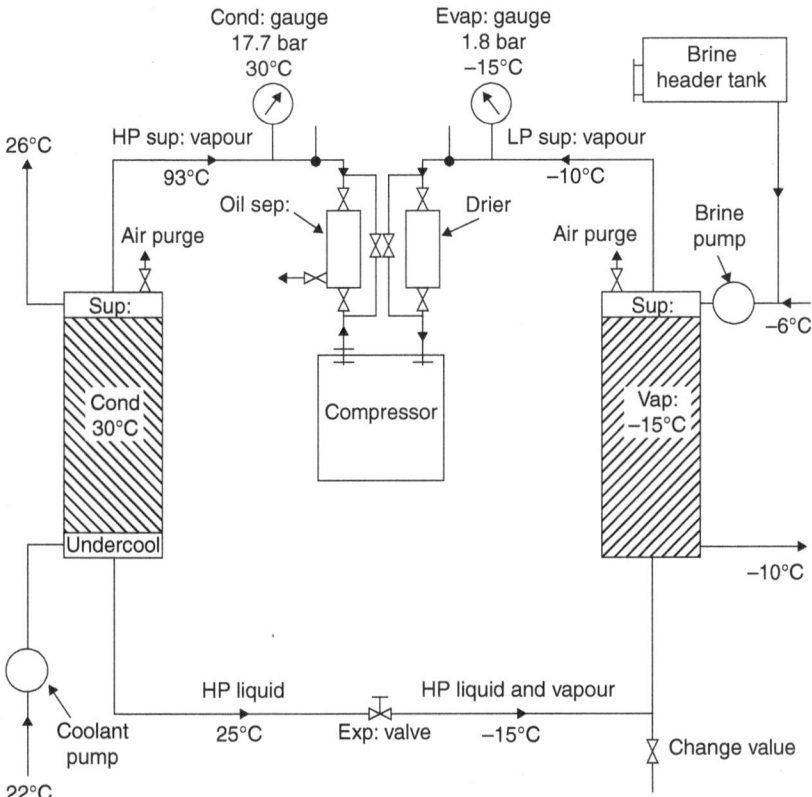

▲ **Figure 7.2** *The vapour compression system*

of 22°C and a particular expansion valve setting that determines that condition. Variations of sea temperature or vaporisation pressure would indicate completely different readings but the basic temperature differentials of 8°C and 5°C should still exist.

The compound pressure gauge shown in Figure 7.3 illustrates the dual scale of saturation temperature and pressure. Ammonia is shown for illustrative comparison purposes only, that is, normally only one refrigerant pressure and temperature is on the scale. Commonly all readings are taken in temperatures only. Correct differentials are an indication of correct working with sufficient vapour charge.

Under correct running conditions the compressor discharge pipe should be fairly hot to the touch and the suction pipe should be just frosting up near the compressor. The compressor discharge and suction lines are commonly provided with cross-over valves in addition to the stop valves. These valves allow the pumping out of the high-pressure side to the low-pressure side for overhauls and allow an easy discharge for starting. Refrigerant is added, with the machine running normally, at the charging position.

Many of the circuits employ a liquid receiver after the condenser and CO_2 types commonly have intermediate liquid cooling receivers. The capacity of a liquid receiver is usually sufficient to cover the outlet to the liquid line. Methods of control of the flow of refrigerant are (a) low-side float, (b) high-side float, (c) hand manual control, (d) capillary, (e) direct expansion with constant pressure, (f) direct expansion with constant superheat. These are discussed later.

The system should always be kept clear of water, air and dirt. Appropriate filters are fitted in the systems and these should be checked on a regular basis.

The correct operation of the refrigeration plant is an area for the engineering staff to concentrate upon to ensure an energy efficient ship. The plant can so easily come out of adjustment and this would cause additional energy to be consumed. All ships must keep a 'technical file' containing a Ship's Energy Efficiency Management Plan, and the correct operation of the refrigeration plant will be part of that plan. Engineers must keep a close watch on the refrigeration plant and to help with this, some simple faults are listed below.

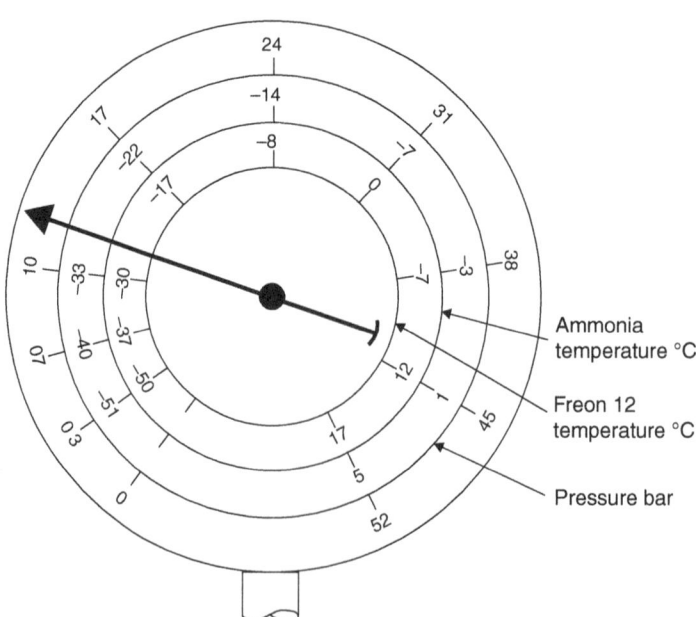

▲ **Figure 7.3** *Compound pressure gauge*

Common faults and simple detection techniques

1. Undercharge: low discharge gauge reading, lack of frost on suction pipe, lengthy running time, of the compressor, under load.
2. Air in system: high-discharge gauge reading (assuming sufficient vapour), jumping of gauge pointers, inefficient operation of the plant.
3. Dirty condenser or insufficient cooling water: high-discharge gauge reading and incorrect condenser temperature differentials.
4. Overcharge: this is unlikely, but it does give a high discharge gauge reading and very sensitive working of the expansion valve.
5. Oil on cooling coils: incorrect condenser and evaporator temperature differentials (oil is an insulator), excess frost on the suction pipe.
6. Choked expansion valve: caused by dirt or freeze-up by water, gives starving of evaporator and rapid condenser pressure rise.
7. Short cycling: condenser coolant restriction causing high-pressure discharge, compressor cut-outs, choked expansion valve giving low pressure suction cut-outs, etc.

Thermodynamic cycles

The circuit appears on the theoretical charts as shown in Figures 7.4 and 7.5. Entropy being a theoretical property of a fluid that remains constant during frictionless adiabatic operations.

$$\text{Heat energy received from cold chamber} = \text{Area under AB}$$
$$\text{Heat energy rejected in the condenser} = \text{Area under CD}$$
$$\text{Heat energy equivalent of work done} = \text{Heat energy rejected} - \text{Heat energy received}$$
$$= \text{Area under CD} - \text{Area under AB}$$
$$= \text{Area of figure ABCDA} + \text{Area under throttle curve DA}$$

$$\text{Coefficient of performance} = \frac{\text{Heat energy received}}{\text{Heat energy equivalent of work done}}$$

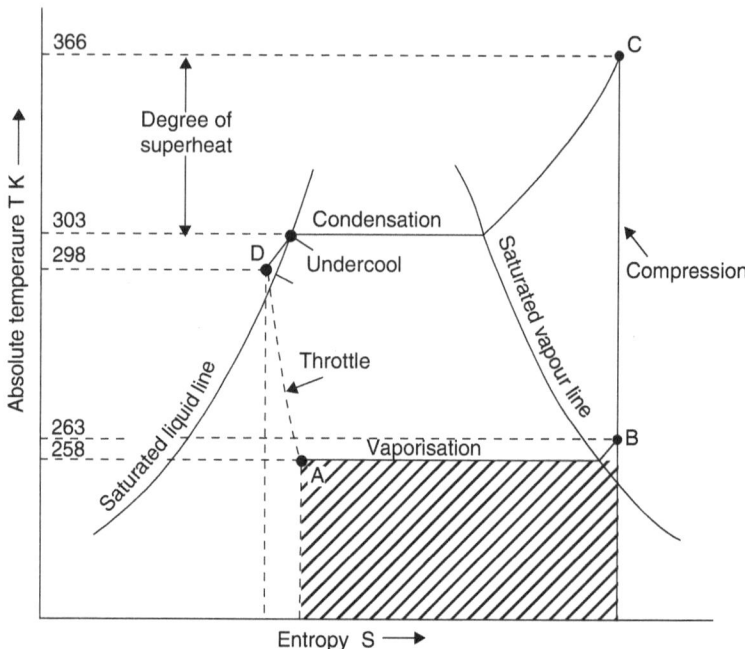

▲ **Figure 7.4** *Absolute temperature-entropy*

The compression is taken to be isentropic (frictionless adiabatic) for calculation work, this means the compression line is vertical (constant entropy). Unit mass of refrigerant is the usual basis. Coefficient of performance for Freon is approximately 4.7.

It should be noted how undercooling, in moving point A to the left, increases the heat received from the cold chamber thus increasing the refrigerant effect.

Refrigeration cycle p ~ v diagram

This diagram is shown (Figure 7.5) to allow comparison of this refrigeration cycle with other more familiar cycles covered in theoretical work on $p \sim v$ diagrams. In practice the $p \sim v$ diagram is rarely used in refrigeration.

Refrigeration cycle p ~ h diagram

Once basic theory has been established by using $T \sim s$ charts, the emphasis shifts in practice to the $p \sim h$ (Mollier) chart (Figure 7.5). This diagram has the big advantage that heat extracted, heat rejected and work done heat equivalent can be read off directly from the h axis in kJ/kg.

Intermediate liquid cooling

It has been mentioned previously that there is a loss in refrigeration effect due to flash-off to vapour when the liquid is being throttled through the expansion valve.

Undercooling before the expansion valve reduces flash-off after throttling, so lowering quality, and increasing refrigeration effect in the evaporator. (Although this applies to all refrigerants the loss would only be taken as *serious* with CO_2 because of its very low liquid specific enthalpy to specific enthalpy of vaporisation ratio.) The practical flash-off loss is about 20% in terms of refrigeration effect for Freon.

This does not justify the complexity of fitting two expansion valves and an intermediate liquid valve between them.

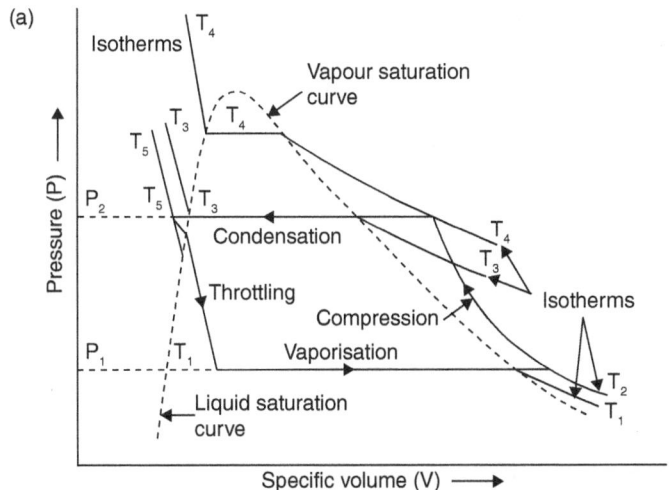

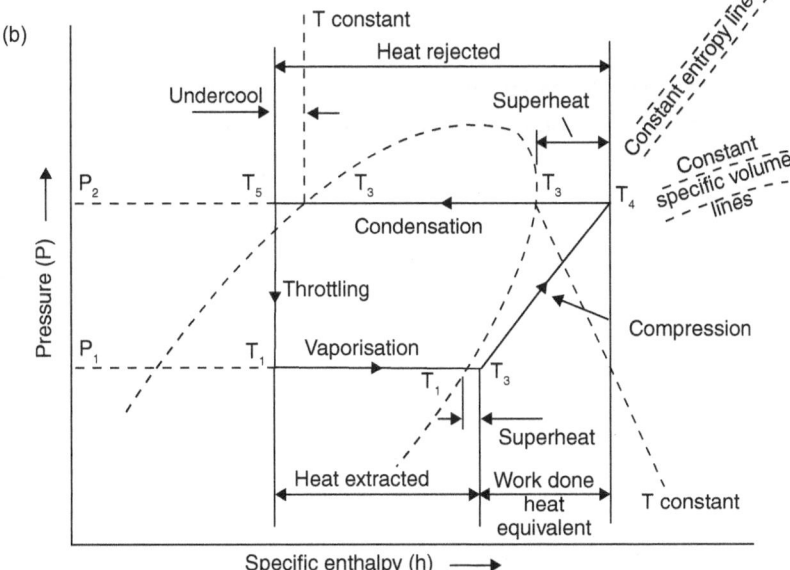

▲ **Figure 7.5** *(a) p ~ v diagram of refrigeration cycle; (b) p ~ h diagram of refrigeration cycle*

Critical temperature

Critical temperature is that temperature beyond which the gas cannot be liquefied by isothermal compression, that is, as a gas, no amount of compression will liquefy if the temperature remains above the critical temperature for that substance. CO_2 has a low value (31°C) and once the sea temperature (coolant) reaches 23°C the critical had been reached (8°C differential) and from this point the efficiency of the CO_2 plant steadily decreases. The critical temperature for most refrigerant vapours is, however, well above the normal condensing temperatures 96°C for R22 and 73°C for the HFC R404A.

Temperature glide

The modern refrigerant gases are a blend with each individual gas having its own characteristics. Due to this the boiling points of each of the gases can be different as can the condensing temperatures at a given pressure. R134a for example is a single refrigerant gas and therefore it will all evaporate and condense at single temperatures. With the chlorine-free HFC blends (Zeotropic blends) this temperature glide could be between 3°K and 7°K. The Zeotropic blends are the 'R4 series' of refrigerants and being a mixture, any top-up of gas in the system must be completed in the correct way and the percentage of mix must be maintained. The gas manufacturer's instructions must be followed or the fully topped-up system might not work well enough to cope with the heat load that needs to be removed.

Newer Azeotropic gas blends are being used where the component gases have the same condensing and evaporation temperatures, which also means zero temperature glide. These gases are designated the R5 series of refrigerants.

Compressor

There are four main types: reciprocating, rotary, centrifugal and screw.

Reciprocating compressors are in the majority in marine applications as they are most suited to low specific volume vapours and large pressure differentials, characteristics of all the main refrigerants.

Reciprocating

Almost all modern machines are motor-driven, high-speed (up to 30 rev/s), single-acting types that have adopted many improvements in line with the automobile industry. The only gland seal here is the crankshaft seal where the shaft emerges from the crankcase, such seal being mainly subject to suction pressure (Figure 7.6).

General

The actual machine itself needs little practical description to the trained engineer as most of the construction is standard reciprocating practice. Multi-cylinder in-line types are popular but there is an increasing usage of Vee and modified *W* designs. Pistons are usually of the trunk type, two or three compression rings and a lower oil seal ring, the suction valve may be located in the head of piston or in the cylinder head, the most modern arrangements have the suction and discharge valves in a valve plate in the cylinder head. Compressor bodies are close-grained castings of iron or steel. The crankshaft can be eccentric drive type or more conventional crank and connecting rod, lubrication being either dip and splash or forced, with pump.

Modern valves are of the reed or disc type mounted in the head and are of high-grade steel on stainless-steel seats with a usual lift of about 2 mm in average sizes. The discharge valve retainer is normally held down to a set position by heavy springs, and if oil or liquid is discharged the retainer lifts give extra valve lift so reducing over-pressure; cylinder relief valves and over-pressure cut outs are also standard-practice fittings.

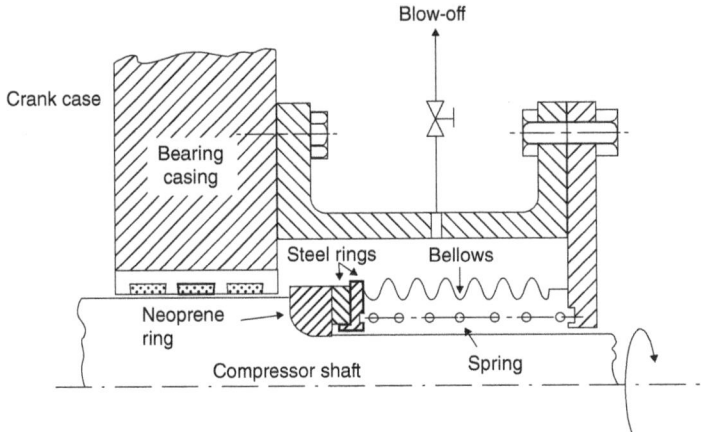

▲ **Figure 7.6** *Shaft gland*

With valve in piston head types the piston is often long and cut away at the side to the centre so that suction vapour enters here and there is no connection through to the crankcase, which reduces oil pumping effects. Screw-type service valves are double seated, full open or full shut, which allows easy gland packing changes. All reciprocating compressors should have the minimum reasonable piston clearance, 1.5 mm as a maximum usually, so as to give maximum efficiency.

Veebloc

In general there are four, six or eight cylinders radially round the upper half of the cast-iron crankcase with two to four connecting rods from each of two crank throws (see Figure 7.7 for four-cyinder *V* and Figure 7.8 for eight-cylinder *W* types).

The aluminium piston is fitted with two compression rings and one scraper ring. Piston and gudgeon details are given in Figure 7.8.

A differential oil pressure switch and overload electrical switch protect the machine from low oil or high vapour pressure. In addition, the discharge valve cage is spring loaded to lift in case of liquid carry-over and there is an overpressure nickel bursting disc to relieve excess discharge pressure to the suction side of the machine. Connecting rods are aluminium with steel-backed, white-metal bearings, the crankshaft is SG iron.

Provision is made for reducing the capacity of the machine either manually or automatically. Capacity reduction gear lifts and holds open the alloy steel suction valves of a specified number of cylinders, which is operated by oil pressure on a servo piston in the automatic type. This can also provide total or partial unloading for easier starting.

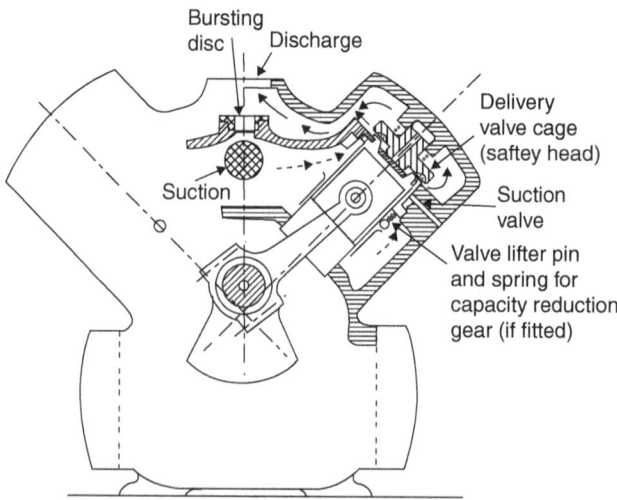

▲ **Figure 7.7** *Veebloc compressor detail*

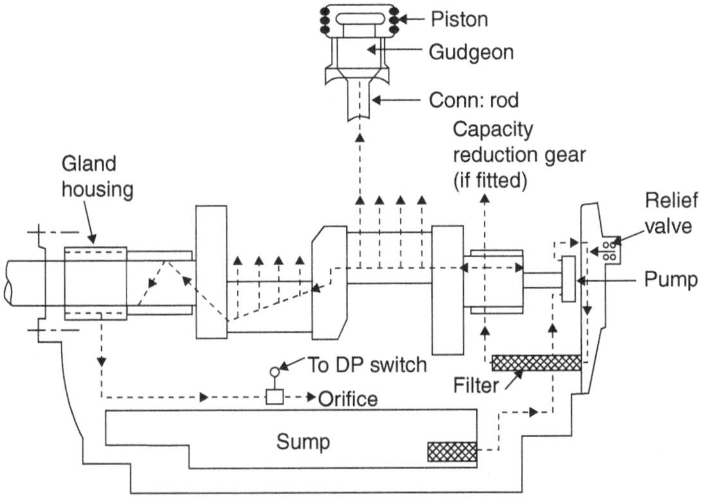

▲ **Figure 7.8** *W type compressor lubrication*

The lubrication should be clear from Figure 7.8. Oil is supplied by a rotor type of pump in which the inner rotor has one less tooth than the outer rotor and oil is induced to flow between the two rotors.

An eight-cylinder machine of 178 mm bore and 140 mm stroke running at about 12.5 rev/s would require a drive of about 90 kW for a refrigerating capacity of about 320 kJ/s.

For low-suction temperature operation (say −20°C or lower) and high temperature of discharge (say 30°C or higher) excessive temperatures may be reached in the reciprocating compressor. This is even more liable to occur in the unloaded state than in the loaded state. It is most often found in the fast-running smaller bore-stroke size of compressor. In certain cases an oil cooler – operation direct expansion, thermostatic – must be used, particularly when automatic unloading is required, and the above conditions apply. Compound compression units must be provided or special changeover valves can be fitted so that an eight-cylinder unit will operate in a single stage down to −20°C and for temperatures below this six cylinders can perform the initial compression and the remaining two cylinders perform the final compression.

For suction pressures below atmospheric level, the risk of air leakage is an important consideration.

Crankcase oil heaters are usually fitted for use with the machine stopped. This prevents formation of liquid refrigerant and oil frothing on starting. Auto compressors should be fitted with solenoid-operated liquid stop valves (see magnetic liquid stop valve, Figure 7.15 later).

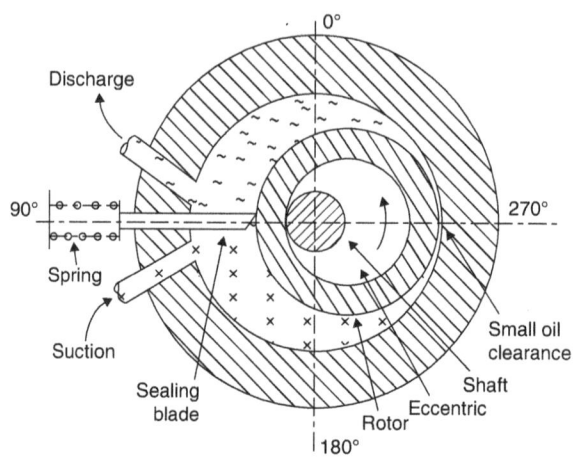

▲ **Figure 7.9** *Rotary compressor*

Rotary

These types are usually of the form shown in Figure 7.9.

At the position shown, the discharge and suction strokes are half completed, 270°C. At 0° discharging at compression stroke, induction at suction stroke. At 90° start of compression and end of suction. At 180° compression taking place and the suction stroke has just started. Thus the leading flank of the rotor acts as the discharger and the lagging flank acts as the inductor.

Such compressors mainly find application in household and domestic units but modern practice is extending their use to cargo purposes.

A variation on the above is a multi-blade type whereby the eccentric rotor contains spring-loaded blades (or relies on centrifugal force). When any rotary compressor is not in use the oil film between eccentric rotor and cylinder is broken, which means pressure equalisation and easy starting but requires the fitting of a non-return valve in the suction line. To reduce sizes these machines are direct drive from the motor.

Centrifugal

These machines work on a similar principle to the centrifugal pump whereby discharge velocity energy is converted to pressure head. For high-pressure differentials, as normally exist, a series of impellers are required on a fast-running rotor, each impeller feeding to the next in series to build up pressure. These machines are best suited to low differential pressure, high-volume capacity work such as air conditioning. Capacity reduction is effected by directional blades at the rotor inlet port. Efficiency is increased if interstage flash vapour formed during liquid expansion is returned to an appropriate stage of the compressor.

Screw

These compressors can be visualised as a development of the gear pump. A male rotor with say four lobes on the shaft meshes with a female rotor of say six lobes on a parallel shaft. Clearance between lobe screws and casing is kept to a minimum with sealing strips and oil film. As the space between two adjacent lobes of the female rotor passes the inlet port at one end of the compressor, a volume of gas is drawn in. With rotation, a lobe of the male rotor progressively fills this space so compressing the vapour and, due to the helical screws, forces it axially to the outlet port at the other end. To reduce capacity, sliding sleeves around the barrel can be moved axially to bring the outlet port nearer to the inlet port.

Lubricant

The first essential for such oil is that it should have a low pour point, that is, must remain fluid with good lubrication properties at low temperatures. Oils that are miscible with the refrigerant can be carried round the circuit and could congeal on the evaporator coils so drastically reducing heat transfer rates. The oil should be free from moisture under all conditions to prevent plant corrosion and freezing at the expansion valve. The viscosity should not be seriously affected at low temperature. Typical analysis:

Density 900 kg/m^3
Flash point 235°C
Viscosity 12cSt at 50°C
Pour point −42°C

A pure mineral oil is advised (Arctic, Seal oil, etc.). The above is suitable for reciprocating compressors.

Compressors should not be run too hot otherwise there is a danger of oil vaporisation and subsequent ignition by the heat of compression.

Hall screw compressor

These compressors are arranged so that there is one main screw and two balancing wheels that are diametrically opposite. The star-shaped rotors are turning by the main screw. The gaps in the fingers of the star and the helix of the screw form the chambers for pumping the refrigerant. The advantage of the single screw is as follows:

1. The outer casing is of a simpler design and therefore easier to make.
2. The simpler casing makes internal fixing points for the moving parts more accurate.
3. Easier assembly of the compressor.
4. Ease of maintenance as the sub-assemblies can be removed separately.

Heat Exchangers

Condensers

Condensers used in water-cooled marine plants are virtually all of the shell and tube type. The shell is welded construction of mild steel with vapour inlet, purge, drain and liquid outlet connections on the main body. The vapour condenses on the outside of the tubes and falls to the lower part of the condenser, which commonly acts as the liquid receiver. The water flow is multi-pass (usually 2, 4 or 8 flow types so keeping inlet and outlet branches at one end) through cast-iron end covers. Tube plates are ferrous, of welded mild steel, with steel tubes expanded into place, or non-ferrous, of muntz metal with

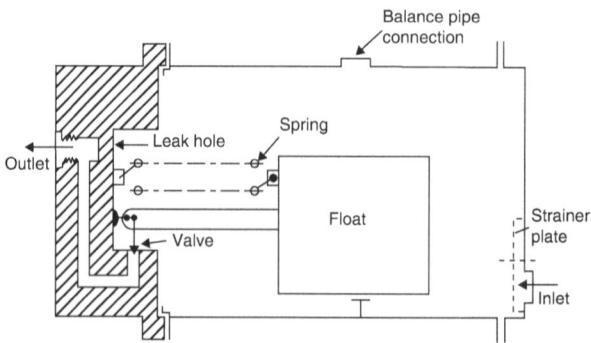

▲ **Figure 7.10** *High-pressure float valve*

aluminium brass expanded tubes. Galvanic protection blocks (zinc or iron) should be provided and steel tube plates are best treated with chromium for corrosion resistance.

Temperature differences are not high and the little expansion can be taken up by metal resilience. (It should be noted that non-ferrous metals are attacked by ammonia refrigerant, which will mean all jointing of lead or soft iron and use of steel tubing.) Air-cooled condensers are only used for small domestic units; they usually have finned tubes and air circulation may be fan assisted.

Evaporators

Modern evaporator grid types are only used on small plants and the distance between supply and return headers is very short so giving quick maximum extraction of vapour formed.

Large evaporators are invariably of the shell and tube type, almost identical to condensers in design and construction. Brine circulates through the tubes in multi-pass flow and the vapour–liquid mixture enters at the bottom at one end. The evaporated vapour leaves at the top of the other end so that speedy vapour extraction, full heat flow and full evaporation are achieved.

Heat transfer (fluids)

The two examples given here serve to revise basic theory:

1. A liquid refrigerant evaporates at 3°C and cools water from 11.5°C to 6.4°C in a heat exchanger of cooling surface area 360 m² for which the overall heat transfer coefficient is 100 W/m²K. Evaluate the log mean temperature difference and heat transfer rate.

$$\theta = \frac{(11.5 - 3.0) - (6.4 - 3.0)}{\ln(8.5 - 3.4)} = 5.566K$$
$$Q = 100 \times 5.566 \times 300 = 200376\,W$$

2. Calculate the effectiveness of a heat exchanger that cools air from 25°C to 15°C with refrigerant evaporating at 5°C.

$$\eta = \frac{25 - 10}{25 - 5} \times 100 = 75\%$$

Liquid level control

Hand-operated expansion valves have the disadvantage that they require fairly regular manual adjustments. The float type of valve automatically maintains a controlled liquid level. A diagrammatic part sectional, view of such a valve is as shown in Figure 7.10.

There are two types of float valves fitted, high-pressure and low-pressure. The high-pressure type as sketched is the more usual and is fitted with the float operating in high-pressure liquid after the condenser, with the object of draining the condenser or liquid receiver of liquid to the float level and feeding the liquid to the evaporator. The level is adjustable by altering the spring tension and to prevent gas locking an equalising or balance pipe is usually led to the condenser top.

The low-pressure type is to maintain a constant level of liquid in the evaporator. The valve is located with the float operating in low-pressure liquid in the evaporator or in a separate float chamber connected to the evaporator by balance pipes. As the liquid evaporates and is drawn off, the liquid level falls and more liquid flows in to take its place.

Some of the most up-to-date refrigeration systems use sensors to detect the refrigerant level and alter the expansion valve setting to keep the level constant. Some systems will also have a receiver before the expansion valve and a 'liquid separator' after the valve, which gives a greater degree of control of the refrigerant flowing through the evaporator.

Control by capillary tube is sometimes applied in small hermetically sealed units. The small-bore capillary tube, between the cooling unit and receiver, controls the point of pressure drop between the high- and low-pressure sides by its length. The system is strictly tied to refrigerant quantity and capillary tube bore and length. Thus once fixed for a set loading this cannot be changed without altering the capillary tube. These limitations reduce its application.

Small domestic units are usually of the direct expansion type where the refrigerant coil is in the cold room. Such types work on constant pressure or constant superheat expansion valve control and are described later in detail.

Direct Expansion Units

The control automation required is (1) start, (2) stop, (3) expansion valve, (4) emergency cut out, (5) cooling air or water circulation, (6) oil separation, (7) liquid in suction line, etc. Obviously this is closely related to electrical control. A few of the components have been simplified and are presented here:

Referring to Figure 7.11: the compressor is started and stopped by a thermal element pressure switch. An emergency pressure cut-out is provided and the expansion valve is of the thermostatic control type. The function of the solenoid liquid stop valve is to isolate different circuits and to shut just before the machine cuts out so that the compressor clears the suction line before it stops. This prevents liquid knock when restarting.

When there are a number of circuits, meat, fish, veg. rooms, ice tanks, ready-use chambers, etc., each circuit is a tapping off the main line.

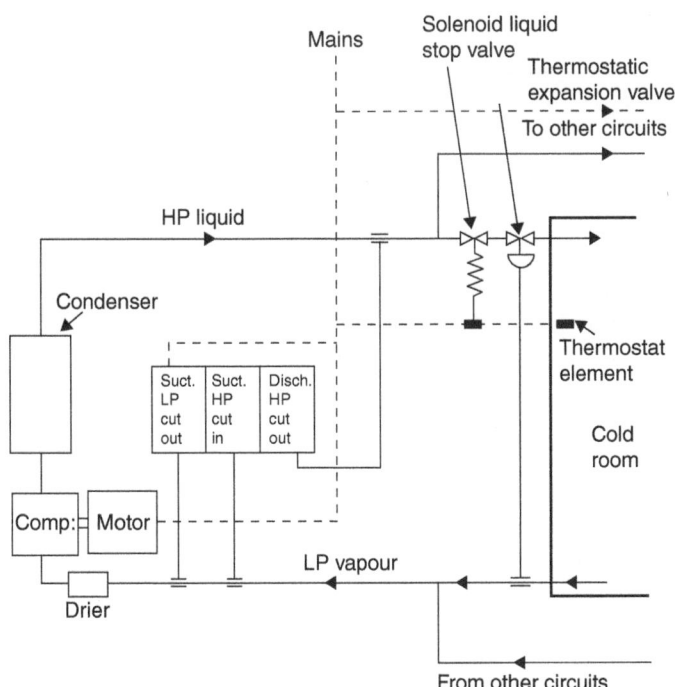

▲ **Figure 7.11** *Direct expansion unit*

Each circuit has its own thermostatic expansion valve and solenoid stop valve. When one chamber is cooled the thermostat shuts the liquid stop valve and cuts out that chamber only. This happens progressively to all circuits and when all circuits are cut out the rapid drop in suction pressure will cut out the compressor.

Electrical control switch

Referring to Figure 7.12: the capillary tubes are usually filled with a volatile liquid (or the refrigerant itself) so that temperature variations cause pressure variations on the flexible metallic bellows. The motion of the bellows operates the trip switches. As the temperature of the suction line *increases*, the bellows pressure *increases* against the spring compression upwards and *closes* the selector switch (HP cut-in at centre) at say 2 bar, so cutting in the motor. With the compressor running the suction temperature *falls* and hence the bellows pressure *falls*. This action against a tensile spring will eventually *open* the other selector switch (LP cut-out at left) at say 1 bar so cutting out the motor and compressor. The differential between these two is set for reasonable running.

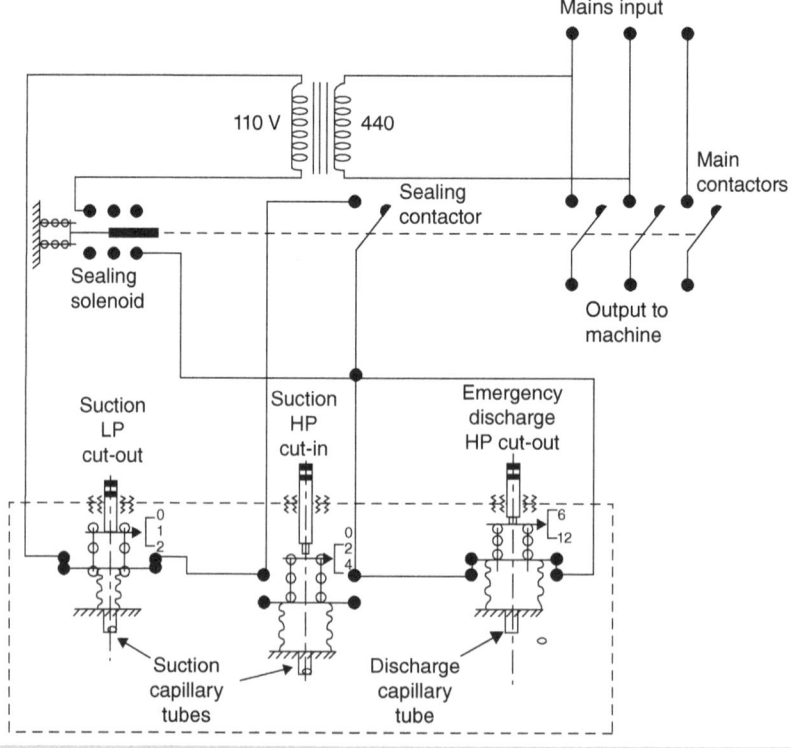

▲ **Figure 7.12** *Electrical control switch*

Again, here the principle of operation is best seen with the slightly older equipment. However, the modern systems will have the pressure sensing 'bellows' replaced by solid-state sensors and the electro/mechanical devices replaced with transducer and electronic control circuits.

The emergency high-pressure cut-out, to operate if cooling failure and pressure build up occurs, works to *open* the switch as for cut-out action.

The sealing contactor maintains the electrical circuit when the high-pressure suction cut-in opens with the machine running.

For manual operation, push buttons (start and stop) would replace the two suction bellows.

Electrical protection is by thermal trips operating the main contactors, which have an inherent time delay during heating.

Thermostatic expansion valve

The expansion valve shown is the thermostatic type, which is designed usually to give a vapour just superheated leaving the cold room.

Considering Figure 7.13 in relation to forces: *down* forces are high-pressure liquid valve force, adjustable compression spring force and power bellows force. *Up* forces are evaporator vapour pressure upon body bellows and tension spring force. For given spring settings the spring forces and high-pressure liquid valve forces balancing gives equilibrium, then force tending to *close* valve depends on evaporator saturation temperature and force to *open* valve depends on temperature of bulb at location, that is, valve operation is controlled by difference of these two temperatures, that is, superheat. Degree of suction superheat is controlled by the adjusting nut and is usually set at about 5°C. If the room temperature tends to *rise* the superheat *increases*, bellows down force from capillary tube *increases* and the valve *opens* to increase flow of liquid refrigerant until the temperature equilibrium is restored. The valve has a certain equilibrium setting at one pressure and another setting at another pressure, that is, the valve has no control over suction pressure.

The pressure control-type expansion valve works to maintain a fixed evaporator coil pressure. This type is similar to the lower part of the valve sketched but is more simple. It consists of the liquid valve and body bellows, the latter being loaded by an adjustable compression spring. *Up* forces on the bellows (evaporator pressure) balances high-pressure liquid valve force *down* for a given setting. Evaporator pressure reduction *opens* valve, pressure increase *closes* valve and when compressor stops the rising pressure causes closure.

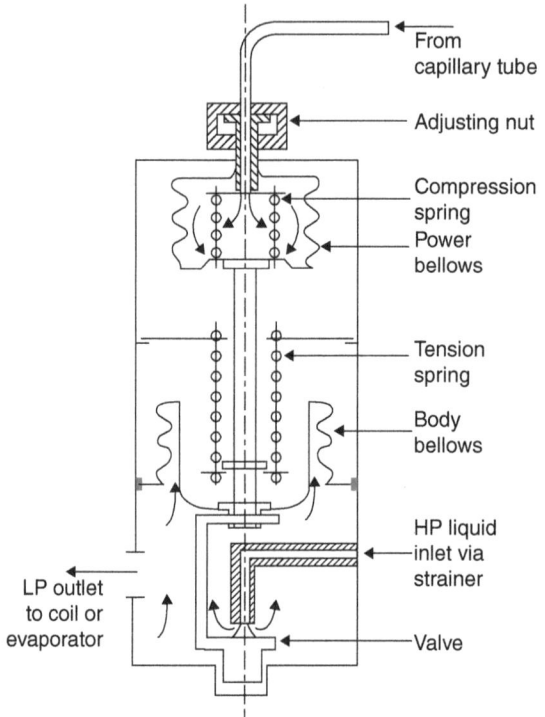

From capillary tube

Adjusting nut

Compression spring
Power bellows

Tension spring

Body bellows

HP liquid inlet via strainer

LP outlet to coil or evaporator

Valve

▲ **Figure 7.13** *Thermostatic bellows-type expansion valve*

Automatic water valve

See Figure 7.14. When the condenser temperature builds up, the capillary bellows pressure increases and the valve opens to permit circulation. This device operates almost immediately after compressor cut-in and serves to reduce water usage. The device can easily be modified to operate a starter for fan air circulation.

Magnetic liquid stop valve

See Figure 7.15. When current flows in the motor circuit, the solenoid attracts up the valve and holds it open provided the room thermostat allows the solenoid current to flow, that is, high temperature. When the room cools the current fails and the valve shuts and due to a time delay the compressor will pump out the room coil and so avoid uncompressible liquid entering the compressor on restarting.

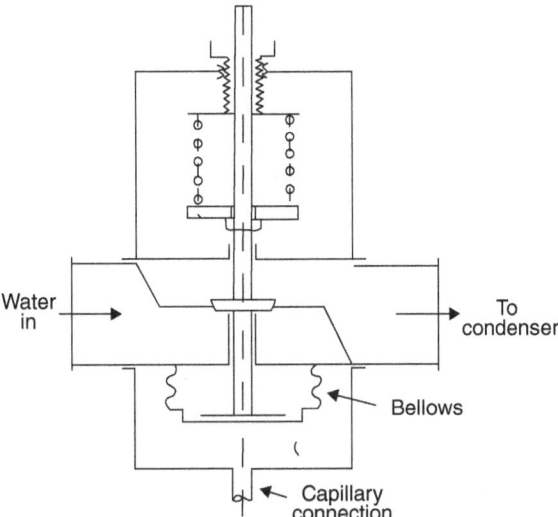

▲ **Figure 7.14** *Automatic water valve*

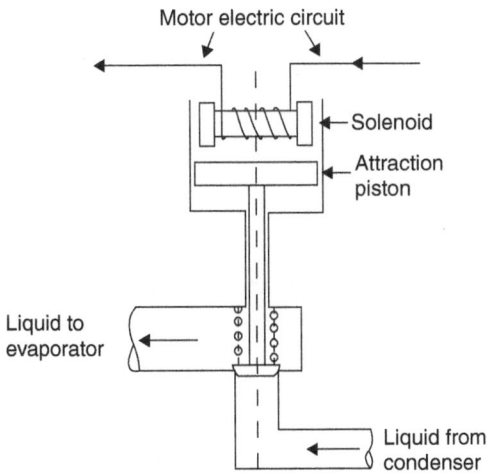

▲ **Figure 7.15** *Magnetic liquid stop valve*

In multiple circuits this valve serves to cut out or cut in the particular chamber to which it is thermostatically connected. Each room has its own stop valve (thermo-electric) and thermal expansion valve as well as a hand-isolating valve. The various circuits will cut out in *sequence* as the temperatures fall and the coolest chamber closure will then serve to stop the machine.

The thermostat switch is simply an electrical contact, working with a spring action against the bellows pressure caused by the room temperature and employing a certain temperature differential.

Sight glasses for liquid observation are commonly fitted at various points in the domestic-type systems.

Automatic oil separator

The oil is retained in the separator until the oil level rises and unbalances a float causing the vapour pressure to blow the oil back to the compressor.

The float is dead loaded so that a considerable blow back is possible before the valve reseats, to avoid fluctuating action.

Absorption-type Refrigerator Unit

This device has no moving parts and is continuous in operation when provided with a heat source, such as a town gas burner or electric element. The total pressure, which is the sum of the partial pressures, is constant through the system. Consider Figure 7.16:

1. Hydrogen vapour, which is insoluble in water, leaves the absorber and rises until it meets ammonia liquid falling into the entry to the evaporator. The hydrogen pressure causes a lowering of the ammonia pressure (two media exerting the same pressure as previously exerted by only one) and this assists in vaporisation of the ammonia. Ammonia and hydrogen vapour are carried down to the absorber where water absorbs and dissolves the ammonia and the hydrogen vapour re-cycles.

2. Ammonia vapour, which is highly soluble in water, rises with the water vapour from the generator to the separator where the water vapour and some ammonia vapour condenses. Ammonia vapour then rises, is liquefied in the condenser, reduced in pressure and vaporised in the evaporator, and falls to be absorbed in the absorber. Ammonia, dissolved in water, falls down into the lower pipe to the generator.

3. Water vapour leaves the generator, is condensed in the separator, falls through the absorber dissolving the ammonia vapour and returns to the generator.

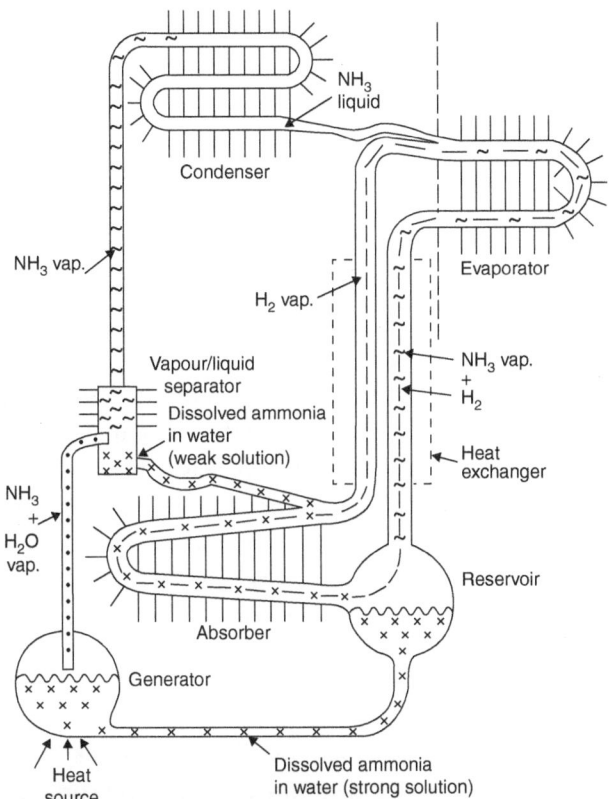

▲ **Figure 7.16** *Absorption-type refrigeration unit*

The unit requires no compressors, or pumps, and is silent and vibration-less. It has been used ashore in domestic units, but rarely on board ship, as a correct and steady level is critical for correct working. It is worth students taking note of this system as by using renewable energy sources this may become a popular system again in the future.

Condenser, evaporator and vapour–liquid separator are air-cooled, with fins welded or brazed on to the piping to give extended surface heat transfer.

In many designs the hydrogen vapour rises up through the absorber to the *underside* of the evaporator. It then passes up through the evaporator and carries ammonia vapour down to the reservoir where ammonia and hydrogen separate out, ammonia condensing and hydrogen rising through the absorber.

The precise method adopted depends on the dimensions of the equipment and the thermodynamics and flow pattern of the media in the unit.

Brine Circuits

Properties of brine

It is an advantage if the coolant coil through the cold chamber contains a fluid that is virtually non-harmful to the contents of the space in the event of leakage.

Small domestic units circulate the refrigerant through the evaporator in direct expansion and have the evaporator arranged in the space that is to be cooled. Larger systems, cooling cargo holds or container units usually employ a secondary refrigerant (brine), which is circulated through the evaporator and then circulated through the pipework to the cold chambers and back. A big advantage is that the brine pipes have a much larger reserve of cold than refrigerant coils when the plant is stopped, also having the advantage that various different circuits can easily be arranged, for example, cooling (+4°C), chilling (–10°C) and defrosting if required.

The brine as used is a mixture of distilled water (preferably) and calcium chloride ($CaCl_2$). The colder the brine circuit the more dense the brine in circulation has to be to avoid any freeze up. Table 7.3 gives the densities and corresponding freezing points.

Under certain conditions sodium chloride (NaCl) could be used with water but an alkali such as caustic soda (NaOH) would be required as an addition as about 1% of the solution. The brine should be maintained in an alkaline state under all conditions; this can easily be checked by the use of litmus paper, phenolphthalein, etc. Brine density should also be taken regularly by standard hydrometer test at 15.5°C and a

Table 7.3 *Brine density and corresponding freezing points*

Freezing point (°C)	Density at 15.5°C (kg/m³)
–3	1,050
–7	1,100
–13	1,150
–21	1,200
–32	1,250
–46	1,290

regular check should also be taken for brine leakage at the brine header tank, which serves to keep a head on the system.

There is a possibility that the air content of brine rooms could become explosive or inflammable under conditions of hydrogen gas liberation due to corrosive action, hence it is advisable not to allow naked lights.

The brine circuit consists of a brine room, containing distribution headers, mixing tanks, evaporators, pumps, etc., then the various piping systems to cold storage spaces maintained under pressure. The piping is usually tested to 7 bar, or 2.5 × WP, whichever is the greater, pipes commonly of mild steel externally galvanised and painted, about 40 mm bore.

It is usual to regulate the flow of brine by the return valves on the distribution and return headers. For chilling chambers about 3m³ of chamber would require about 1 m² of pipe cooling surface, increased to 4.5 m³ if air circulated. For freezing chambers the ratio is about 1.5 m³/m² (2.2 m³/m² air circulated). Normally 1,250 kg/m³ density would be satisfactory for most brine circulation with a pH value of 8.5. Non-freezing solutions can also be based on organic fluids; ethylene and propylene glycol are in general use.

Battery system

This system is to blow air across a brine or direct expansion grid and circulate the storage space. It is well suited to higher temperature storage, for example, shellac, as there is no dripping from overhead grids on to the cargo. Also this system gives some control over the humidity as moisture will be deposited on the cooling coil. The supply of air circulation to any storage room will reduce the brine cooling surface required by as much as 50%. Direct expansion grids employ only about 40% of brine-cooled grid pipe surface but do not have the same large reserve of cold.

Ice making

The ice tank is usually wrought iron and contains lead-coated sheet steel ice moulds. The moulds are immersed in a brine bath and a cooling coil, and brine or refrigerant lowers the bath temperature until the water in the moulds is converted to ice. The tank is insulated and coil supply and return valves are fitted. A direct expansion coil would be approximately 120 m² of surface area per kilogram of ice per second.

Hold ventilation control

Shown in Figure 7.17 is a method of air-delivery temperature control very suitable for fruit cargos. The sensor bulb is situated in a bypass pocket in the air trunk and senses air-delivery temperature whether fans operate normally or reversed. The diaphragm-operated control valve can be supplied as direct acting (fail in the open position) or reverse acting (fail in the shut position). For fruit cargo where frost damage could occur, reverse-acting valves that would fail in the shut position on air failure are used. For chilled meat cargo where failure would mean a long period of time would pass before the temperatures could be again reduced to the correct value, direct-acting valves are preferred. Refrigerated containers may each have their own refrigerator or the containers may be connected to the ships' ducts.

The refrigerated units that fit on to containers are called 'clip-on' units. They are self-contained refrigeration units and will need connecting to the ship's AC supply. Careful checks must also be made to these units as they can fail in various ways and if they do then the cargo in the container may perish.

Basic Principles of Air Conditioning

Air conditioning is the control of humidity, temperature, cleanliness and air motion. Winter conditioning relates to increasing temperature and humidity while summer

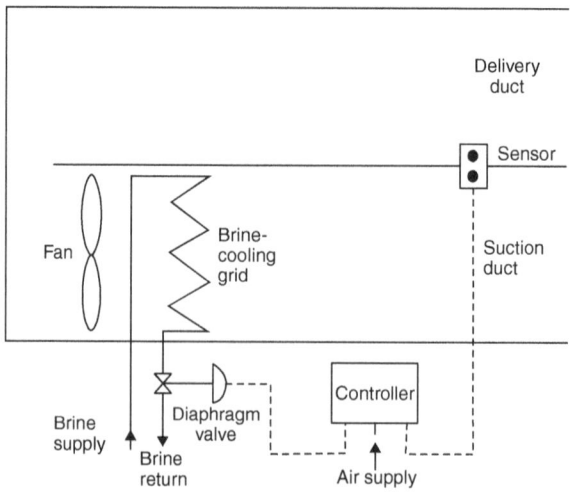

▲ **Figure 7.17** *Hold ventilation temperature control*

conditioning relates to decreasing temperature and humidity. Basically the practical difference is dependent upon whether the air fluid is passed over a hot grid (steam) or cold grid (brine or direct-expansion refrigerant). Water can be used instead of brine.

Comfort cooling for accommodation will only briefly be described although the field covers in addition much industrial usage as well as heating applications in both cases.

Specific humidity

Is the ratio of the mass of water vapour to the mass of dry air in a given volume of mixture.

Per cent relative humidity

This is the mass of water vapour per m^3 of air compared to the mass of water vapour per m^3 of saturated air at the same temperature. This also equals the ratio of the partial pressure of the actual air compared to the partial pressure of the air if it was saturated at the same temperature, that is.

$$\frac{m}{m_g} = \frac{p}{p_g}$$

Partial pressures, Dalton's Laws

Barometer pressure = partial pressure N_2 + p.p. O_2 + p.p. H_2O, from Dalton's Laws, viz:

1. Pressure exerted by and the quantity of the vapour required to saturate a given space (i.e. exist as saturated steam) at any given temperature are the same, whether that space is filled by a gas or is a vacuum.
2. The pressure exerted by a mixture of a gas and a vapour, of two vapours, or of two gases, or a number of same, is the sum of the pressure that each would exert if it occupied the same space alone, assuming no interaction of constituents.

Dew point

When a mixture of dry air and water vapour has a saturation temperature corresponding to the partial pressure of the water vapour it is said to be saturated. Any further reduction of temperature (at constant pressure) will result in some vapour condensing. This temperature is called the dew point; air at dew point contains all the moisture it can hold at that temperature, as the amount of water vapour varies in air then the partial pressure varies, so the dew point varies.

It can be seen from Figure 7.18 that cooling at constant pressure brings the low-pressure, superheated vapour to the dew point after which condensation occurs. It can also be noted that cooling at constant temperature increases the partial pressure until the saturation point is reached and thus relative humidity can be found.

$$\%R.H. = \frac{m}{m_g} \times 100 = \frac{p}{p_g} \times 100$$
$$= \frac{p\,\text{dew point}}{p_g} \times 100$$

where g refers to the saturation condition. This means dry air contains maximum moisture content (100% RH) at the saturation condition.

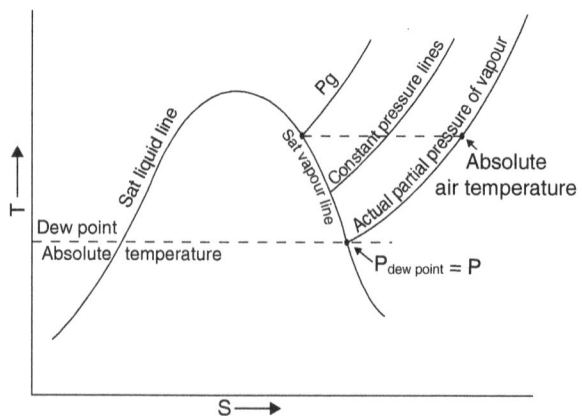

▲ **Figure 7.18** *Dew point*

Dry bulb and wet bulb temperatures

The hygrometer (or psychrometer) consists of an ordinary thermometer, which gives the dry bulb temperature, and a wet bulb thermometer (wetted gauze cover). The wet bulb reading will be less than the dry bulb reading, the difference is quoted as the wet bulb depression. The drier the air, the more rapid the moisture evaporation from the gauze, giving a cooling effect. Thus the greater the difference between the readings the drier the air and the less the % RH. Still air thermometers are inaccurate and hand sling types to secure air motion across the wick until equilibrium conditions exist are preferred.

Psychrometric chart

Cooling air at constant pressure gives constant moisture content, increasing in relative humidity until saturation (dew) point is reached.

Cooling air in practice gives some pressure drop due to fluid friction but this is not high in a correctly designed plant. If the cooling rate is kept in line with the pressure drop then the relative humidity will stay constant; if the cooling rate is slower the relative humidity will reduce; if the cooling rate is faster (as it will be usually in practice) then the relative humidity will increase.

When dealing with air mixtures, for example, Z 17 m³ of air at 35°C DB and RH 40% mixed with X 83 m³ of air at 27°C DB and RH 50%, set off on Figure 7.19 and proportion XZ off so that:

$$\frac{XY}{XZ} = \frac{\text{mass of 17 m}^3}{\text{mass of 83 m}^3}$$

the masses being found from specific volumes (ignoring small water mass) then Y is condition of 100 m³ of mixture (100 m³ 28°C DB RH 48%).

Points X, Y and Z on Figure 7.19 as shown are merely illustrative. The chart is drawn for a pressure of 1 bar but is sensibly accurate between 0.9 and 1.1 bar.

From wet and dry bulb readings the various properties of the air–vapour mixture can be estimated. Enthalpy is a function of the wet bulb temperature, and moisture content and vapour pressure are functions of dew point. The chart gives a quick performance check on the air entering and leaving the cooling coil, dew point, temperature, humidity, enthalpy, etc.

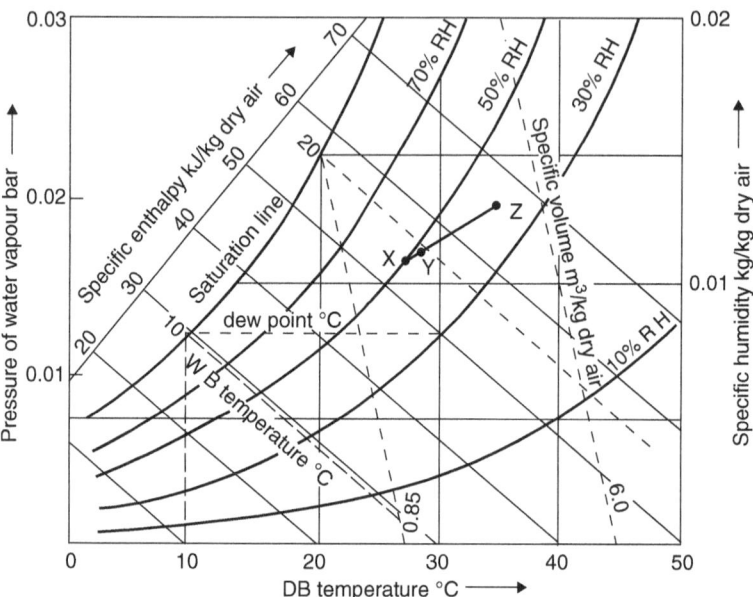

▲ **Figure 7.19** *Psychrometric chart*

Comfort under summer conditions is dependent on dry and wet bulb readings and relative humidity as well as air motion. For a given degree of air turbulence (75–127 mm/s), relative humidity between 30% and 70%, average 50%, and thermometer readings 19–25°C, average 22°C gives the best degree of summer comfort. Air at low temperature and high humidity can be as comfortable as air at high temperature and low humidity.

A differential of about 7°C between inside and outside conditions is usually aimed at but this is variable with the outside conditions, as a coil can extract large amounts of heat from warm dry air, so reducing temperature appreciably, or large amounts of moisture from humid air with little temperature reduction.

Air conditioning circuit

A typical circuit is as illustrated in Figure 7.20.

The average temperature differential is about 11°C between fan discharge and room temperature. The amount of air recirculated depends on the installation, space conditions (smoking, etc.), degree of air motion (draughts, etc.), and so number of air changes per day is a balance of quantity and temperature. Thus temperature, humidity and air motion are interrelated and the designer must correlate correctly.

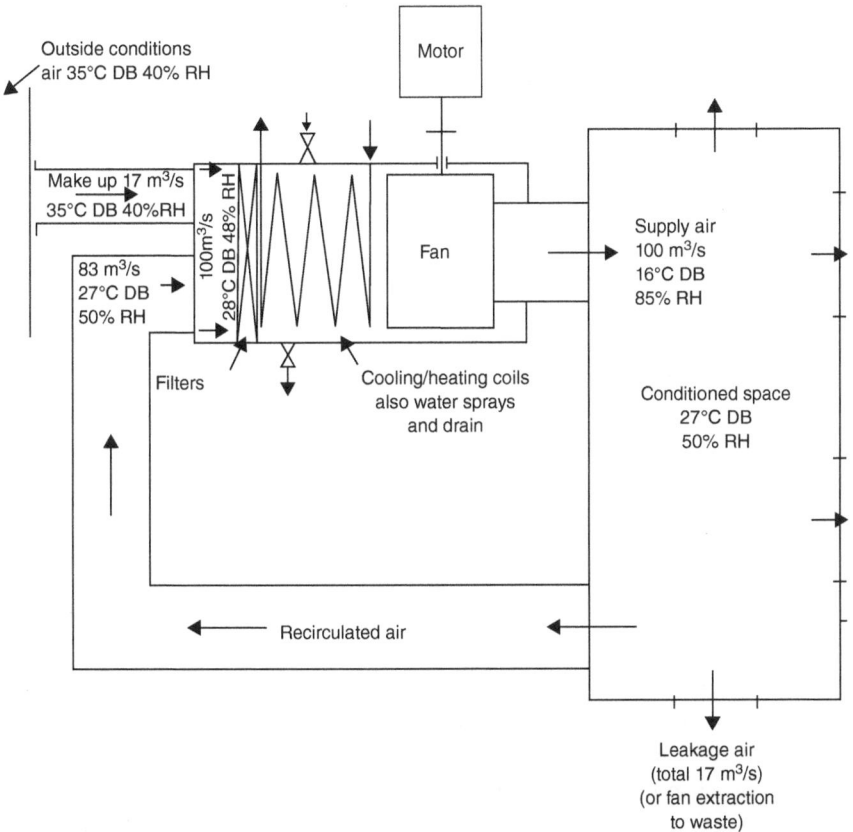

▲ **Figure 7.20** *Air conditioning circuit*

About 0.1 m³/m² floor space (accommodation) to 1.33 m³/m² floor space (kitchens) may be regarded as typical maximums, air motion about 100 mm/s.

The air conditioning unit (compressor, evaporator, condenser, etc.) will usually be independent from the rest of the refrigerating plant, although located often in the same space. The brine supply will be distributed to the cooling grids incorporated in a unit. The number of units would depend on the number of accommodation circuits necessary, say at least one unit per accommodation deck. Size of the plant would depend on the type of vessel.

Each unit would exist in a similar circuit to that shown in Figure 7.20. Leakage air must be cut to a minimum by closure of ports and doors. Air circulation would be through the normal louvre system to the various spaces, and when heating is required the air would bypass the shut-down cooling grid and be passed over heating elements, in this case a controlled water spray controls humidity before leaving the unit.

The temperature and humidity are controlled at the grid, drainage condensation being led away from the unit. Air motion will be determined by the initial design of the fans, ducts and louvres. The flow is usually by centrifugal or propeller-type fans and the humidistat or thermostat controller is situated at the unit together with fan controls. Cleanliness and purity depend on the filters. In marine practice viscous-type filters are used in which the filter medium (glass wool, fibre, compressed cardboard) is inserted between metal grids (about 50 mm apart) and the assembly mounted as a removable case. The cartridge is immersed in an odourless oil and then dried. Such filters are usually arranged to be cleaned by steam, alkalis, etc.

Other types are available that are in cardboard cases and are destroyed when dirty but they have little marine usage as yet. Similarly, dry filters (paper compressed screens) and electric precipitator types are available but also have no great application in marine services as yet.

Contamination by Legionella bacteria

This bacteria was first identified in 1977 following the deaths of a group of people in one hotel. The bacteria had probably been responsible for the death of people before that date but the symptoms are much like pneumonia and therefore any previous illness was not attributed to the bacteria.

The bacteria is found in showers and air conditioning systems, mostly in buildings but the potential exists for the bacteria to be found in marine systems as well. The UK Marine Guidance Note MGN 38 gives information about how the bacteria flourishes in stagnant water or areas that are constantly wet such as air inlet arrangements, filters or cooling units. It is recommended that these areas are treated with super chlorinated solution (50 p.p.m.) at least every three months.

Simple heat pump circuit

The circuit in conventional diagrammatic form for the air source heat pump is shown in Figure 7.21. The circuit has been set up to show how the accommodation would be heated during the winter. The heat can be transferred to the circulating 'refrigerant' from the heat exchanger outside in the open. The compressor raises the temperature and the pressure further. The hot gas is used to heat the accommodation and then as it gives up some of its heat the vapour condenses and is passed to the expansion valve as

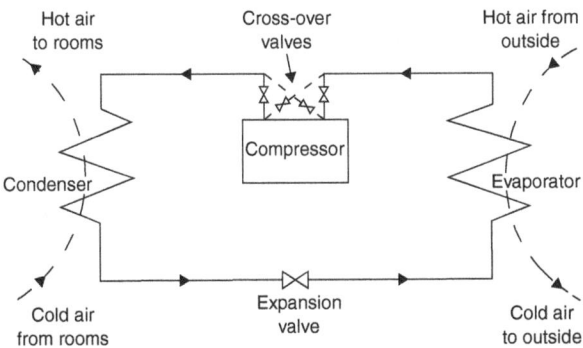

Hot air to rooms | Cross-over valves | Hot air from outside

Compressor

Condenser | Evaporator

Cold air from rooms | Expansion valve | Cold air to outside

▲ **Figure 7.21** *Simple heat pump circuit*

a liquid. At the expansion valve the temperature is greatly reduced and passes through the outside heat exchanger where the cycle begins again.

By reversing the direction of flow, by means of the cross-over valves, the heat pump can cool the accommodation in summer. Coefficient of performance averages about 4 or 5, which means that for every kW of electrical power that is spent on driving the compressor and fans about 4–5 kW of heat is gained from the air outside, which is used to warm the accommodation.

Dehumidifier

Moisture is removed from the humid air by passing the air through a fluted rotating drum of asbestos-type material surface impregnated with water-absorbing salts (Figure 7.22). A separate stream of heated air passes through the drum over a 1.6 radius sector in the opposite direction to continuously remove the moisture deposited on the salts. The drum is mounted on two rollers covered with a high-frictional coefficient material. A motor drive to the rollers causes rotation of the drum in the two air streams.

Heat input of 6.7 W would remove 21.4×10^{-3} kg/s of moisture from humid air at 27°C and 60% RH and flow rate of 0.77 m³/s.

The dehumidifier prevents moisture damage to cargo from direct moisture deposit or moisture deposited on internal ship surfaces.

One unit is normally provided for each hold and an electrical control and recording panel is provided on the unit as well as at a centralised station.

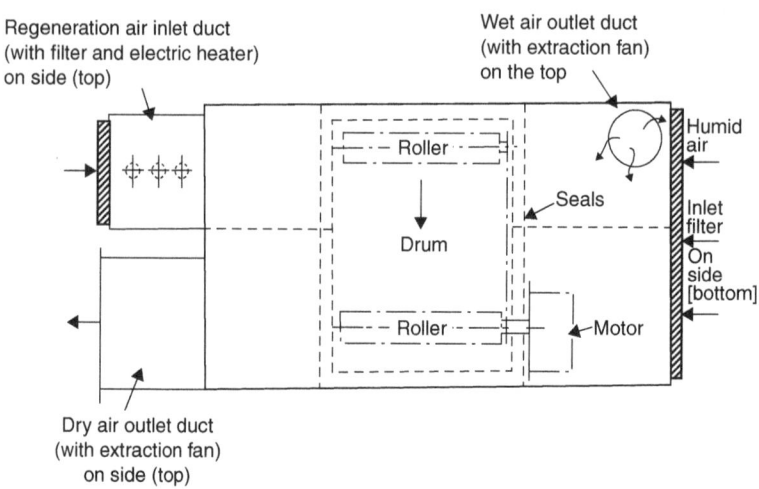

Regeneration air inlet duct
(with filter and electric heater)
on side (top)

Wet air outlet duct
(with extraction fan)
on the top

Roller

Humid
air

Seals

Inlet
filter

Drum

On
side
[bottom]

Roller

Motor

Dry air outlet duct
(with extraction fan)
on side (top)

▲ **Figure 7.22** *Dehumidifier (diagrammatic plan view)*

The fan units would normally each provide about 70 air changes in one day for the full hold volume and damper units are arranged to give re-circulation or ventilation with outside air.

Two recorders are normally utilised per unit: one recorder with sensors for ship skin moisture, that is, weather temperature, weather dew point, sea water temperature and hold dew point; and the other for cargo moisture with sensors for weather dew point, cargo temperature and hold dew point.

Most refrigerated cargos are now carried in containers. The design must take into account temperature control, air exchange, humidity levels as well as the correct loading of the container. Chilled meat will be at −1.4°C and frozen meat at −18°C or −22°C. Fruit and vegetables are mainly stored at high humidity (90–95%) but meats generally must be in drier air. Fruit could be carried at temperatures ranging between 0°C and +13°C to ensure that they are delivered in the best possible condition. Bananas, for example, are reduced to +4°C within 48 hours of loading and carried at this temperature until discharging. This means that they do not ripen during the voyage.

Insulation

The traditional ship-side insulation is shown in Figures 7.23 and 7.29. An alternative space saving method is to put the granulated cork in the air space so moving the vertical tongued and grooved board with zinc sheathing right up to the insulating paper.

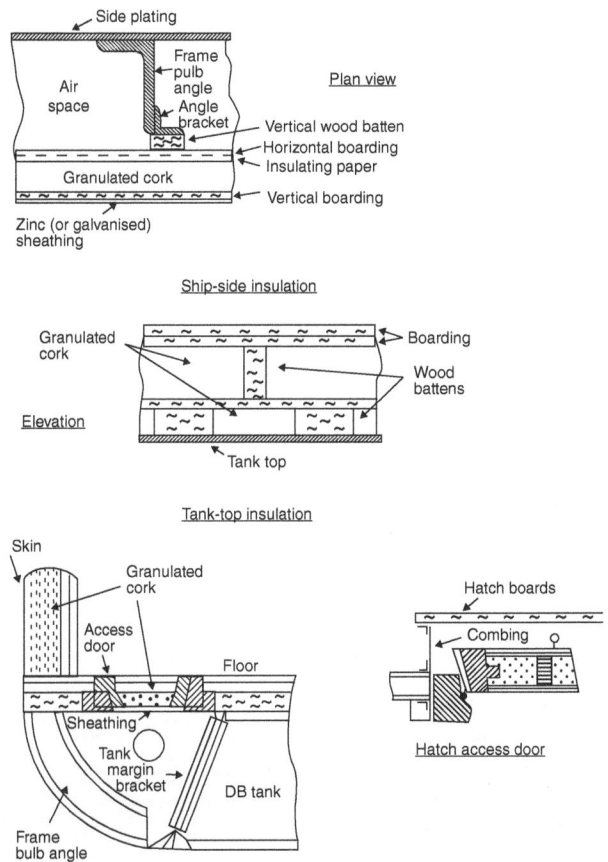

▲ **Figure 7.23** *Insulation details (1)*

Tank-top insulation is as shown although in some cases the lower battens and bottom tongued and grooved board can be omitted. In the case where the insulation extends right across a bilge, with a tank margin plate form of construction, access doors must be provided as shown.

Deckhead insulation is virtually the same as side-shell or bulkhead insulation, the fastenings being to the deck beams.

All insulation should be rat and vermin proofed and vapour proofing is also required. Air flow through walls will reach dew point and moisture will form, which destroys the insulating value and rots the material. Vapour-proof papers are commonly used, and all joints and concrete plaster over wire mesh covers should be coated with odourless asphalt (bitumen), tar or oakum.

Table 7.4 *Thermal conductivity of different materials*

Insulating material	Thermal conductivity w/mk average values
Polyurethane foam	0.026
Expanded polystyrene	0.034
Corkboard	0.043
Fibreglass	0.043
Rockwool	0.036
Kapok	0.035
Ground cork	0.043
Celotex	0.052
Building brick	0.346–0.692
Concrete	0.737–1.373

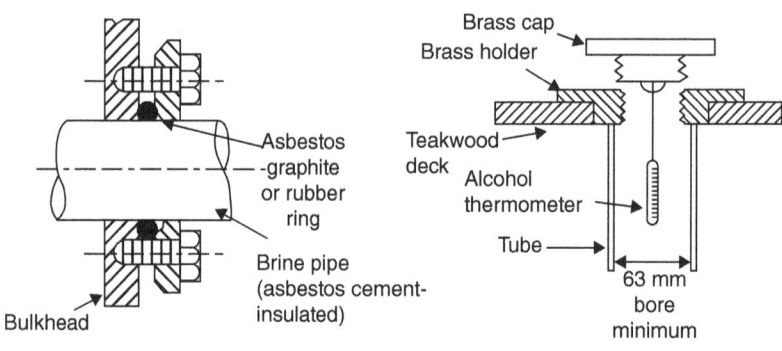

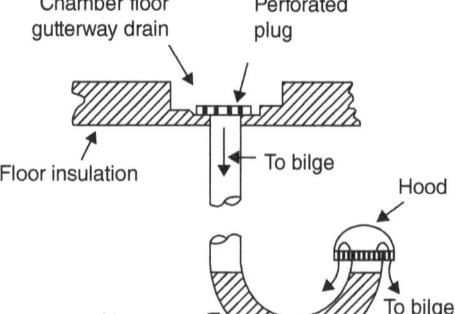

▲ **Figure 7.24** *Insulation details (2)*

Hatch covers, inspection doors, etc. are of heavy construction, usually of the wedge type, the taper being lined with felt or having ball rubber joint inserts.

Brine rooms and the like are usually concreted on wire mesh on the inside to give a smooth, clean appearance. Piping to and from the compressor to the brine or evaporator room should also be effectively insulated. Silicate cotton and also sawdust are packed 192 kg/m^3 and granulated cork about 128 kg/m^3. Plastics that flow easily and form firm hygienic mouldings are now in more general insulation use. Fibreglass and cork slabs are also used (see Table 7.4).

In the past 'asbestos' has been used as an insulation material and although this material has been banned from being used in any new buildings for some time, there may be older ships being scrapped that still have asbestos in them. When building a new ship a set of plans is required together with a list of the materials that are used in building the vessel. (Samples may also need to be inspected before installation.)

DNVGL in their published rules state that the insulation material should have the following qualities.

1. Should be non-absorbent and should not give off odours or gases that may affect the cargo.
2. Should have good mechanical resistance to vibrations and deformation at the design working temperatures.
3. Should not disintegrate or change structurally.
4. Should be resistant to decay and be chemically neutral.
5. Should have high insulating properties. When insulation materials with low resistance against moisture transmission and air movements are used, the integrity and completeness of the lining and vapour barrier shall be given special attention.
6. Organic foams will have flame-retardant properties.

Gas Containment Systems

Liquid natural gas (LNG) and liquid petroleum gas (LPG) are becoming more important as a source of energy for the modern industrialised world. However, the places where the gases are formed are not necessarily the same and therefore they need transporting around the world. The best method is by ship and the development of the LNG/LPG gas carriers has been a success story of the late 20th century.

Classification societies have now updated their rules for the construction of these vessels in light of the development of new materials for their construction.

A further development in this area is the use of LNG/LPG as a fuel, especially for ships other than the ones that are carrying gas as a cargo i.e. container ships and passenger ferries.

Cargo containment-piping systems for LNG and LPG utilise aluminium-stainless steel with membrane double-skin tank, construction and perlite insulation (boxed) – alternative polyurethane foam inside and/or outside. Cascade re-liquification plants are now being used as an alternative to burning the 'boil-off' in the ship's main engines and generators.

The design concept needs to have a primary and secondary means of containment when the temperature of the cargo is not lower than -55°C. Here the hull structure is allowed to be regarded as the secondary barrier. However, below this temperature a separate or partial secondary barrier will be required.

The effects of vibration on the completed structure will need to be considered carefully and values checked during sea trials.

The materials used for the constructions of the gas vessel's containment systems also need to be carefully considered. Stress and/or corrosion-induced cracking can occur due to the temperature differentials and the chemical composition of the cargos. Therefore, the chemical composition of the steel used in the construction process is carefully specified and quality checked according to the classification societies rules. The (stainless) steels used are:

- carbon-manganese steel;
- fully killed, aluminium-treated fine-grain steel;
- 1.5% nickel steel (minimum design temperatures -60°C and 2.25% nickel steel for minimum temperatures of -65°C;
- 9% nickel steels and specially treated austenitic steels are used for temperatures of -165°C and lower.

A full set of specifications for all the steels used is published by Lloyds Register and other classification societies on their websites

Tensile, toughness and yield stress specifications are also detailed for the steel to be used. The common factor of safety used is 2.0 but this could be changed if the vessel's design team submit suitable alternative plans. The pipework used in the transfer and bunkering system will be of the 'seamless and welded' type.

Non-metallic materials used in the construction of gas carriers can be used for:

- joining different materials and/or sections of the same material;
- containing liquid gas or the vapours from the gas;
- contributing to the load-bearing strength of the containment system;
- limiting the heat flow to the cargo (thus preventing over-pressurisation of the cargo)

All materials used for insulation must conform to the ISO 8301 and 8302 standards for determining the steady-state resistance to heat flow. In addition, the insulation material chosen must be able to retain its design characteristics during the environmental variations (thermal cycling and vibrations) that will be present during the vessel's duties and life cycle.

Some LNG carriers utilising aluminium and/or stainless-steel construction for the membrane double-skin cargo tanks have boxed perlite insulation or alternatively polyurethane foam inside and/or outside the metal.

Cascade re-liquification plants are also now being used as an alternative to burning the 'boil-off' in the ship's main engines and generators.

DNV, Daewoo Shipbuilding & Marine Engineering Co., Ltd. (DSME) and MAN Diesel & Turbo continue to develop the cryogenic systems required for burning LNG as a fuel in ships other than gas carriers. The design concepts Triality and Quantum 2000 were launched in 2011 and some of the concepts are now finding their way on to ships that are being built.

General information about reducing heat transfer (insulation)

The following two examples serve to revise basic theory:

1. A brick wall 0.225 m thick with a thermal conductivity of 0.6 W/mK, measures 5 m long and 3 m high and has a temperature difference of 25 K between inside and outside faces. What is the heat conduction rate?

$$Q = \frac{0.6 \times 5 \times 3 \times 25}{0.225} = 1,000\,W$$

2. A brick wall 0.24 m thick has a concrete surface facing 0.015 m thick. Thermal conductivities are 0.6 and 1.5 W/mK, respectively; surface heat transfer coefficients, inside and out, are 3.333 and 20 W/m²K, respectively. Evaluate the overall heat transfer coefficient needed to estimate the heat conduction rate through the wall.

$$R_o = R_i + R_e + R_b + R_o$$
$$\frac{1}{U} = \frac{1}{h_i} + \frac{x_c}{k_c} + \frac{x_b}{k_b} + \frac{1}{h_o}$$
$$= \frac{1}{3.333} + \frac{0.015}{1.5} + \frac{0.24}{0.6} + \frac{1}{20}$$
$$U = 1.316 \text{ W/m}^2\text{K}$$

Students should think about the effect of increasing the thermal conductivity of the insulation material.

8

FIRE AND SAFETY

One of the most dangerous hazards that seafarers may encounter on board ship is that of fire; the phrase 'Prevention is better than cure' was never more appropriate. Crew training and general awareness about the risk of fire is generally better than it was thanks to better on-board training media from companies such as Videotel. However, seafarers can never become complacent and they must know how the 'first aid' firefighting extinguishers work and what type of fire they can be used upon. The Manila Amendments to STCW will help tremendously in this respect. This is because for the first time in the history of the industry there is a specific requirement to have refresher training in fighting live fires.

However, informal staff development should not be overlooked. Realistic drills on board are still an excellent form of training. Firefighting and safety awareness discussions should also take place at team meetings as well as safety meetings and briefings. Safety signs and training videos and DVDs all help to promote a 'safety culture' among the ship's company.

Everyone – not just senior staff – should be on the lookout for danger. The obvious things such as the removal of oily rags from places of work are fairly easy to spot and rectify and now that smoking is severely restricted or banned on most ships, it does help fire safety. Modern 'fire retardant' materials are also an important addition to fire safety. However, the less obvious should not be overlooked, such as obstruction in front of the doors to the breathing apparatus (BA) locker and the testing of fire dampers. One of the biggest dangers is complacency; an 'it will never happen to me' thought process will only be overcome by changing the on-board culture to one of 'safety first'.

Principle of Fire

A fire is the rapid oxidation of combustible elements or compounds with oxygen, resulting in the liberation of heat. The heat, oxygen and fuel coming together start the process and then the resultant heat produced keeps the chain reaction going until one of the 'four' (heat, oxygen, fuel and the chemical reaction) is removed. NOTE: if only the chemical reaction is tackled then the fire could easily restart when the chemical inhibitor disappears, as the other three are still present. (Dry powder acts as a chemical chain breaker.)

Materials such as coal, wood, paper and any goods that are manufactured from them contain carbon, which is a combustible element. Any of these materials if subjected to heat will ignite in the presence of oxygen, in other words, the carbon will combine readily with the oxygen from the atmosphere in the correct ratios, and heat energy will be liberated.

Vapours given off from oils are full of hydrocarbons, that is, they contain the combustible elements hydrogen and carbon, which, again, if conditions are correct – that is, they are in the correct ratio for the given temperature and pressure – will ignite in the presence of oxygen.

Fire Prevention and Precautions

Cleanliness, vigilance and common sense are the principal weapons with which to prevent fire. Training, experience and the promotion of a 'safety culture' are also extremely important.

Tank tops should be kept clean and well lit. It is recommended that tank tops be painted a light colour so that any oil leakages from drip trays, pipes, joints, filters and valves may be easily spotted and the leakage dealt with promptly before any dangerous accumulation of oil arises.

Bilges must be kept clean and the pumps and strainers for the bilges maintained in good working order.

All firefighting appliances must be kept in good working order and tested regularly. Emergency pump and fan stops, collapsible bridge (quick-closing) oil valves and watertight doors should all be tested frequently and kept in a good operating condition. All fire detection devices should be regularly tested and any faults rectified.

All engine room personnel should be fully conversant with the recognised procedure for dealing with a fire aboard ship and should know the whereabouts and method of operating *all* firefighting equipment. Regular realistic drills will be very helpful in keeping the ship's crew conversant with the firefighting procedures. However, to reinforce the message all drills should be accompanied by a debrief giving all members of the exercise a chance to input their ideas about how the event unfolded and to suggest any improvements for the future.

When flammable materials are carried, as cargo, the compartment or compartments where it is situated should be well ventilated and the coal should, as far as possible, be stacked in such a way that it presents as large a surface area as possible to the atmosphere. This will reduce risk of an outbreak of fire due to spontaneous combustion of the coal. Personnel should be thoroughly familiar with the problems associated with any special cargo the vessel may be carrying, for example LPG, LNG and chemical carriers.

Classification of Fires and Extinguishing Methods

Knowledge about some specific information on fires and fighting fires is important for all seafarers. If a fire occurs on board then the ship's staff must deal with the situation because unless the ship is close to or berthed at a port then there will be very little in the way of external help.

The characteristics and properties of fires are not all the same and a fire does not occur unless all *three* essential features are present: oxygen, fuel and heat. They also have to be in the correct ratio. For example, a lighted taper/match can be thrown into a bath of cool diesel oil and the fire will be extinguished because the diesel is too concentrated and the ratio of oxygen, fuel and heat is not correct. The name for this combination is the fire triangle. Some teaching includes a fourth side and the combination is then called the fire tetrahedron (Figure 8.1). The fourth side represents the chemical reaction or oxidation process that is part of the mechanics of the fire. If this is broken then despite the presence of the other three properties the fire still goes out. This is an important aspect because using extinguishers that are 'chain breakers' is an efficient way of fighting fires. However, in the past this involved the use of halogenated gases, which are ozone depleting substances (see Chapter 7 pages 285–286) and are now prohibited from being used as a firefighting medium.

To distinguish between different types of fire they are classified into different groups. Each group has similar characteristics.

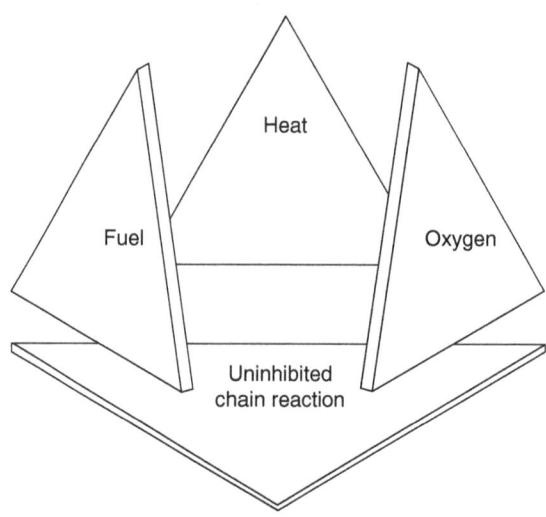

▲ **Figure 8.1** *Fire triangle and tetrahedron*

Class A fires

These fires involve free-burning material such as wood, paper, waste, bedding, fabrics and plastics. If these materials catch fire they are most effectively extinguished by cooling, principally with water or foam. It is now common for portable water extinguishers to have an additive included in the water to make them more effective.

Class B fires

These fires involve flammable liquids and solids such as fuel oil or waxes. Fires are started because the oxygen, heat and fuel come together in the correct ratio. If heat is introduced to liquid oil the combination will be too rich for combustion therefore the oil must give off vapour and it is the vapour that will support combustion as it is able to mix with oxygen in the correct ratio. Igniting the oil vapour causes a rise in temperature in the oil so that more oil vapour is readily given off from the oil to replace that already burnt. It can be seen that the temperature of oil can be raised to a point where the heat from the oil can ignite the vapour without another external heat source. This is called the auto ignition temperature. The methods of extinguishing oil fires are as follows:

1. Sand, used for small oil fires, serves as a blanket so excluding the atmosphere.

2. Water spray or fog must *completely* cover the surface of the burning oil. The water has a cooling effect that will reduce the rate at which vapour is given off from the oil. Hi-fog is very effective as the water droplets take away the heat as well as smothering the fire and taking away the oxygen. Hi-fog is gaining considerable favour as a replacement for 'Halon' in the engine room.

3. Foam should be used as a blanket to smother the fire.

4. Dry powder extinguishers fires in two ways: the first is by the application of a fine cloud over the surface of the fire, which excludes the oxygen from the fire. The second way is by interfering with the chemical reaction that allows the fuel vapour to combine with the oxygen. This action stops the combustion process. This double action gives dry powder the ability to knock down the flames very quickly.

   Dual application – extensive practical tests have shown that the dry powder extinguisher is extremely effective when used in conjunction with water/foam.

5. Inert gas, for example, carbon dioxide, is heavier than air hence it falls over the fire replacing the oxygen-bearing atmosphere.

6. Steam can be used to smother a fire.

7. Asbestos blanket is used for smothering small fires, especially chip pan fires.

Class C fires

Class C fires involve flammable gases such as propane, butane and methane. Extreme care and caution must be taken with class C fires. They can be extinguished by using dry powder if the initial heat source in not still close by. However, if the flame is extinguished but the gas is still escaping then the fire fighter might be making the situation more dangerous. If the gas source is removed by isolating the supply then of course the fire will go out.

Class D fires

Class D fires involve burning metal such as magnesium, aluminium, titanium, lithium and their alloys. The class D extinguishers are highly specialised and use types of dry powder as the extinguishing medium. Due to the very high temperatures of class D fires it is very important that water does not come into contact with the fire because the heat can bring about the disassociation of the water into hydrogen and oxygen that would obviously feed the fire instead of putting it out. However, the temperature can be reduced by boundary cooling techniques.

Electrical fires

Electrical fires are not a separate class of fire under European standards. This is because the equipment may catch fire due to the overheating of some part of the circuit or a malfunction of one or more of the components. This is a cause of a fire and not a different type of fire. The fuel for the burning will not be the electricity itself but the surrounding material, oil or metal that has become heated by the electrical fault. If it is possible to isolate the electrical supply and know that the fire area is isolated then the fire could be extinguished by using water. Dry powder is effective for fires involving live electrical equipment, being non-conductive and safe to use. However, the most popular medium is an inert gas such as CO_2. The advantage of CO_2 is that it does not make a mess of the equipment like dry powder does.

Class F fires

Class F fires are specific to cooking oils and fats. One of the best and most effective methods of extinguishing these types of fire is with a specially made fire blanket. Water must never be used for these fires: as the water turns into steam it expands rapidly to about 1,700 times its volume taking the burning oil with it. In the confined space on board this is disastrous as the burning oil is carried to the ceiling as a fire ball.

Fire Detection Methods

Fire prevention is better than having to deal with an established fire. If fire prevention has not worked then early detection is essential to be able to tackle the fire successfully.

The strength of fire patrols should not be underestimated and it should also be remembered that regardless of the fire detection system, the responsibility still rests with the watchkeeper.

Fire (good housekeeping) patrols

Formal fire patrols are not normally carried out on a regular basis upon modern cargo vessels. However, the watchkeeper should be on the continual lookout for any potential danger. A watchkeeping 'round' should be conducted:

1. Immediately prior to or upon sailing. A thorough inspection of the vessel being made especially in hold compartments, stores, engine and boiler rooms, etc.

2. When the vessel has been vacated by shipyard personnel while the vessel is in port undergoing repair. Someone may have been using oxy-acetylene burning or welding equipment on one side of a bulkhead totally unaware that the beginnings of a fire were being created on the other side of the bulkhead.

The patrol should be looking for fire but in addition the staff should also assess and correct any possible dangerous situations, for example loose oil or paint drums, incorrectly stored chemicals, etc.

SOLAS Chapter II-2 regulation 7 states that effective fire patrols should be completed on a passenger vessel carrying more than 36 passengers. Each member undertaking this task should be trained to be familiar with the ship, the equipment and how to use the equipment in the event of a fire.

Fire detection and alarm systems

It should be noted by the student, presenting themselves for the flag examinations, that all marine equipment must be manufactured to a recognised standard or specification. MSN 1734 gives details of the UK Type Approval of Marine Equipment. The standard covering the construction of the latest fire detection and alarm systems is BS EN 54. Reference should also be made to Merchant Shipping Notice MSN 1666 (M). The UK's 'M' notice gives significant details about the specifications required and the performance expected of all sections of the firefighting equipment on board.

Traditional or smaller systems consist of an alarm panel, which when covering the machinery space alone is situated outside but fairly close to the machinery space in a position that is easily accessible for the staff who have command and control responsibility. The alarm panel will show the rough location of the fire by giving an indication of the fire-affected zone. Zone circuits, audible alarms and auxiliary power supply details are shown in Figure 8.2.

Circuits

The less sophisticated type of circuit has each zone serviced by two wires. The contacts in a detector head are open under normal conditions, when they short the circuit then the operation of the audible fire alarm is activated. The supervisory current from the supply or the battery flows through the lines of each zone. The voltage drop is provided by the

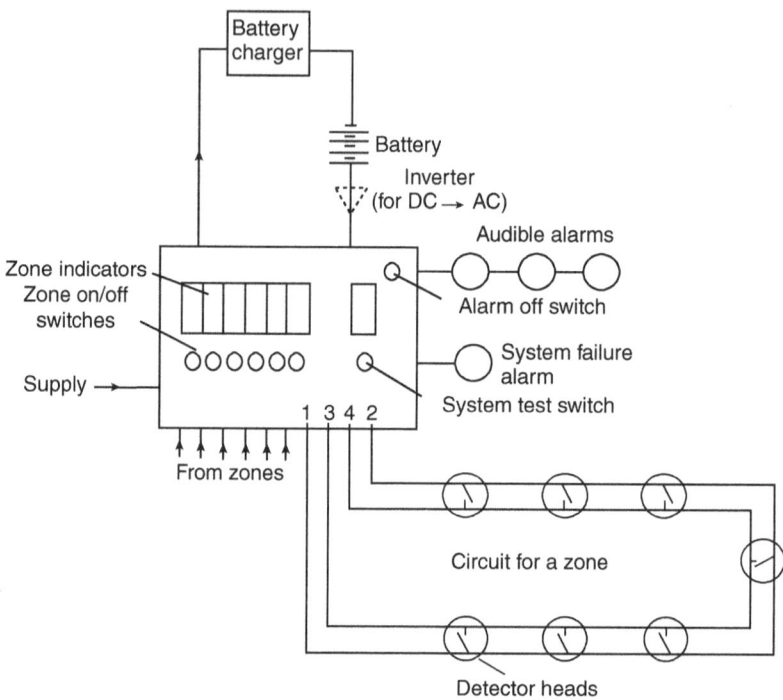

▲ **Figure 8.2** *Fire alarm circuit*

current flowing through an end of line (EOL) resistor and is continuously monitored by the voltage-sensing circuit inside the control panel (1–2 and 3–4 shown in Figure 8.1), a change in voltage due to a fault in cabling or detector causes the system failure alarm to sound.

Activation of the detector due to an alarm, however, will 'short circuit' the EOL resistor and the voltage-sensing circuit will register a 'Fire' alarm condition. It must be remembered that these systems need very careful installation and it has been known that poor installation has not been realised until the system has been required to perform in a real fire situation.

The most sophisticated systems could be up to 320 loops with 120 detectors per loop. They use multiplex electronic circuits and a system of 'logic' gates to monitor the field wiring or some form of computer area network (CAN) circuitry and devices.

The systems can give a graphic display of deck plans showing the positions of all detectors, manual call points and other features such as watertight doors. The graphics will show the progression of a fire and give valuable information to the ship's staff. The operation of the system is via keyboard and mouse accessing the computer software.

The new systems all have 'intelligent' detectors that have built-in electronic logic circuits. These detectors are addressable by the main control board and they can include multiple types of detection devices such as smoke and heat or carbon monoxide. Some will include fast-acting circuitry to give detection of a fire that has a very fast rise in temperature while other detectors will be designed to detect a slow rise in temperature. These new designs need the electronics built in to be able to determine the local conditions that go to make up a fire.

Power failure

The main power supply is arranged to keep the 'stand-by' batteries fully charged. In the event of a failure of the main power supply, the auxiliary power is automatically supplied, from the fully charged batteries, for up to 6 hours. Most systems operate on 24 V DC. For those systems operating at the mains supply of 220 V AC an inverter converts the 24V DC to 220 V AC

Audible alarms (sounders) covered by BS EN 54–3

Due to the extreme variety of different sounds the new standard does not specify what the sound signals are to be made by the sounder. The level of sound required may well be different in different situations: for example, the level of sound on the bridge will be different from the main part of the engine room. However, manufacturers need to declare the levels produced and the standard does specify how these levels are declared. The standard does specify construction details. The fire alarm in marine practice is usually an intermittent audible signal whereas fault and manual test are normally a continuous audible signal. However, some cruise line companies have found it necessary to have a two-stage process with the first stage alerting the crew only. If the fire became more serious then the passengers would be alerted and required to move to their muster station.

General rules for fire detection systems on board

The SOLAS chapters, relating to the guidelines for the construction and use of ships, are being updated on a regular basis as new information comes to light. The classification societies, especially DNV and ABS, make information available on the website about those changes. Recently, SOLAS has included a requirement that where it is designed that a detector or zone can be temporally taken out of service then there must be a specific indication showing that this has been done. There must also be a feature where the detector or zone will be reactivated after a pre-set time period. The Fire

System Safety (FSS) Code Chapter 9 of SOLAS also says that where the sensitivity of the detectors can be adjusted the 'set point' must be clearly displayed.

Fault monitoring is an essential feature of the EU standard. All critical signal paths should be monitored and faults indicated including:

- alarm circuits;
- detector circuits;
- all power supplies;
- battery backup systems;
- connections to any remote communication equipment.

Any fire or fault in one zone should not affect the operation of any detector in another zone.

Fire detector heads

Various types are available for fitting into an alarm circuit. The choice of head is dependent upon fire risk, position, area to be covered, volume and height of compartment, atmosphere in the space, etc. To economise and simplify, standard bases are generally used in the circuit into which different types of detectors can be fitted.

The construction of fire detection systems must be to EU standards where the units are of a 'fail-safe' construction, which is where the circuit integrity is monitored at all times. Any break in the system will trigger the alarm.

Heat sensors

These may be fixed-temperature detectors, rate-of-rise detectors or a combination. Rate-of-rise detectors do not respond and give alarms if the temperature gradually increases, for example, moving into tropical regions or heating switched on.

Figure 8.3(a) Pneumatic type: increase in temperature increases the air pressure inside the hemispherical bulb. If the bleed of air through the two-way bleed valve from the inside of the bulb is sufficient the diaphragm will not move up and close the contacts. If, however, the rate of rise of temperature causes sufficient pressure build-up inside the bulb to close the contacts, alarm will be given. In either case a bi-metal unit will at a predetermined temperature close the contacts on to the fixed temperature adjustment screw, giving alarm.

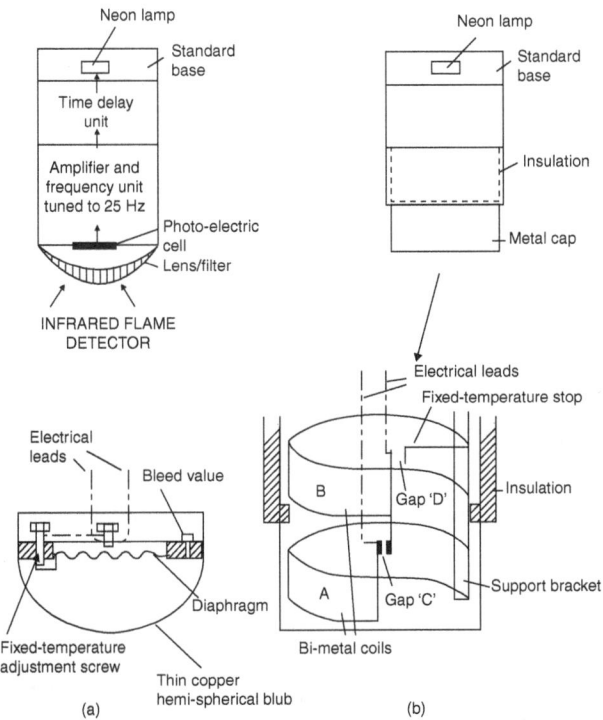

▲ **Figure 8.3** *Fire detectors*

Figure 8.3(b) Bi-metal coil type: two bi-metal coils attached to a vertical support bracket are encased in a protective metal cap. When the temperature increases A will move to close the gap C at a faster rate than B moves to maintain the gap. This is due to B being better insulated from the heat than A. If the rate of rise of temperature is sufficient, gap C will be closed and alarm given. At a *fixed* temperature gap D, then gap C will be closed, giving alarm.

Quartzoid bulbs of the type fitted into a sprinkler system are fixed-temperature detectors used for spaces other than engine and boiler rooms.

Relevant points

Sensitivity: a typical response curve for a rate-of-rise detector is shown in Figure 8.4. The greater the heat release rate from the fire, the poorer the ventilation, and the more confined the space, the quicker will be the response of the detector and the sooner an alarm sounds.

Fixed-temperature setting depends upon whether the detector is in accommodation or machinery spaces and can vary from 55°C to 70°C.

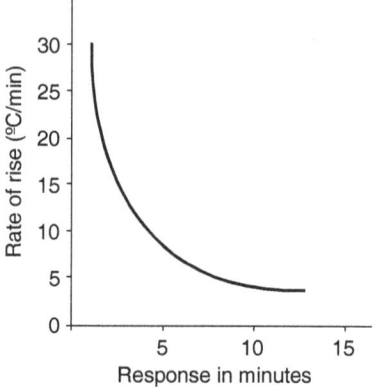

▲ **Figure 8.4** *Detector response curve*

The detector is useful for dusty atmospheres as it is completely sealed but it does not give as early a warning of fire as other types of detectors. It can be tested by a portable, electric hot-air blower or muff.

Infrared flame detector (incorporating ultraviolet detection)

Figure 8.3 also shows in simplified form this type of flame detector. Flame has a characteristic flicker frequency in the range of 5–30 Hz and use is made of this fact to trigger an alarm, which can be done in less than 50 ms. However, units usually have a delayed response of between 3 s and 30 s: the latest European standards require less than 10 s. Flickering radiation from flames reaches the detector lens/filter unit, which only allows infrared rays to pass and be focused upon the cell. The signal from the cell goes into the selective amplifier, which is tuned to the required selection frequency, then following the time delay unit it triggers the alarm circuits. The detector is designed so as not to give false alarms for steady hot body radiation, arc welding, artificial lighting or sunlight.

Where UV detection is incorporated in the detector it is highly resistant to false alarms from the sources mentioned because a microcontroller monitors the signals from each sensor to identify a variety of flame conditions, and only when the condition from the combined signals is an alarm is the warning triggered.

Relevant points

Very early warning of fire is possible and therefore this type of detector is suitable for areas where fire risk is high, that is, machinery spaces or modern boiler rooms that will normally be free from naked flame.

Some problems with these detectors are obstruction from solid objects such as machinery that is between the fire and the detector, and obscuration by smoke, water, dirt or dust. The detector can be tested with a naked flame or with a UV/IR test lamp.

Photo-electric cell smoke detectors

Three types are in use: those that operate by light scatter, those that operate by light obscuration and a type that combines scatter and obscuration.

Light-scatter type

A photo-cell separated by a barrier from a semi-conductor intermittently flashing light source are housed in an enclosure whose containment allows smoke but not light inside. When smoke is present in the container, light is scattered around the barrier on to the photo-cell and an alarm is triggered (see Figure 8.5).

Relevant points

Smoke may be present without much heat or any flame; therefore this detector could give early warning of fire. Photo-cells and light sources are also vulnerable to vibration and dirt whereas this one is not. Testing can be done with smoke from a cigarette.

The light-obscuration type is used in oil mist detectors for diesel engine crank cases and the obscuration/scatter type is to be found in the detecting cabinet of the carbon dioxide flooding system shown in Figure 8.17 (page 380).

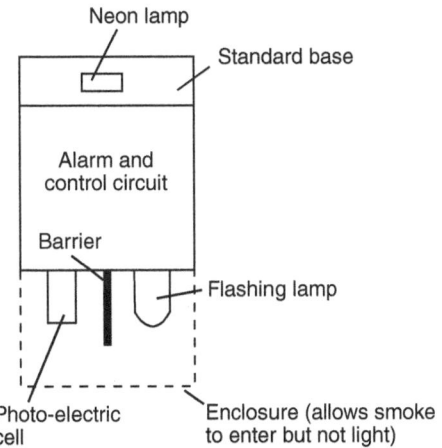

▲ **Figure 8.5** *Smoke detector (light-scatter type)*

Standard bases

The standard bases shown in the figures for the various detector heads have a neon light incorporated that flashes to indicate which detector head has operated. Detector heads can be simply unplugged from the base and tested in a portable test unit that has an adjustable time delay, audible alarm and battery.

Combustion gas detector

A circuit diagram of a combustion gas detector is shown in Figure 8.6. Two ionisation chambers connected in series contain some radioactive material, which emits a continuous supply of ionising particles.

The detecting chamber is open, the reference chamber closed and operating at a constant current since it contains air that is being ionised and the applied potential ensures that saturation point is passed. Current strength is dependent upon the applied potential, since if the potential is low not all the ionised particles reach the electrodes – some will combine with electrons and thus be neutralised.

When the potential reaches a certain value all the ions formed reach the electrodes giving saturation. Beyond this, the current will remain approximately constant irrespective of any further increase in potential. In this way the reference chamber has a constant resistance.

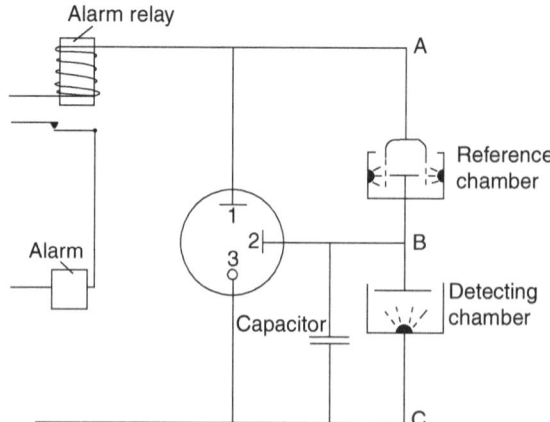

▲ **Figure 8.6** *Combustion gas detector*

If combustion particles, visible or invisible, pass through the open detecting chamber the current will drop since the combustion products are made of larger and heavier particles than normal gas molecules. When ionised, the particles are less mobile than ionised air particles and because of increased bulk and lack of mobility can readily combine with particles of opposite charge and hence be neutralised. The effect is to greatly increase the resistance of the detecting chamber, and this change in resistance produces a substantial change in the potential at the centre point B.

Normal voltage A to C is 220 V A to B 130 V, B to C 90 V. When voltage shift, due to increasing resistance in the detecting chamber, reaches 110 V across BC this is sufficient to trigger a discharge in the valve from 2 to 3, the capacitor then unloads itself across 2 to 3 encouraging a discharge from 1 to 3, bypassing the chambers and causing heavy current flow through the alarm relay and the alarm to sound.

It can be tested by cigarette smoke or the use of butane gas delivered from an aerosol container. It is a very sensitive fire alarm and a time-delay circuit may be incorporated to minimise the incidence of false alarms.

Critical Analysis of Fire Extinguishing Media

Water

It has high latent heat, 2,256.7 kJ/kg at atmospheric pressure, hence it has a very large cooling effect. If it absorbs this heat it expands to 1,700 times its liquid volume to produce steam, which is a smothering atmosphere.

It is plentiful, non-toxic, safe to use on most fires and can be easily directed over considerable distances.

When used on oil fires all of the liquid surface should be covered by the water spray, and surrounding hot metal should be cooled to prevent re-ignition. If water droplets enter the hot oil they will be converted to steam – this rapid expansion from water to steam leads to spluttering and possible spread of the fire. The water droplets should be fine enough (mist or fog) so that they cool by taking heat from the burning vapours. This is especially necessary in the case of oils with low fire points, for example, crude, petrol, etc., as direct cooling of these oils is not possible.

Steam

It has a very limited cooling effect. Its higher temperature makes the control of smouldering fires somewhat protracted. It is not always available and large quantities are necessary. Steam should not be used in conjunction with carbon dioxide for hold compartments, since to use it after carbon dioxide would be to replace a good firefighting medium with a relatively poor one. Steam smothering is not recommended. Some classification societies specify that steam could be used in lieu of foam, water or gas on vessels under 1,000 tons in boiler rooms or main engine rooms.

Foam

Foam is an accumulation of gas-filled bubbles surrounded by an aqueous solution. The production of the foam solution of modern vessels is by mixing foam concentrate and water. The expansion ratio is a measure of the volume of the final foam solution compared to the volume of the concentrate.

Aqueous film forming foam (AFFF) is a low-expansion foam (about 12:1) containing a free-flowing agent, such as a fluorinated surfactant, which alters the surface tension and thereby enables the foam to flow rapidly and evenly across large areas. This is popular for use on product carriers and chemical tankers for use in the event of a fire on deck. In addition, AFFF will have other additives for use when cargos such as methyl alcohol are being carried, thus preventing breakdown of the foam due to water loss.

Other foam concentrates are 'alcohol-resistant' (AR), film-forming fluoroprotein foam concentrate (FFFP), which is specially formed for use on oil fires. Protein foam concentrate (P), synthetic foam concentrate (S) and finally Types A and B foam concentrates. The MSC.1/Circ. 1312 from the IMO sets out the revised guidelines for the performance and testing criteria, and surveys of foam concentrates for fixed fire extinguishing systems.

Foam is generated either by mixing, using a suitable agitator, a protein foam concentrate or a synthetic detergent concentrate with water and air. Foam concentrate has been found to deteriorate in storage and hence there is a reduction in performance; regular testing and recharging are required. The 'throw capability' depends upon the expansion ratio, that is, the volumetric ratio of the amount of foam to the amount of water used in its formation. Low-expansion foam (8:1) has a reasonable throw but medium- and high-expansion foams (about 150:1 and 1,000:1, respectively) have very little throw capability. The foam can be pumped to foam monitors (e.g. foam guns at suitable deck stations, from which it can be thrown considerable distances, depending upon the pumping pressure).

Foam extinguishes oil fires by providing a heat radiation blanket from the flames burning above the oil and this prevents vapour forming above the surface thus cutting off the supply of fuel to the fire. The water content of the foam may in part be converted to steam and this together with any water not converted produces a cooling and smothering action.

Carbon dioxide

This has relatively low latent heat and hence it has a limited cooling effect. When released it expands to some 450 times its liquid volume to produce a cold gas that is heavier than air and has a penetrating three-dimensional action. The gas displaces the atmosphere, lowering the oxygen level and smothering the fire.

Its vapour pressure is approximately 40 bar at 0°C, hence if it is liquid at ambient temperature its pressure must be greater than 40 bar. Containers (except for bulk systems) are heavy, which limits the size of portable extinguishers.

Carbon dioxide is non-corrosive, toxic, does not deteriorate and does no damage. It is a non-conductor of electricity, clean and relatively inexpensive firefighting medium suitable for most fires except those that liberate oxygen while burning.

Inert gas

These systems must generate enough gas equal to at least 25% of the gross volume to be protected every 72 hours. The gas has a smothering effect but very little in the way of cooling. However, they are a useful back-up system where inert gas could be used if the initial charge of CO_2 was, for example, escaping and the danger was not passed. Inert gas is also a very efficient 'preventative' medium that is used in the free space over oil cargo in tankers to ensure that the vapours do not become a flammable mixture.

Vaporising fluids

Halogenated hydrocarbons BCF and BTM are now banned for any use on board international ships. If by any chance this gas still exists in a marine system the fact should be reported to the ship's flag administration immediately.

These gases are actually quite effective as a firefighting medium. With a higher latent heat than carbon dioxide, they have a better cooling effect and extinguish fires by interrupting the chemical oxidation process, the so-called 'breaking the fire chain', and they extinguish a flame in milli-seconds. Therefore only 5–5.5% saturation of a compartment is required. The Halon gas is very expensive but the weight required is only one-third to one-half that of carbon dioxide and they are not good for deep-seated smouldering fires. Hence for large spaces and holds carbon dioxide remains the most popular.

In the United Kingdom, MGN 258 gave the time-scale for the removal of Halon on all EU flagged vessels. The deadline for removing this gas from UK ships was 31 December 2003.

There are no 'drop-in' replacements for this system. The chlorine-free hydrofluorocarbon (HFC) gas is a close alternative but the further use of this gas is in some doubt. As a HFC gas it is not subject to restrictions in some parts of the world but in others it is. HFCs have a zero ODS number (Chapter 7 page 289) but it does contribute to global warming therefore any leak (although remote) would contribute to the ship's registry for total release of global warming gas. The alternative 'fixed installation' system choice is between inert gas, carbon dioxide, water-based systems such as Hi-Fog and low- and high-expansion foam (Figure 8.8).

Portable Fire Extinguishers

The European standards – BSEN3 – introduced a change to the UK practice of distinguishing different types of fire extinguishers by their colour. The UK colour coding involved having the whole extinguisher a specific colour depending upon the type of firefighting medium. This has now been replaced. The new standard requires the following:

- 95% of the extinguisher is to be red.
- There can be areas of a different colour indicating the contents.
- All markings must follow a specific layout.

- Extinguishers must be marked with the standard symbols.
- A minimum body-shell thickness.
- Minimum fire performance rating for the size of extinguisher.
- Operating temperatures have been increased.
- Discharge times have been increased.

The old colour-coded systems did not inform the user about the type of fire that the extinguisher was suitable for, it was there to inform only about the contents. The new system of pictures informs about the type of fire class (A, B, C or F). The concession to the older systems was for a band representing 5% of the total area to be allowed showing the content.

The colours are:

- red for water
- cream for foam
- blue for powder
- black for carbon dioxide
- canary yellow for wet chemical

The standard is not retrospective and therefore students may come across older types such as the possibility of coloured or even chrome extinguishers. Students should be mindful of this as they might come across a vessel with a combination of types. The assumption could then be made that the red extinguisher contains water. Always look at the instructions before use. The wise and experienced seafarer will always check his/her places of work or cabin area as soon as they step on board when joining the ship.

MSN 1665 issued by the UK government states that portable fire extinguishers (other than carbon dioxide or halogenated hydrocarbon fire extinguishers) shall be between 9 litres and 13.5 litres in capacity and have an extinguishing capability at least equivalent to a 9 litre extinguisher. The portable CO_2 extinguishers should be at least 3 kg.

Extinguisher type

In the past, ships have used extinguishers such as the soda-acid type. When a plunger was depressed and sulphuric acid was released to react with sodium bicarbonate solution, this resulted in the production of CO_2. The CO_2 built up pressure and the firefighting medium, usually water, was then driven out.

The inverting type foam extinguisher was another popular extinguisher where the inner container is filled with a solution of aluminium sulphate and the annular space formed by the inner and outer containers is filled up to the level indicator with a solution of sodium bicarbonate and foam stabiliser. The proportion of solutions is approximately 1:3 in inner and outer containers respectively, total solutions 9 litres.

By inverting the extinguisher the lead seal will fall, clearing the ports in the inner container and the two solutions can then freely mix. As the solutions mix they react, generating foam under pressure that is discharged through the nozzle.

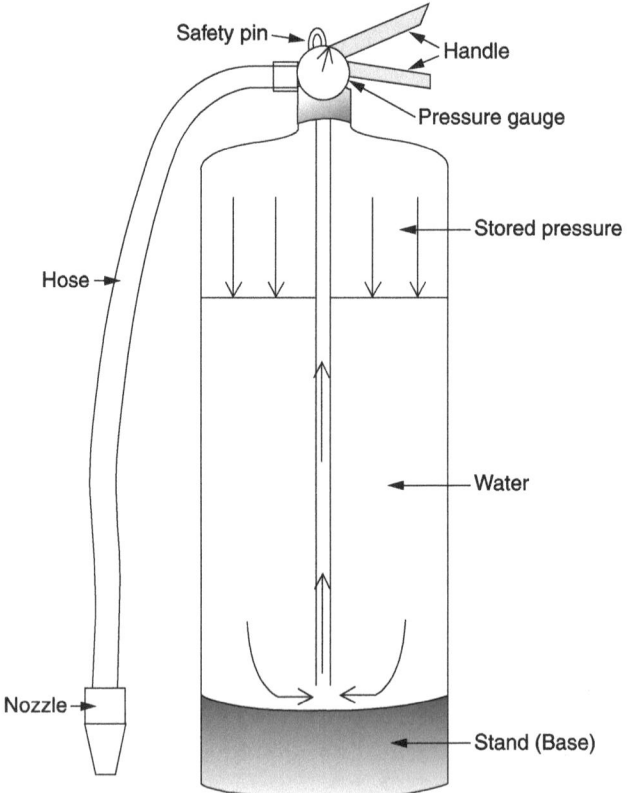

▲ **Figure 8.7** *Construction of a stored pressure fire extinguisher*

The modern extinguishers now delivered to a ship are one of two types:

• stored pressure (this is the most common type) (Figure 8.7) or
• generated pressure (cartridge type).

All extinguishers on board British registered vessels must be of an approved type meeting the requirements of British Standard BS/EN or ISO standard for the construction and use of portable fire extinguishers. With the stored pressure type, the propellant and the firefighting agent are both stored in the single chamber in the container. The pressure is about 10 bar which is indicated on a simple pressure gauge showing green or red depending upon the actual pressure and if any has leaked away. With dry powder or other dry chemicals nitrogen could be used as the propellant and with water and foam the extinguisher is usually pressurised with air. Inside the extinguisher is a tube that extends to the bottom of the cylinder. This enables the extinguisher to be used upright by removing the safety pin and squeezing the handle. The extinguishing agent is released through the nozzle pushed out by the pressure in the top of the cylinder.

The cartridge type contains the propellant in a separate container which has to be punctured prior to discharge. The small cartridge contains CO_2 at 130 bar rather than air or nitrogen. The cartridge has to be ruptured by a pin piercing the container that then releases the gas and pressurises the extinguisher forcing out the extinguishing medium. Dry powder extinguishers are regularly fitted with a cartridge extinguishing system so that the medium is not affected by moisture in the air with the stored pressure type.

Construction

The standard extinguishers are of steel construction with a plastic stand fitted to the bottom. They must be of sufficient strength to contain the pressure and they must be manufactured to BE/EN standard 3–1 to 7 with the testing completed to BS/EN 3–8. However, units are now appearing that are made of lightweight 'composite' material. According to the manufacturers these will not corrode and therefore should not need servicing as long as they remain pressurised. However, the marine regulators have not yet given 'type approval' to these units for use at sea.

The design of the nozzles will differ slightly depending upon the type of medium being used to extinguish the fire. The foam extinguisher has a special nozzle to assist with aerating the mixture to form the foam jet that can be aimed at the fire. CO_2 has a shallow tapered horn that slows the velocity and concentrates the gas in the right direction as well as preventing it from mixing with air. Water extinguishers have a rounded end that helps to form a jet that can be directed toward the seat of the fire and the dry powder extinguishers have a nozzle that helps the powder to be directed in the correct way and does not capture too much air.

Servicing

All ship's staff should be aware of the importance of fire extinguishers and MGN 276 gives details of the maintenance requirements for both steel- and aluminium-bodied extinguishers. The M Notice was published in 2006 following two fatal accidents associated with the operation of portable fire extinguishers.

Foam fire extinguisher

The 9 litre low-expansion foam extinguishers use a protein- or synthetic-based foam concentrate in a separate container from the water, although the stored pressure of CO_2 cartridge is still used as the propelling agent. They are designed so that the extinguishing agent will float on oil, forming a break between the oil and the flames. The AFFF type has a significant water content that will also help to reduce the heat. However, due to the water content foam should not be used on electrical fires.

Foam fire extinguishers would be useful in the engine room where oil-based products are in abundance. The technique is to aim the foam jet at a convenient structure close to the fire so that the foam can then spread to cover the area providing the fuel for the fire. If there is a distinguishing colour band near the top of the extinguisher it will be cream.

CO_2 fire extinguisher

The CO_2 extinguisher can be used on B- or C-class type fires. The action is to cover the fire area with the CO_2 so that the oxygen is excluded. This type of extinguisher can be used on electrical fires. However, it must be remembered that if CO_2 is released in a confined space it could endanger personnel.

Water fire extinguisher

Water has an effective cooling effect and is the best medium to use on class A fires. When fighting fires in organic materials such as cloth, paper or wood the water jet should be aimed at the base of the fire and kept moving over the whole area involved.

Dry powder fire extinguisher

Dry powder acts as a break between the fuel and the flames. These portable extinguishers should be used by pointing the discharge jet/cloud at the fire and driving it towards the far corner of the area that is alight. Some extinguishers will have a shut-off control lever; if so then wait until the initial cloud of powder has settled and attack the fire again if necessary.

The fire blanket

Made of fire-resistant materials, these are very good for smothering small fires involving flammable liquids. They will be clearly marked to show if they can be used again or not. Care must be taken to ensure that the whole area is covered and the person fighting the fire must also take great care to ensure that s/he is protected by the blanket while moving towards the fire.

Fixed Firefighting Installations

Chapter II-2 of SOLAS

This chapter of SOLAS sets out the regulation for the construction and use of all firefighting systems on board all SOLAS approved vessels built after 1 July 2002. All the flag states that are signatories to IMO will have implemented the SOLAS regulation through their own domestic law. In the United Kingdom, this chapter came into law through the Merchant Shipping (Fire Protection) Regulations 2003. These regulations amend all the previous regulations that applied in the past and came into force on 11 December 2003. The original regulation is also updated as required, due to new thinking and improved technology, via amendments.

Regulation 2 has the following objectives:

1. Prevent the occurrence of fire and explosion.
2. Reduce the risk to life caused by fire.
3. Reduce the risk of damage caused by fire to the ship, its cargo and the environment.

4. Contain, control and suppress fire and explosion in the compartment of origin.

5. Provide adequate and readily accessible means of escape for passengers and crew.

The aim is to achieve these objectives by:

1. The appropriate vertical and horizontal division of the vessel.

2. The use of thermal and structural boundaries.

3. Restricted use of combustible material.

4. Detection of any fire.

5. Containment of fire.

6. Protecting escape routes.

7. Availability of firefighting equipment.

8. Minimising the possibility of vapour ignition.

In addition to the regulation above, IMO consolidated the information into the International Code for Fire Safety Systems, which first came into force on 1 July 2002, and amendments have been made to this code ever since as a result of the decisions taken at IMO's Maritime Safety Committee.

Foam-spreading installations

When fitted, permanently piped foam-spreading installations, operated external to boiler or machinery space, which supply foam to boiler and/or engine room tank tops, must

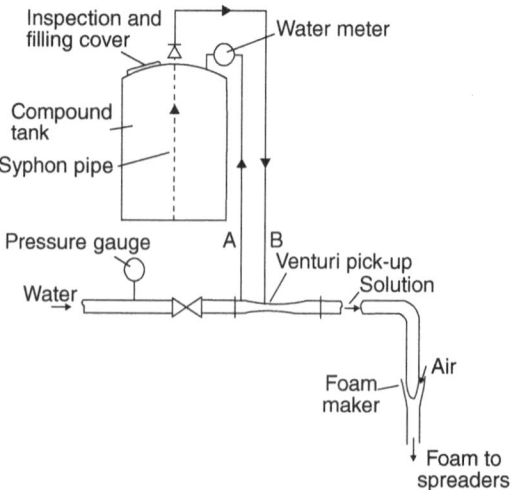

▲ **Figure 8.8** *Mechanical foam installation*

have sufficient capacity to give a depth of foam of at least 152 mm over the whole tank top. Water at a pressure of at least 6 bar (0.6 MN/m²), supplied from the ship's mains, passes through the water control valve into the Venturi fitting. Two small-bore pipes are connected to the Venturi fitting, pipe A is the high-pressure pipe led through a water meter to the top of the foam compound tank, pipe B is the low-pressure pipe that permits a controlled quantity of foam compound to be entrained into the Venturi fitting. The protein foam compound and water pass along the main delivery pipe to the foam makers situated in the boiler or engine room. As the mixture passes through the foam maker, air is entrained, which then produces a stable foam that is delivered to the foam spreaders. A diagrammatic representation of a foam maker is included with Figure 8.8.

Pre-mixed foam

This type of mechanical foam installation is self-contained, that is, it does not require motive power from the ship's pumps. To put the system in operation it is only necessary to pierce the CO_2 bottle seals by means of the operating gear provided. The CO_2 is then delivered at a pressure of approximately 42 bar (4.2 MN/m²) to the metering valve. As the CO_2 passes through the orifice plate it falls in pressure to 8 bar (0.8 MN/m²) or less. The solution in the storage tank is driven out via the delivery pipe to the foam makers situated in the boiler or engine room space, wherein air enters the system and foam is produced for distribution to the foam spreaders.

For tanker installations the foam may be delivered to the pump room, the engine or boiler room sprayers, or to a hydrant system to which hoses can be connected, which has foam-making nozzles attached (see Figure 8.9).

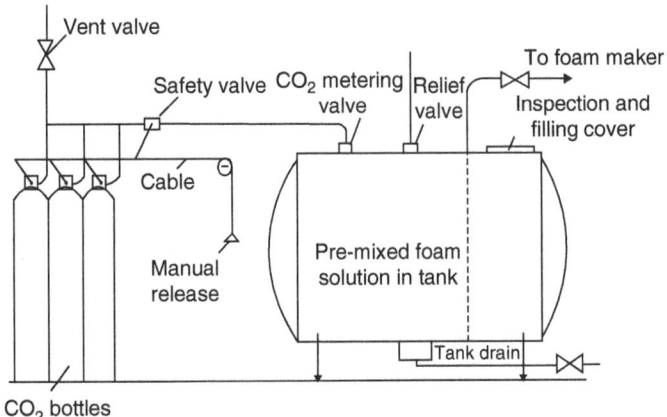

▲ **Figure 8.9** *Pre-mixed foam installation*

Foam compound injection system

Figure 8.10 shows diagrammatically the compound injection system often found on tankers for deck and machinery spaces. Tank and pumps may be situated wherever it is convenient.

Foam compound is drawn from the sealed tank by the compound pump, and air enters the tank through the atmospheric valve (this being linked to the compound valve). Both open simultaneously and delivery is to the automatically regulated injector unit. The injector unit controls the amount of water to compound ratio for a wide range of demand by the foam spreaders etc. A fire pump delivers the foam-making solution at sufficient pressure to the deck monitors (multi-directional-type foam guns) so that foam can reach any part of the deck, or to the spreaders for machinery spaces.

To bring the system into operation it is only necessary to open the linked air/foam compound valves and start the pumps. After use the system must be thoroughly flushed through and recharged.

In chemical foam installations the principal disadvantage is the deterioration of the chemicals and chemical solutions, hence regular checking is necessary to ensure the system is at all times capable of effective operation. However, with the chemical foam system good-quality uniform foam is capable of being produced.

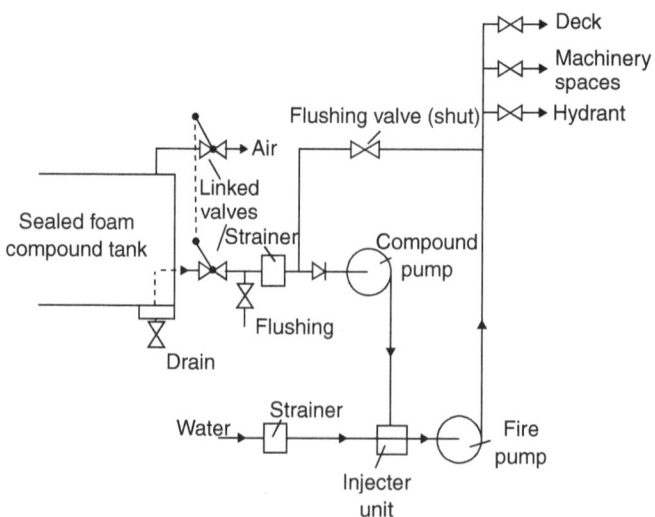

▲ **Figure 8.10** *Foam compound injection system*

With mechanical foam systems, storage and deterioration of the foam compound presents no difficulty, which is one of the reasons why this particular type of system is generally preferable.

High-expansion foam system

High-expansion foam firefighting systems shall be approved by the flag administrations to the guidelines developed by IMO. Any system must be capable of discharging a quantity of foam sufficient to fill the greatest space to be protected, through fixed outlet to cover at least 1 m depth per minute. The total quantity in reserve should be sufficient to fill the largest space five times. The expansion ratio shall not exceed 1,000:1. The Merchant Shipping Notice MSN 1666 issued by the Maritime Coastguard Agency (MCA) has more details about the regulations for the installation of this alternative fire extinguishing medium for boiler and engine room compartments.

The generators are large-scale bubble blowers that are connected by large-section trunking to the compartments (Figure 8.11).

A 1½ m long, 1 m square generator could produce about 150 m³/min of foam, which would completely fill the average engine room in about 15 minutes. 1 litre of synthetic detergent foam concentrate combines with 30–60 litres- of water (supplied from the sea) to give 30,000–60,000 litres of foam.

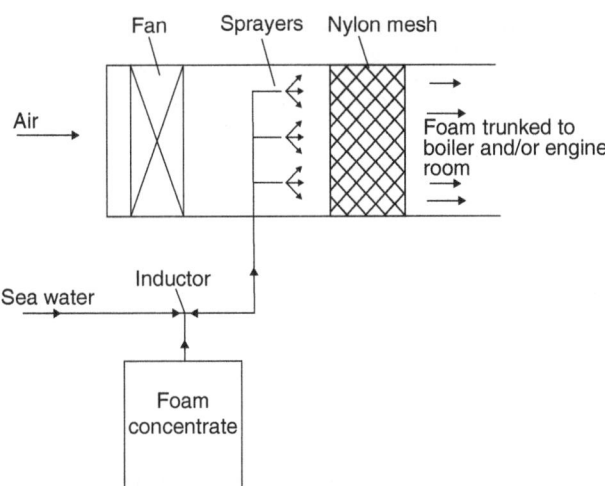

▲ **Figure 8.11** *High-expansion foam system*

The advantages are it is (1) economic; (2) can be rapidly produced; (3) could be used with the existing ventilation system; (4) personnel can actually walk through the foam with little ill effect.

The disadvantages are it is (1) persistent, it could take up to 48 hours to die down in an enclosed compartment; (2) large trunking is required; (3) should be trunked to the bottom of a compartment to stop convection currents carrying it away.

CO_2 flooding of holds and pump rooms

CO_2 is an effective medium for protecting paint stores, holds, cargo pump rooms and engine rooms for the following reasons:

- It does not react with other metals or electrical cables.
- There is no residue left after use.
- Modern systems react very quickly (following activation and safety checks).

Some vessels still in operation have a system of smoke detection and fire smothering that uses pipes of 20 mm diameter. These sampling pipes were led from the various hold compartments in the vessel to a cabinet on the bridge. Air drawn continuously through these pipes to the cabinet by suction fans would deliver the air through a diverting valve into the wheel house. Class and flag requirements are that alarms are also used.

If a fire started in a compartment, smoke would issue from the diverting valve into the wheelhouse, warning bridge personnel of the outbreak. Simultaneously, an electronic smoke detector in the cabinet sets off audible alarms, hence if the bridge is unoccupied (e.g. in port) the notice of outbreak of fire is still obtained.

Changeover valves are generally situated inside the lower portion of the cabinet, one for each of the sampling pipes. To flood an affected compartment with CO_2 gas, the operator would first operate the appropriate changeover valve and second release the requisite number of CO_2 cylinders for the compartment. CO_2 gas would then pass through the sampling pipe to the space in which the fire exists.

When a smoke detection system is to be used for the hold compartments of a refrigerated cargo vessel, the lines to the refrigerated holds will be blanked off in the detector cabinet. These blanks can be removed once per watch as a test (for a few days after loading cargo) and removed altogether when the hold is open and defrosted.

Note

When an outbreak of fire in a compartment is detected, the fire may be of small proportions and be capable of being extinguished by means other than flooding with the CO_2 equipment provided. In this event it would be necessary for personnel to enter the compartment in order to extinguish the fire. However, after inspection, the fire may be such that CO_2 flooding is necessary. Before this is done, an audible alarm should first be operated warning personnel that CO_2 flooding of the compartment is about to be used.

After the fire has been extinguished, the compartment must be well ventilated before entry for damage inspection. As CO_2 gas is heavier than air and does not support human life great care must be taken before re-entry into the hold.

CO_2 total flooding system for machinery spaces

The International Code for Fire Safety Systems, which is part of SOLAS, gives the constructional and performance details for fixed firefighting CO_2 flooding systems. For machinery spaces containing diesel main propulsion, or auxiliary machinery whose total power is 746 kW or more, a fixed firefighting installation has to be provided.

If a CO_2 total flooding system is fitted than there must be enough gas to fill 40% of the gross volume excluding the casing or 35% of the gross volume including the casing. The piping must be of a size to allow 85% of the gas to enter the space within 2 minutes of activation. Classification societies also require that modern systems protecting engine rooms or boiler rooms should discharge approximately 90% of the CO_2 above floor level and 10% below the floor level. CO_2 flooding is often used for tanker engine rooms and pump rooms even if the machinery used is steam turbine.

Operation

See Figure 8.12. First ensure that the compartment is evacuated of personnel and sealed off. This could be achieved by calling everyone to their muster station. Sealing off the compartment means closing all doors to the engine room, shutting down skylights, closing dampers or vents and stopping ventilation fans. Pumps should also be stopped and quick-closing valves closed. In a modern vessel the sealing off can be done by remote control from the fire control station generally using a compressed air or hydraulic system.

The door of the steel control box, situated at the fire control station, would then be opened, which operates a switch that may be arranged to have a dual purpose:

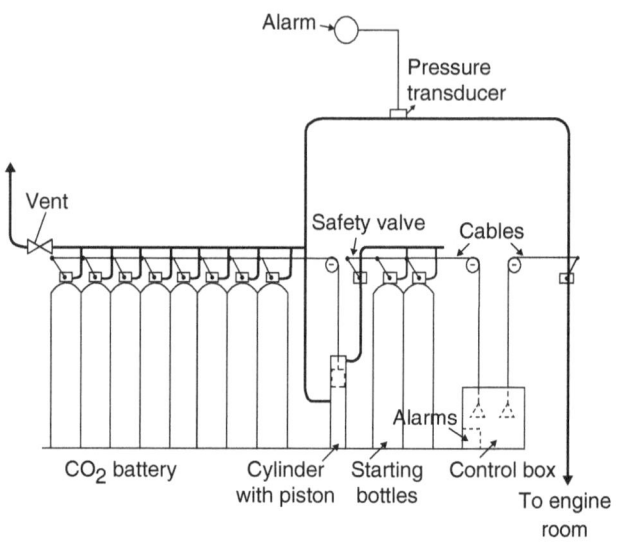

▲ **Figure 8.12** *CO₂ total flooding system*

1. To operate audible and visual alarms in the engine room spaces.
2. To initiate the shut off signals to the ventilation fans.

The CO_2 release handles would then be pulled; the first opens the main discharge valve to the engine room and the other operates the valves to the starting or pilot cylinders. The gas from these cylinders in turn operates the gang release mechanism to open the main bank of cylinders at the same time.

Reference must be made to MGN 389 (M + F) as these are the instructions to all seafarers relating to the Operating Instructions and signage for fixed gas fire extinguishing systems. In bold wording the note says '**Know YOUR system; don't be caught out in an emergency**'. It is very important that the engineer presenting him/herself for examination is able to describe the contents of MGN 389.

Maintenance and testing

Ensure that all moving parts are kept clean, free and well lubricated. Wires must be checked for tightness, toggles and pulleys must be greased. With the use of compressed air the CO_2 distribution pipes could be blown through periodically. CO_2 bottles must be weighed regularly to check contents (an ultrasonic or radioactive isotope unit detector could be used to check liquid level).

Note

The CO_2 storage bottles have seals that also act as bursting discs; should there be a CO_2 leakage from one or more of the starting bottles this cannot result in CO_2 discharge into the engine room from the battery because of the cable-operated safety valve. When leakage occurs either in the starting section or main battery, a pressure switch in the lines will cause alarms to be sounded, vents to the atmosphere can then be opened.

Bulk carbon dioxide system

This system was designed to replace the flooding systems for machinery and hold spaces that use a considerable number of carbon dioxide bottles.

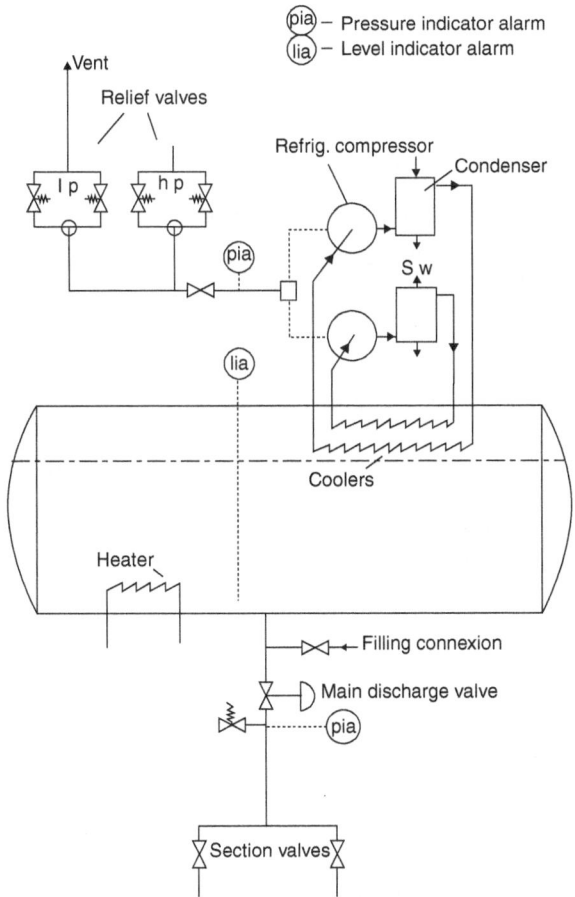

▲ **Figure 8.13** *Bulk carbon dioxide system*

It consists of a large, well-insulated container (older systems may have two containers) that holds carbon dioxide at a working pressure of about 21 bar, temperature about −20°C. In order to maintain this low temperature, duplicated refrigeration units automatically controlled by the pressure of the carbon dioxide are used. One refrigeration unit would be in operation and in the event of failure, the other would cut in automatically and warning would be given (see Figure 8.13).

Since it is essential that pressure in the container be maintained for fire extinguishing within a set range, a heater cuts in if required to increase the pressure of the carbon dioxide.

Two sets of relief valves are fitted: low-pressure (LP) valves are set at around 24.5 bar, high-pressure (HP) valves at around 27 bar. It is a requirement that the HP valves vent into the compartment in which the container is situated – this venting would occur in the event of fire in the compartment where the container is situated.

Alarms

These are provided for:

1. Loss of 5% of contents (low level).
2. Increase up to 98% of free volume (high level).
3. Leakage past main discharge valve.
4. Opening of section valve.

Balloons fitted over open ends of waste pipes give indication of relief valve leakage and two level indicators are provided, one remote the other local.

Advantages

Lower initial cost, reduced filling cost and filling is simplified. About a 50% saving in weight compared to a multi-cylinder system.

Disadvantages

This is a relatively complex system, which can reduce reliability. A power supply is required.

Operation

The appropriate section valve is opened (alarm sounds) and the main discharge valve is opened. The main discharge valve is usually fitted with an actuator for remote control, carbon dioxide is then delivered for a specified period (which depends upon the size of the compartment) and the main valve is closed.

Inert gas generator

This was originally developed to supplement CO_2 flooding systems. If a fire occurred on board a ship at sea and the fire was extinguished through using all the CO_2 available and a further outbreak of fire occurred, the situation could be dangerous.

In a compartment wherein there is an outbreak of fire, the minimum percentage of oxygen in the atmosphere in the compartment that will allow combustion to proceed varies with different materials between 12% and 16% approximately. Hence if the oxygen content of the compartment can be reduced below 12%, it would be insufficient to allow combustion to continue. This reduction in oxygen content can be achieved by employing a generator that will supply inert gas, which is heavier than air, so displacing the atmosphere in the compartment.

The generator consists of a horizontally arranged brick-lined furnace that is cylindrically shaped and surrounded by a water jacket. This is connected to a vertical combustion chamber in which water spray units and Lessing rings (cylinders of galvanised metal arranged to baffle the gas flow) are fitted. A water-cooled diesel engine, usually fitted alongside the generator, drives a fuel pump, a constant-volume air blower and an electric generator. The electric generator supplies current to an electric motor, which in turn drives the cooling water pump; the motor and pump are usually situated at the forward end of the shaft tunnel. By fitting the cooling water pump in the shaft tunnel and having it connected to the wash deck line, this pump can also be used as an emergency fire pump. Cooling water for the gas generator can also be supplied by ballast and general service pumps in the engine room, the amount of water required is approximately 545 litres per hour for every 27.7 m³ of inert gas produced.

The oil fuel burner is initially lighted by means of high-tension electrodes, the electrical supply being through a small transformer. A constant pressure regulator is fitted to the oil supply line to the burner along with a control valve.

A control panel for the gas generator incorporates an O_2 recorder, water, oil and fuel alarms as well as pressure gauges. In the gas piping system leading from the combustion chamber, condensate traps and drains are fitted. The following is an approximate analysis of the gas generated:

Oxygen	0–1%
Carbon monoxide	Nil
Carbon dioxide	14–15%
Nitrogen	85%

Under some circumstances there could be some unburnt hydrocarbons and oxides of nitrogen mixed in with the composition shown above.

Inert gas installation for tankers

Figure 8.14 shows, diagrammatically, a system of inert gas supply for the cargo tanks of an oil tanker. Gas from boiler uptakes passes through two pneumatically operated, remote-controlled, high-temperature valves. It then passes through a scrubber into which sea water is sprayed for cooling the gases to about 3°C above water temperature and scrubbing out soot particles and most of the sulphur oxides.

The gas then passes through a plastic demister, which can be cleaned by back flushing.

After the scrubbing the gas analysis would be about 13% carbon dioxide, 4% oxygen, 0.3% sulphur dioxide, remainder nitrogen and water vapour.

Two centrifugal blowers are provided, only one would normally be operated the other being a stand-by unit.

The supply of cleaned, dry inert gas at a pressure of 1.2–6 kN/m^2 gauge pressure is regulated by the automatically controlled bypass valve that is linked to the main supply valve. When the main valve starts to close the bypass begins to open and vice versa.

Safety features control and alarms

1. High-temperature gas valves: open/shut indication on control room panel, close automatically when soot blowing of the boiler is put into operation.
2. If inert gas temperature gets too high, automatically delivery valves are closed and fans stopped.
3. Low water flow to seals and scrubbers.
4. High gas temperature.
5. Scrubbing tower overflow.
6. High oxygen content in the gas.

The system is suited to boiler installations as gas from diesel engines contains large quantities of excess air. An inert gas generator may be installed for gas delivery to the system in the event of no gas being available from the boilers.

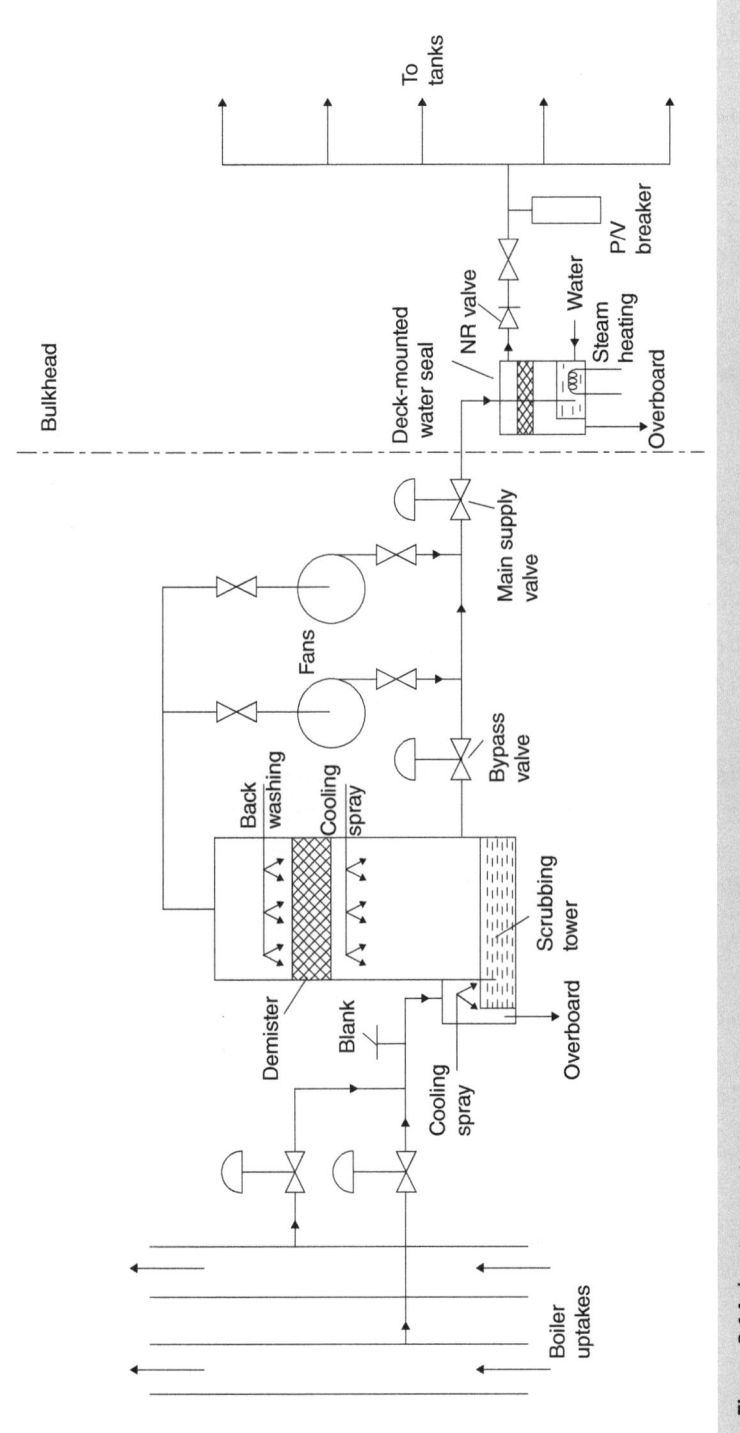

▲ Figure 8.14 *Inert gas system*

Advantages of the system

1. No explosive mixtures can form in the tanks.
2. Reduces corrosion.
3. Voyage cleaning of tanks is unnecessary.
4. Reduces pumping time because of the positive pressure in the tanks at all times.

Water Spray Systems

Automatic sprinkler system

The sprinkler system is an automatic fire-detecting, alarm and extinguishing system that is constantly 'on guard' to deal quickly and effectively with any outbreak of fire that may occur in accommodation or other spaces.

The system consists of a pressurised water tank with water pipes leading to various compartments. In these compartments the water pipes have sprinkler heads fitted that come into operation when there is an outbreak of fire and the sprinkler head becomes heated. Different heads will 'burst' at different temperatures depending upon the positioning and design of the units.

Figure 8.15 is a diagrammatic arrangement of the system. The pressure tank is half-filled with fresh water through the fresh water supply line. Compressed air delivered from the electrically driven air compressor raises the pressure in the tank to a predetermined level; this should be such that the pressure at the highest sprinkler head in the system is not less than 4.8 bar (0.48 MN/m^2).

Sprinkler heads are grouped into sections with not more than 150 heads per section and each section has an alarm system. Each sprinkler head is made up of a steel cage fitted with a water deflector. A quartzoid bulb, which contains a highly expansible liquid, is

Rating	Colour
68°C	Red
80°C	Yellow
93°C	Green

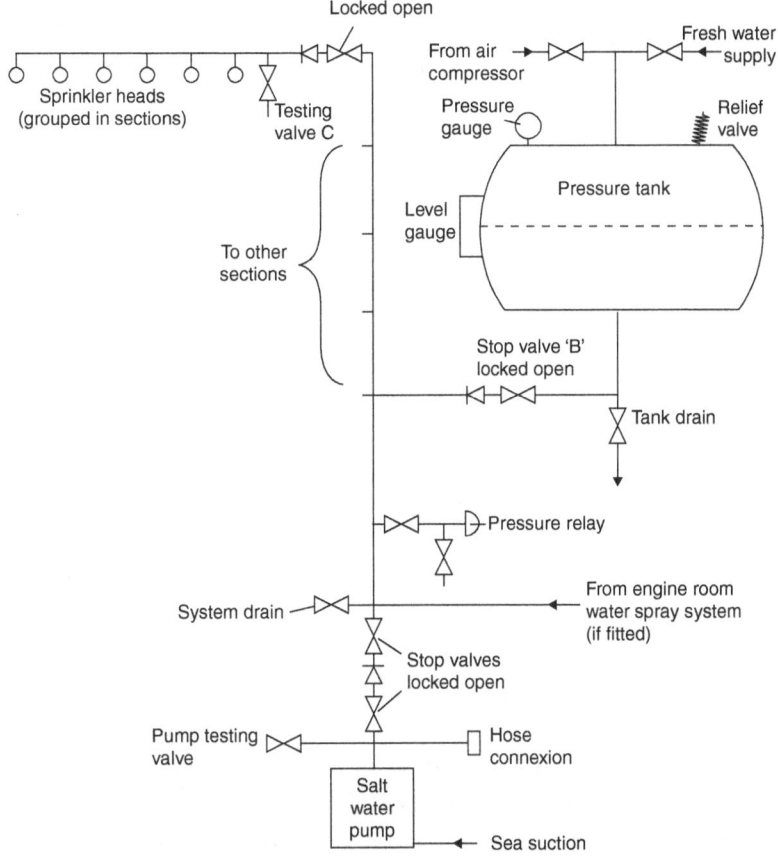

▲ Figure 8.15 *Automatic sprinkler system*

retained by the cage. The upper end of the bulb presses against a valve assembly that incorporates a soft metal seal.

When the quartzoid bulbs are manufactured, a small gas space is left inside the bulb so that if the bulb is subjected to heat, the liquid expands and the gas space diminishes. This will generate pressure inside the bulb and the bulb will shatter once a predetermined temperature (and hence pressure) is reached. Generally, the operating temperature range permitted for these bulbs is 68–93°C but the upper limit of temperature can be increased, which would depend upon the position of the sprinkler head or heads. Quartzoid bulbs are manufactured in different colours; the colour indicates the temperature rating for the bulb.

Once the bulb is shattered, the valve assembly falls permitting water to be discharged from the head, which strikes the deflector plate and sprays over a considerable area.

When a head comes into operation the non-return alarm valve for the section opens and water flows to the sprinkler head. This non-return valve also uncovers the small-bore alarm pipe lead and water passes through this small-bore alarm pipe to a rubber diaphragm. The water pressure acts upon the diaphragm and this operates a switch which causes a break in the continuously live circuit. Alarms, both visible and audible, fitted in engine room, bridge and crew space are then automatically operated.

Stop valves, A and B (Figure 8.16), are locked open and if either of these valves is inadvertently closed a switch will be operated that brings the alarms into operation. The alarm system can be tested by opening valve C, which allows a delivery of water similar to that of one sprinkler head to flow to drain.

An electrically operated pump with a direct suction to the sea comes into operation when the fresh water charge in the pressure tank has been used up. This is arranged to operate automatically through the pressure relay. A hose connection is also provided so that water can be supplied to the system from the shore when the vessel is in dry dock.

High-pressure water spray system

These systems are used as firefighting provision for machinery spaces. The design should be such that at least 5 litres/m²/min of water is supplied into the space being protected. The system incorporates an air vessel, fresh water pump and salt water pump all connected to piping that is led to sections, each section having its own shut-off valve and sprayer heads, which unlike the sprinkler system have no quartzoid bulbs or valves but are open (Figure 8.16). Precautions must be taken to ensure that the pipe work, nozzles and pumps do not become blocked or corroded.

With all section valves closed, the system is full of fresh water under pressure from the compressed air in the air vessel. When a section valve is opened water will be discharged immediately from the open sprayer heads in that section, and pressure drop in the system automatically starts the salt water pump, which will continue to deliver water to the sprayers until the section valve is closed.

After use the system should be flushed out and recharged with clean, fresh water.

The air vessel is incorporated into the system to prevent the pump cutting in if there is a slight leakage of water from the system.

It should be tested at weekly intervals. Open A, close B, open C; the pump should automatically start and discharge from A. This avoids having to refill the system with fresh water.

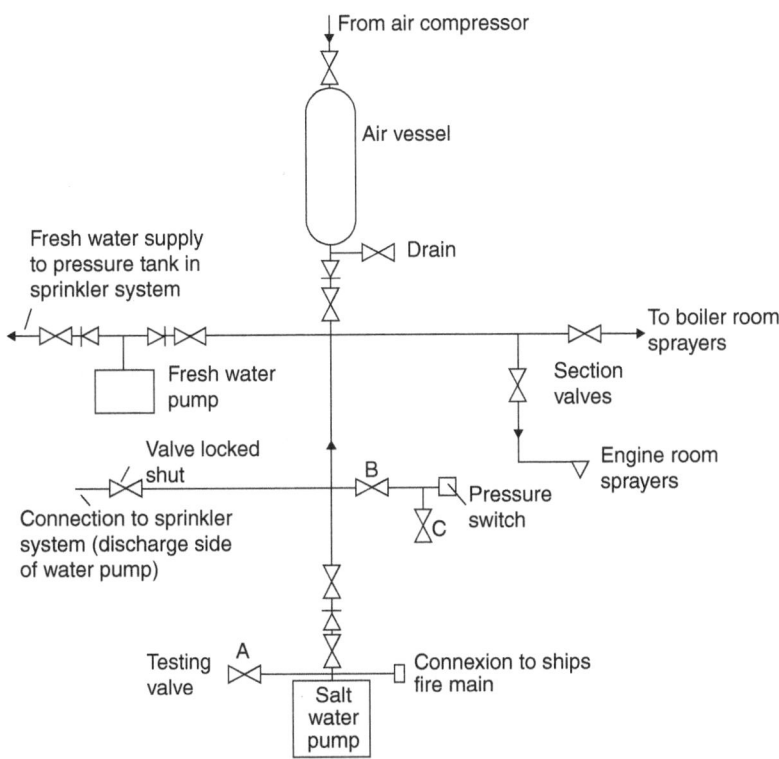

▲ **Figure 8.16** *High-pressure water spray system*

High-pressure water mist systems for engine room protection

The flag state is responsible for approving these systems. They have been developed for use instead of gas flooding systems for the fire protection of machinery spaces. The system uses fresh water pushed through a nozzle at high pressure. The advantage is that as well as blocking the oxygen to the fire locally the water mist is very effective at cooling the fire and blocking radiant heat making it much more effective than traditional sprinkler systems, and by using considerably less water.

Merchant shipping (fire protection: large ships) regulations (1998)

Additional details for the UK legislation are set out in MSN 1666 (M). The 1998 regulation is the original legislation and should be studied alongside SOLAS Chapter II-2 and the International Code for Fire Safety Systems as well as MSN1666/MSN1665.

It is strongly recommended that students study these regulations that are available on the various websites.

General information

All seagoing vessels (registered in the UK) are assigned to a specific 'class'. A ship of Class I is a passenger ship engaged on voyages any of which are long international voyages. The ship must have fire pumps, a fire main, hydrants, hoses and nozzles. The equipment and performance specifications are detailed in the Merchant Shipping Regulations.

Machinery space of category A

Every machinery space of category A (which is a space with internal combustion power of 373 kW or above or has any oil-fired boiler) must be fitted with a fixed firefighting system, which could be one of the following:

1. Water spray system.
2. Gas fire extinguishing system.
3. High-expansion foam fire extinguishing system.

Every ship should have the following:

1. A fixed foam fire extinguishing installation operated from outside of the space and capable of giving a depth of foam of at least 150 mm in not more than 5 minutes over the largest single area over which oil fuel is liable to spread.

   Such installations shall include mobile sprayers ready for immediate use in the firing area of the boiler and in the vicinity of the oil fuel unit. A pressure water spray system or fire-smothering gas installation may be used as an alternative.
2. A 136-litre foam fire extinguisher (or 45 kg CO_2) capable of delivering foam to any part of the compartment.
3. Two portable fire extinguishers suitable for extinguishing oil fires.
4. A receptacle containing at least 0.3 m^3 of sand and a scoop.
5. Two fire hydrants, one port, one starboard, with hoses and nozzles (spray nozzles must also be provided).

Machinery spaces containing internal combustion machinery

In every ship of Class I the following should be provided:

1. A fixed foam fire extinguishing, water spray or smothering gas installation operated from outside of the machinery space if the machinery space is Category A, which is a space with internal combustion power of 373 kW or above or has any oil-fired boiler.
2. One foam fire extinguisher of at least 45 litres capacity or a CO_2 fire extinguisher of at least 16 kg capacity.
3. One portable fire extinguisher suitable for extinguishing oil fires for each separate space but not less than two such extinguishers or more than six.
4. Two fire hydrants, one port and one starboard, with hoses and nozzles (spray nozzles must also be provided).

Machinery spaces containing steam engines

1. Foam fire extinguishers each of at least 45 litres capacity or CO_2 fire extinguishers each of at least 16 kg capacity, sufficient in number to enable foam or CO_2 to be directed on to any part of the pressure-lubrication system and on to any part of the casings enclosing pressure-lubricated parts of turbines, engines or associated gearing. These would not be required if a fixed firefighting installation similar to (1) for internal combustion machinery were provided.
2. One portable fire extinguisher suitable for extinguishing oil fires for each 746 kW (or part thereof) but not less than two such extinguishers or more than six.
3. Two fire hydrants, one port and one starboard, with hoses and nozzles (spray nozzles must also be provided).

Portable fire extinguishers

1. Those discharging fluid shall have a capacity of not more than 13.5 litres and not less than 9 litres.
2. CO_2 extinguishers shall have a capacity of not less than 3.2 kg.
3. Dry powder extinguishers shall have a capacity of not less than 4.6 kg.

International shore connection

An international shore connection must be provided to enable water to be supplied from another ship or from the shore to the fire main, and fixed provision shall be made to enable such a connection to be used on the port side and starboard side of the ship.

Cargo spaces and store-rooms

1. Every ship of Class I shall be provided with appliances whereby at least two powerful jets of water can be rapidly and simultaneously directed into any cargo space or store-room.

2. Every ship of Class I of 1,000 tons or over shall be provided with appliances whereby fire-smothering gas can be rapidly conveyed by a permanent piping system into any compartment appropriated for the carriage of cargo. The volume of free gas shall be at least equal to 30% of the gross volume of the largest hold in the ship, which is capable of being effectively closed. Provided that steam may be substituted for fire-smothering gas in any ship in which there are available boilers capable of evaporating 1.3 kg of steam per hour for each 1 m^3 of the gross volume of the largest hold in the ship.

Fire pumps

The regulations relating to the capacity of fire pump power is complex and set out in MSN 1665. The latest regulations relate to the total suction head of the main pumps, their ability to pressurise the fire hoses and the relative power of the emergency fire pump. It does, however, make sense to have at least two alternatives as the pumps may be doubled up as the ballast or general service pump.

In any passenger ship each fire pump must be able to deliver the following pressures to any fire hydrant.

1. 4,000 tons and upwards at a pressure of 310 kPa.
2. 1,000 tons and upwards but under 4,000 tons at a pressure of 270 kPa.
3. under 1,000 tons at a pressure of 210 kPa.

In any ship other than a passenger ship each fire pump must be able to deliver the following pressures to any fire hydrant.

1. 6,000 tons and upwards at a pressure of 270 kPa.
2. 1,000 tons and upwards but under 6,000 tons at a pressure of 250 kPa.
3. under 1,000 tons at a pressure of 210 kPa.

In every ship of Class I fitted with main or auxiliary oil-fired boilers or internal combustion propelling machinery, the arrangements of sea connections, pumps and the sources of power for operating them shall be such as will ensure that a fire in any one compartment will not put all the fire pumps out of action.

Water pipes, hydrants and fire hoses

Every ship of Class I shall be provided with water pipes and hydrants. The diameter of the water pipes shall be sufficient to enable an adequate supply of water to be provided for the simultaneous operation of at least two fire hoses and for the projection thereby of two powerful jets of water. The number and position of the hydrants shall be such that at least two such jets may be directed into any part of the ship by means of two fire hoses each not exceeding 18 m in length, each jet being supplied from a separate hydrant. At least one fire hose shall be provided for each hydrant.

Fire-fighter's clothing (SOLAS – fireman's outfit)

Every SOLAS-approved ship shall be provided with at least two sets of protective clothing for the staff designated to fight fires on board. The outfit will include:

1. A safety lamp.
2. A fireman's axe.
3. (i) a breathing apparatus; or (ii) a smoke helmet; or (iii) a smoke mask. The outfits shall be kept in at least two separate places on board.

Some owners recognise the need to equip their ships with more that the minimum required by SOLAS. Flash hoods for example are regarded by regular firefighters as essential protection especially against burns from the hot buckles and clips that are fitted to the helmets.

Fire patrols

Every Class I ship will ensure that an efficient patrol system is in place to detect any outbreak of fire promptly. In addition there will be:

1. Adequate call points.
2. Trained personnel.
3. Sufficient communication including two-way portable radios for patrol members of vessels carrying more than 36 passengers.

Breathing Apparatus

Emergency escape breathing device

The Fire Safety Systems Code, which is part of SOLAS, gives the minimum number of emergency escape breathing devices (EEBDs) that should be carried on board cargo ships and passenger ships. The EEBD is a supplied air or oxygen emergency device that is used only to escape from a hazardous area and is not to be used for fighting fires.

Traditionally a smoke mask with a foot-operated bellows connected by hose to the face mask was used by crew to fight fires. A harness with lifeline attached accompanies it, and it is essential that the signal code used must be fully understood by all personnel.

MSN 1665 gives details of the requirements for the smoke helmet and hose if carried on UK ships. The main advantage with the smoke mask was that everyone could familiarise themselves with its operation by wearing and using the apparatus, no bottles have to be re-charged or maintained. In large vessels, however, the excessive hose length made the unit unsuitable and the self-contained breathing apparatus (BA) would have been preferred. Another major disadvantage is that the combination of hose and life-line could make things dangerous if sharp complex obstructions have to be negotiated, cutting or fouling of the hose and line could occur.

Self-contained breathing apparatus (BA) sets

Every self-contained BA shall be of the open-circuit, compressed-air type that has a certificate of assurance issued by the Health and Safety Executive The minimum statutory capacity requirement is 1,200 litres of free air, and most sets contain from 1,200 liters to 1,800 litres at a pressure of about 140 bar to 210 bar.

The basic set consists of an air cylinder mounted on a plastic back plate fitted with a harness. A moulded rubber face mask incorporating a demand valve, exhalation valve/speech transmitter, head harness and visor is connected by high-pressure (HP) reinforced hose from the demand valve to the air manifold. A pressure gauge and low-pressure warning whistle, which gives audible warning to the wearer when 80% of the air has been used, completes the assembly.

Means shall be provided for the automatic regulation of air supply to the wearer of the apparatus in accordance with his breathing requirements when he is breathing any volume of free air of up to 85 l/min at any time when the pressure in the supply cylinder or cylinders is above 1.05 MPa. Means shall be provided for overriding the automatic air supply to increase the volume of air available to the wearer if required. If the ship is of a type where toxic chemicals are supplied then the BA sets should be of the positive-pressure type.

Familiarity with the apparatus is essential and to facilitate this it would be useful to have on board a compressor that delivers oil free air to recharge the bottles.

Emergency fire pump

This independent pump with its own prime mover, generally a diesel engine with its own low-flash-point fuel supply, must be situated outside of the engine room and connected into the fire main.

In the event of fire in the engine room and subsequent evacuation and sealing, the emergency fire pump must be started and the engine room isolating valve in the fire main closed.

Figure 8.17 shows a completely independent emergency fire pump system. The centrifugal fire pump and hydraulic motor would be completely submersible and irrespective of the height and draught of the vessel the pump would not require a priming device as it would be below the water level. Such an arrangement may also be used as a booster/priming device for a main fire pump situated on deck.

Safety is an important and far-reaching topic and students presenting themselves for flag state examination must be mindful of the common sense aspects of safety. This is at the heart of the examination process and searching questions will arise that require a thorough technical knowledge such as in the following examples:

- How does oil vapour pressure affect flammability?
  - Oils with high vapour pressure are those that are extremely volatile are such as crude oil and petroleum. This means that they have a higher vapour pressure and a greater fire hazard.

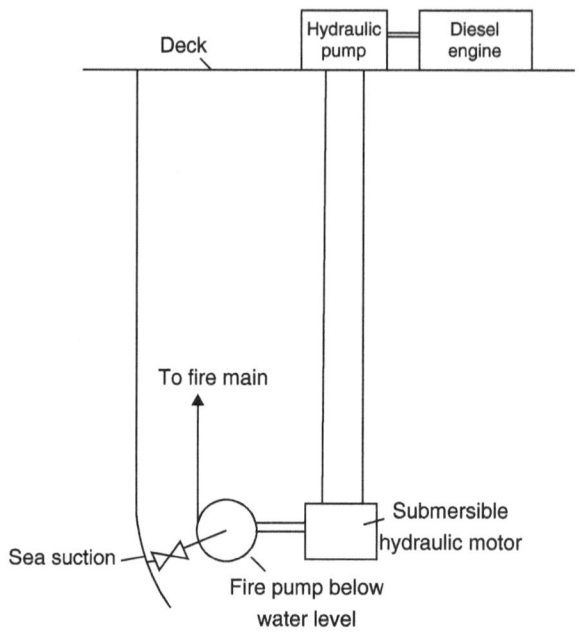

▲ **Figure 8.17** *Emergency fire pump system*

- How does the explosive limit of oil affect flammability?
  - o Explosive limits are classified as (1) lower explosive limit (LEL) and (ii) higher explosive limit (HEL). They are the percentage of oil vapour by volume with air, usually 1–10%, respectively for most hydrocarbons. If the mixture is below the LEL or above the HEL the mixture will not ignite or explode. Explosive limits are related to vapour pressure. If vapour pressure is 0.05 bar then percentage by volume with air is $0.005 \times 100\% = 0.5\%$ (below the LEL).
- How does oil vapour density affect safety?
  - o Most petroleum vapours are heavier than air, hence in still air if a tanker is loading cargo some of the vapours could gravitate into the lower recesses of the vessel – for example, accommodation and machinery spaces. PV valve arrangements usually vent the vapours at high velocity and to considerable height in order to minimise the risk.
- How does heating of high-viscosity oil affect safety?
  - o For good atomisation fuel viscosity should be about 150 seconds Redwood N°1 or less. With high-viscosity oil it is necessary to heat the oil in order to reduce its viscosity. *Note:* Pressurised heavy fuel oil may be heated to a temperature above its flash point to achieve the necessary viscosity and obviously if a leakage of this

hot pressurised oil occurred it could be extremely dangerous. Fuel lines carrying this oil should not pass near high-temperature surfaces, for example, exhaust manifolds or electrical equipment. Ideally they should be jacketed, with jacket drains leading to a safe place – a drain tank with level alarm.

- How does the auto-ignition temperature of fuel and lubricating oil affect safety?
  - The auto-ignition temperature of the vapours of fuel and lubricating oil are much lower than those of vapours from the volatile petroleum liquids. Hence they are more likely to ignite if sprayed on to a hot surface. Remember it is the vapour that burns.

9

PUMPS AND
PUMPING SYSTEMS

A ship's engine room, and on some ship types the pump room, contain a number of complex pipe and pumping systems. Some of the most important systems are listed below:

- bilge and ballast;
- fuel and lubricating oil;
- cooling water;
- fresh water and grey water;
- as well as other systems, such as hydraulic pressure systems and compressed air systems.

These pumping systems will involve a large number of pipe leads, cross-connections, valves and connecting systems that will be constructed in different ways. The pumps and equipment provided on a modern ship for these various duties are also of different types and sizes.

With the introduction of the mandatory Ship's Energy Efficiency Management Plan, pumps and pumping systems have become the focus of attention for the potential to save energy. Components made from lightweight materials have the added advantage of being non-corrosive as well as reducing weight. Variable speed control of AC electric motors means that pumps can be operated at their optimum setting for greater proportions of their duty thus saving considerable energy but still carrying out the same task. This area of marine engineering is developing fast but the basic engineering concepts remain, therefore the descriptions within this chapter will be a broad overview and concentrate on the *essential* information that will be of importance to the student preparing for the flag state examinations.

Initially the chapter will consider the main types of pumps in regular use together with any relevant points and the latest development. Associated equipment is then considered, such as heat exchangers, pneupress fresh water systems, sewage systems oil–water separators and feed water injectors. The chapter will conclude with some pipe arrangements and fittings and a concise grouping of the rules and regulations pertaining to pumping systems. It should be stressed that these rules do not have to be memorised and are given to allow the student to have an appreciation of the minimum requirements. The International Association of Classification Societies (IACS) and the classification societies have extensive rules for the design of pumping systems and the student wishing to become involved in this area can refer to the related class society for additional detailed information.

Types of Pumps

Classification

See Figure 9.1.

1. *Positive displacement pumps.* One or more chambers are alternately filled then emptied, usually by mechanical means. These include reciprocating, screw, gear and water-ring types, etc. They do not require a priming device; in fact they may be used as priming devices. In general, they would be used for low to medium discharge rates, they can pump fluids of a wide range of viscosity and can develop – especially in the case of the reciprocating pump – high-pressure differentials if required.

2. *Dynamic pressure pumps (or roto-dynamic pumps).* A tangential acceleration is imparted to the fluid, which is then changed into pressure energy by the shape of the casing. These include centrifugal, axial and mixed-flow (part axial, part centrifugal) types. Depending upon supply head they may require a positive displacement pump as a priming device. In general they would be used for medium to high discharge rates and are usually confined to low-viscosity fluids, generating low to moderate pressure differentials, although special application, multi-stage pumps can be used to generate higher pressure differentials.

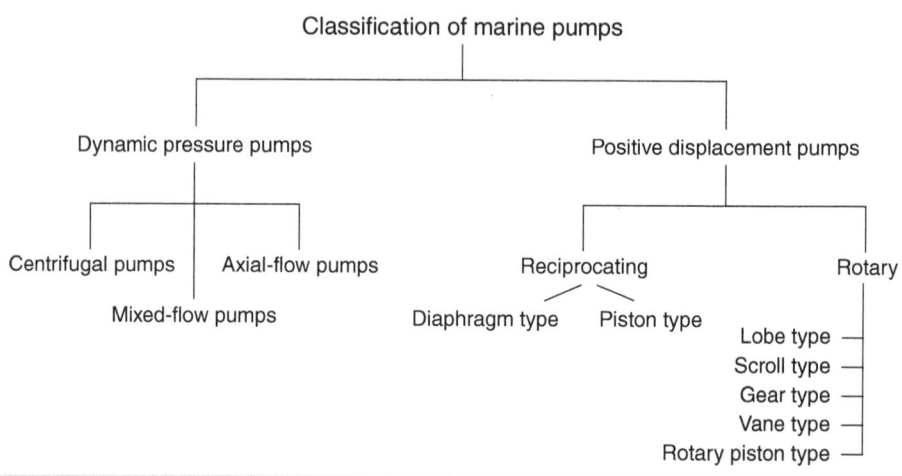

▲ **Figure 9.1** *Classification of pumps*

Priming

Priming pumps is a very important concept for the marine engineer (and deck officers working on tankers) to understand. It is at the very heart of problem-solving faults with pumping systems, especially in bilge systems and other important systems such as fuel oil transfer or feed water delivery to main or auxiliary boilers.

The fundamental concept pertaining to pumping liquids is that they are not compressible. Therefore the forces imparted to the liquid means that the liquid reacts straight away by moving in the desired direction. As long as the force is maintained, the liquid will continue to move in the way that the designer intended.

However, if air is allowed to take the place of the water (which could be for a number of reasons) then it will have to be removed before the pump will operate correctly. This is the action that is called 'priming'.

The positive displacement pumps will move the air on through its normal action, which means that if there is a liquid available at the suction the pump will start pumping again. This type of pump is called a 'self-priming' pump. The roto dynamic pumps are not in themselves self-priming but some do have priming devices adjacent to them, which makes the two coupled together a self-priming pumping unit. If the centrifugal, axial-flow or mixed-flow pump loses suction and air gets inside the pump it will stop working and the air will have to be removed.

Identifying the malfunction and problem-solving the causes of pumping failures takes considerable thought and marine engineers will need to be mindful of several factors. Suction and discharge pressures will be an important indicator as will the type of service that the pump is performing. For example, if a reciprocating bilge pump that should be self-priming is found to have no discharge but a high suction pressure then the suction valve could be shut or the suction filter could be blocked. If there is little or no suction pressure then the pump will still be drawing in air from somewhere.

Air might enter the pump via the drive shaft seal or gland and therefore it is important that these are all checked by the engineer during normal watchkeeping duties.

Feed water pumps might fail to pump on occasion due to gassing up inside where steam bubbles start to form in the water, which as they build up will reduce the ability of the pump to discharge the water into the boiler.

The majority of the pumps working in the engine room will be placed at the bottom of a closed system such as the engine fresh water cooling system or the domestic water supply system. In these cases, the pumps will have a level of liquid above the pump and the simple action of opening the suction valve will allow the oil or water to enter the pump and any air inside will be displaced. Under these circumstances the pump, independent of type, will stop and start without the need for intervention by the engineer or without any priming device installed.

When the pump is above the end of the suction pipe, as with a fuel transfer pump taking fuel from a double-bottom tank, then upon completion the pump can be stopped and the suction valve shut with liquid still inside the pump. This action might result in the pump retaining its suction pressure on the next operation.

Central priming system

Figure 9.2 shows diagrammatically a central priming system arranged to give automatic priming to four pumps. The system can be used for as many pumps of the centrifugal type that would be used in an engine room.

Water ring exhausters maintain a vacuum condition between pre-set limits in the vacuum tank. Opening the priming cock, or SDNR valve, for a pump causes priming to take place. To prevent water entering the vacuum tank after priming, float-operated 'air release' valves will automatically close.

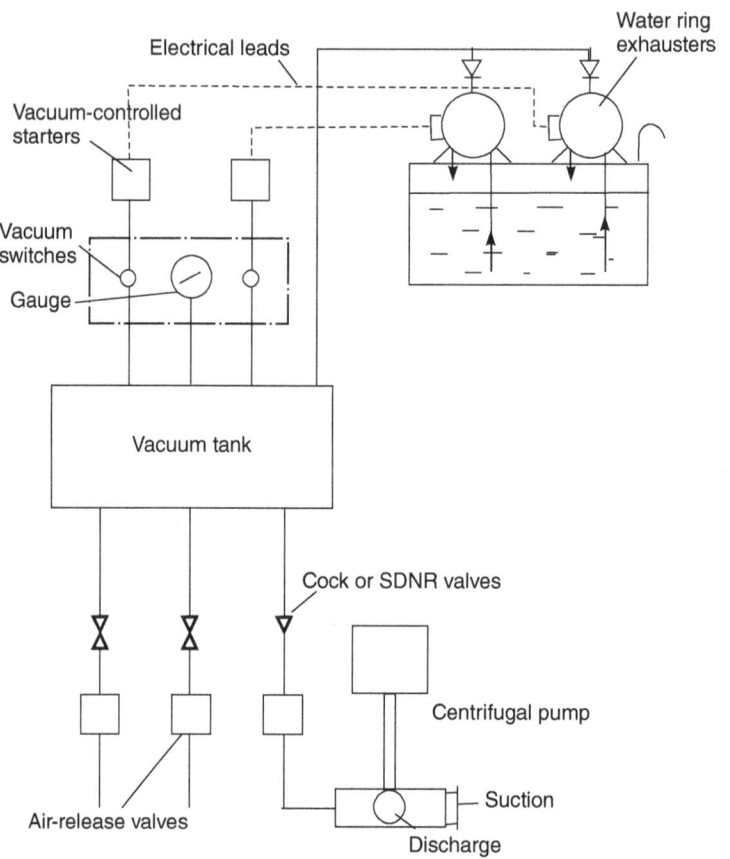

▲ **Figure 9.2** *Centralised hydraulic system*

For essential services a SDNR valve would be fitted instead of a priming cock so that if the valve is inadvertently left open, and due to mal-operation or a defect the vacuum in the tank is lost, air is not drawn into the pump and its suction lost with possible serious consequences to plant.

These are the advantages of the system:

1. Saving in total power since each pump does not have its own exhauster or priming unit operating all the while the pump is operating.
2. Reduced capital cost.
3. Simplified maintenance.
4. Automatic – takes care of any minor leaks that may be present in the suction side of a centrifugal pump.

Two water ring-type of exhausters mounted on top of a water supply tank are shown, one would act as a stand-by unit, but both could operate together in the event of heavy demand.

Positive Displacement Pumps

Reciprocating pumps

Pumps are becoming more and more sophisticated with the advancement of material science and there is now a good variety of such pumps; the piston or plunger type is still the most common type in the marine environment and can be arranged in both horizontal and vertical configurations. They are used for all types of duties on ships but find most favour as bilge pumps. The most common form is currently to produce the reciprocating motion through a coupling and connecting rod from an electric drive motor.

The piston pumps are a simple concept where the piston moves up and down in the cylinder and liquid (usually water) is moved through the pump under the action of the piston. Suction and delivery valves are arranged to open and close under the differential pressure set up by the movement of the pistons. The reasonably tight tolerances mean that the pump has the ability to push air in the same way that it pushes the fluid. Therefore the air is removed, which makes the pump a self-priming pump. This self-priming ability gives this type of pump the advantage of pumping the last drop of liquid from a tank or bilge. Other advantages are its ability to create a high suction, which means that liquids could be lifted from a large distance below the pump. The pumps also complete their pumping task with a low shear effect on the liquid being pumped. Therefore, for example, the pumping of oily water means that there is only a small amount of emulsifying effect, which means that the oil–water separation process is more effective.

Gear pumps

These positive displacement pumps are usually driven, by an electric motor, through a flanged coupling or gears. The lubricating oil gear pumps inside an engine might also be driven via a chain drive. The pumps are usually arranged to supply a constant volume at the correct pressure and therefore the pump is self-regulating and will not need controlling. However, for applications where control is required this will either be by the use of a bypass valve for conventional single-speed AC drives or with the use of electronically controlled variable-speed AC or DC motors. Often these pumps are also arranged so that they are full of liquid at all times and therefore there are no suction or discharge valves, similar to the centrifugal pumps arranged in fresh water cooling systems.

Gear pumps are often arranged so that one of the gears inside the pump is driven by the external power source and this gear in turn drives the second gear. The two intermeshing toothed wheels shown in Figure 9.3 are a close fit in the casing. The fluid is carried through the pump by travelling around in the space between the teeth and the casing. The oil is NOT, as some students will think, pushed through the teeth that are meshed together. The advantage of these pumps are that they are a positive-displacement, smooth-running pump that has a constant output and are well suited to pump oil, particularly for the lubricating oil pressure feed to diesel engine bearings where the slightest hesitation in flow could cause a bearing failure.

Screw displacement pumps

Considering Figure 9.3, it is seen that the fluid enters the outer suction manifolds and passes through the meshing worm wheels, which are gear driven from a motor, to the central discharge manifold. Such pumps are quiet and reliable and are particularly suited to pumping all fluids, in particular oil. The pump can deal with large volumes of air while running smoothly and maintaining discharge pressure. It is well suited to tank draining and intermittent fluid supply such as may occur in lubricating oil supply systems to engines, with the vessel rolling.

Timing gears are fitted to some screw pumps to ensure that correct clearance is maintained at all times between the screws, thereby preventing overheating and possible seizure. Modern designs of screws preclude the use of timing gears, ensuring smooth, efficient, simple operations that eliminate turbulence and vibration. It is interesting that scroll pumps supplying oil to a gearbox can operate for many years and still have the original machining marks on the scroll as the oil does a good job of keeping the internal parts from touching each other.

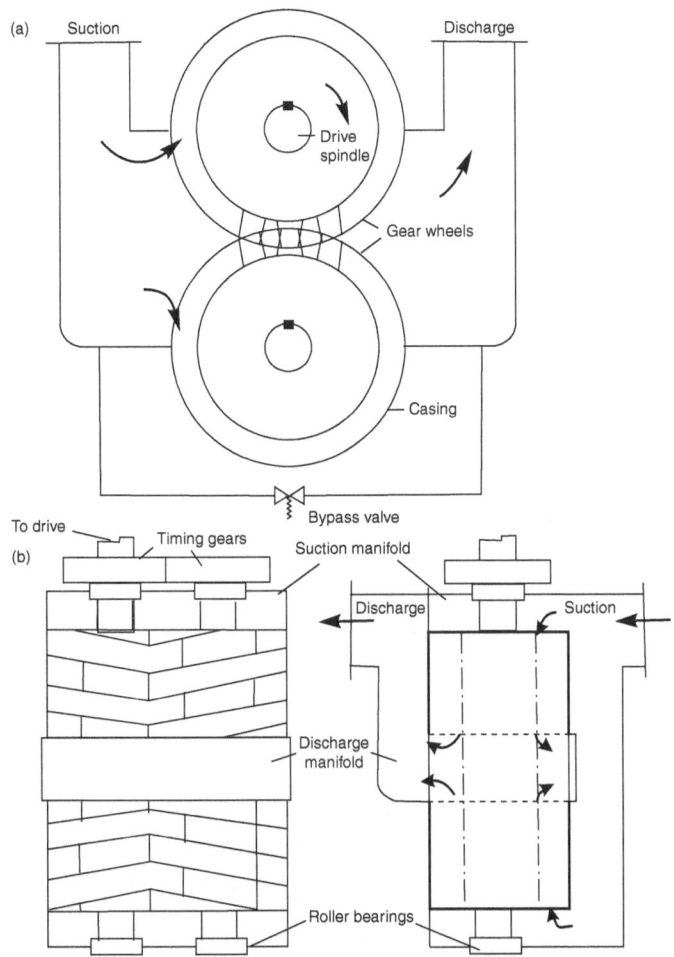

▲ **Figure 9.3** *Screw displacement pump*

Roto-dynamic Pumps

Centrifugal pumps

Impeller and volute casing

In single-stage pumps a single impeller rotates in a casing of spiral or volute form and in multi-stage pumps two or more impellers are fitted on the same shaft. Fluid enters the impeller axially through the eye then by centrifugal action continues radially and discharges around the entire circumference. The fluid in passing through the impeller receives energy from the vanes giving an increase of pressure and velocity. The kinetic (velocity) energy of the discharging fluid is partly converted to pressure energy by the suitable design of impeller vanes and casing. In some pumps, for example, turbine-driven boiler feed pumps, diffusers are used. These consist of a ring of stationary guide vanes surrounding the impeller, and the passage through the diffuser vanes is designed to change some of the velocity energy in the fluid to pressure energy. In double-inlet pumps, fluid enters from two sides to the impeller eye as if there were two impellers back to back giving twice the discharge at a given head. In multi-stage pumps the fluid from one impeller is discharged via suitable passages to the eye of the next impeller so that the total head developed (or discharge pressure) is the product of the head per stage and the number of stages, such a pump is often used for high-pressure discharge at moderate speed (e.g. the turbo-feed pump for main or auxiliary boilers).

A double-inlet impeller and single inlet impeller for comparison, together with a volute casing, are shown in Figure 9.4.

The impeller and volute casing design will depend on the required duty, for example, head to lift, head to discharge (pressure), quantity, etc. A typical centrifugal bilge pump would give an output of about 30 kg of water in 1 second, consume about 12 kW of

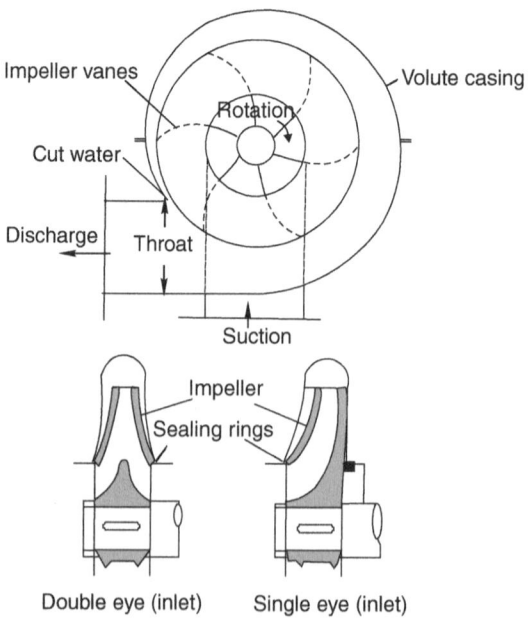

▲ **Figure 9.4** *General arrangement of the centrifugal pump*

power and discharge up to 5 bar running at about 17 rev/s. The casing usually has the suction and discharge branches arranged at the back so impeller and spindle can be removed from the front without breaking pipe joints. The discharge branch is usually on the pump centre line so that the pump is not 'handed'. The number of impeller vanes is not fixed but usually there are six to ten. The volute casing is like a divergent nozzle that is wrapped around the impeller and serves two main functions:

1. It enables velocity energy to be converted into pressure energy, the degree of conversion is governed mainly by the degree of divergence.
2. It accommodates the gradual increase in quantity of fluid that builds at the discharge from the circumference of the impeller.

For the velocity to be constant the volute is made so that the cross-sectional pipe area increases uniformly from the cut water position to the throat of the casing (see Figure 9.4). With an impeller having six vanes then the cross-sectional area of the volute at No. 1 vane will be one-sixth of the throat area as one vane is pumping one-sixth of the water quantity, similarly one-third at No. 2 and so on, taking vanes in turn from cut water to throat. If the discharge were choked or blocked then the pump would merely churn water so that the fitting of a relief valve is not essential. However, if it is run in this way for a length of time it will cause over-heating. A common fault for repair with these pumps is the increase of clearance due to wear at the bearing rim (or sealing ring) faces. This allows connection between suction and discharge so drastically reducing efficiency. On the larger pumps these faces are often brass strips on liners secured by countersunk screws, clearance adjustment is effected by adding further liners. On the smaller pumps the faces are made by sealing rings, which are renewable. After any overhaul or clearance adjustments, care must be taken on re-assembly to make sure motor or impeller is not pulling on each other at the junction coupling.

In the smaller designs the shaft gland seal is by an ordinary stuffing box, water cooled, usually employing lead foil-type packing. Great care must be taken with the packing because if it is not adjusted properly it is very prone to nip and score the shaft severely.

For larger types, a rotary packing is used. This consists of a fixed clamp ring on the shaft driving another ring cup, with packing rings on to the shaft, through driving pins. Ring cup and rings are free to slide along the shaft under the action of axial springs from the clamp ring. The cup ring presses on to a fixed ball ring that in turn sits in a ball socket joint in the back plate that bolts to the pump casing. Grease lubrication is provided to the face between ring cup and fixed ball ring, worked by spring or water pressure.

Figure 9.5 shows diagrammatically a vertically arranged single inlet centrifugal pump. This is arranged with the casing split vertically and one-half has suction and discharge

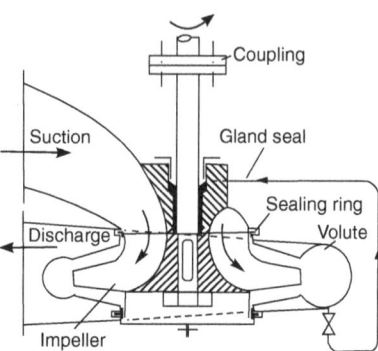

▲ **Figure 9.5** *Vertical, single inlet, centrifugal pump (drysdale)*

branches so that the impeller and shaft can be removed without breaking pipe joints. The impeller has a single eye (inlet), upward facing so that air locking is eliminated under operating conditions. Pressure in the space under the impeller ensures hydraulic balance.

Centrifugal pump characteristics selection

This depends mainly upon duty and space available. Duty points: (1) Flow and total head requirements. This will govern the speed of rotation, impeller dimensions, number of impellers and type, for example, single or, double inlet. (2) Range of temperature of fluid to be pumped. If suction capability is insufficient to accommodate supply conditions due, for example, to high inlet temperature cavitation can occur. (3) Viscosity of the medium to be pumped. (4) Type of medium, for example, corrosive or non-corrosive; this would affect the choice of material (although for salt and fresh water the difference is often just the casing).

If the vessel is expected to be running extensively in or close to rivers then the sea water pumps may experience considerably more wear of its internal components due to the abrasive material, such as sand and stones, that will be present in the water.

Materials for salt water could be casing – gunmetal (cast iron for fresh water), impeller – aluminium bronze, shaft – stainless steel, casing bearing ring seals – leaded bronze. Space points: with vertically arranged pumps less floor space is required. This usually means that no hydraulic balance is necessary, impeller access is simple and no pipe joints have to be broken.

A typical engine room pump could be a vertical, in-line, overhung (i.e. suction and discharge pipes are in a straight line and the impeller is supported, or hung, from above), either base or frame mounted, from which the impeller can be removed without splitting the casing, breaking pipe joints or removing the electric motor.

Losses

Power supplied to the pump must take into account the various losses:

1. Friction loss in bearings and glands, surfaces of impeller and casing. Some impellers are highly polished to minimise friction loss.
2. Head loss in pump due to shock at entry and exit to impeller vanes and eddies formed by vane edges.
3. Leakage loss in thrust balance devices, gland sealing and clearances between cut water and casing and bearing seals.
4. The force required to keep the heavy metal impeller rotating.

A characteristic curve for a centrifugal pump is obtained by operating the pump at rated speed with the suction open and the discharge valve shut. The discharge valve is then opened in stages to obtain different discharge rates and total heads (can be measured by discharge pressure gauge, suction head constant) corresponding to them. A typical curve is shown in Figure 9.6, reciprocating and axial pump curves given for comparison.

Failure to deliver caused by loss of suction may be due to: insufficient supply head, air leakage suction pipe (e.g. valve open on empty bilge, etc.), loss of priming facility or leaking shaft gland. Capacity reduction could be the result of: a damaged sealing ring, leaking gland, obstruction (valve partly closed), incorrect rotational speed. Excessive vibration may be caused by: loose coupling, loose impeller, damaged bearing, or impeller imbalance.

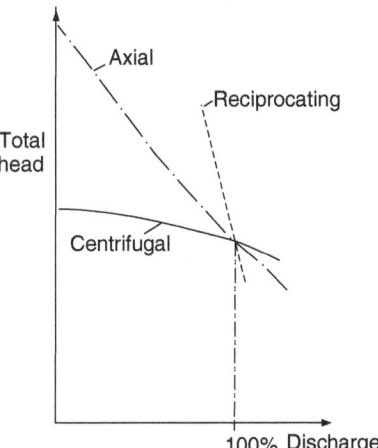

▲ **Figure 9.6** *Characteristic curves*

Alternating current (AC) electric motors have been the preferred method of driving marine centrifugal pumps for many years. The problem has been that until recently the AC motor has been a single-speed machine. Some have been fitted with star-delta changeover electrical connections, which enabled a start-up mode that required less energy than the direct-on-line starter. This made the start-up process more efficient but after that the pump was still a single-speed unit, which meant that if the pump's delivery needed changing it had to be completed by means other than speed control.

Therefore, for example, if a reduction in flow was called for by the system being serviced by a pump, then the energy input to the pump could not be reduced and the flow would have to be throttled or a bypass used.

The modern techniques of thyrister control systems means that the speed of AC motors can now be controlled to any speed while the pump is in operation. A pump is only required to operate at full load for small proportions of the time in service. Therefore any reduction in energy will bring savings. ABB Ltd are leaders in this field and their information claims up to 60% energy saving by using a variable-speed drive.

Another energy-saving technology is the use of lightweight material for the construction of centrifugal pump impellers. These components save energy in two ways. The first is due to their light weight, which will require less energy to drive the impeller. The second method is due to the fact that the impeller stays at its design dimensions and does not corrode and lose its profile, which reduces efficiency.

Axial flow pump

When large-capacity, wide variation of low lift head at constant speed conditions have to be met the horizontal or vertically arranged axial pump is the most suitable.

The pump is efficient, simple in design and is available in a wide range of capacities. It can, if required, be reversible in operation (a friction clutch between motor and pump would be required) and is ideally suited to scoop intake for condensers as it offers very little resistance when idling.

Figure 9.7 shows such a pump. Its casing would be cast iron or gun metal, and its impeller of aluminium bronze. Guide vanes, of gunmetal, guide water without turbulence to the discharge. Pump shaft, of stainless steel, with solid and flexible couplings is driven, if low head, by a relatively small prime mover at higher speeds than a comparable centrifugal pump. A water cooled thrust of the tilting pad type is required because of the considerable thrust generated (consider a propulsion system).

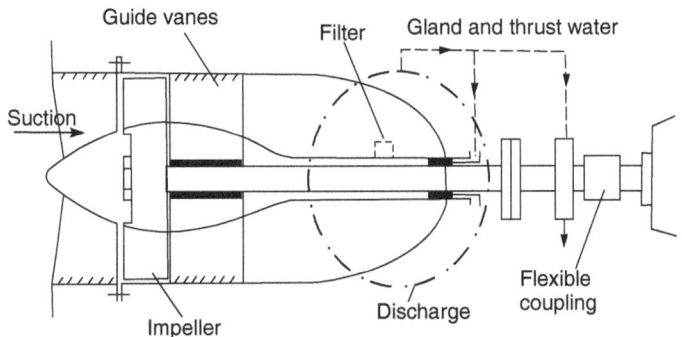

▲ **Figure 9.7** *Axial flow pump*

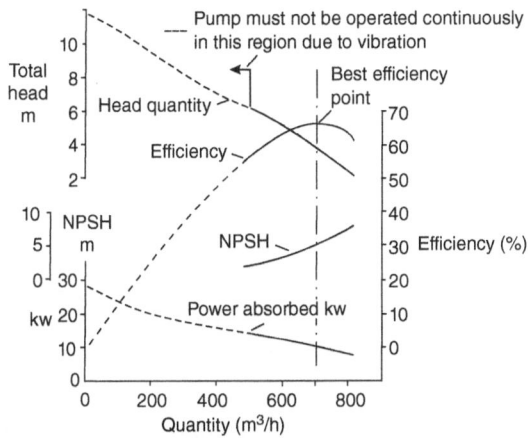

▲ **Figure 9.8** *Characteristic curves for an axial pump*

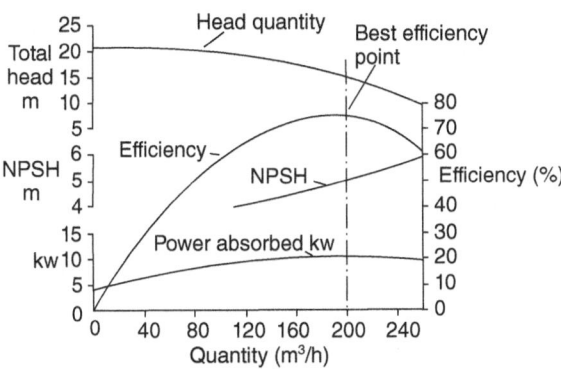

▲ **Figure 9.9** *Comparison of charateristic curves*

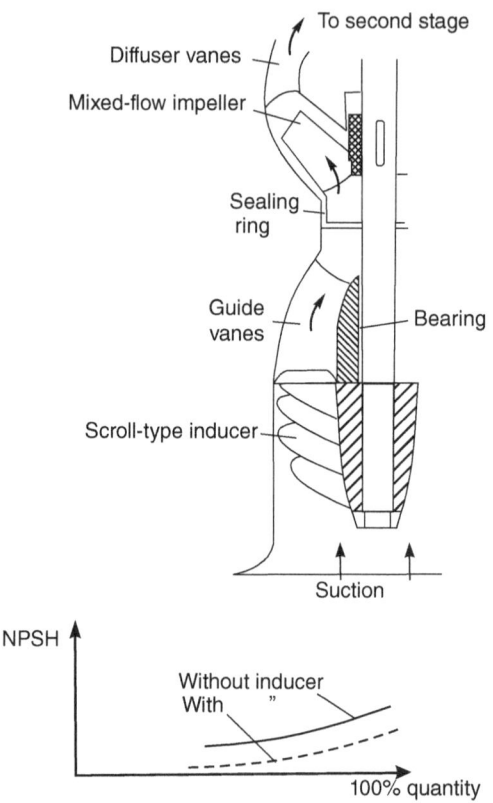

▲ **Figure 9.10** *Cargo pump*

The mechanical seal is water cooled as is the composition bush for the shaft. The latter is via the multi-leaf filter, in the case of condenser circulating, because of the possible ingress of sand.

Characteristic curves for an axial flow pump are given in Figure 9.8. Careful study of these curves and comparison with those for a similar-speed centrifugal pump given in Figure 9.9 will greatly assist the reader to answer some of the questions asked, for example, throttling the discharge valve, its effect on pressure, efficiency and power.

A mixed-flow pump, that is, part centrifugal part axial, for cargo duty with cryogenic carriers is shown in Figure 9.10.

It is fitted, in this case, with a scroll-(or screw-) type of inducer to reduce NPSH (net positive suction head) requirement and eliminate the need for a stripping pump. Only one stage is shown in the diagram, in practice two or more vertically arranged stages would be used, operated by the prime mover on the deck while the bell mouth suction at the bottom of the tank and the pump casing act as a long discharge pipe.

Cargo pumps

Centrifugal cargo pumps differ according to type of cargo, for example, product tankers (crude oil, etc.) would have a separate pump room with conventional centrifugal pumps, probably vertical overhung impeller, sometimes called barrel-type cargo pump installed. This double-eye inlet pump with either a straight-through or 90° suction discharge angle with pipe connections in the bottom half of the casing has two external bearings above the impeller, the upper one takes all the hydraulic thrust and the lower acts as a radial load bearing. This pump has certain advantages over its counterparts: (1) impeller can be sited lower in the pump room thus improving suction conditions and reducing stripping time; (2) impeller can be removed without disturbing pipe joints; (3) there is easier access to bearings and shaft seal without removal of rotating elements.

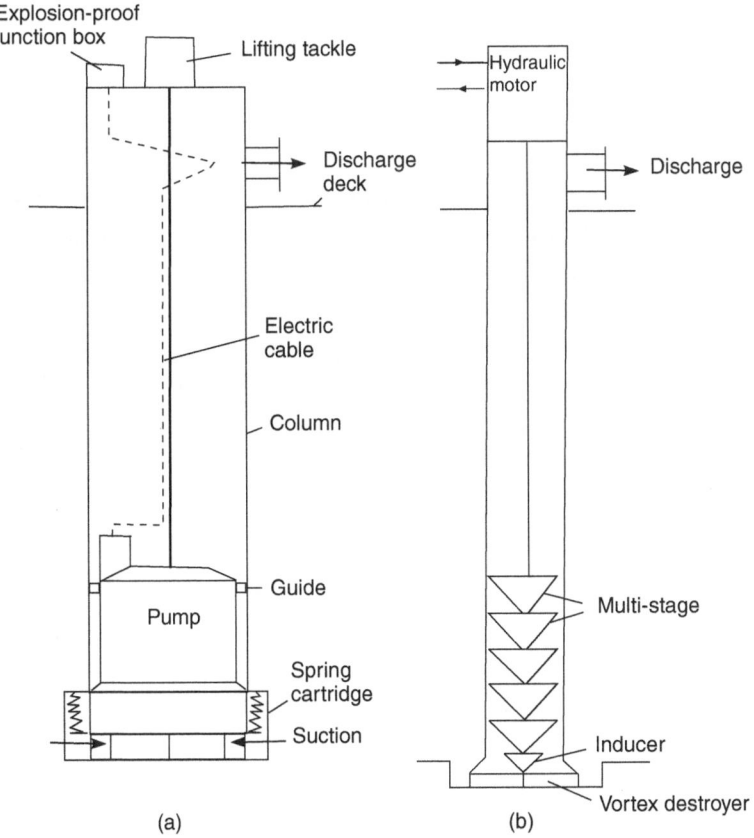

▲ **Figure 9.11** *(a) Submerged cargo pump; (b) deepwell cargo pump*

In chemical, LPG or multi-product tanker, a separate pump is sited in each tank. Pumps driven through line shafting coupled to hydraulic motors on deck would be deep well, single- or multi-stage with radial or mixed-flow impellers, respectively. Or, submerged pumps electrically or hydraulically driven with, usually, single elements. The line shaft pump, despite some bearing problems, is proving to be more popular, especially for LPG carriers.

Submersible pumps eliminate line shaft bearings and gland problems, but expensive problems could occur due to hydraulic fluid leakage into the cargo and vice versa.

Figure 9.11 shows diagrammatically two of the cargo pump arrangements just described. The submerged electric motor-driven pump rests on a spring cartridge that closes when the pump is raised and seals off the tank from the column.

Air extraction on most pumps is required, especially on all bilge pumps. Early designs of circulating pumps employed a steam ejector on the volute casing together with a steam jet into the casing to condense and prime, or a direct water priming valve. Later designs of centrifugal pump incorporated a separate air pump. In the first types the air is separated from the water in the suction chamber, it rises and is withdrawn by the air pump via a float-operated valve. Twin single-acting air pumps are fitted, driven by worm and wheel from the pump spindle, and are crank driven. The pumps are capable of operating flooded should the float gear break down but in normal operation the flooded water suction closes the float valve and the air pumps idle. This design can be sketched fairly easily for examination requirements and is shown in the emergency bilge pump diagram in Figure 9.12. In the more modern designs the reciprocating air pumps are usually replaced by rotary types. In these designs the usual suction separating chamber and ball float are provided but the air connection from the top of the ball float chamber is taken to the rotary air pump, which is directly driven by an extension of the motor spindle on top of the pump. The rotor revolves in a special, variable-shaped chamber that is supplied with fresh water from a reservoir in the air pump casing.

Due to the casing shape the water is made to flow from and towards the rotor centre during each revolution. The water motion is utilised to act as suction and discharge for the

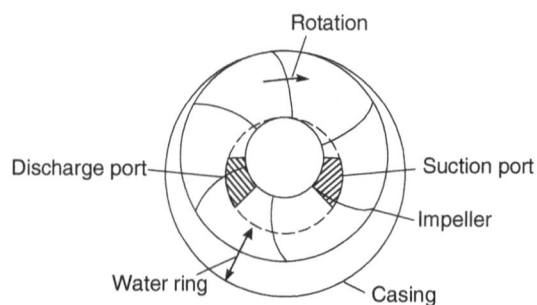

▲ Figure 9.12 *Water ring air pump principle*

air through appropriate sets of ports. The rotor casing is continuously cooled by a closed-water circuit from the pump discharge round the air pump jacket and returns to the pump suction. The air pump can be placed in or out of operation by a control cock on the front of the air pump casing. The principle of operation is referred to as the 'water ring principle'. Figure 9.12 shows this in simplified form. As the impeller vanes pass the suction port, air is drawn in and trapped between the water ring and the pump shaft. This 'slug' of air is carried around and delivered to the discharge port; hence this pump is a positive displacement type. In some ship plants the priming connections for all pumps, etc., are led to a central exhausting system. This system, under the operation of auto compressors, functions to give priming from a central control station to all units in the engine room as required.

Emergency bilge pump

The function of this pump (Figure 9.13) is to drain compartments adjacent to a damaged (holed) compartment. The pump is capable of working when completely submerged. The pump is a standard centrifugal pump with reciprocating or rotary air pumps. The

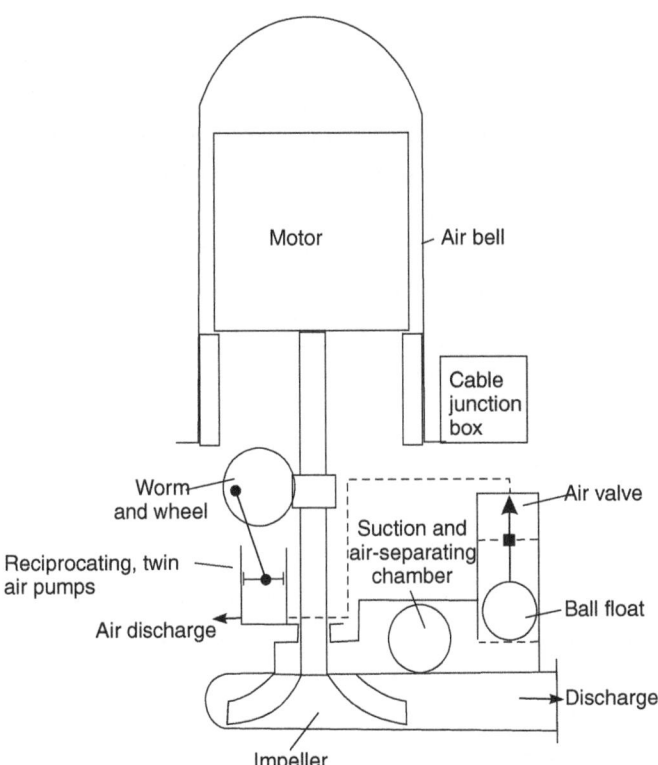

▲ **Figure 9.13** *Emergency bilge pump*

motor is enclosed in an air bell so that even with the compartment full of water the compressed air in the bell prevents water gaining access to the motor. The motor is usually DC operated by a separate remote-controlled electric circuit, which is part of the vessel's emergency essential electric circuit. The pump is designed to operate for long periods without attention and is also suitable for use as an emergency fire pump. This design is particularly suited for use in large passenger vessels giving outputs of about 60 kg/s.

Comparison of Pumps, Suction Lift, Cavitation

Consider first the performance of a reciprocating pump. If the pump *could* create a perfect vacuum in the barrel it should *theoretically* be able to lift cold fresh water from a height of 10.3 m above the suction valve.

1.013 bar ~ 760 mm mercury ~ 10.3 m water

Thus the pump lift depends on the barometer reading (for vacuum attainable) and also the fluid pumped, that is, oil below a density of 1.0 will be capable of being lifted a greater amount. Also the fluid pumped should be cold as warm fluids tend to vaporise and destroy the vacuum. In *practice* a good reciprocating pump will lift cold water from about 8 m with a high barometer. As the temperature of the fluid rises the suction lift decreases so that at 94°C the pump will not draw water. Above this critical temperature water must be supplied from a head to increase the pressure on the suction valve and prevent vapour lock. The following figures give an indication of the above points:

Barometer 750 mm figures for cold water		Practical lift	7.5 m
Temperature	64°C	Practical lift	3 m
Temperature	77°C	Practical lift	2.1 m
Temperature	94°C	Practical lift	0 m
Temperature	110°C	Head required	3 m
Temperature	123°C	Head required	6.7 m

Air vessels are usually fitted to reciprocating pump discharge lines to ensure uniform water flow velocity in discharge lines so reducing the inertia head required. The vessel is merely a cylinder forming an air space damping cushion with fluid entry at one side and discharge over a weir at the other side, or via an internal pipe.

In pumps carrying liquids a phenomenon known as cavitation occurs. Low-pressure regions occur in the flow at points where high local velocities exist. If vaporisation occurs due to these low-pressure areas then bubbles occur; these expand as they move with the flow and collapse when they reach a high-pressure region. Such formation and collapse of bubbles is very rapid and collapse near a surface can generate very high-pressure hammer blows, which results in pitting, noise, vibration and fall-off in the pump efficiency. This phenomenon is usually not very pronounced in reciprocating pumps. Incipient cavitation, that is, cavitation that is just beginning, can occur when suction lift capability cannot meet supply requirements and the output reduces until the two coincide. Under these conditions of operation the pump runs noisily and cavitation damage can occur. By throttling the discharge, or reducing pump speed, rough running of the pump and possible damage can be avoided.

Supercavitation occurs when the vapour bubbles collapse within the liquid after the impeller.

Inducers

These are sometimes fitted to centrifugal pump impeller shafts at suction. Their purpose is generally to ensure that supply of fluid to the impeller is at sufficient pressure to avoid cavitation at impeller suction, or it enables the pump to operate with a lower net positive supply head. Different types are used: scroll, screw or propeller.

The propeller (like a stub-bladed fan) inducer is fitted to super-cavitating pumps, that is, pumps where the cavitation occurs between the inducer and the impeller. Such pumps can operate at about one-third of the net positive supply head normally required for conventional centrifugal pumps and they are suitable for LPG and LNG carriers.

Considering the previous remarks with respect to centrifugal pumps, the following points are applicable: (1) The same remarks apply for pump suction head but as clearances in a fast-running centrifugal pump are difficult to maintain then even with the good pump with facing clearances of a few millimetres it will probably lift about 7.3 m of cold water with a high barometer. (2) Air vessels are rarely fitted as steady flow and air extraction is usual. (3) Such pumps are very prone to cavitation especially at inlet to impellers and it may be advantageous to reduce the suction lift to prevent the formation of bubbles due to low-pressure regions; incorrect attention to this point may cause severe cavitation and very poor pump performance.

The gear and screw displacement pumps are also affected by changes of barometer pressure and fluid temperature. A reasonable mechanical clearance must be provided and any clearance will of course reduce the vacuum efficiency and hence suction head available. Summarising for these two types on the above three points, it may be said that such good pumps will probably lift cold water from about 6.7 m with a high barometer, rarely need air vessels and are not specially prone to cavitation when correctly designed.

In a modern vessel most pumps would probably be motor-driven centrifugal, with reciprocating, gear, screw displacement or turbo pumps only fitted for specialist individual duties. The discharge head attainable (or pressure of discharge) is virtually unlimited for a reciprocating pump. Provided a good steam pressure is available, the principle of area differentials gives very high discharge pressures. For the other forms of pumps, rotational speeds are increased to obtain higher discharge pressures up to a reasonable maximum, for really high pressures, impellers or wheels running in series are required. For example the maximum peripheral impeller speed is best fixed at about 105 m/s from the stress viewpoint (although cavitation may be appreciable). A 200 mm impeller at 167 rev/s, a 120 mm impeller at 275 rev/s, or three compound 330 mm impellers running at 60 rev/s would all produce 50 bar at peripheral speeds within 105 m/s, although the latter is preferable. The maximum head for series impellers is often fixed at about 170 m (16 bar) per stage, with a maximum of say nine stages, but these figures are by no means rigid.

Associated Equipment and Systems

A short selection of units and systems are considered on which examination questions have been set.

Heat exchanger

Thermodynamic characteristics

$Q = U\theta A$ is the rate of heat transfer from one fluid to another in a heat exchanger, where Q is in watts.

U is the overall coefficient of heat transfer in W/m²K, this depends upon the properties of the fluids, their speeds and the form of the heat-exchanger surface.

θ is the logarithmic mean temperature difference in °C between the two fluids. θ is a maximum with counterflow.

A is the area of heat-exchanger surface in m².

Figure 9.2 shows some of the different flow patterns used in heat exchangers, counterflow is the best thermodynamically of the basic patterns. In practice most heat exchangers use mixed flow to obtain the best possible characteristics.

In the selection of a heat exchanger, certain points have to be considered:

1. Quantity of fluid, maximum to minimum, to be cooled.
2. Range of inlet and outlet temperature of fluid to be cooled.
3. As above for the cooling medium.
4. Specific heat of the mediums.
5. Type of medium, corrosive or non-corrosive, as well as safety.
6. Operating pressures.
7. Maintenance, fouling, cleaning, access.
8. Position in system and associated pipe work.
9. Cost, materials, streamline or turbulent flow.

Streamline and turbulent flow

In Figure 9.14 simple diagrams show: (1) The laminar, streamline flow of a fluid whose velocity variation is approximately parabolic, being maximum at the centre and zero where the fluid is in contact with the pipe or plate surface. (2) The turbulent flow of a fluid.

Whether flow is streamlined or turbulent depends upon certain factors, which are summed up by Reynold's number.

$$\text{Reynold's number} = \frac{\text{Velocity of fluid flow} \times \text{Pipe diameter}}{\text{Kinematic viscosity}}$$

If the number is less than 2,000 the flow is streamlined. If the number is more than 2,500 the flow is turbulent. (Kinematic viscosity is the ratio of absolute viscosity to relative density.) Obviously pressure difference is a hidden factor in the calculation, the greater its value the greater the velocity.

For efficient heat transfer turbulent flow is best, but erosion of metal surface will be greatest. For little erosion of metal surface, streamline flow is required but heat transfer will be relatively poor.

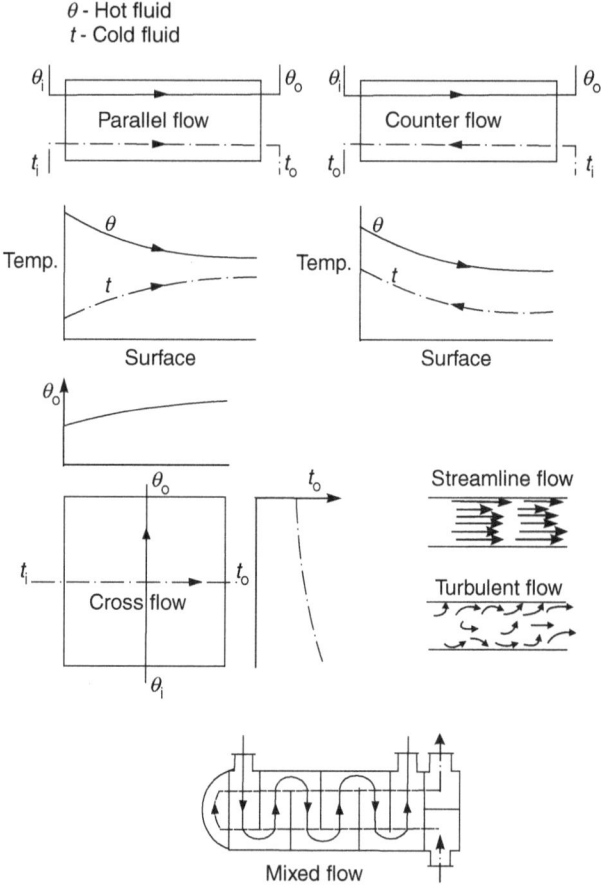

θ - Hot fluid
t - Cold fluid

▲ **Figure 9.14** *Characteristic of streamline and turbulent flows*

Shell and tube type heat exchanger (serck)

The shell (or cylinder) is usually made of close-grained cast iron, with surfaces machined as required. Gun-metal or fabricated steel may be used as alternatives depending upon requirements. Inspection doors are fitted in the distribution belts (Figure 9.15).

End boxes with end access covers are of the same material as the shell. Sacrificial anodes in the rod or plug form and an electrical contact strip are fitted to minimise corrosion.

The tube stack is made up of stress-relieved aluminium brass tubes expanded into naval brass tube plates; one plate is fixed as shown and the other is free to allow for expansion of the stack. Brass circular baffles give radial flow to the fluid and support to the tube stack.

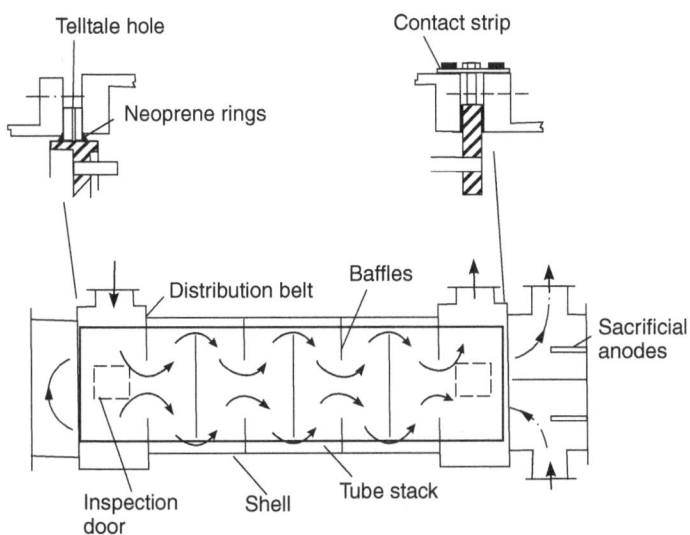

▲ **Figure 9.15** *Shell and tube heat exchangers (Serck)*

In more recent designs of tube-type heaters and coolers the guided flow concept has been introduced, that is, a secondary heating, or cooling, surface in the form of radial fins integral with the tubes between which flow is guided radially, alternately out and in from section to section. This gives: (1) greater heat transfer surface, (2) better heat transfer, (3) lower metal surface temperature, (4) in the case of oil heaters less risk of oil cracking and hence fouling.

Plate-heat exchangers

Figure 9.16 shows diagrammatically the Alfa-Laval plate-type heat exchanger. It consists of a variable number of titanium (stainless steel or aluminium brass) plates, clamped together, between a closing pressure plate and a frame. The plates are sealed by the use of rubber seals that sit in a groove. The surface of the plates is corrugated to give strength and additional heat transfer surface. The most recent development is to use a herringbone pattern with the 'V' pointing alternately up and down, touching in a criss-cross pattern. This gives additional support, allows pressure to be increased and plate thickness to be reduced.

Seals are usually nitrile rubber bonded to the plate and arranged so that in the event of failure the two fluids cannot mix.

Principal advantages of the plate heat exchanger are listed here:

1. Compact and space saving, virtually no head room is required.
2. Easily inspected and cleaned, all the pipe connections are at the frame plate hence they do not have to be disturbed when plates are dismantled.

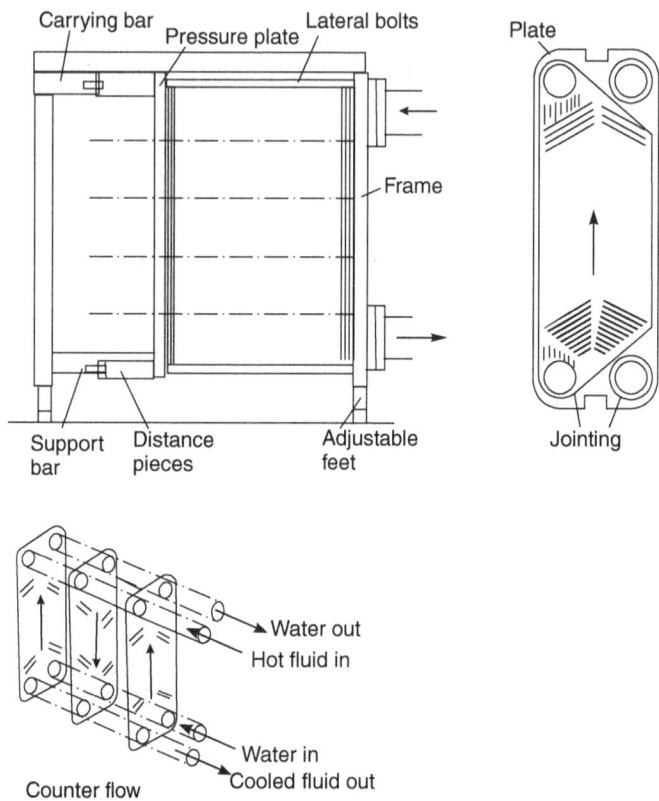

▲ **Figure 9.16** *Plate-type heat exchanger (Alfa-Laval)*

3. Variable capacity. Plate numbers can be altered to meet capacity requirements.
4. With titanium plates there is virtually no corrosion or erosion risk and turbulent flow (which is erosive), which takes place between the plates, will increase heat transfer and enable fewer plates to be used.

Central cooling systems

These have been designed for diesel and steam plants. Figure 9.17 shows diagrammatically the arrangement for a diesel engine installation.

Large, sea water-cooled heat exchangers, one in operation the other standby, are the 'central coolers', which will have excess cooling capacity to allow for fouling. A controlled bypass of the fresh water to be cooled maintains it at a steady temperature of 35°C up

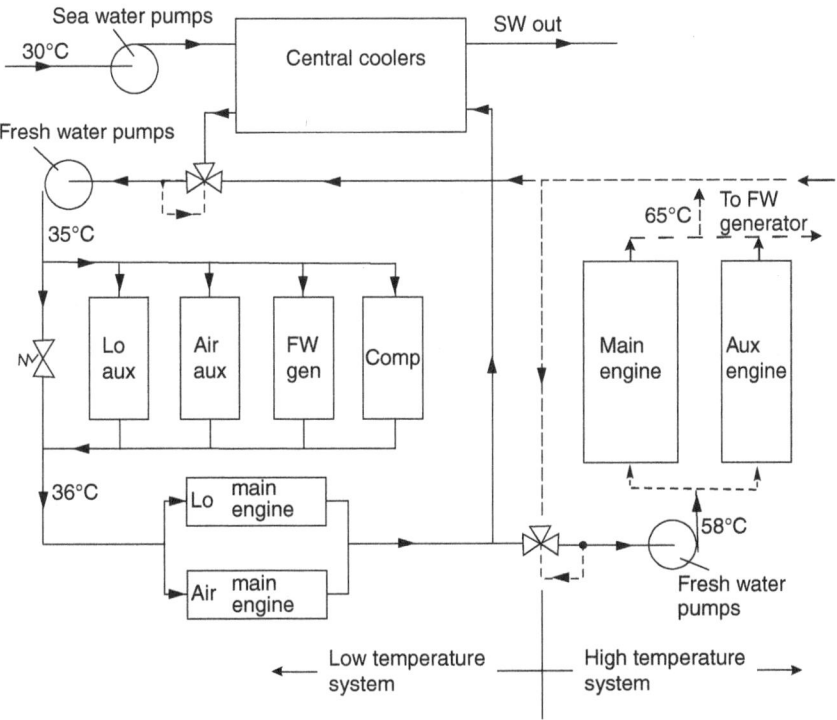

▲ **Figure 9.17** *Central cooling system*

to a maximum sea water temperature of 33°C. Sea water temperature above 33°C will result in an increase in fresh water temperature.

The system is divided into low- and high-temperature zones. The low-temperature zone contains the coolers, which can be arranged in different ways to suit requirements. Automatic bypass valves are arranged across each cooler unit that control the upstream water pressure keeping it constant irrespective of the number of coolers in use. The main advantages of using a central cooling system are as follows:

1. Reduced maintenance, due to the fresh water system having clean, treated water circulating. The cleaning of the system and component replacement is reduced to a minimum.

2. Fewer salt water pipes with attendant corrosion and fouling problems.

3. With titanium-plate heat exchangers used in the central coolers, cleaning of the coolers is simplified and corrosion reduced.

4. The higher water speeds possible in the fresh water system result in reduced pipe dimensions and installation costs.

5. The number of valves made of expensive material is greatly reduced. Also, cheaper materials can be used throughout the fresh water system without fear of corrosion/ erosion problems.

6. With a constant level of temperature being maintained, irrespective of sea water temperature, this gives stability and economy of operation of the machinery, for example, no cold starting since part of the cooling system will be in operation. Reduced cylinder liner wear, etc.

Modular systems for auxiliary plant

Modular systems are used for items such as lubrication, fuel, boiler feed water, cooling, etc.

To decide what items have to be included in the module, we need to know what performance is required of it. For example: fuel to a diesel engine should be clean, free of water, at the correct pressure and viscosity. Hence in the module we require filters, centrifuges or coalescing filters, pumps, heaters and sensors.

These are the advantages to be gained by using modular techniques:

1. The engine room layout will be simplified.
2. Pipe runs will be simple, external to the module, consisting only of supply and return.
3. Module is assembled in the workshop – this in itself has considerable advantages:

    (a) The environment is easier to control and it should be clean, dry and oil free. There should be reduced risk of damage to plant when the module is installed because no rust, scale, oil, waste, weld spatter, etc. would be present inside the module, which would have all open-ended pipes blanked after satisfactory testing and examination, and they would remain blanked until they have to be connected to piping on board ship. This has the added advantage of reducing pre-commission cleaning time on board.

    (b) The best possible arrangement of integral components for ease of maintenance coupled with shortest pipe runs can be achieved. This would be accomplished by designers and assemblers working in close collusion.

    (c) The module can be easily tested and inspected.

4. Installation time at the shipbuilding yard would be reduced.
5. Standardisation with the least amount of material used, together with the best possible design for access, maintenance, reliability, etc. results in economy.

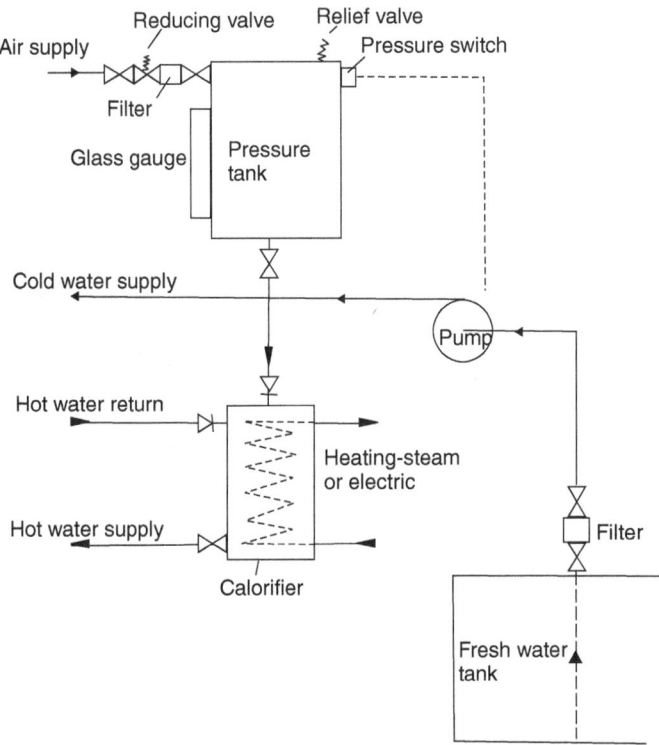

▲ **Figure 9.18** *Pressurised fresh water system (pressure 1.3–2 bar at highest point in system)*

Standardisation does have its limitations. Units would be made in capacities of standard incremental quantities, and the standard unit may not provide exactly for requirements. Also, the unit must not become so large that transport becomes a problem (Figure 9.18).

Efficiency increases by variable pump-drive motors

Currently pumps supplying a service on board ship are driven by AC synchronous motors. These motors run at one speed, which is determined by the frequency of the alternating current that is being supplied.

However, this will not be the most efficient for a significant percentage of the operating life of the pump. With the advent of modern power electronics, variable speed control of AC motors is now a real possibility and will bring significant efficiency gains to the operation of pumping systems.

The energy consumption of the pump can be reduced by lowering the speed to match the demand. At present the valves to the coolers are either throttled or the cooler is bypassed. Either way the pump's power consumption hardly varies from the maximum.

Automatic domestic water supply systems

For sanitary and fresh water supply, modern vessels now usually employ automatic systems from a tank or reservoir. The pump discharge is led in and out of the bottom of the tank on its way to the piping system. The tank containing the water has an air space provided above the water. As the water is used up the pressure of air will drop. A pressure switch is connected to the tank – a switch that is almost identical to that described in the refrigeration section, so that when air pressure falls to say 2 bar the lead from the tank to the bellows serves to operate the switch so starting the pump. The pump builds up water quantity in the tank until the air pressure is say 4 bar when the pressure switch serves to shut off the pump. The differential for cut-in and cut-out can be adjusted for reasonable running periods while maintaining a satisfactory pressure on sanitary and/or fresh water fittings (Figure 9.19).

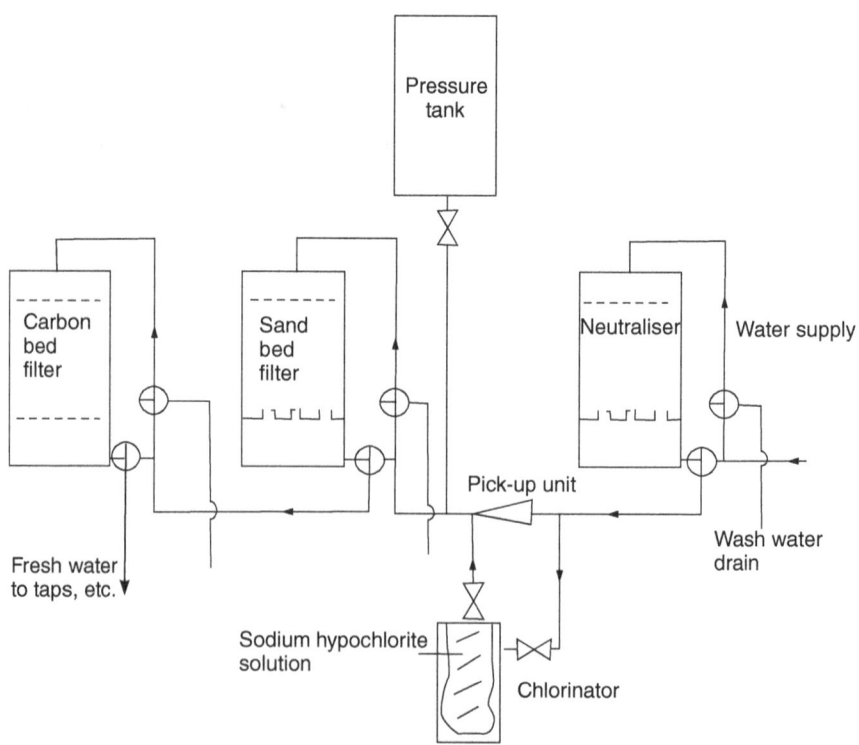

▲ **Figure 9.19** *Fresh water treatment plant*

Water purification

For domestic purposes the water used must be slightly alkaline, sterilised, clear and pleasant tasting.

1. To give alkalinity and to improve the taste of insipid distilled water, carbonates of calcium and magnesium are used as a filter bed in a neutraliser.
2. To sterilise the water chlorine is used, this would normally be solutions of hypochlorite or possibly the powder calcium chloride. About 0.25–1 kg of chlorine would be required for every 1,000,000 kg of water.
3. To produce clear water it can be passed through a sand bed filter.
4. To improve taste a de-chlorination process is used. Chlorinated water is passed through an activated carbon filter bed, which will absorb excess chlorine.

Neutraliser, sand bed filter and carbon bed filter can all have their flows reversed for cleaning purposes.

Hydraulic system

A centralised hydraulic system consisting of duplicated oil pumps, usually rotary reciprocating, accumulators, filters and an oil reservoir, fitted with pressure regulators that govern the pressure in different lines for different purposes is an economic, reliable and safe power distribution system.

Items that can be operated by such a system include: pumps, for example, submersible or line-shaft driven from deck motors; deck machinery, for example, winches, windlass, cranes, derricks, hatch covers, ramps, water tight doors, bow thruster, etc.

The main advantages of a centralised hydraulic system are: (i) smooth operation, (ii) infinitely variable speed control, (iii) self-lubricating, (iv) intrinsically safe therefore useful for hazardous cargos, (v) centralised for ease of control.

Figure 9.20 shows in simple diagrammatical form how a controlled unit, pump, fan, etc. would be connected into the return and pressure lines that are connected to the centralized hydraulic system. The stop/start speed control may be fitted with a pilot line for remote operation and appropriate sensors would telemeter measurements to the control station.

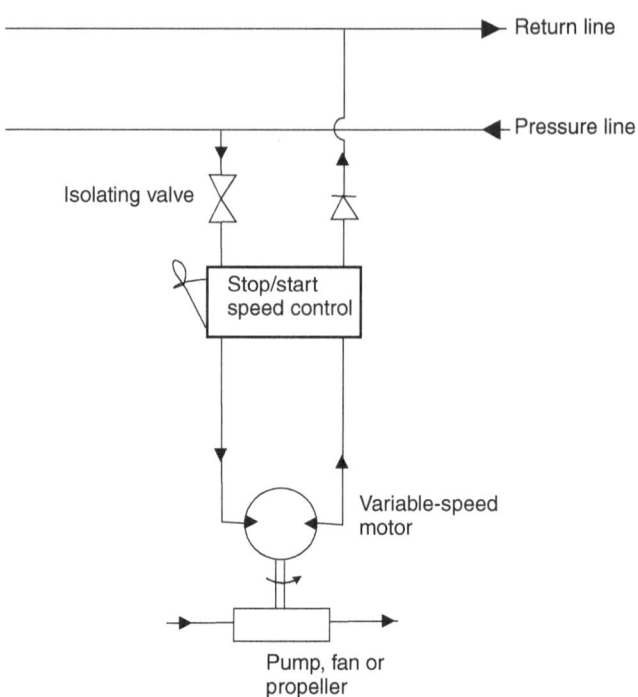

Return line

Pressure line

Isolating valve

Stop/start
speed control

Variable-speed
motor

Pump, fan or
propeller

▲ Figure 9.20

Prevention of Pollution of the Sea by Oil

MARPOL, Merchant Shipping Acts and Flag administration, requirements means that allowing oil to enter the water is prohibited by law. The legal maximum oil particle discharge quantity is currently 15 parts per million of water, but this may be reduced further in the future. However, it may be said that from the operating engineer's viewpoint, he is not allowed to discharge any oil or contaminated water overboard. This means that unless a holding tank is utilised, which is quite feasible in smaller vessels, which can then be cleaned in port, then a separator suitable for extracting oil from bilge or ballast water must be provided, for all vessels. The rules require such a separator to be of sufficient design, size and construction. Provision must also be made to prevent overpressure and discharge into confined spaces. Prison sentences are given for the illegal fitting of the so-called magic pipe, which could be used to bypass the oil–water separator.

Avoidance of pollution of the sea with oil

The merchant shipping notice M1196 gives details of a manual on the avoidance of pollution of the sea that has been produced by the H.M.S.O in the United Kingdom. This together with current, relevant M Notices are essential reading for marine engineers before attempting the examination. Some precautions are to be observed when bunkering: See Figure 10.1.

- All scuppers to be plugged so that in the event of a small spillage on to the deck it is contained and can be dealt with.
- Drip trays must be placed under the ship-shore connection.
- Good communication between ship and shore must be established and checked to regulate flow as desired.
- Personnel operating the system must be fully conversant with the layout of pipes, tanks, valves, etc.
- Moorings and hose length should at all times be such that there is no possibility of stretching or crushing the hose.
- Ensure blank at opposite end of cross-over pipe is securely in place.
- Air pipes should be clear, soundings checked and depth indicators tested.

When transferring oil within the ship it should be ideally done during daylight hours. Under the Merchant Shipping Act 1995, oil is not to be transferred ashore at night unless agreed first with the harbour authorities, and the overboard discharge connections should be closed and secured, overflow alarm should be tested and soundings taken at frequent intervals.

The MCA produce their own rules for the construction of ships based on the SOLAS regulation. These rules cover fuel oil and ballast water tank construction and use. Features such as the position and use of sounding pipes are all covered by the MCA rules.

Merchant Shipping Acts and MARPOL

It is an offence to discharge oil into UK waters under the Merchant Shipping Act 1995. This will include bilge water contaminated with any oil. Furthermore IMO have established emission control areas (ECAs) of which North America is the latest addition.

When bilge water is discharged via an approved oil–water separation system in any waters, an entry must be made in the oil record book consisting of:

- the quantity discharged;
- the source of the bilge water;
- the time of discharge;
- the ship's position;
- the date.

It must then be signed by the Master and Chief Engineer.

MARPOL and flag states require that every vessel of 400 gt must be fitted with an oil–water separator. MSN 1823 states that no ship shall discharge oil into UK waters and that suitable provision must be made to retain oil on board until such time that it can be discharged ashore. A new form of oil record book formed in two parts must be on board and in use by UK ships from January 2011.

Oil–water separator

Various types of oil–water separators have been produced over the years but most of the gravity types fall far short of the modern requirements. IMO have laid down requirements for separators:

1. Oil-water separators for bilge and ballast applications should be capable of treating waste water containing less than 100 p.p.m. of oil irrespective of the oil content (0–100%) of the feed supplied to the equipment.
2. Filtering systems are further required to provide a discharge of no more than 15 p.p.m. under all inlet conditions.

The original gravity separators that have been installed in vessels for many years will remove solids and oil from waste water down to 15 p.p.m. if working in a system containing coalescing filters. Tougher requirements introduced as part of IMO MEPC 107(49) will require treatment to between 5 p.p.m. and 15 p.p.m. The bilge water content in modern vessels is an ever-changing cocktail, containing not only diesel oil and water, but also lube oil, hydraulic oil, heavy fuel oil, oil additives, chemicals and detergents. This unpredictable mixture has to be separated into three distinct phases: oil, water and sludge. The process is further complicated by the presence of emulsions, which are even mixtures of immiscible liquids such as tiny oil droplets mixed into the water phase of the bilge water. Although gravity would normally cause these droplets to separate from the water, particles or surfactant chemicals from cleaning products used on board can prevent this process from happening. Alfa-Laval have now introduced the PreBilge system that will treat bilge water before the ship's standard OWS, meaning that the systems as a whole could be used to tackle some of the problems that exist on modern ships.

The type of pump used for delivering the oil–water mixture governs considerably the degree of contamination in the effluent. A large number of bilge pumps are centrifugal and they are often used as the supply pump to the separator. They churn the supply and produce small oil droplets (<200 μm) dispersed throughout the water so that the 100 p.p.m. requirement cannot be met.

A positive displacement pump, for example, slow-running double vane, screw, reciprocating or gear enables a much better performance to be achieved from the separator as they do not produce large quantities of small oil droplets.

The pumping mode is becoming important since it is claimed that with any kind of pump operating in the suction mode (i.e. pump after the separator) the IMO requirement of 15 p.p.m. or less can be met without the use of second- and third-stage filters or coalescers.

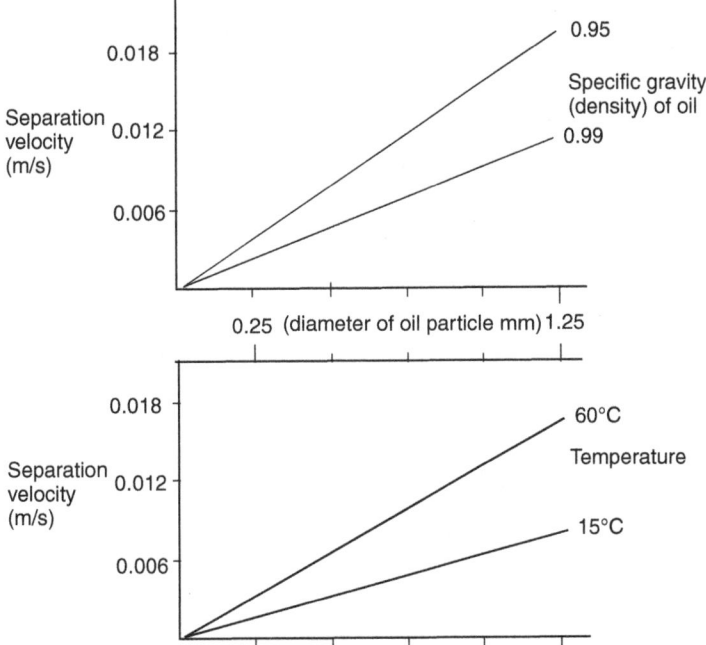

▲ **Figure 9.21** *Oil particle size verses separation velocity*

The graphs in Figure 9.21 show clearly the effect of oil particle size and separation velocity thus further emphasising the importance of pump selection and mode, presence of oil coalescers (gather oil into larger droplets) and controlled flow within the separator. Oil density and mixture temperature also govern speed of separation and hence separator throughput.

Automatic oil–water separator

Figure 9.22 shows the essential parts of an automatic separator. The operation is as follows.

Clean water is delivered to the separator through the oil–water inlet until discharge takes place out of the vent valve which is then closed. Oil–water is now delivered to

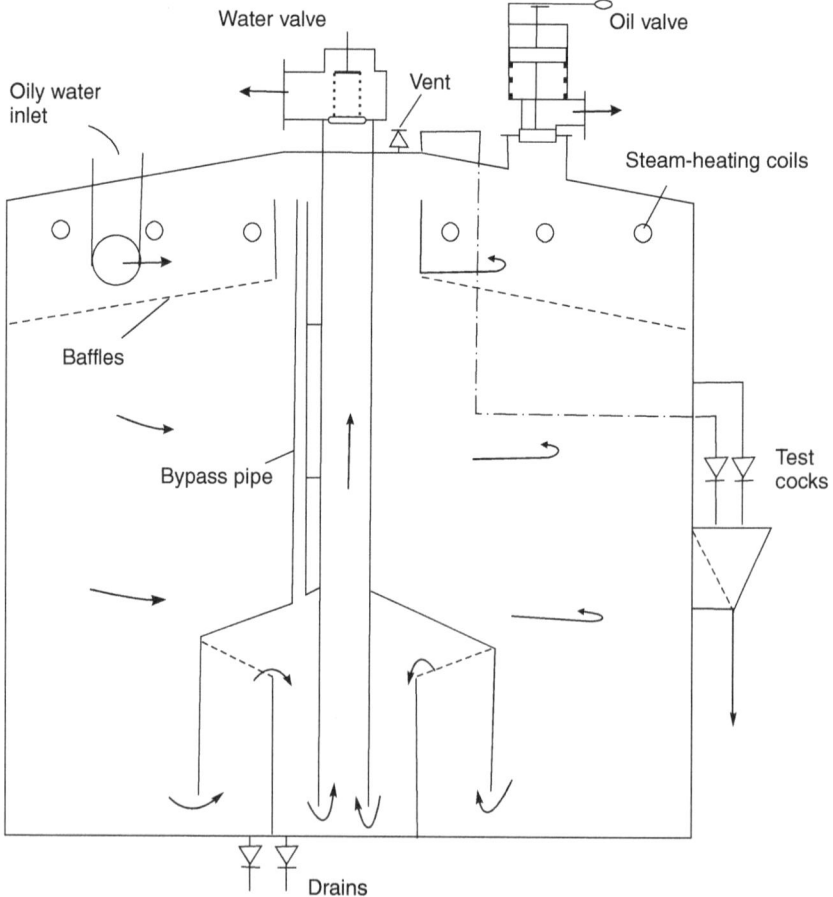

▲ **Figure 9.22** *Automatic oil–water separator*

the separator and when the pressure inside the separator reaches 2 bar approximately the water discharge valve automatically opens. The mixture circulates and flows across weirs and perforated baffles, which assist in separating the oil and water. Oil will now accumulate at the top of the separator, and as the oil–water interface gradually moves down, the oil discharge valve is automatically opened up (see later for details).

A bypass pipe takes the remaining traces of oil from the last separator stage up to the top of the separator.

Steam-heating coils are provided in the oil space to reduce viscosity and assist separation, test cocks can be used to ascertain the levels of oil and water approximately as a check for the automatic detection. A spring-loaded valve is usually fitted on both discharges but it is essential that a relief valve is provided on the shell or incoming mixture line to prevent overpressure and accidental discharge to a confined space or overboard under all working conditions. Such a relief valve should preferably be led back to the suction side of the supply pump or to an overflow tank. The usual working pressure for the separator is in the region of 2 bar, that is, the pressure at which the spring-loaded water discharge valve is set. The relief valve is set at about 2½ bar approximately.

Figure 9.23 shows a three-stage separator that complies with IMO requirements. The first stage is as previously described for the automatic separator, the second and third stages are coalescers. The effluent from the first stage enters the bottom of the second stage and passes up the middle of the coalescer, the coalesced oil collects at the top and the water discharges at the bottom and then goes to the next stage.

Electric separator probe

This is a rather complex AC circuit and only the simplest operating principle is considered. Two sheathed probes are provided at the highest and lowest levels of the oil–water interface and a third probe is fitted low down in the clean water space, the latter acts as an emergency cut-out should priming-over occur. The probes are connected to an AC circuit very similar to the DC Wheatstone Bridge principle. Two coils are energised from the supply and two condenser circuits complete the bridge, one condenser connected to the probe being the variable in the circuit. The probe and tank form two electrodes of the variable condenser. The capacitance depends on the dielectric constant of the material between (for given distance apart and electrode size). Thus the value of the capacitance depends on the material between probe and tank. The bridge when balanced in air would become unbalanced by change of capacitance in oil or water and the electrical signal could be magnified and relayed, similarly balance in oil would react to water, etc.

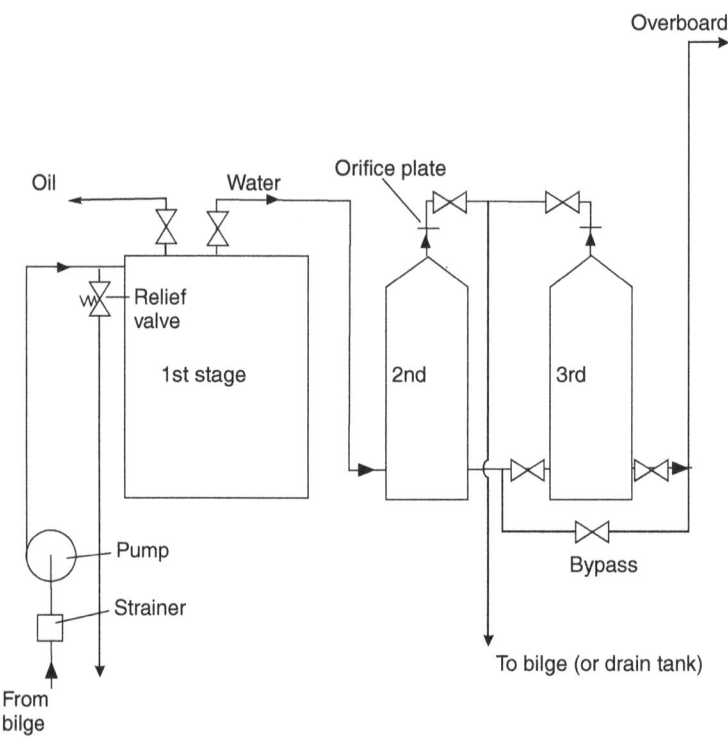

▲ **Figure 9.23** *Three-stage oil–water separator (Victor)*

Automatic valve operation

The relay can be arranged to operate lights, alarms, etc., or preferably solenoid operation via air, steam or hydraulic servos for the operation of the appropriate oil or water discharge valves. With the lower probe in oil, the imbalance due to capacitance change will function to open the oil valve and close the water valve, similarly the upper probe in water will serve to shut the oil valve and open the water valve.

The actual operation of automatic valve control combined with alarm and protection circuits can be considered by referring to Figure 9.24.

Oil has the same effect as air. With the separator empty (both probes in air) the two probe indicator lamps will be out, oil discharge lamp on, main contactor energised, solenoid energised, pilot valve up, alarm bell ringing, pump tripped. When water enters the separator and reaches the lower probe the bell stops ringing and the lower probe indicator light comes on. When water reaches the upper probe its indicator light comes on, the water discharge light comes on and the oil discharge lamp goes out. The solenoid is de-energised and the pilot valve moves down. These conditions are reversed when oil pushes the interface down.

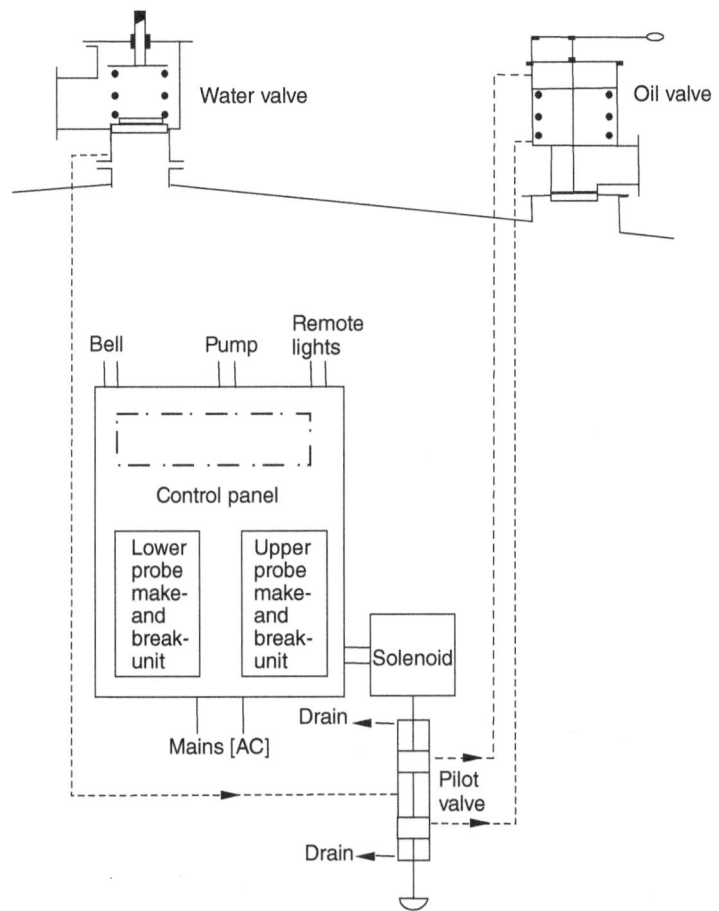

▲ **Figure 9.24** *Automatic oil–water separator system*

When oil build-up occurs and pushes the oil–water interface down to the lower probe the solenoid is energised and the pilot valve moves up. Clean water at 2 bar acts on top of the oil valve piston causing the valve to open. Pressure reduces, the water valve closes and pressure reaches about 1⅔ bar with the interface rising to the upper probe. The solenoid is now de-energised, the pilot valve moves down, clean water at 1⅔ bar acts under the oil valve piston causing the valve to close, and pressure starts to increase again.

With the main isolator shut, the lower probe unit can be made to isolate and cut out the pump if oil reaches the danger level. With the main isolator open this action is shorted out so that the pump can be used for other duties. A float-controlled air-release arrangement can be fitted to give automatic air release from the shell.

The suction mode automatic oil–water separator shown in Figure 9.25 can, it is claimed, reduce effluent level to 2 p.p.m. of oil in the mixture or less. In order to achieve this

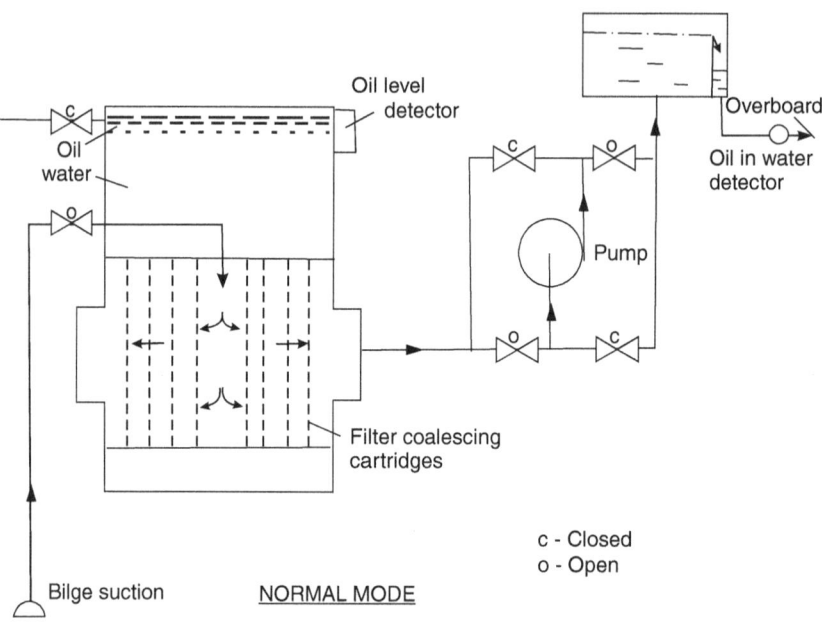

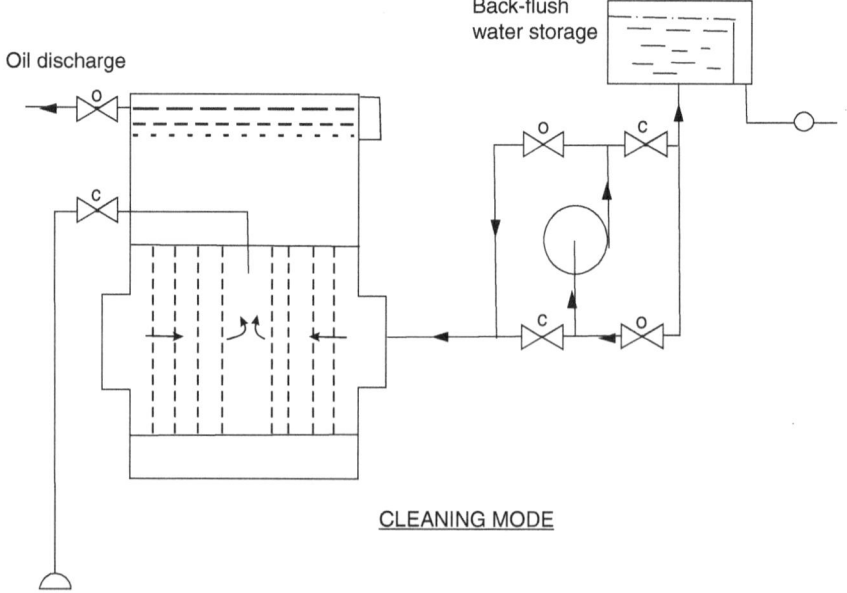

▲ **Figure 9.25** *Automatic oil–water separator*

low level the separator incorporates concentric, cylindrical oil coalescing cartridges through which the oil–water mixture is drawn by a positive displacement pump. The coalesced oil rises to the top of the separator where its accumulation is detected by an oil–water interface probe. In the normal mode a controller is constantly monitoring the oil–water interface level and the overboard discharge. In the event of the effluent exceeding set limits, the process is stopped and alarm given.

When the oil–water interface reaches its lower level the controller changes the operation to one of cleaning by back-flushing and oil discharge. The oil–water interface will then rise to the higher level when reversion to normal mode takes place.

By using an oil–water separator in the suction mode rather than the delivery mode (i.e. the pump after the separator not before) disintegration of the oil–water mixture prior to separation is achieved, thus improving separation efficiency (Figure 9.26).

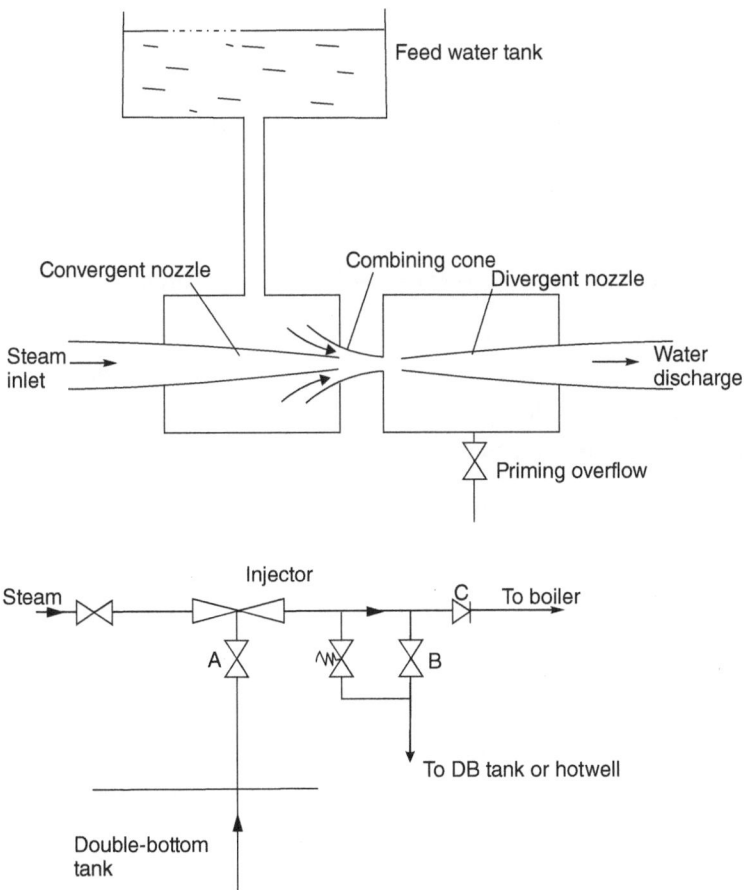

▲ **Figure 9.26** *Feed water injector*

Injectors and Ejectors

The feed water injector could be used in place of a feed pump. It is rarely seen in marine practice but it is sometimes provided as a standby feed supply device. Compared to a pump the injector has the advantage of no working parts but it has the disadvantage of being restricted to fairly cold water. Consider Figure 9.26 for the theory of the injector.

The working steam expands through a convergent nozzle, losing pressure and gaining velocity so that it emerges at high velocity from the nozzle. It contacts the cold feed water from the feed tank and condenses and the resulting jet of entrained water is guided through the combining cone and has its maximum velocity at entrance to the diverging nozzle. The kinetic (velocity) energy of the jet is now converted to pressure energy again and as it passes along the divergent nozzle it loses velocity and increases in pressure so that at exit the pressure is arranged to be higher than the boiler pressure so water will enter the boiler. Thus it can be seen that the principle of operation is conversion of energy. Ignoring temperature (energy) changes and losses and considering no rise of discharge pipe then the pressure and velocity energy of the entering steam together with the energy due to the head of entering water appears as pressure and velocity energy of the combined mass at discharge.

The feed tank need not be supplying from a head, that is, the injector will lift water but for given orifice sizes this reduces the discharge pressure from the injector. The orifice sizes are variable so that velocity and pressure energies are variable and the discharge pressure can be adjusted as required.

Figure 9.26 also shows typical connections to a feed water injector. To put the injector into operation, valves A and B would be opened, valve C closed and steam would then be supplied to the injector until it is primed and water is being delivered to the double-bottom tank or the hot well. Delivery would then be changed over to the boiler by opening C and closing B. This arrangement simplifies priming when the injector has to lift water and ensures that all the steam used is condensed.

Ejectors are used for bilge systems, evaporators and gas-freeing systems on tankers, etc. Their principle is similar to the injector but water is used instead of steam as the pumping medium. Ejectors consist of a convergent–divergent nozzle arrangement, similar to a Venturi, with a connection for pick-up of the fluid to be discharged at the throat. Ejectors are simple, reliable, inexpensive, effective and virtually maintenance free.

Sewage and Sludge

MARPOL Annex. IV was approved in September 2003 and entered into force on 1 August 2005. The Annex. requires new ships, above 400 gt, engaged on international voyages and capable of carrying more than 15 persons, to be equipped with either an approved sewage treatment plant or an approved sewage comminuting and disinfecting system or a sewage holding tank. Therefore, since September 2008 the discharge of untreated sewage into the sea at a distance of up to 3 miles from land is prohibited and between 3 miles and 12 miles sewage can only be discharged through an approved system from a disinfected holding tank. Further updates are being discussed and the draft MEPC resolution on Guidelines on Implementation of Effluent Standards and Performance Tests for Sewage Treatment Plants was proposed for approval at MEPC 64.

Retention systems

Their main advantage is simplicity in operation and virtually no maintenance. They comply with present regulations within the limit of their storage capacity. Since no sewage can be discharged in port, prolonged stays create a problem, which could be reduced by the use of a vacuum transportation system for toilets where only about 1 litre of water/flush is used compared to about 12 litre/flush for conventional types. Vacuum systems use smooth, small-bore plastic pipes (except in fire hazard areas) which, are relatively inexpensive, and because of the small amount of water used they are usually supplied with fresh water, which keeps salt water out of accommodation spaces with obvious advantages.

With some retention systems the sewage is passed first through a comminutor, which macerates the solids giving greater surface area. The mix is then passed into a chlorine contact tank where it must remain for at least 20 minutes before discharge overboard.

Biological treatment plants

Plant description: raw sewage passes through a comminutor into the collection compartment. When the level in this compartment rises sufficiently, overflow of the liquid takes place into the treatment – aeration – compartment where the sewage is broken down by aerobic activation. Fluid in this compartment is continuously agitated by air, which keeps the bacteriologically active sludge in suspension and supplies the necessary oxygen for purification.

The effluent is then pumped to a settlement compartment where the sludge settles leaving treated effluent, which passes over a weir into the final compartment for chlorination before discharge overboard. The settled sludge is continuously returned to the aeration compartment by an airlift pump.

Excess sludge builds up in the settlement chamber and this must be discharged at regular intervals. In port this may not be possible hence it should be pumped to a sullage tank for disposal later. By using an incinerator to deal with excess sludge a sullage tank may not be required.

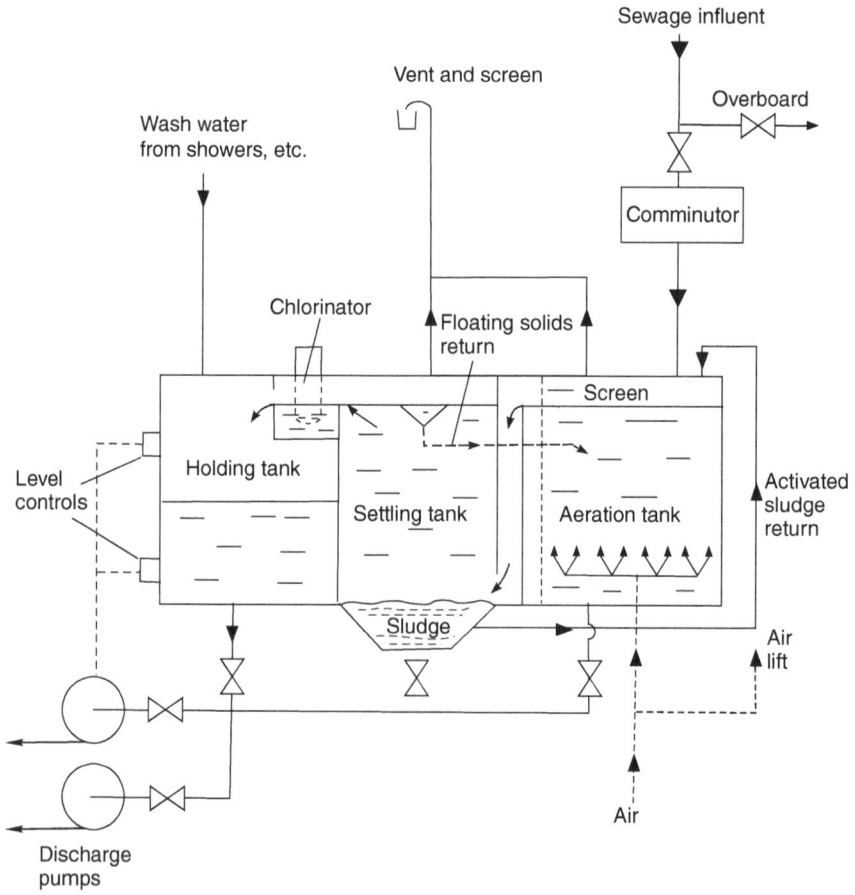

▲ Figure 9.27 *Extended aeration sewage plant*

With the extended aeration system it can be 5–14 days before the plant is fully operational because of the prolonged aeration of sewage necessary to produce the bacteria that carry out the purifying process. Hence, the plant should be kept operational at all times. It should be noted that oil or grease entering the system kills useful bacteria.

Chemical treatment plants

These are recirculation systems in which the sewage is macerated, chemically treated and then allowed to settle. The clear, sterilised, filtered liquid is returned to the sanitary system for further use and the solids are periodically discharged to a sullage tank or incinerator. The main advantages are: (i) no necessity to discharge effluent or sludge in port or restricted waters, and (ii) relatively small compact plant. However, chemical toilets are not always what they should be and with this relatively complex system increased maintenance is something that does not endear itself to engineers.

Sludge incinerators

These are capable of dealing with waste oil, oil and water mixtures of up to 25% water content, rags, galley waste etc., and solid matter from sewage plants if required.

Figure 9.28 shows a small water tube type of boiler combined with an incinerator plant in order to provide economy.

Homogenous oil–water mixtures that have been formed by passing them through a comminutor – a kind of grinder, macerator, mixer that produces a fine, well-dispersed emulsion – are supplied to the rotating cup burner. Solid waste from the galley and accommodation, etc., would be collected in bags and placed in the chamber adjacent to the main combustion chamber, the loading system of which is self-evident in the diagram. The loading arrangement incorporates a locking device that prevents the doors (loading and ash pit) being opened with the burner on. The solid waste goes through a process that may be described as pyrolysis, that is, the application of heat. Hydrocarbon gases are formed, due to the low air supply to this compartment, which pass into the main chamber through a series of small holes and burn in the furnace. Dry ash remaining in the chamber has to be removed periodically through the ash pit door.

Solid matter from sewage systems could be incinerated in this unit, for which a connection would have to be made from the sewage plant to the pyrolysis chamber of the incinerator.

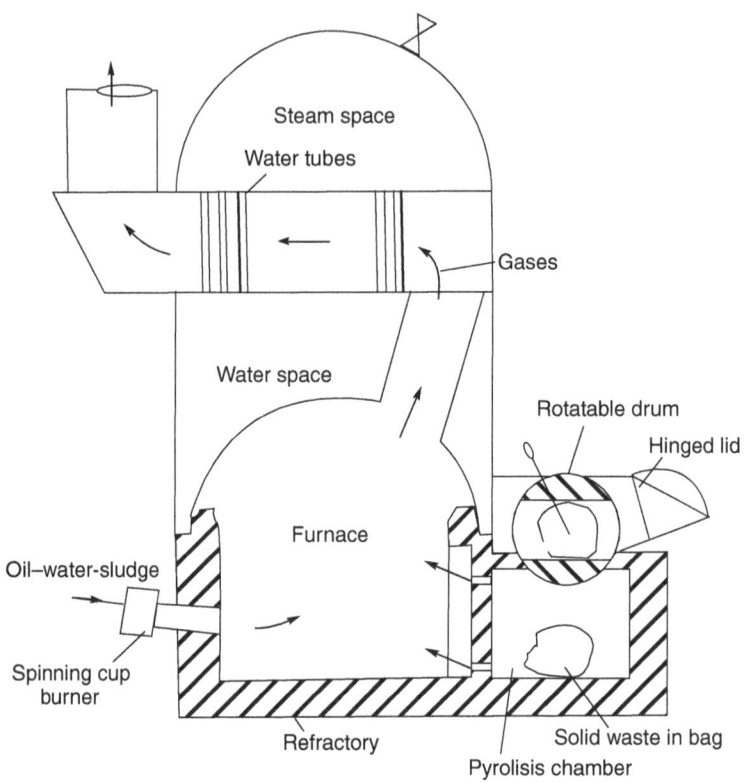

Steam space
Water tubes
Gases
Water space
Rotatable drum
Hinged lid
Furnace
Oil–water-sludge
Spinning cup
burner
Refractory
Solid waste in bag
Pyrolisis chamber

▲ **Figure 9.28** *Incinerator*

Pipe Arrangements and Fittings

Most of the details given are for cargo vessels as the tanker is virtually a whole pumping system within itself.

A typical bilge pipe arrangement

The arrangement as sketched satisfies the conditions required. The diagram should be considered in conjunction with the rules given later. For examination purposes it is advised that the student should, if asked, sketch and describe the bilge system and fittings with which he has been associated on a vessel at sea. In so doing he should be able to sketch the arrangement from memory and there should be no need to consider Figure 9.29 further, it is presented merely for guidance.

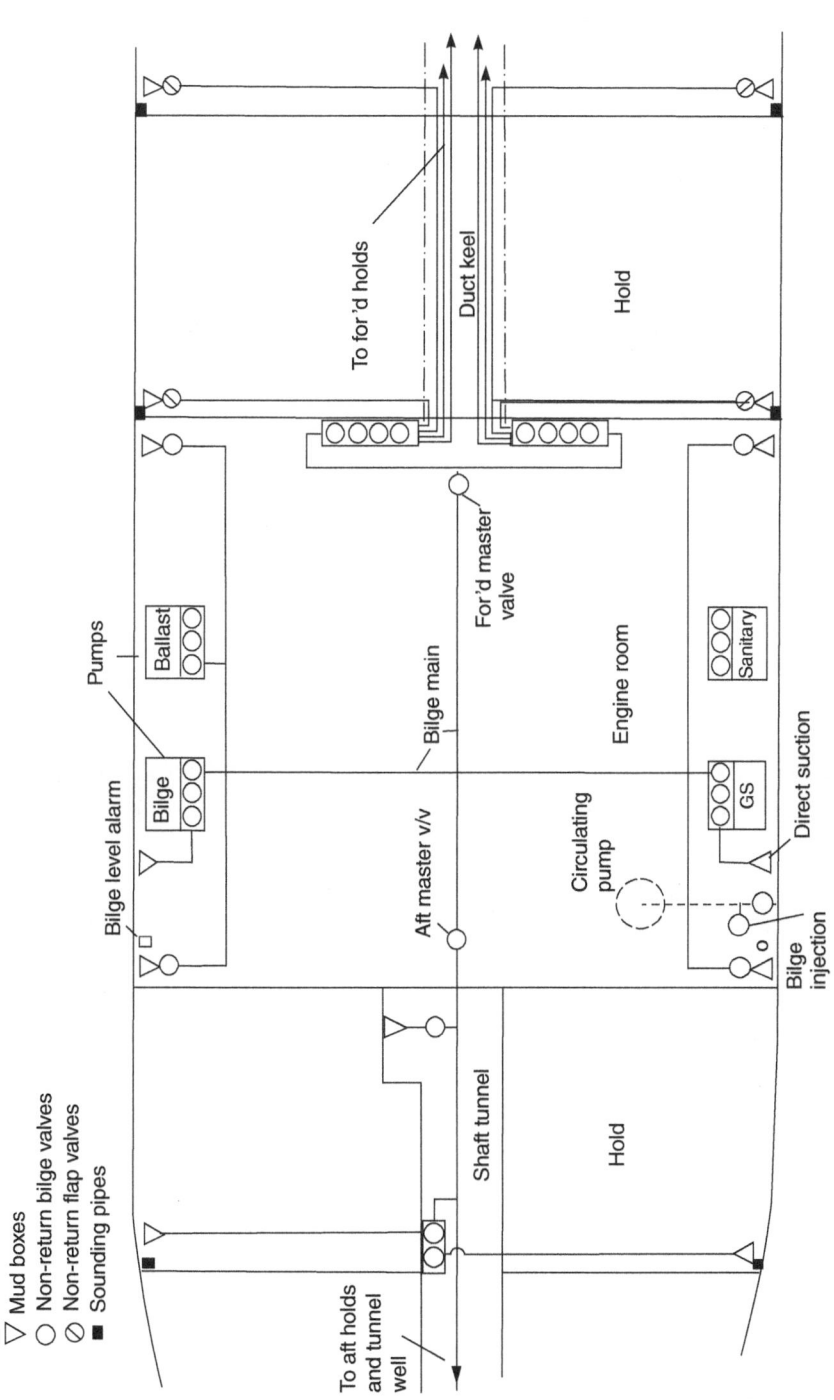

To for'd holds

Duct keel

Hold

Pumps

Ballast

For'd master valve

Bilge main

Engine room

Sanitary

Bilge level alarm

Bilge

Aft master v/v

Circulating pump

GS

Direct suction

Bilge injection

Shaft tunnel

Hold

To aft holds and tunnel well

Mud boxes
Non-return bilge valves
Non-return flap valves
Sounding pipes

▲ **Figure 9.29** *Bilge pipe arrangement*

Deep tank piping arrangement

Referring to Figure 9.30 it will be seen that the tank is arranged for filling with water ballast. In the early stages filling will be by means of the ballast pump discharge and later by gravity through the ballast pump suction chest, although if time is available gravity throughout is probably best. Final filling before closing the tank lid is best carried out by hose from the deck service line. When the tank is full the screw lift ballast valve is shut and the line is blanked off until the tank requires to be pumped out. When the tank is to be used for dry cargo the ballast line is blanked and the bilge line is open. Great care is necessary to avoid any mistakes being made and a rigid routine is advised. Clear explanatory notices are to be provided and all valves and fittings should be in good order and easily accessible.

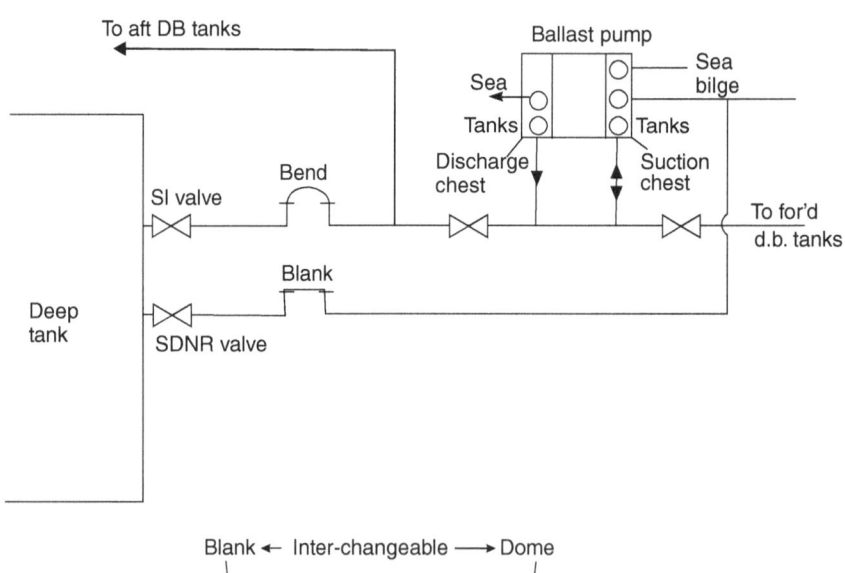

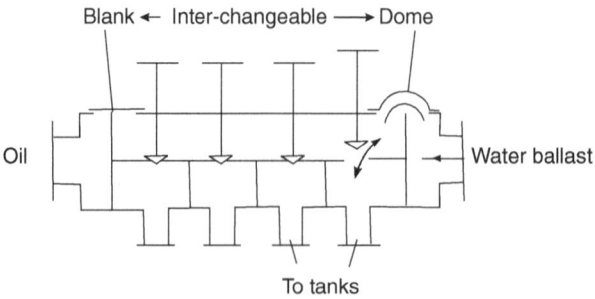

▲ **Figure 9.30** *Oil–water ballast chest*

Bilge injection valve

The bilge injection valve as sketched in Figure 9.31 is one of the most important fittings in the machinery space. It is provided for use in the event of serious flooding in the machinery space. By closing in the main injection valve and opening up the bilge injection valve, the largest pump (or pumps) in the engine room are drawing directly from the lowest point in the space; this action can remove large quantities of water. A doubler plate is welded to the skin, and machined usually after welding operations, the chest flange being bedded to the doubler and then studded in place. The joint is either spigot and jointing compound, or flat with a joint of canvas and red lead putty.

The diameter of the bilge injection valve is at least two-thirds of the diameter of the main sea inlet. Valve spindles should be clear of the engine platform and valves and operating gear require regular examination and greasing, with cleaning of strum or strainer.

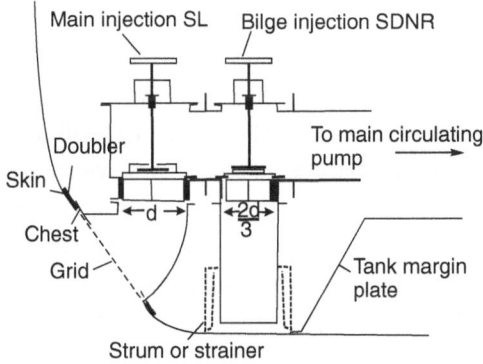

▲ **Figure 9.31** *Main and bilge injection valves*

Oil–water ballast chest

This chest is a standard fitting on most cargo vessels, on the double-bottom piping system. Normally all chests are open to oil fuel (bend) and blanked to water ballast. For ballast, or ballast prior to cleaning purposes, the bend and blank are as shown in the diagram (Figure 9.31). This means that an error in opening the wrong valve would not in itself allow crossing of circuits.

As an alternative to this fitting, hollow one-way discharge plug cocks or a system of interlocking valves would be acceptable. Any system employed must prevent easy joining of oil and water circuits by accident.

Materials used in pipework construction

Chapter 4 of this volume explains much more about the corrosion of metals. The precautions and treatment required to protect the main plant such as boilers and diesel engines is also covered in Chapter 4.

However, the whole vessel is working in a marine environment and as such all the metals used in the construction of the vessel will need to be chosen with the cost/life expectancy equation carefully considered.

The most problematic area is with the sea water system and until recently the solution has revolved around choosing a low-cost metallic material and a more frequent replacement programme or a higher initial cost for metal that will last longer before requiring replacement.

Recently a third option is gaining popularity especially for lines where more corrosive fluids are to be transported around the vessel. The developments in design using composite material have reached the stage where classification societies and flag administrations are able to approve their use for this role, bearing in mind that any failure of a seawater line low down in the vessel could have a fatal effect and the composite material under consideration is glass reinforced plastics or carbon fibre. These have the advantage, over steels and other metals, of not corroding in the presence of moisture and/or seawater.

A composite material is the combination of two different materials that are joined together to form a third structure where the new material has more useful characteristics than the two original components.

The composite differs from an 'alloyed metal' in so much as the physical, chemical and mechanical characteristics of the original components remain unchanged in the new structure. The increased strength that is gained is achieved with a low-density material and therefore another advantage of the modern composite material is that it is lightweight when compared to most metals undertaking the same job.

However, like all things the low volumes of production have, until now, not allowed this material to compete with the metallic competition. There have also been concerns about the durability of composites against the traditional material but the most important characteristic to consider is the composite's mechanical properties.

When metals are manufactured the chemical composition of the mix in a furnace is carefully monitored to ensure that the final product will have the correct mechanical strength and other properties that were intended. The process of making alloyed metals is well known and the even consistency of the final product is very important, therefore the builders of ships will have confidence in the quality of the material being used.

The classification societies will still have rules for the material manufacturers to follow. Lloyds Register for example have a 300+ page document called 'Rules for the Manufacture, Testing and Certification of Materials', which gives details about the quality assurance process.

The manufacturing process for plastic 'composite' material covers things such as the resins, gel coat, binder and density of the fibre itself. Therefore the manufacturing process can be more exacting and close monitoring is required to produce a product of consistent quality.

Some Rules Relating to Pumping Systems' Bilge System

These rules pertain to pumping systems bilge in vessels over 90 m long.

1. A piping system and pumping plant should be provided to pump out and drain any adjacent-to-damage water tight compartments (including tween decks) under all reasonable damage conditions. Efficient drainage should be provided especially to unusual-form compartments and the piping system design should not allow flooding under damage conditions.

2. Vessels shall have at least four independent power pumps connected to the main line. Ballast, sanitary, etc. are acceptable, also engine-driven pump, provided they are of sufficient capacity and are connected to the main line.

3. One such pump should be of the remote-controlled submersible type *or* the power pumps and controls should be so placed so that one pump is always available under all reasonable damage conditions. Each pump should *where possible* be located in a separate watertight compartment.

4. Pumps should be of the self-priming type unless efficient priming devices are provided. The capacity of the pumps should give a water speed in the main line of not less than 2 m/s, and the capacity may be determined from a given empirical formula.

5. Each pump should have a direct suction to the space in which it is situated, such suction to be at least the same bore as the bilge main. Not more than two such suctions are required and in the machinery space such suctions should be arranged one on each side.

6. Main engine circulating pumps shall have a direct suction (with non-return valves), draining the lowest level in the machinery space, with at least two-thirds of the diameter of the main sea inlet. In motor ships this should apply but direct suctions on other suitable pumps of equivalent capacity is acceptable.

7. Bilge pipes should not be led through oil tanks or DB tanks. Joints should be flanged, and pipes well secured and protected against damage. The pipes should be independent to the bilge systems only.

8. Collision bulkheads should not be pierced below the margin line by more than one pipe, such pipe to be fitted with a screw-down valve operated from above the bulkhead deck, valve chest being secured to the forward side of the collision bulkhead (divided peaks may have two pipes).

9. Valves and cocks not forming part of a pipe system are not to be secured to watertight bulkhead. Pipes, cables, etc. passing through such a bulkhead are to be provided with watertight fittings to retain the integrity of the bulkhead. Connections attached to such bulkheads are to be made by screwed or welded studs, not by tap bolts passing through clearance holes.

10. The bilge piping system is to be separate from cargo and oil fuel systems. Spindles to all master valves, bilge injection, etc. should be led above the engine room platform. All valves, extended spindles, etc. to be clearly marked and accessible at all times.

11. Diameter of bilge suction lines in millimetres to be determined from the given empirical formula. No bilge main should be below 65 mm bore and no branch should be below 50 mm or need be over 100 mm bore.

12. Bilge valves should be one of the non-return types. Valves, blanks, lock-ups, etc. must be provided to prevent connection between sea and bilges or bilges and water ballast, etc., at the same time.

13. Emergency bilge pumping systems if provided should be separate from the main system.

14. Bilge pipes to be provided with mud boxes. Suction pipe ends should be enclosed in easily removable strum boxes, the holes should be approximately 10 mm diameter and their combined area not less than twice the area of the suction pipe.

15. Sounding pipes where provided are to be as straight as possible, easily accessible, normally provided with closing plugs, and machinery space pipes to have self-closing cocks.

Note

One explicit rule, covered by the generalisation summary in rule 1 given above, is considered worth repeating, in full, in view of a recent casualty.

Provision is to be made in every vessel to prevent the flooding of any watertight compartment served by a bilge suction pipe in the event of the pipe being severed or damaged, by collision or grounding, in any other watertight compartment. Where any part of such a pipe is situated nearer to the side of the ship than one-fifth of the mid-ship breadth of the ship measured at the level of the deepest subdivision load water line, or in any duct keel, a non-return valve shall be fitted to the pipe in the watertight compartment containing the open end of the pipe (see Figure 9.30).

Ballast system

Ballast water has been and still is one of the most important environmental issues currently under discussion within the industry. Until fairly recently ships have taken on ballast water, usually at a discharge port, and released the water at the loading port, which could be many thousands of miles away. This practice, however, was transporting micro-organisms, including bacteria and viruses as well as the adult and larval stages of many coastal plants and animals, to very different parts the world. Some of these suited their new environment, became 'invasive species' and caused devastation to the local inhabitants, in some cases causing a risk to human health. IMO has now developed regulations to greatly reduce the risk of introducing non-native species into areas from the ballast water of ships.

Ships are now required to have ballast water management plans in place that will include completing appropriate action to comply with the IMO convention. Ballast water exchange, several times during a long voyage, was allowed until 2014 for vessels that have between 1,500 m^3 and 5,000 m^3 of ballast capacity and until 2016 for all other vessels. After these dates the vessels were required to be fitted with appropriate and approved ballast water treatment technology. These could be physical separation systems or disinfection systems using chemical treatment or physical treatment, such as UV, gas injection of ultrasonic treatment.

One of the most difficult problems is taking a 'representative sample' and owners will have to think carefully how this is to be achieved. (See Chapter 12 for more details.)

Oil fuel installations

Best practice has evolved from many years of experience and taking guidance from several sources such as:

- the code of safe working practices;
- classification society rules;
- flag administration rules;
- SOLAS;
- on-board ism plans;
- Chief Engineers' standing orders.

Some will have been repeated under sections such as boilers, fuel testing and ship construction. However, it is worth considering guidance relating to instructions to ships' engineers.

Instructions and best practice

1. A plan and description of the oil piping arrangement should be clearly displayed.
2. Escape of oil heated to or above the flash point is most dangerous, and may result in explosion or fire.
3. After lighting burners, the torches *must* be fully extinguished by means of the appliances provided for that purpose.
4. Cleanliness is essential to safety; no oil or other combustible substances should be allowed to accumulate in bilges or gutter ways or on tank tops or boiler flats.
5. Before any oil tank that has contained oil fuel is entered for any purpose the oil should be removed entirely, all oil vapour must also be carefully removed by steaming and efficient ventilation. Tests of the atmosphere in tanks or bunkers should be made to ensure safety before inspection or work in them is begun.

Boiler, settling tank and oil fuel unit spaces, etc. must be clean, have no combustible material and have good access. Oil tanks, oil pumps, etc. should be fitted as far from boilers as is practicable and should be provided with trays and gutters, drain cocks, etc. They should be self-closing and efficient sounding or indicating devices provided. Relief valves should be fitted to discharge to an overflow tank fitted with level alarms, and filling stations should be isolated, well drained and ventilated. Every oil tank should have at least one air pipe, such air pipe or any overflow pipe system provided (preferably returning to an overflow tank with visual and sight returns) should have an aggregate area at least 1¼ times the aggregate area of the filling pipes. All means should be considered to prevent discharge of oil overboard.

Oil pipes and fittings should be of steel, suitably hydraulically tested. Oil units should be in duplicate and any oil pump should be isolated to the oil system only, provided with relief preferably back to the suction side, capable of being shut down from a remote control position, and provided with shut-off isolating valves. Heating coil drains should be returned via an observation tank. Valves or cocks fitted to tanks in the machinery and boiler spaces should be capable of being operated from a remote position above the bulkhead deck. Ample ventilation and clearance spaces for circulation should be provided and no artificial lights capable of igniting oil vapour are allowed. Ventilator dampers, etc., must have reliable operating gear clearly marked for shut and open positions.

Note: Essential features are care and cleanliness together with reliable overflow and isolating equipment. Particular care is advised during bunkering to avoid overflows (gravitating is always a safer process where practicable) and during tank cleaning, venting or inspection periods.

10

ON-BOARD MANAGEMENT OF FUEL AND LUBRICATING OIL

One of the most important tasks on a ship is to ensure that it does not run out of fuel. The fuel consumption calculations are very carefully completed by the Chief Engineer who will then determine how much fuel is required and at which port it will be taken on board. The procedure of taking on fuel and lubricating oil is called 'bunkering'. This name originated with older ships that used to load coal into the coal bunker and the term 'taking on bunkers' has stuck, but today it means taking on all types of fuel as well as lubricating oil. The term is also applied to water in the unlikely event of a ship having to take on water instead of making their own.

The choice of the bunker port is important because it must fit in with the ship's operation and it must also be the most cost-effective port that the vessel calls at. A quick search of the internet will reveal that the cost of all the different grades of fuel will differ from port to port and if a port does not have a bunkering facility it might be cost-effective to bring in the fuel by bunker barge from another port.

It is also very important to check that the vessel has received the correct amount and specification because the fuel taken on board will have to last until the next bunkering opportunity. If the fuel is taken on board when it is not up to specification, then the ship runs the risk of breaking down in the middle of the ocean.

The development of the MARPOL regulations and the increased willingness by flag state administrations to impose severe penalties on any ship leaking oil into the water has brought the bunkering procedure into sharp focus.

The bunkers will usually come as bulk delivery from a road tanker or a barge. It will be pumped on board through a pipe using the contractor's equipment. Before the operation all the ship's officers should be aware that bunkering is due to take place and all the engineering staff should be aware of their role and responsibilities throughout the process. It is important that the following points are covered by the Chief Engineer's instructions to the staff:

1. Ship's engineer to ensure that the hoses supplied by the contractor are free from any defects.
2. Ship's engineer to be present as the final flange from the contractor's hose is connected to the ship's pipework – including the flange joint.
3. All holes in the final flange should have a suitably tightened nut and bolt.
4. All the deck scuppers should be blocked off and there is sufficient absorbent material available to soak up any oil spill.
5. Savills under the pipework flanges should be intact.
6. Hoses have sufficient slack and are supported so that there is no weight on the flange.
7. Efficient communication needs to be set up between the bunker station and the barge or road tanker as well as between the bunker station and the ship's engineering staff controlling internal oil pathways.

During the bunkering operation the fuel should be sampled so that the oil may be sent to a laboratory for analysis. A lot of development work is being carried out to design suitable equipment so that fuel can be analysed on board instead of being sent to the laboratory, which takes time and before the results are received back on board the bunkers may already be in use.

Samples should be kept on board for at least 12 months and the ship should have a system in place to keep track of the retained samples.

Figure 10.1 shows just where these precautions will be found in relation to the bunkering process. Guidance should also be made to MARPOL Annex. 1 – regulations for the prevention of pollution by oil, MEPC.1 /Circ. 508 Bunker Delivery Note and Fuel

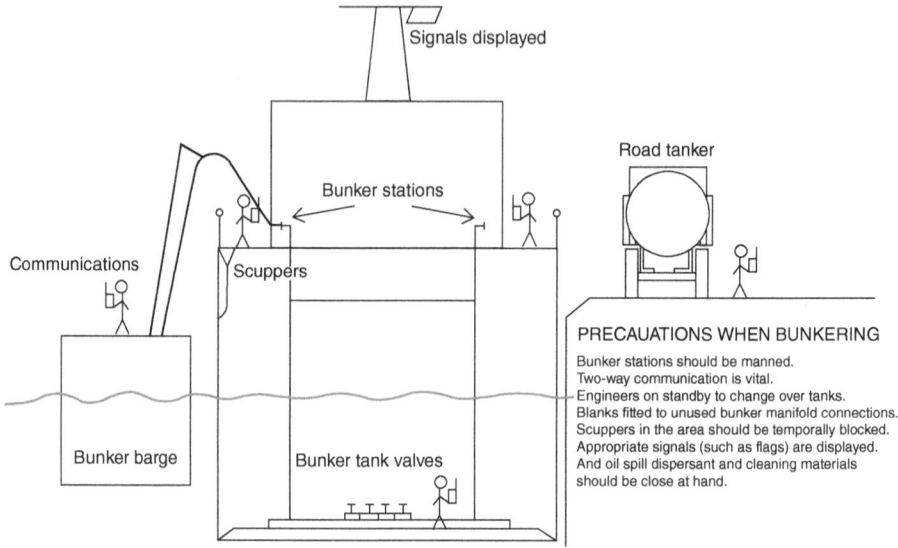

▲ **Figure 10.1** *Precautions to be taken when bunkering*

Oil Sampling, MARPOL Annex. VI – Guidelines for Sampling of Fuel Oil and ISO 8217 fuel standard 2010.

The reason for such care of fuel on a modern ship is that it is easy to receive fuel that is not up to the specification ordered. There is so much that can go wrong and if it does the ship could be placed in a potentially dangerous position. More detail about the refining process and the production standards can be found in Chapter 1 and information about mass flow meters for measuring bunkers can be found in Chapter 11. It is important when bunkering heavy fuel oil that it is kept at a temperature suitable for pumping and storage. If the oil cools down wax can form, which may clog up filters and affect the operation of pumps. Therefore the steam plant should be ready for service.

Careful records should be kept about which storage tanks are holding the bunkers loaded in which port. Also it is good practice not to mix bunkers from any one place. This means that if defective fuel is loaded there will be as little cross-contamination as possible. Special care should be taken in cold climates to ensure that the correct temperature is maintained. When in the storage tank the heat will help to separate out any water that has been delivered with the fuel. However, if this is a double-bottom tank it will be difficult to remove the water at this stage as it cannot be drained.

Fuel and its on-board treatment – gravity separation

Fuel oil will be transferred from the 'in use' bunker tank to the settling tank where it is allowed to stand for a while allowing the water, to move to the bottom of the tank where it can be drained off.

On board the three processes used to clean fuel and lubricating oil are gravitation, filtration and centrifugal purification. The use of gravity, where water and particles of dirt are allowed to separate out of the oil and fuels, is carried out in the settling tanks. When the oil is allowed to stand undisturbed in the tank, elements of higher relative density than the oil gravitate to the bottom of the tank where they are discharged periodically through a manually operated sludge cock. The process of separation in a settling tank can be speeded up to a certain extent by heating the tank contents. If heating of the contents is possible, steam-heating coils are generally used, but care must be taken not to heat the oil to too high a temperature. Figure 10.2 shows a settling tank with the usual fittings provided. SOLAS requires that marine fuels have a flashpoint of 60°C but the classification societies recommended a storage temperature, in the bunker tanks, to be above 45°C, which preserves the quality of any blend. This means that an ideal temperature for heavy oil in a bunker tank would be about 50–55°C. The next stage in on-board fuel management is in the settling tank where there needs to be sufficient heating capacity to raise the heavy fuel oil temperature ready for the final preparation phase prior to use by the main engine. Figure 10.2 shows the arrangement of a typical oil storage tank showing the internal and external fittings. In an examination these fittings should be itemised and a brief description of their function should be given as shown below.

Sludge valve or cock. Used for draining water and sludge from the bottom of the tank. It must be self-closing; if it were not, and it was left unattended, a dangerous situation could arise whereby the tank content could be drained into the oily bilge or sludge tank.

Dumping valve. This fitting can be used in the event of fire to dump the oil from an elevated settling tank to a double-bottom tank, which could possibly be below the level of the fire.

Exhaust steam. From heating coils this would be led to a steam trap, which ensures maximum utilisation of the heat content in the steam, then to an observation tank where any defect in the pipe work can be diagnosed due to oil crossing over into the steam-heating circuit.

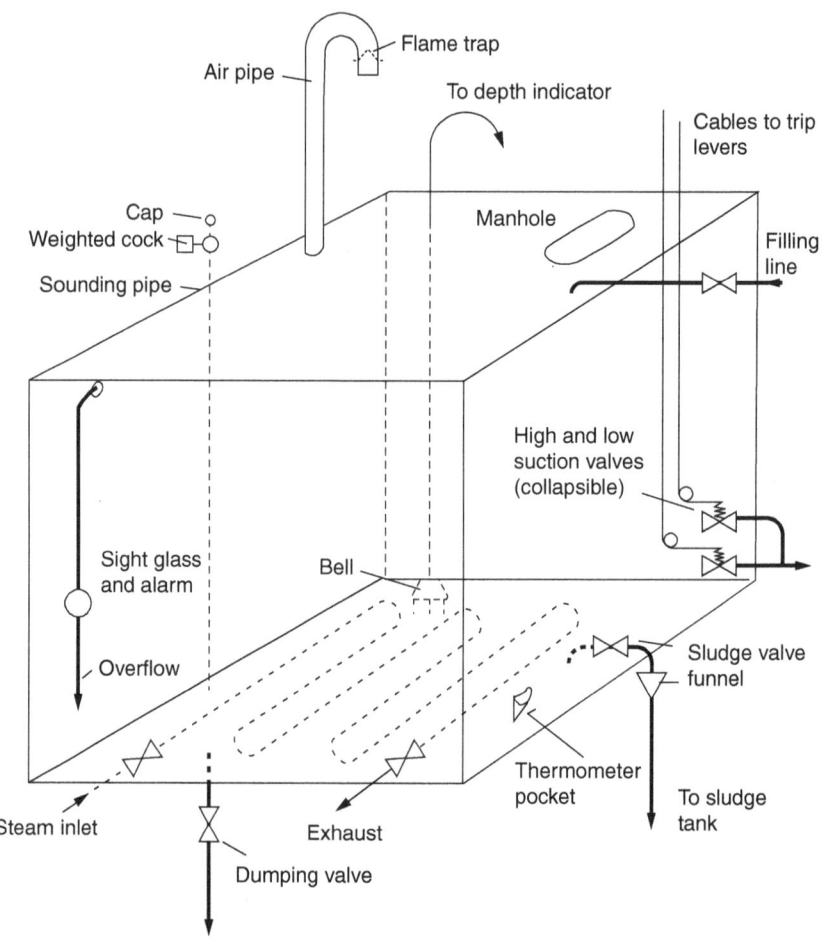

▲ **Figure 10.2** *Settling tank including fittings*

Overflow pipe. This is an important feature of the tank that could stop it from being overpressurised. However it has caught engineers out in the past. If a tank is filled right up to its maximum with fuel that is cooler than it should be, then when the fuel is raised in temperature it will expand. This could lead to the oil being forced up the overflow pipe, or indeed the sounding pipe, and end up flowing over the deck and into the water.

Sounding pipe. This is a tube extending from a platform above the tank through the top of the tank to the bottom. At the foot of the tube are two important features. The first is a hole so that the liquid in the tank can enter the pipe and the level in the pipe and the tank can then be measured. The second feature is a flat metal plate placed at the bottom of the pipe, which is called a striker plate and is there to take the force of the weight on the end of the sounding tape hitting the bottom. The weighted cock on the top of the sounding pipe is important and should be checked for correct operation by the watchkeeper during his/her tour of the engine room. The reason for the weight is to ensure that the cock is closed following each use and because of this it can be awkward to used; therefore staff have sometimes been tempted to wedge it open in some way to make taking a sounding easier. The cock restricts any liquid travelling up the sounding pipe from spilling out over the deck or into the engine room. In the event of a rupture in the tank this could be a potential position for ingress of water.

Remote cables. Tanks containing fuel and oil will have high- and low-suction valves with remote cables fitted. In the event of a fire in the engine room, the tanks can be isolated from a safe position. This is an essential feature associated with settling and service tanks.

Manholes. These are provided to give staff access to the tank for cleaning and repair. To gain access the manholes will have to be unbolted after ensuring that the tank in empty. However, it is vitally important that before any person is allowed to enter the tank a strict procedure is followed to make sure that the tank is safe for human occupation. An example of the procedure to follow for entry into enclosed spaces appears in the code of safe working for merchant seamen that is produced by the UK administration. The procedure should also be clearly set out in the ships safety management system (SMS). This point is so important that students should expect, and be prepared, to answer questions every time in their examinations.

With the introduction of the need to use different grades of fuel it will be very important for the engineers to keep track of where the different oils are stored. Ships may have low-sulphur marine diesel oil (LSMDO), intermediate fuel oil (IFO) and heavy fuel oil (HFO). These oils should not be mixed in the storage tanks and will only come into contact with each other for short periods during a changeover phase from one grade to another. Don't forget also that there may be the bunkers from different ports to allow for as well.

Filtration

The process of filtration of lubrication and fuel oils removes unwanted particles of material such as cotton threads, paint chippings, small pieces of metal, etc., which could cause damage to pumps and engines, if left to circulate with the oil. Filtration does not separate much water from the oil; however, by pumping heated lubricating oil into a vacuum chamber, vaporisation of water can be achieved. Also, water repellent and water coalescing filter cartridges can be used, which will cause some separation of water from oil.

Many different types of filters are manufactured, the simplest being the wire mesh type that are fitted in pairs in the lubricating oil piping system. One filter is used at a time and this arrangement enables the operator to clean the filter not in use without shutting down the oil system. Others of a more complex nature can be cleaned while in operation and may be fitted singly in the oil piping system or again in pairs.

Wire gauze-type filters are made with coarse or fine mesh depending upon the positioning of the filter unit in the oil system. An example of this are the hot and cold oil filters fitted in oil burning and pumping installations: the coarse-mesh suction filters are used for cold oil and the fine-mesh discharge filters are used for the heated oil. The wire mesh-type filter, however, is rarely made to filter out particles below 125 μm in size. If finer filtration is required, other types of filter unit are used: one such being the well-known Auto-Klean strainer.

Auto-Klean strainer

See Figure 10.3. This type of filter is an improvement on the wire gauze strainer. It can be cleaned while in operation. It can filter out particles down to 25 μm in size. The dirty oil passes between a series of thin metal discs mounted upon a square central spindle. Between the discs are thin, metal, star-shaped spacing washers of slightly smaller overall diameter than the discs. Cleaning blades, fitted to a square stationary spindle and the same thickness as the washers, are between each pair of discs. As the oil passes between the discs, solid matter of sizes larger than the space between the discs remains upon the periphery of the disc stack.

The filter is cleaned by rotating the central spindle, which rotates the disc stack and the stationary cleaning blades scrape off the filtered solids, which then settle to the bottom of the filter unit. Periodically the flow of oil through the filter unit is interrupted and the sludge in the bottom well is cleaned out. To facilitate the cleaning operation the filters are generally fitted in pairs.

Pressure gauges are fitted before and after the filter unit to indicate the condition of the filter. The pressure difference (DP) across the filter is low when the filter is clean and the pressure difference becomes progressively higher as the filters start to become clogged up. The DP should not be allowed to increase too high as the flow of oil could then be restricted.

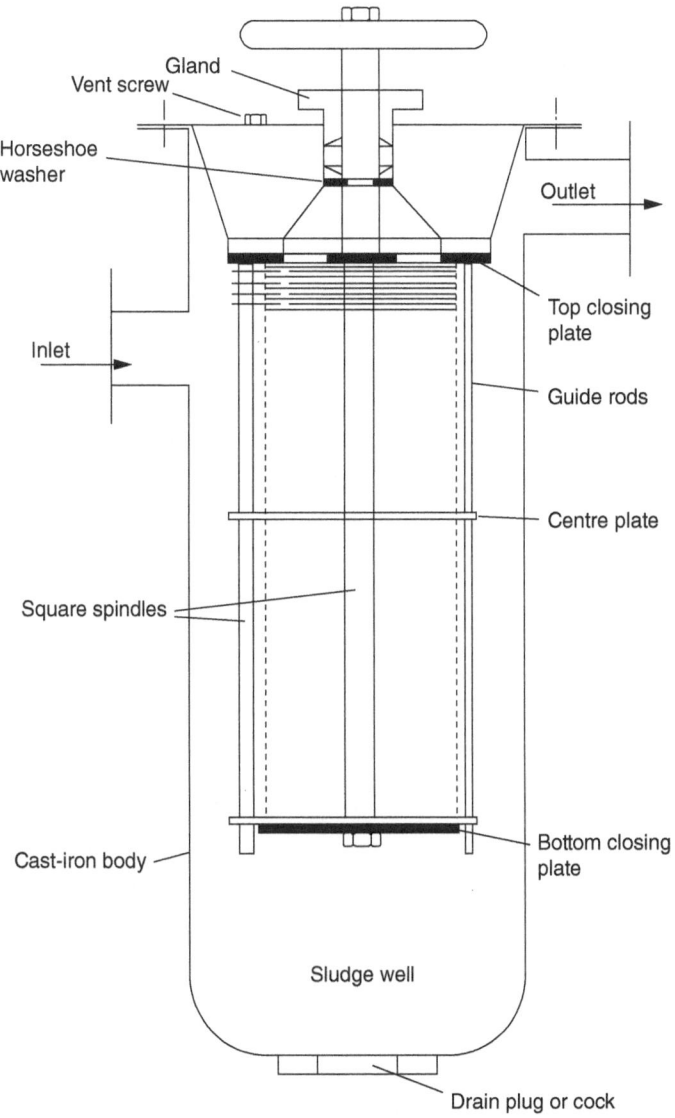

▲ **Figure 10.3** *Auto-Klean strainer*

Figure 10.3 illustrates the internal arrangement of an Auto-Klean filter unit, which can be arranged for automatic or manual operation. The standard unit was capable of filtering out particles down to 200 μm in size, this type can also be made to filter particles of under 75 μm, but the mechanical strength of the cleaning blades will be low. The current Auto-Klean strainer has the modified disc stack and cleaning blade arrangement shown in Figure 10.4. With this modified disc stack, particles down to 25 μm in size can be filtered out without impairing the mechanical strength of the cleaning

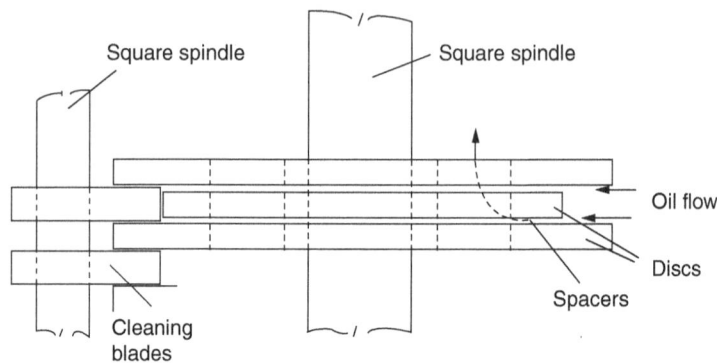

Square spindle   Square spindle
Oil flow
Discs
Spacers
Cleaning
blades

▲ **Figure 10.4** *Modified disc stack (Auto–Klean)*

blades. The 'fluxflo' self-cleaning wire gauze filter can be used to filter out particles of between 10 p.p.m. and 15 p.p.m.

Streamlined lubricating oil filter

The streamlined filter consists of a two-compartment pressure vessel containing a number of cylindrical filter cartridges. Each section rod is held in longitudinal compression. The discs can be made from a wide variety of materials, for lubricating oil special paper discs are generally used. The oil can flow from the dirty to the clean side of the filter via the small spaces between the compressed discs then up the spaces formed by the hole in the disc and the rod. In this way, the dirt is left behind on the periphery of the disc stack and it is claimed that particles of the order of 1 μm can be filtered out, which means that this type of filter maintains the oil in a very good condition without the need for other treatment plants.

For cleaning, compressed air is generally used. Closing A and B and opening D and C results in reversal of flow (see Figure 10.5).

Oil mist eliminators

The Vokes oil mist eliminators can be used to remove the oil mist from air flows that have been contaminated with oil from areas such as lubrication oil tanks or engine crankcases and car decks on ro-ro ships.

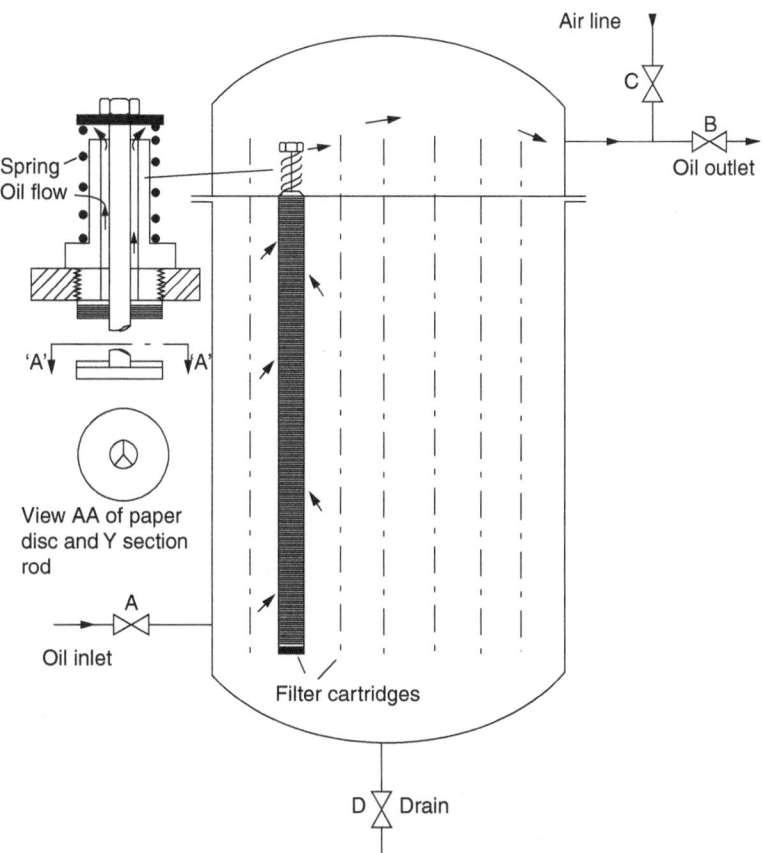

▲ **Figure 10.5** *Streamline lubricating oil filter*

Filter coalescers

These have been designed to replace the centrifugal method of particulate and water removal from fuel and lubricating oils.

The unit consists of some form of pre-filter for particulate removal followed by a compressed, inorganic fibre-coalescing unit in which water is collected into larger globules.

Coalescing action is relatively complex but briefly, the molecular attraction between the water droplets and the inorganic fibres is greater than that between the oil and the fibres. As the number of droplets increases they join together to form a layer of water.

When the water globules are large enough they will drop to the bottom and out of the coalescing unit.

Downstream of the coalescing cartridges are PTFE-coated, stainless-steel, water-repelling screens that act as a final water-stripping stage. Water gravitates from them and from the outlet of the coalescer cartridges into the well of the strainer body from where it is periodically removed.

In modular form these units would have pumps, motors, alarms, indicators, water probes with automatic water dumping, heaters to lower the viscosity of the oil, together with the filtration system described above.

Lubricating oil filter-coalescer

Lubricating oil in circulation round a closed system, for example turbine, generators, sterntubes, etc. will absorb moisture from the atmosphere that will reduce the lubricating properties of the oil. Figure 10.6 shows a filter-coalescer that will remove solid particles of 3 μm and above and also up to 99% of the water present in the oil.

Lubricating oil is pumped through the water coalescer filter cartridges, which remove solids and coalesce (i.e. gather into larger droplets) the free water droplets held in suspension in the oil. Most of the water then gravitates to the bottom of the body and the oil with the remaining water droplets passes to the water-repelling screens, which permit passage of oil only. Water droplets that collect on the screens eventually settle at the bottom of the body.

To clean the unit, it must first be drained and then the filter cartridges are renewed; the water-repelling screens need not be touched.

A heater would be incorporated in the supply line, which would heat the lubricating oil, thus assisting separation.

Centrifugal filter

Smaller diesel engines are increasingly being fitted with a filter that utilises centrifugal force to remove some of the carbon from the oil thus prolonging the time between overhauls (TBO). The oil is introduced into the unit under the lubricating oil system pressure. The oil is directed out of small holes in the bottom of a rotating drum. Inside the cylinder part of the drum is placed a card and as the bulk of the oil is introduced into the body of the unit the heavy material accumulates on the card. After a period of time the card will need to be changed and this will take out the acclimated carbon as well.

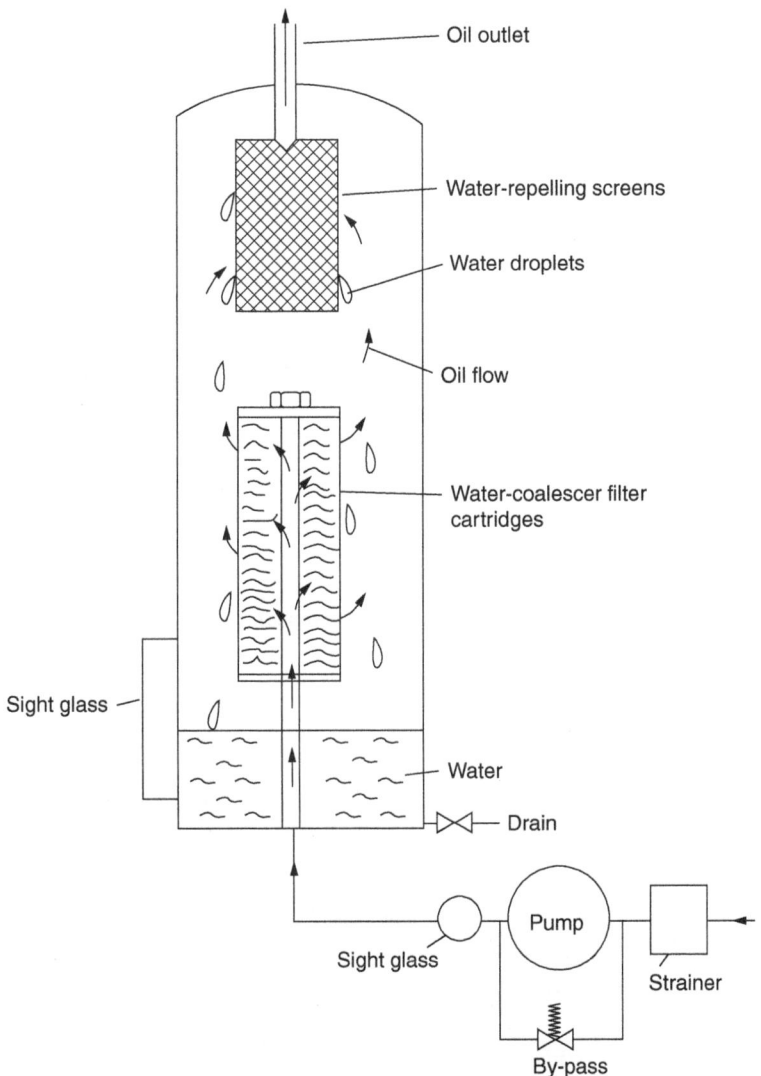

▲ Figure 10.6 *Lubricating oil filter-coalescer*

Oil module (fuel or lubricating oil)

An automatic oil-cleaning module comprising duplicated filter assembly, pumps and controls mounted on a water/sludge tank that serves as a base is shown diagrammatically in Figure 10.7.

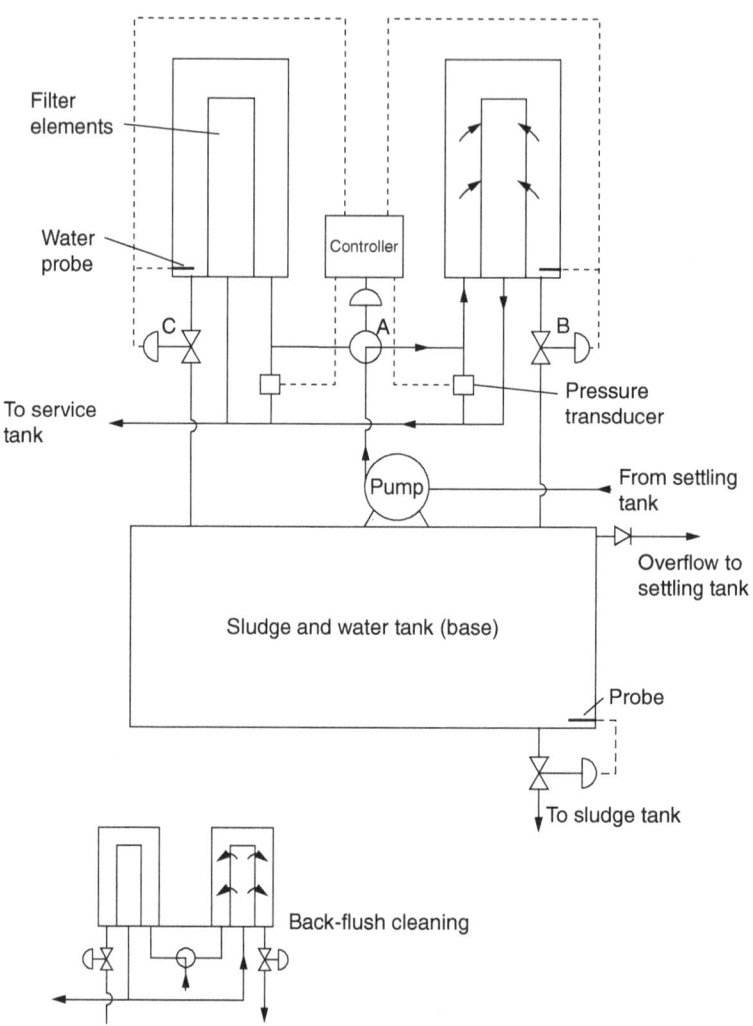

▲ **Figure 10.7** *Automatic oil filter module*

During normal operation, dirty oil from the oil settling tank would be pumped through the filter to the service tank. Impurities collect on the outer surfaces of the filter and this results in an increase in pressure differences across the filter. When this pressure difference reaches the pre-set limit of about 0.4 bar a signal from the differential pressure switch to the controller starts the cleaning procedure. The controller sends signals to valves A and B, which change over and open respectively so that the back-flush cleaning of the dirty filter as shown in Figure 10.7 takes place for about 60 seconds. At the end of the cleaning period, valve B closes and the system is back to normal operation.

If water enters the filter body its presence is detected by a water-detection probe. A signal from the probe causes valves B or C to open, depending upon which filter is in use, and the water is discharged into the sludge tank. When the water is completely discharged the valve automatically closes.

Sludge and water in the base tank are automatically located and discharged to a sludge storage tank. The relatively clean oil from the top of the base tank overflows into the settling tank for recycling.

A differential pressure device that could be used in the module is shown in Figure 10.8. Increasing the spring force increases the pressure differential setting at which the plunger will operate a switch for the timed cleaning sequence. The synthetic rubber diaphragms at each end would resist attack from the oil.

The candle-type back-flushing filter is a popular choice for designers of modern machinery plants. The Boll & Kirch system can have between 1 and 16 candles dividing the filtration process automatically. The candles work continuously but are arranged so that one filter chamber can regenerate while the others are working.

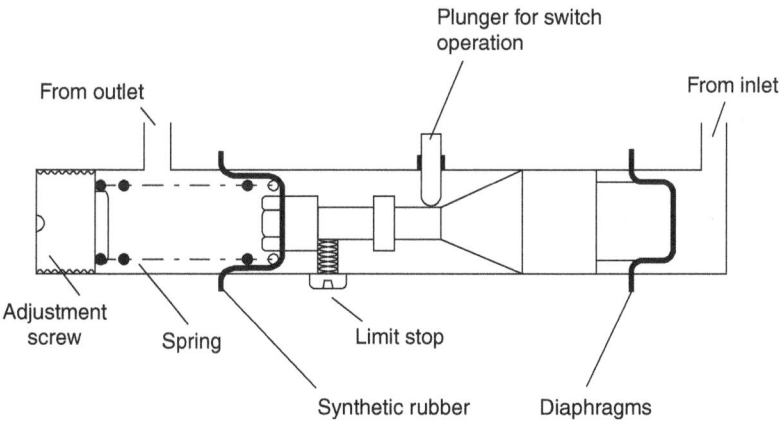

▲ **Figure 10.8** *Differential pressure switch*

Clarification and Separation

Clarification

The term clarification is used to describe separation of solids from a liquid. A centrifuge arranged to discharge a single liquid is called a clarifier.

Figure 10.9 (top) shows clarification taking place in a settling tank by the use of gravity. Oil from the storage tank, containing solids and any water products, is pumped into the tank at (a) and as the oil flows to (b) the solids present in the oil gravitate to the bottom of the tank. The heaviest solids deposit first at (a) and the lighter solids, which are carried forward by the flow of the oil, deposit nearer to (b).

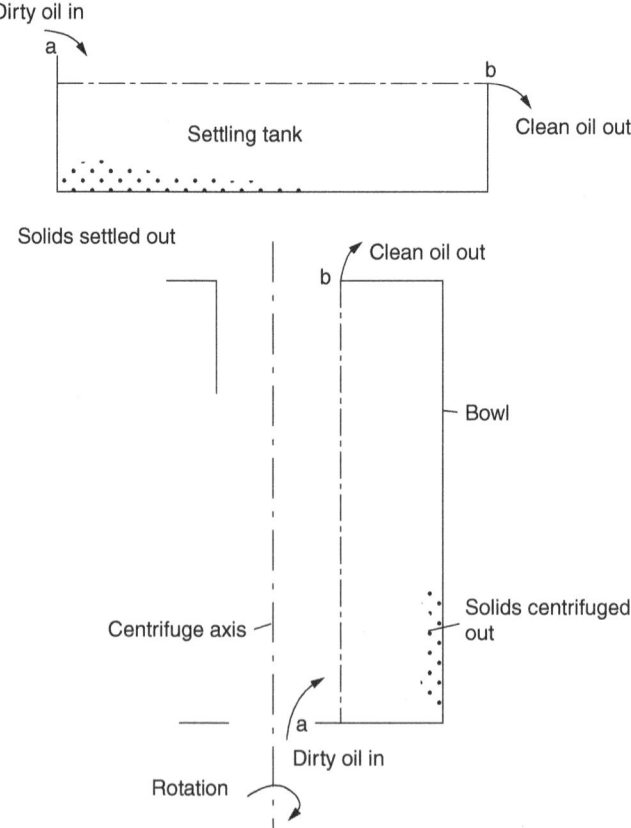

▲ **Figure 10.9** *Clarification*

A similar process takes place in the centrifugal clarifier, shown by rotating the settling tank analogy in Figure 10.9 through 90° to give the lower picture in the diagram. Here, oil is fed in at (a) and is thrown by centrifugal force to the side of the rotating bowl. Solids present in the oil pass through the oil to the bowl side and accumulate there, the heavier solids depositing near the bottom of the bowl and the lighter solids towards the top. The similarity of the two methods of clarification, gravitational and centrifugal, can be seen from the diagram.

Separation

The term separation is used to describe the separation of two liquids and in the maritime industry these are usually oil and water. A centrifuge arranged for the separation and continuous discharge of two liquids is called a centrifugal separator.

In a purifier arranged to operate as a centrifugal separator any solids that are present will deposit upon the side of the bowl, and hence clarification takes place at the same time as separation. For a purifier to operate as a centrifugal separator a water seal is necessary and this operates as follows:

Using the gravitation analogy in Figure 10.10, water is first fed into the tank at (a) until level (b) is reached, overflow of the water will then take place at (b) and when this occurs the water supply is stopped, as any additional water added would not increase the level of the water above (b). Next, oil is supplied into the tank at (a) and the oil will displace some of the water in the tank, the amount displaced depending upon the relative density of the oil. If overflow of oil takes place at (c) separation will occur, that is, water present in the oil will settle causing overflow of an equal amount of water at (b) and hence the quantity of water forming the seal remains constant. If, however, when oil is supplied to the tank of sufficiently high relative density to cause the oil–water interface level (b) to reach (d), oil will be discharged at (b) and no oil will be discharged at (c). This is called loss of seal.

Rotating the settling tank analogy in Figure 10.10 through 90° gives the lower figure in the diagram, which shows separation taking place in a centrifugal separator, the principle being analogous to that of gravitational separation described above. Water is first delivered to the centrifuge and when discharge of water takes place at (b) the water supply is shut off and then oil is delivered at (a), some of the water is displaced and when oil and water are being separately and continuously discharged, the centrifugal separator is operational.

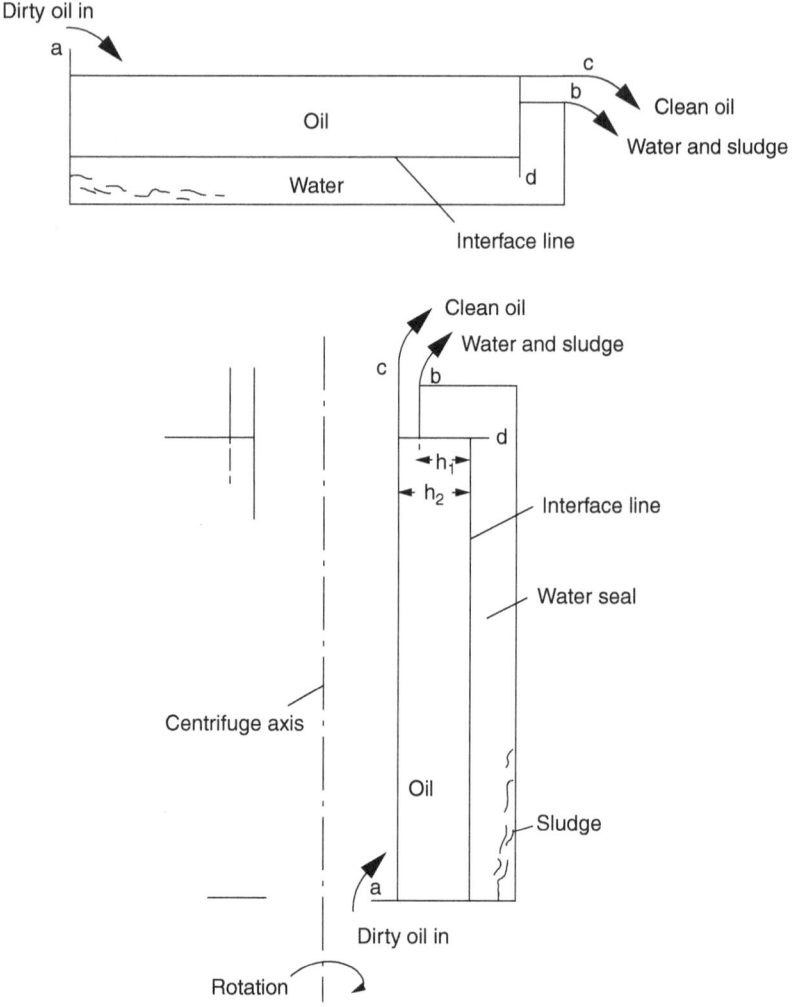

▲ **Figure 10.10** *Centrifugal separation*

The water dam ring (or screws) is used to vary the position of (b) and the choice of dam ring should be such that the oil–water interface is as near as possible to (d) without oil discharge taking place at (b). This ensures as large a quantity of oil in the centrifugal separator as possible. Thus for a given throughput rate the oil will be in the separator for as long a time as practicable, enabling the centrifugal force to give good separation and clarification.

The equilibrium equation for the centrifuge is: $h_1 = h_2 \times$ oil density where h_1 is the head of water and where h_2 is the head of oil.

When oil of high relative density is to be passed through the centrifugal separator the dam ring that would be fitted will bring (b) closer to the axis of rotation increasing h_1 without altering h_2.

If oil of low relative density is to be passed through, the dam ring would have to be such that (b) is moved away from the axis of rotation reducing h_1 without altering h_2.

Control of the oil–water interface line, or equilibrium line, can be achieved in some purifiers by variation of back pressure, in which case no dam ring would be required.

Centrifuges

There are two basic types of oil centrifuges that have been in marine use: the large-diameter bowl type fitted with discs and the thin tubular bowl type without discs. Both types of centrifuge give good separation and clarification although the thin bowl type must be rotated at a higher speed to generate the same centrifugal force as the large-diameter type. The use of AC motors and direct on-line starters means that care has to be taken with the thin bowl centrifuges because the bowl is prone to imbalance as fast acceleration takes place. This could lead to failure of the bearings supporting the spindle.

Action of particles in a disc-type centrifuge

If oil flows in a streamlined condition between two parallel plates its velocity varies between zero where it is in contact with the plate, to a maximum at the mid-point between the plates. Figure 10.11 shows such a velocity variation.

Oil flows between two discs in a centrifuge is radially inward and up towards the clean oil outlet. Any particles in the oil will follow this general flow, but will also be acted upon by a radially outward centrifugal force. The magnitude of the force depends upon the mass of the particle, the speed squared and the radius at which the particle finds itself.

By referring to Figure 10.11 it will be seen that any particles finding their way to the underside of a disc enter a region of zero velocity and they can then move, due to centrifugal force, down the underside of the disc and eventually into the sludge space of the bowl.

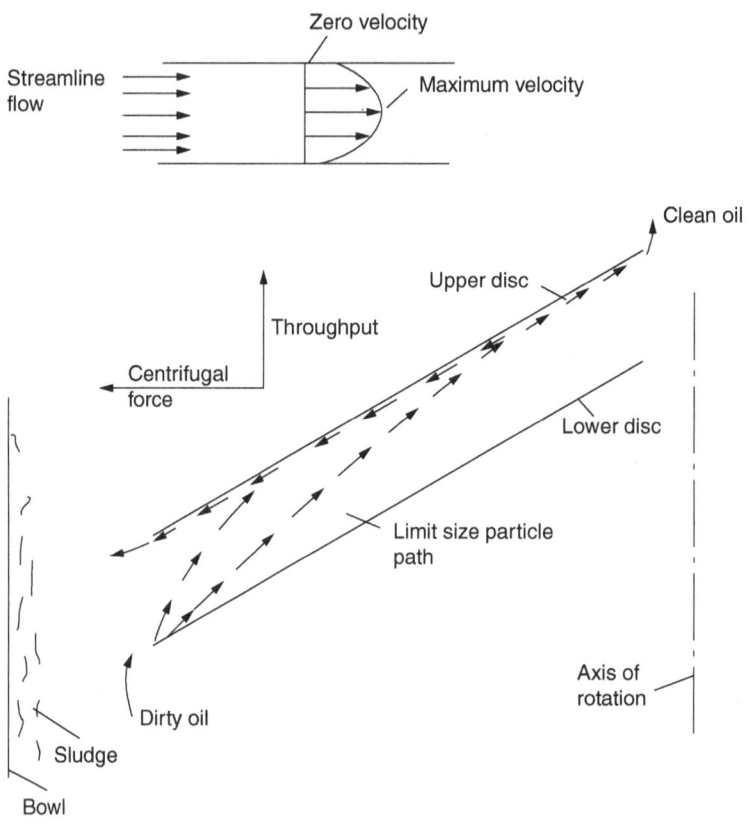

▲ **Figure 10.11** *Velocity variation with laminar flow*

The important path taken by particles is that of the limit size particle, which is the smallest to be removed in the centrifuge, and particles smaller than this will pass out with the clean oil. Some of the factors affecting the limit size particle would be:

1. Viscosity of the oil in the centrifuge; the higher its value the greater will be the viscous drag on the particles. Hence the oil should be preheated to as high a temperature as practicable.
2. Disc spacing, diameter and inclination to the vertical.
3. Speed of rotation of the purifier.
4. Throughput. If this is low the limit size particle will be small and the oil discharged will be cleaner. If throughput is great the limit size particle will be large. However, if the oil contains appreciable quantities of water this would be effectively removed and this form of contamination could be the reason for operating at a high throughput.

Alfa-Laval bowl-type centrifuge

The stainless-steel bowl of up to 0.6 m diameter is mounted on a tapered spindle, the lower part of which is fashioned into a sleeve that passes over a stationary spindle. The stationary spindle carries two ball races that are provided for the rotating spindle. These bearings, the upper one serving as a thrust bearing, give a high degree of flexibility.

A constant-speed electric motor is currently the most common type of motor to supply the motive power for the oil suction and discharge pumps (if fitted) and the wheel and worm drive for the centrifuge bowl. However, there is a growing trend to use variable-speed AC motor control and there is an energy advantage to starting the purifier slowly and gradually increasing the speed to a constant bowl rotation, under normal operating conditions, of between 5,000 and 8,000 rev/min depending upon the size of the machine. This gives a lower centrifugal settling force than would have been found in the older, thin, tubular bowl type of centrifuge. To compensate for the lower centrifugal settling force, stainless-steel conical discs carried by splines on the distributor are fitted to reduce the settling distance.

To operate the centrifuge as a purifier, it is first brought up to operating speed, supplied with fresh water to form the water seal and then the oil to be purified is delivered to the distributor by the inlet pump.

As the oil passes down the distributor it is rapidly brought up to the rotational speed of the purifier by the radial vanes provided for this purpose. The oil passes from the distributor through the space between the bottom plate and bowl to the supply holes.

From the supply holes the oil is fed to the spaces between discs through the distribution holes in the discs. Separation and clarification takes place between the discs, water and sludge moving radially outwards pass along the under surface of the discs and the purified oil moving radially inwards passes over the upper surface of the discs. Water and sludge are eventually discharged at (a) and the purified oil at (b) (see Figure 10.12).

If the centrifuge is to be operated as a clarifier, no water seal is provided and the bottom plate and discs have no supply and distribution holes. Discharge of the clarified oil takes place at (b), and sludge and solids collect upon the bowl wall. Since there is no water seal in a clarifier more bowl space is available for the oil, hence a greater centrifugal settling force will be available due to the increased radii.

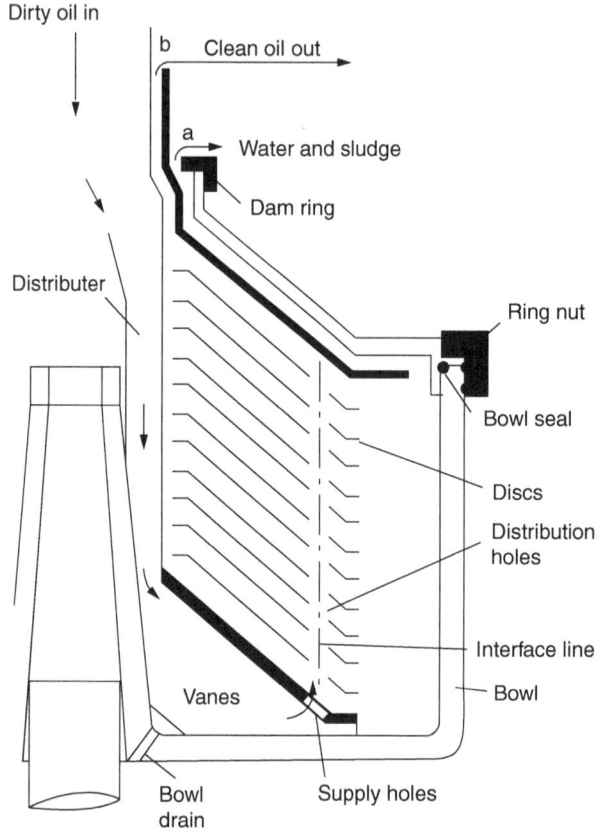

Dirty oil in

b Clean oil out

a Water and sludge

Dam ring

Distributer

Ring nut

Bowl seal

Discs

Distribution holes

Interface line

Vanes

Bowl

Bowl drain

Supply holes

▲ **Figure 10.12** *Purifier*

Figure 10.13 shows a centrifuge arranged for clarification. The degree of diagrammatic simplification between it and the purifier should be noted by the student, and as an exercise in examination sketching it is recommended that the purifier should be drawn in a similar diagrammatic form.

Sharples super-centrifuge

Figure 10.14 illustrates, diagrammatically, the Sharples super-centrifuge with 'one-pass' bowl, designed to replace the purifier–clarifier series combination used for the purification of residual oils. This purifier can also be used for the purification of lubricating and diesel oils. It is an older design concept that is not used for modern machines. There could, however, still be some in service, which is the reason for the brief description here. The problems with them were that if the bowl was not completely

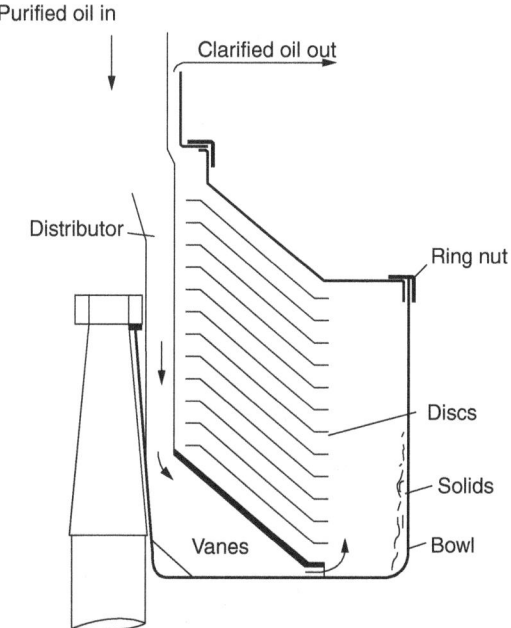

Purified oil in

Clarified oil out

Distributor

Ring nut

Discs

Solids

Vanes

Bowl

▲ **Figure 10.13** *Clarifier*

clean then the machine could start slightly out of balance. This was fine until direct online starters for AC motors, when the acceleration was so fast that the out-of-balance forces could damage the machine as it was trying to run up to speed.

It consists of a stainless-steel tubular bowl about 110 mm diameter and 760 mm long supported at its upper end by a ball thrust bearing and flexible spindle assembly and guided at the bottom by a plain bearing. This arrangement permits the bowl to take up its own alignment.

An electric motor mounted on the top of the purifier framework drives the bowl through a belt drive and this gives a bowl speed under operating conditions of about 15,000 rev/min. Suction and discharge pumps of the gear type driven by the electric motor can be fitted, if required, thereby making the purifier an independent unit.

A three-wing assembly made of tinned or stainless steel is fitted inside the bowl and is retained in position by spring clips. This assembly brings the oil rapidly up to bowl speed, thus ensuring that the oil is subjected to the maximum possible centrifugal settling force that the purifier is capable of producing.

At the top of the bowl a small annular space is provided and this space contains the water seal. By having a small water seal, all of the bowl space is available for the oil. This

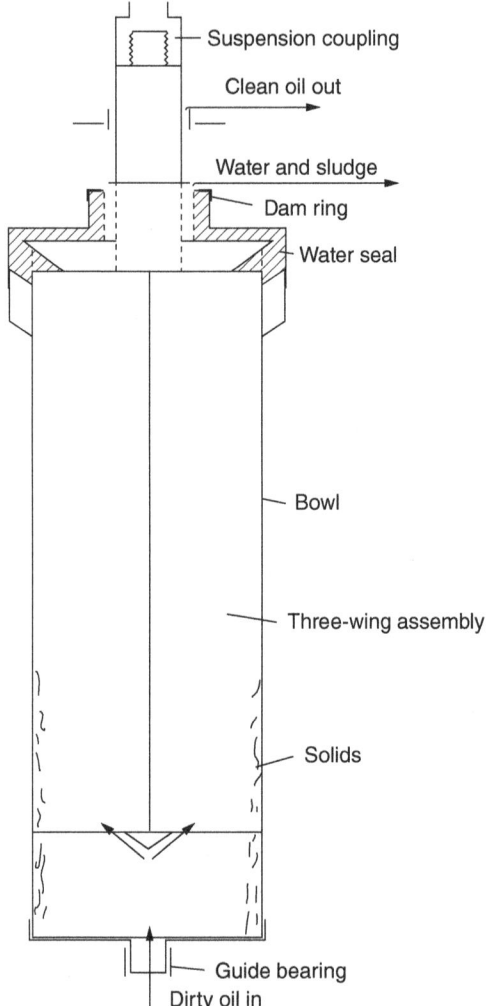

▲ **Figure 10.14** *Super-centrifuge*

means that purification will be improved in this type of purifier compared with the type that has a water seal throughout the length of the bowl.

Self-cleaning purifier

Figure 10.15 shows diagrammatically the method of sealing and sludge ejection for a self-cleaning purifier of the type where the lower section of the bowl is allowed to drop down creating an exit path for the bowl contents. The bowl is closed again, returning the purifier to a condition ready for further purification.

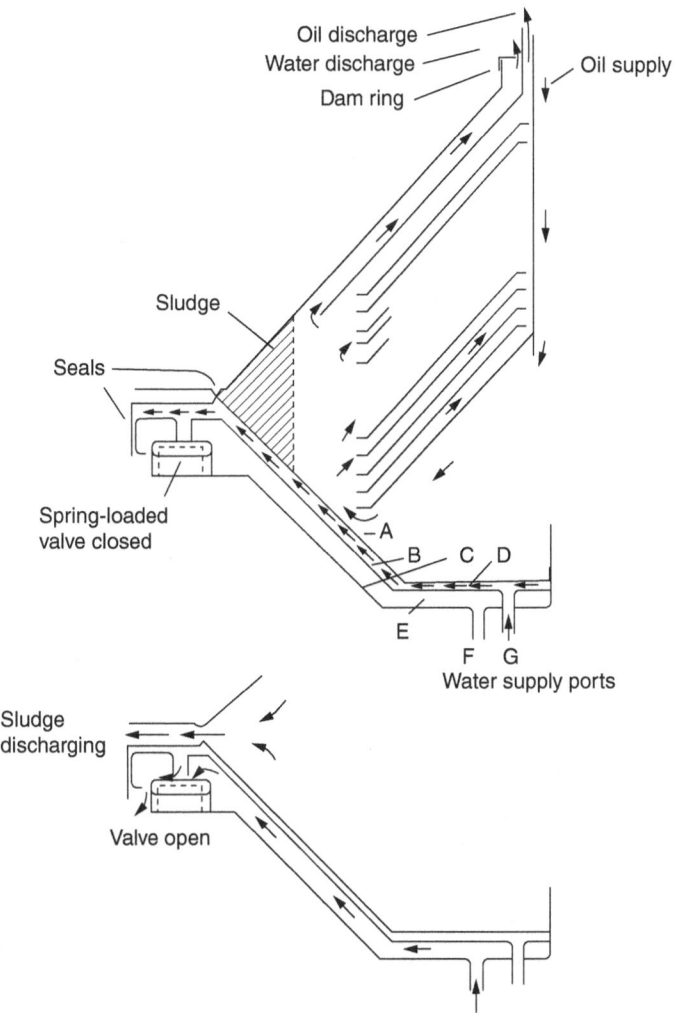

Oil discharge
Water discharge
Dam ring
Oil supply
Sludge
Seals
Spring-loaded
valve closed
A
B C D
E
F G
Water supply ports
Sludge
discharging
Valve open

▲ **Figure 10.15** *Self-cleaning purifier*

Bowl sections A, B and C are all keyed to the central drive spindle. B and C are secured so that they cannot move vertically whereas A is free.

The purifier is first brought up to operating speed and water is then supplied to space D through supply port G. Due to centrifugal force the water pressure in space D moves A vertically to form a seal at the bowl periphery. Water and then oil would next be supplied to the purifier in the usual way.

When the purifier needs to be cleared of sludge the oil supply is shut off and water supply is changed over from G to F supply port. The hydraulic pressure created in space E is sufficient to open the spring-loaded valves and the water from space D will – together with water from space E – be discharged and A will fall, the bowl seal will now be broken and the sludge ejection will take place.

Centrifugal Purification

Centrifugal purification of fuel oils

For the purification of diesel fuel oil the single-stage process is normally used. The diesel oil is delivered to a centrifugal purifier through a heater unit. If the oil is of low viscosity it may be purified efficiently without preheating. Oils of medium or high viscosity should be heated before purification in order to reduce their viscosity. This gives better clarification by reducing the viscous drag upon solid particles moving through the oil.

Purifier capacity depends upon various factors: grade of oil, purification temperature, type of impurities and degree of purification required. For diesel engines developing 3,000–10,000 kW, a purifier having a capacity range of 2,200–8,200 l/h would be used. This would be sufficient to deal with a day's supply of fuel in 8 hours.

For the purification of residual fuel oils the two-stage process is commonly used. The residual fuel is heated in a supply tank to about 50–60°C and is drawn from this tank by the purifier inlet pump. The inlet pump delivers the oil to a thermostatically controlled heater that raises the oil temperature to about 80°C and thence to the centrifugal purifier. The dry purified oil is then transferred to a centrifugal clarifier by the purifier discharge pump. After clarification the clarifier discharge pump delivers the oil to the daily service tank for engine use.

There is a limit of 991 kg/m^3 (at 15°C) for the density of the fuel oil that can be cleaned using conventional purifiers. As the density approaches 991 kg/m^3 the oil–water interface becomes progressively more sensitive until it eventually becomes unstable and the purifier stops working correctly.

Fuel oil between 991 kg/m^3 and 1,010 kg/m^3 at 15°C can be cleaned using a computer controlled system such as the ALCAP system from Alfa-Laval. The system operates with a continual feed to the separator and is not interrupted while the sludge is being discharged. There is no gravity disc and the machine is in effect a clarifier where at a predetermined time the sludge is discharged. There is also a water detection transducer fitted to the clean oil discharge pipe and if water is detected before the pre-set time for sludge discharge then the control circuit will activate the system and the water and sludge will be discharged. If the water reaches the disc stack before the pre-set discharge time then it will discharge through the water drain valve.

Centrifugal purification of lubricating oils

Systems

Lubricating oil purifiers for diesel or turbine lubrication plants are normally arranged to operate on the continuous bypass system. In this system, oil is drawn from the engine sump or drain tank by the purifier feed pump, delivered through a heater to the purifier. Following the purification process the clean oil is then discharged, by the purifier discharge pump, to the engine sump or main lubricating pump suction.

The system layout may vary slightly depending upon the engine and lubricating oil system arrangement, but the best configuration for the 'continuous bypass purification' method of cleaning the lubricating oil is to take oil for the purifier from a point in the lubrication plant where the oil has passed through the engine and has had time to settle. At this point the oil should contain the most amount of water and/or debris picked up from inside the engine. The clean oil should then be delivered back into the system drain tank adjacent to the suction for the main lubricating oil pump. Figure 10.16 illustrates a diagrammatic arrangement of such a system.

Sometimes the layout of pumps, piping, tanks, etc. for the purifier permits operation on the batch system of purification. In the batch system, the contents of the engine drain tank or sump would be discharged to a dirty oil tank and the drain tank or sump would be replenished with clean or new oil. The lubricating oil in the dirty oil tank can then be purified at leisure.

Under normal operation it is recommended that after shutting down the main engine the purifier should be kept running for about 12 hours in order to minimise corrosion due to acid vapours condensing inside the engine as it cools down. Modern practice with centralised cooling systems is to keep the main engine warm with the heat produced by the generators. Moving into the era of cold ironing could change the common practice again as the use of electric heaters, to keep the engine warm, might prove expensive.

Water wash

A feed pipe capable of supplying hot fresh water in a thin stream, intermittently or continuously as desired, should be fitted to the purifier oil inlet pipe. This hot water, fed

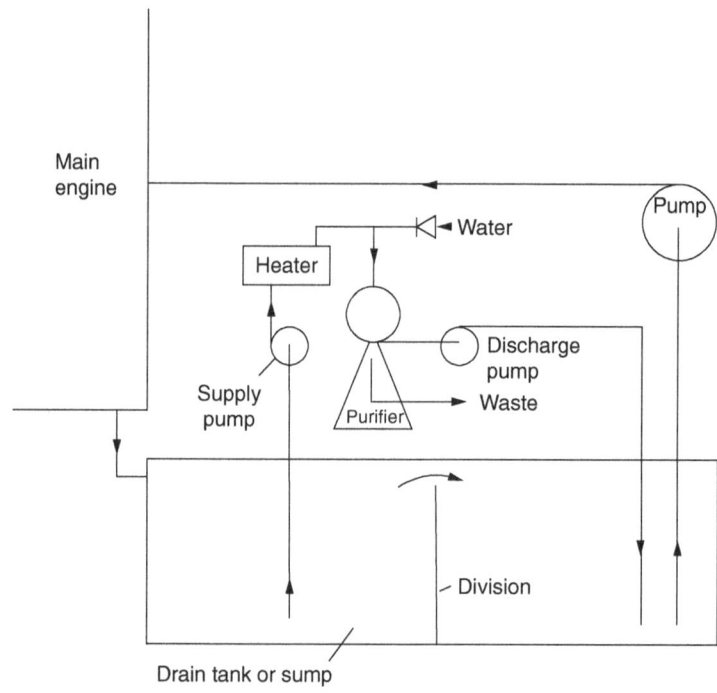

▲ Figure 10.16 *Continuous bypass system*

in at approximately the same temperature as the oil, serves to sludge out some of the lighter dirt from the purifier bowl and wash out any acids.

In IC engines, the lubricating oil can become contaminated with sulphur combustion products that may combine with any water that is present in the oil to form sulphuric acid. Sulphuric acid can corrode cylinder liners and bearings, etc. In turbine installations, the lubricating oil, due to decomposition, may contain harmful acids that can cause corrosion. The acids formed in both cases are more soluble in water than in the lubricating oil, hence if a hot fresh water wash is put in the purifier it has the effect of flushing out the acids present in the oil and reducing risk of corrosion. The purifier itself could be corroded by these acids and the water wash reduces this risk since the water seal is continually being replaced.

Detergent oils

If it is considered necessary to water wash detergent lubricating oil, the wash water supply must not exceed 1% of the total oil flow otherwise excessive depletion of detergent additives will occur, in addition, emulsion of the oil may be created.

Steam jet

By blowing steam into IC engine lubrication oil just prior to purification, coagulation of the colloidal carbon will occur. This enables the purifier to centrifuge this carbon out more effectively. This steam jet arrangement is *not* meant to serve as the preheating system for the lubricating oil, it is in addition to the oil heater provided for that purpose.

Cleaning the purifier

Despite modern purifiers being described as 'self-cleaning' there will be times when they will need to be opened up and carefully cleaned inside, manually. Great care must be taken as this equipment is subjected to high forces and therefore will be prone to any small mistake in re-assembly. The most important point is that all the cones in the stack of discs numbered and it is important that they are re-assembled in the same order. Care must also be taken not to damage the discs when they are being cleaned, but they must be cleaned thoroughly as any debris left behind could result in out-of-balance forces being set up as the machine is restarted.

Lubrication

Manufacture of lubricating oil

Mineral-based lubricating oil base stocks are obtained by fractional distillation of crude oil in a vacuum distillation plant. The reader should refer to Chapter 2 for details.

Crude oils are roughly classified into paraffin base, which has a high lubricating oil content with a high pour point and high viscosity index, and asphalt base, which has a low lubricating oil content with a low pour point and low viscosity index. Lubricating oils refined from these bases would be subjected to various treatments to improve their properties, and they would be blended to produce a wide range of lubricating oils.

Compound oil

From 5% to 25% of a non-mineral animal or vegetable oil may be added to a mineral (or mineral blend) oil to produce a compounded oil.

Oils that have to lubricate in the presence of water or steam are usually compounds of fatty animal oil and mineral oil, and they tend to form a stable emulsion that adheres strongly to metal surfaces. Fatty oils have a high load-carrying capacity and if sulphurised they have extreme pressure (EP) property – used for cutting oils and running in of gearing.

It must be remembered that British Standards recommend that mineral oil only should be used for the lubrication of steam machinery as fatty oils contain acids that can cause corrosion in feed systems and boilers.

Synthetic oil

This is oil that has been manufactured from the best part of the refined crude oil. It is not a natural production of the distillation of crude oil.

Lubricating oil additives

These are chemical compounds that are added for various reasons, though mainly they would be added to give improved protection to the machinery and increased life to the oil by (i) giving the oil properties it does not have and (ii) replacing desirable properties that may have been removed during refining and improving those naturally found in the oil.

The problems that can be overcome by the introduction of fuel additives are discussed below:

- poor release of water – leading to emulsification of the oil;
- reduction in load carrying capacity over the life of the oil;
- reduction in the base number in low consumption oils;
- polishing of the cylinder liner;
- piston ring sticking;
- increased wear in areas of extreme pressure.

Some of the additives that could be used are given here.

Anti-oxidant

Reduces oxidation rate of the oil. Oxidation rate doubles for approximately every 7°C rise in temperature and at temperatures above 80°C approximately oxidation rapidly reduces the life of the oil. Viscosity usually increases due to oxidation products and some of the products can help to stabilise foam, thereby preventing the formation of a good hydrodynamic layer of lubricant between the surfaces in a bearing and reducing the load-carrying capacity. Oxidation products cause lacquering on hot metal surfaces, and they form sludge and possibly organic acids that can corrode bearings.

Corrosion inhibitor

An alkaline additive is used to neutralise acidity formed in the oil and in the case of cylinder lubricants for diesel engines to neutralise sulphuric acids formed from fuel combustion.

The additive will increase the total base number (TBN), and prevent rusting of steel and corrosion of bearings.

Detergents

These keep metal surfaces clean by solubilising oil degradation products and coating metal surfaces, due to their polar nature, hindering the formation of deposits. They also neutralise acids.

Dispersants

These are high molecular weight organic molecules that stick to possible deposit-making products and keep them in fine suspension by preventing small particles forming larger ones. At low temperatures they are more effective than detergents.

Pour point depressant

Added to keep oil fluid at low temperatures. The additive coats wax crystals as they form when temperature is reduced, preventing the formation of larger crystals.

Anti-foaming additive

When air is entrained into the oil, which could be due to low supply head or return lines not running full, etc., foaming could result, which can lead to breakdown of the load-carrying oil film in bearings.

An anti-foam or defoamant, acting like a conditioner in boiler water, is insoluble in the oil and finely dispersed throughout it. It may in time become soluble and the protection is lost.

Viscosity index improver

This is added to help maintain the viscosity of the oil as near constant with temperature variations as possible.

Oiliness and extreme pressure additives

These reduce friction and wear. They may form a film chemically, with the metal reaching welding temperature; the film has a lower shear strength than the base metal and hence welding and tearing of the metal is prevented. These additives would be important during the running in of gearing.

Other additives could include emulsifying and demulsifying agents, tackiness agents and metal de-activators.

In-service experience

It would be easy on board if there was the possibility of a multigrade oil that covered all applications such as there is in the automotive industry. However, the complexity of modern tonnage means that there is a corresponding complexity in the different types and grades of lubricants used.

It is usual that one manufacturer/supply will provide the lubrication scheme for the whole ship. They will provide a listing or 'schedule' showing which type or grade of oil is provided to the vessel for each piece of equipment or application.

There could be different oils for different purposes:

- slow-speed crankcase oil;
- slow-speed cylinder oil;
- medium-speed crankcase oil;
- high-speed crankcase oil;
- turbocharger oil;
- steering gear;
- hydraulic oil;
- gear oils;
- air compressor oil;
- refrigerator and air conditioning compressor oil;
- greases;
- open gear greases;
- oil and grease for deck machinery;
- stern tube oil.

Reference should always be made to the lubrication oil schedule before topping up the oil on any equipment. Oils need particular care when handling. They are expensive and therefore must be used in just the right amounts and not wasted. They are a potential fire hazard and therefore any spill must be cleaned up immediately. The same action must be taken because they are a slip hazard. Barrier creams and or gloves should be worn by all operators working with oils as they can be the cause of dermatitis.

Microbial degradation of lubricating oil

Bacterial attack of diesel engine lubricating oils, crankcase and cylinder, generally resulting in a smelly (not always) emulsion has occurred, in the past, with consequent damage to bearings, crankshafts, cylinders and piston rings.

This problem has been the subject of research but is still under investigation and has prompted many discussions in technical circles and papers. Certain similar points emerge from the cases involved.

1. Infections have usually taken place after water ingress into the oil, which could even occur after condensation.
2. Evidence suggests that the microbes produce long-chain organic acids.
3. Aerobic and/or anaerobic bacteria have been detected.
4. Iron oxides in suspension, probably caused by corrosion, help to produce a tight emulsion in the oil that cannot be effectively removed by centrifuging.
5. Corrosion of the system exhibits as (i) fine golden brown film on steelwork and (ii) bearings and journals that are finely pitted.

Remedy and prevention

1. Burn oil (extreme case), clean out and disinfect system.
2. If oil is just beginning to show water separation difficulty, heat it in a tank for about 2 hours at 80–90°C in order to sterilise.
3. Prevent water entry into the oil.
4. Keep system and engine room clean, use disinfectant wash for tank tops and bilges, etc.
5. Treat P&J water with biocide.
6. Use P&J water additives that do not feed the microbes (nitrogen and phosphorous are, apparently, nutrients).

7. Use biocide in the oil.

8. Test lubricating oil and P&J water with prepared dip slides for the presence of bacteria.

Environmental considerations

Oil, in its original mineral or synthetic form, when released into the sea is not environmentally friendly. Inevitably from time to time oil is spilled into the sea from ships either by accident, faulty equipment or by deliberate action. The introduction of new sections of MARPOL and the introduction of new legislation from individual countries means that the industry has to change. One of the contributions to this subject is the introduction of biodegradable lubricating oils. These are especially recommended as lubricants for deck machinery and for oil-filled stern tube bearings. These are the places where oil is most likely to get into the sea from the ship.

In addition to the environmental consideration of the oil itself, ships are having to consider their overall environmental impact. Ships have evolved to burn residual fuel oil that is left over from the refining of crude oil. This heavy fuel oil, as it is known, is high in sulphur. The sulphur is not affected much by the combustion process but it is introduced into the atmosphere with the flue gases and is harmful to the atmosphere.

However, a change to low-sulphur fuel changes the chemistry that happens inside the engine around the combustion process. With the two-stroke, low-speed engine, the crankcase oil is separated from the scavenge space and the underside of the piston by a division within the engine structure (see Volume 12 fo the Reeds series for more detail). The piston rod is fixed to the piston and passes through a gland called a stuffing box and on to the crosshead bearing. As the piston is segregated from the crankcase there is no lubrication to the piston from the crankcase oil as there is in the four-stroke engine. Therefore, the cylinder and piston rings on the two-stroke, slow-speed engine are lubricated by the 'total loss' cylinder oil system.

The cylinder oil used for this system is specially formulated to combat the acids that are found in an engine burning high-sulphur fuel oil. When the ship changes over to low-sulphur fuel a different set of circumstances will exist and therefore there is a different requirement for the oil. Some oil majors have produced a cylinder oil that is suitable for both applications while others have kept with the requirement to use two oils. This means that the engineers must change the cylinder oil at the same time as the fuel is changed from high-sulphur to low-sulphur.

Lubrication fundamentals

A lubricant is designed to carry out the following:

- reduce friction and thus reduce or eliminate wear;
- keep metal surfaces clean by carrying away possible deposits;
- provide a seal to keep out dirt
- carry away the heat generated thus preventing overheating seizure and possible breakdown.

A lubricant can be a liquid as with oil, semi-liquid as with greases and solid as with graphite or PTFE. Each is well suited to the task that it was designed to complete. For example, a grease application would stay in one place better than oil, which would tend to run off the item being lubricated.

Bearing lubrication

The addition of even the slightest trace of lubricant to a moving bearing modifies the friction force appreciably. The two most important properties of a lubricant would be *oiliness* and *viscosity*. Oiliness is a form of bond between molecules of lubricant and material surface in which the lubricant is adsorbed by the material. The adsorbed film is very thin and once formed is very difficult to remove, which is most advantageous, in this respect colloidal suspension graphite is a very successful additive. If a layer of finite thickness lubricant exists without material contact, then friction is determined by viscosity, if the layer is only a few molecules thick then oiliness is the main factor.

$$F = \eta A \frac{dv}{dy}$$

where F is the viscous force required to move one plate over another with a velocity dv when the area of the plate is A, thickness of lubricant between surfaces dy, η is viscosity coefficient.

Boundary friction is the condition where the high spots of a microscopic surface are in contact, while the low areas in between are separated by a finite lubricant layer. In this state the thickness of the oil film is so small that oiliness becomes the predominant factor. This lubrication condition is only successful in low-pressure situations and in the past it could be said to exist in some top end bearings and crosshead guides. However, increases in lubricating oil system pressure have increased the efficiency of the lubrication in these difficult areas of the engine.

Film lubrication, or hydrodynamic lubrication, is the condition whereby the bearing surfaces are completely separated by an oil layer. The load is taken completely by the oil film. The film thickness is greater at inlet (initial point in direction of rotation) than at outlet, and the pressure at inlet increases quickly, remains fairly steady having a maximum value a little to the outlet side of bearing centre line, and then decreases quickly to zero at outlet. This form of lubrication is ideal but can only be satisfied in certain types of bearing. Simple examples are high-speed journal bearings, turbine bearings, or plane surfaces that can pivot to allow wedge oil film to allow for load, speed and viscosity effects, as in Michell thrust bearings.

Using the variables of oil viscosity, relative speed of the bearing surfaces and pressure, Figure 10.17 shows how the friction and form of lubrication alters in a journal bearing.

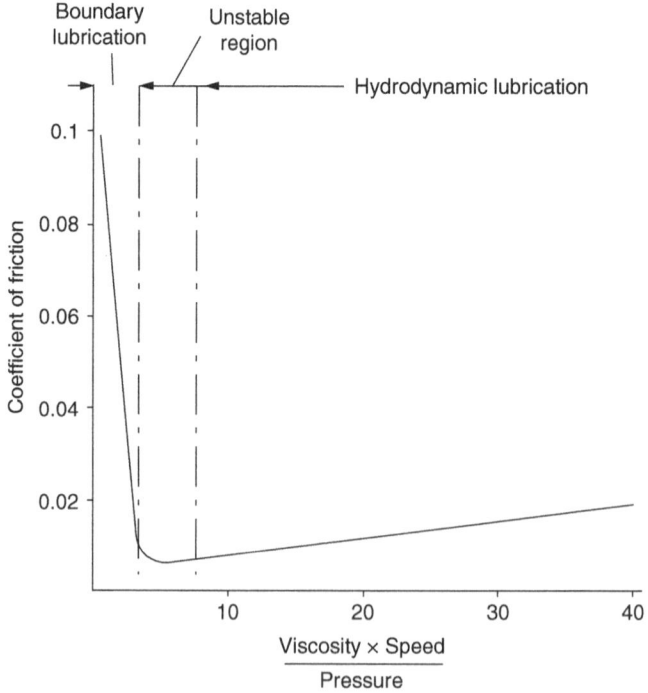

▲ **Figure 10.17** *Diagram for a journal bearing*

Factors affecting hydrodynamic lubrication

Viscosity of the lubricant

The higher the viscosity the greater the tendency towards hydrodynamic lubrication. Obviously the type of lubricant – oil, water or grease – and the temperature are important. Temperature can be increased by insufficient lubricant circulating to remove the heat generated in a bearing – this could be caused by clearances being too small and/or insufficient supply of oil.

Relative speed of the surfaces

The higher the relative speed the greater the tendency towards hydrodynamic lubrication. Increasing a journal or crankpin diameter and retaining the original rotational speed will increase the speed of the journal's surface, which will have a beneficial effect on the move to hydrodynamic lubrication.

In two-stroke reciprocating engines the oscillatory motion of the crosshead and guide shoe means that there is a tendency in these units towards boundary lubrication as the relative speed goes from a maximum to zero. This is one of the reasons why crosshead lubrication may be a problem. The latest engine designs have overcome this problem by introducing the lubricating oil into the engine at the crosshead bearing. This means that the full systems pressure is introduced to the bearing instead of the oil inlet being at the main of big end bearing as in the previous engine types.

Bearing clearance

Oil can withstand a certain amount of pressure before it breaks down and a bearing is carefully designed with just the right amount of clearance to give good hydrodynamic lubrication and also be able to withstand the pressures from the forces produced from the combustion process. A bearing knocking sound will be heard if the bearing becomes worn and the clearance too large. The increase in movement increases the loading and the pressure between the surfaces, and this squeezes out the oil and will allow contact between the journal and the bearing. If the clearance is too small, there could be overheating of the oil, a breakdown in all lubrication and possible seizure could result.

Pressure

If pressure, that is, bearing load per unit area, is high it can lead to boundary friction. If peak loads are high in the cylinder of a diesel, due to incorrect fuel injection timing or other reason, bearing pressure will increase.

Journal bearings

Consider Figure 10.18 in which the amount of clearance and pin shift movement has been much exaggerated for clarity. When movement first begins the pin climbs up the journal bearing *against* the direction of rotation, the friction angle is Φ. The layer of lubricant tends to be scraped off so boundary lubrication exists. As speed increases the oil is dragged behind the pin by viscous action until the oil film breaks through and separates the surfaces, the line of contact having moved *in the direction* of rotation, film lubrication exists. The variation of oil pressure circumferentially and longitudinally is as shown.

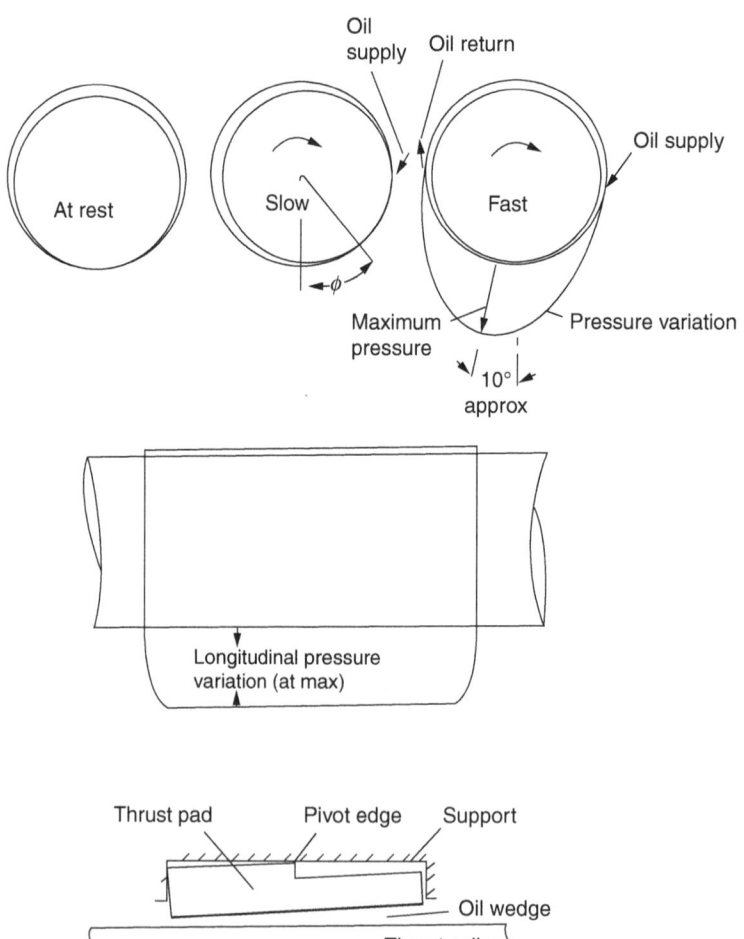

▲ Figure 10.18 *Hydrodynamic lubrication*

Michell bearings

The bearing surface is divided up into a number of kidney-shaped pads extending part or all the way round the surface, this principle being utilised in tunnel and thrust bearings.

The pads are prevented from moving circumferentially but are free to tilt and incline to the direction of motion. Such tilt allows a self-adjusting oil film wedge giving full film lubrication. This film fully carries the load and allows pressures of 30 bar and shows a coefficient of friction value of 0.003.

Certain definitions and general points are now considered.

Scuffing

Breakdown of the oil film between surfaces causes instantaneous microscopic spot welding of the high points on the two surfaces that should be kept apart by the lubrication. Further movement causes tearing out of the material and the resultant condition is known as scuffing. It is most liable to be found when the lubrication film is difficult to maintain, for example, on turbine gear teeth and in IC engine cylinder liners.

Extreme pressure lubricant

This refers to special additives to the oil to maintain oil film under most severe load conditions and where film is difficult to maintain. Molybdenum disulphide (moly slip) additive is often used. Such lubricants are used to prevent scuffing.

Pitting

More a fatigue or a corrosion fatigue phenomena, usually the result of too high contact pressures giving minute cracking at contact surfaces.

Emulsion

Oil that is contaminated or has deteriorated in service will not separate easily from water and may cause an emulsion, in whole or in part. Emulsification is associated with precipitation of sludge at an increasing rate; such sludges are formed from accretion of resins and asphaltenes. The oil should have a good demulsibility when new and should retain this in service.

Oxidation

A bearing oil subject to oxidation due to a high 'heat load' on the oil in circulation forms products in the oil that include polar compounds, for example, the fatty acids such as oleic in which the acidic group is polar. Severe shaft and bearing corrosion can result. Polar substances have a molecular structure such that one part of the molecule is electrically negative with respect to the other part. This polar form tends to disperse one fluid in the other and stabilise the emulsion and tends to favour orientation at interfaces. Oxidation and corrosion products, such as oxides of iron etc., stabilise emulsions. Anti-oxidation additives or inhibitors restrict polar molecule formation. Pure mineral oils normally have a high resistance to oxidation.

Typical bearing pressures

Crankpin bearings	91 bar (max)
Top end bearings	138 bar (max)
Guide shoes	5 bar (max)
Michell thrust bearing	30 bar (max)

Note that fluid film lubrication applies for most bearings of high-speed engines but a guide shoe is a case of boundary lubrication.

Lacquering

Oxidation and corrosion products plus contamination products lead to deposits. In high-temperature regions, hard deposits form thin lacquer layers on pistons or heavier deposits form, for example, on the upper piston ring grooves of IC engines. Lacquer varnishes also form on piston skirts. On cooler surfaces sludge of a softer nature is more liable to be deposited.

Shipboard lubricating oil tests

Qualitative oil tests carried out on board ship do not give a complete and accurate picture of the condition of the oil; this could be obtained only from a laboratory. However, they do give good enough indication of the condition of the oil to enable the user to decide when the oil should be replaced, or if some alteration in the cleaning procedure is considered necessary. The usual tests for alkalinity (or acidity), dispersion, contamination, water and viscosity apply.

Samples of oil for analysis should be taken from the main supply line just before entry into the engine since it is the condition of the oil being supplied to the engine that is of the greatest importance.

Alkalinity test

A drop of indicator solution is placed on blotting paper and this is followed by a drop of sample oil placed at the centre of the drop of absorbed indicator. A colour change takes place in the area surrounding the oil spot: if it is red – acid, if blue/green – alkaline, if yellow/green – neutral.

Dispersiveness, contamination and water

A drop of oil is placed on blotting paper and the shape, colour and distribution of colour of the spot gives an indication of the oil condition. An irregular shape indicates water is present.

A uniform distribution of contaminants indicates good dispersion. If they are concentrated at the centre of the oil spot, dispersion is poor. If the colour of the spot is black, heavy contamination is the cause.

Viscosity test

Four equal-sized drops of oil – one used, one of the same grade unused, one with viscosity higher than and one with viscosity lower than the unused oil – are placed in a line along the edge of an aluminium plate.

When sufficient time has elapsed so that they are all at room temperature the plate is inclined from the horizontal and when one of the oils has run down about 7.5 cm the plate is returned to the horizontal.

By comparing the distances travelled by the sample of used oil with the three reference oils an estimate of viscosity is possible. Obviously, if the distances travelled by used and unused oils of the same grade are equal there is no change in viscosity.

If the viscosity is reduced this could be due to dilution by distillate fuel. Heavy contamination due to carbon and oxidation would cause the viscosity to increase, as would contamination by heavy fuel oil. If variations in viscosity of 30% from initial viscosity are encountered the oil should be renewed.

A simple viscosity test of a similar nature to that described above known as the 'Mobil Flostick' test uses equal quantities of used and unused oils of the same grade in a testing device. Equal-capacity reservoirs are filled with the oils, which are allowed to reach room temperature, then the device is tilted from the horizontal and the oils flow down parallel channels. When the reference oil reaches a reference mark, the device is quickly returned to the horizontal and the distance travelled by the used oil in comparison to the unused oil gives a measure of viscosity.

Crackle test for water in oil

If a sample of oil in a test tube is heated, any water droplets in the sample will cause a crackling noise due to the formation of steam bubbles – this test indicates that small amounts of water are present. A simple settling test would be sufficient to detect large quantities of water in the oil.

Corrosion of white-metal bearings

White metals are tin based, that is, they have a larger proportion of tin in the alloy than any other metal. A typical composition could be 86% tin, 8.5% antimony, 5.5% copper.

In the presence of an electrolyte, corrosion of the tin can occur, forming extremely hard, brittle, stannous and stannic oxides (mainly stannic oxide SnO_2). These oxides are usually in the form of a grey to grey-black-coloured surface layer on the white metal, either in local patches or completely covering the bearing. The hardness of this brittle oxide layer could be as high as twice that of steel and if it became detached, possibly due to fatigue failure, serious damage to bearing and journal surfaces could occur.

The formation of the oxide layer is accompanied by an upward growth from the white metal, which can considerably reduce clearance and could lead to overheating and seizure, etc.

Some factors that appear to contribute towards the formation of the tin oxides are as follows:

1. Boundary lubrication, for example starting conditions.
2. Surface discontinuities.
3. Concentration of electrolyte, for example fresh or salt water or other contamination.
4. Oil temperature.
5. Stresses in the bearing metal.

Additives to the lubricating oil seem to offer some degree of protection, as does centrifuging and water washing of the oil.

Grease

Description

It is a semi-solid lubricant consisting of a high-viscosity mineral oil and metallic soap with a filler.

Soaps are compounds of a metal base – calcium, sodium, aluminium – with fatty acids obtained from animal or vegetable fats.

Fillers are lead, zinc, graphite, molybdenum disulphite. Fillers enable grease to withstand shock and heavy loads.

What it does

1. Will stay put.
2. Will lubricate.
3. Will act as a seal.
4. Useful for inaccessible parts.

What to use

Calcium soap greases are water resistant and have a melting point of about 95°C and are suitable for low speeds. Sodium soap greases have a high melting point, about 200°C, and are suitable for high speeds but emulsify in water. Aluminium soap grease has a high load-carrying capacity.

11

INSTRUMENTATION AND CONTROL

Developments in instrumentation and control along with advancements in material science are probably two of the most important areas of learning for the modern marine engineer. The need for ships to become more efficient will inevitably place a focus on control technology. Instrumentation and control is the subject of a complete Volume 10 in this Reeds series. Therefore, this chapter will cover more about the application of control as well as new technology and very importantly how the human element responds to increases in sophisticated automation systems.

An important function of the marine engineer's role is one of systems engineering or, as the industry call it, 'watchkeeping'. This important function is about taking responsibility for the safe and efficient operation of the machinery that is required to do the following:

1. Propel the vessel in the correct direction.
2. Protect the crew and the cargo.
3. Provide life support for the crew.
4. Complete the tasks in an as environmentally friendly way as possible.

For the machinery to be able to complete these tasks it will have to convert energy into work to produce mechanical and electrical power. During this action heat will be produced in various quantities. The job of the systems engineer is to ensure that the machinery operates within the correct parameters, such as keeping the values for temperature and pressure correct around any number of different systems. This means applying just the right amount of cooling at the right time and to complete this function information about the process needs to be gathered.

For example, most modern ships are propelled by a diesel engine driving a propeller. The diesel engine will burn fuel and become hot. Therefore it will need cooling – but only the correct amount of cooling – overcooling can be nearly as bad as undercooling. A fundamental requirement for any system is the ability to measure the condition of the process being undertaken. The measurement function will need to be coupled to other devices depending upon the sophistication of the control system. The parts of a control system for a modern ship can be divided into the following:

- Instruments are devices or sensors that are used for detecting and measuring the properties of elements.
- Telemetering; the remote signal transmission and conversion (transduce) of a signal from one place or form to another. This description has been in existence in control engineering for some time; however, more recently, the adapted word telemetry is used to describe remote signal transmission usually by wireless means. These wireless transmissions are common within the world of Formula One motor racing, when monitoring drones, and in the space industry; international shipping is now expanding the use of this technology but it has the potential to grow much further as the use of autonomous underwater (and surface) vehicles (ships) become much more widespread.
- Control theory is the basic concept of hydraulic, pneumatic and electronic control actions that are completed in response to predetermined variables. The modern systems will have mathematical algorithms completing the logic processes and deciding the control actions required.
- Marine control systems are the application of interdependent control loop functions to marine processes.

The measurement of temperature and pressure has been available to marine engineers from a very early point in the development of mechanical propulsion systems. It is a tribute to the technology that some of these instruments are still being used today. The early control logic, application and interaction with other systems, however, were carried out by the engineer. It was his/her thought processes skill and experience that would know just how much to open coolers, close heaters, increase flow rates and ensure correct levels in header tanks. Today this function is *still the responsibility* of the watchkeeper. Control systems and data logging functions may have taken away the physical tasks of opening and closing valves, recording temperatures and pressures, etc., but it does not take away the responsibility for these tasks and the student must be ready to express this point to the flag state examiner.

Instruments

Accurate readings of the plant's operation are essential to the marine engineer. Temperatures and pressures have to be taken local to where the process is being controlled, for example, to know if a charge air cooler is working correctly the temperature of the air before and after the cooler needs to be measured as does the pressure differential across the cooler. The local readings will be measured accurately by 'liquid in glass'-type thermometers but it will also be essential on modern vessels to have the same temperatures and pressures sent in electronic form to local controllers and also to a remote monitoring point-such as the machinery control room. It will be essential for the students to understand the point that local and remote readings should be the same and may need checking from time to time. Therefore it is *always* good practice to repair or replace defective local instruments. Despite the fact that the automation cannot read a 'liquid in glass'-type thermometer, there is the need to encase machinery to make it as thermally efficient as possible, hence it will become more difficult to fit local instruments but even so the sensors will still be sending information to a local panel as well as to the MCR.

The need for accurate measurements means that instrumentation is very important. However, this equipment is working in a relatively hostile environment and needs to be constructed to withstand the high level of vibration, temperature and sometimes corrosive environment. With this in mind, marine instruments might have to be constructed more robustly than their land-based counterparts. It also means that when specifying replacement parts it is essential that the supplier understands that the replacement is for marine use. It is also the job of the watchkeeper to be careful in his/her inspection of the instruments. With reference to Figure 11.5, for example, it can be seen that the pressure gauge has some delicate internal components for transmitting the movement of the bourdon tube to the pointer, which will then indicate the pressure. Normally, a pressure gauge will spend most of its working life indicating pressure within a narrow band, which means that the internal components will also be working over a narrow range. That action will increase wear within that area of operation and as there are some brass components inside the gauge it means that the pointer may become stuck at or near the point of the 'correct' reading. The actual pressure may not be at the point that is showing on the gauge.

This illustration carries an important word of warning to the marine engineer, which is that the instrumentation may not always be reporting the correct value. The flag state examiner will be very interested to ensure that the student presenting him/herself for examination understands this point and also has a range of strategies to identify faulty readings and know how to work safely knowing that a fault exists.

Temperature measurement

One of the most common requirements of control engineering is to measure temperature. The design and construction of sensors that carry out this function will be covered later in this chapter but the more traditional technology includes liquid in glass, filled-system and bi-metallic types. Students studying for marine engineering qualifications will be aware of the different temperature ranges. The System International (SI) unit for the measurement of temperature is degrees Kelvin. However, most engineers and designers will ensure that thermometers are calibrated in degrees centigrade.

Liquid in glass thermometers

These are simple devices and their use has stood the test of time. The principle is that when a substance is heated it expands and when it cools it subsequently contracts. Depending upon the temperature to be measured different liquids can be used but for the marine environment mercury is preferred as it can be used within *liquid in glass* thermometers from −38°C to 366°C; if pressurised and contained in specially resistant glass the temperature range can be increased up to 600°C. Alcohol can be used for low-temperature measurement (−80°C to 70°C) and pentane can be used down to −190°C. Sometimes the mercury in the capillary tube can become separated, leading to false readings. It might be possible to rectify the problem by shaking or if possible centrifuging. It may also be possible to place the thermometer in a cold freezer room, which will allow the mercury to contract enough for it to collect back together as one liquid.

Filled system thermometers utilise a bulb sensor, connecting capillary and bourdon tube measure element. The system is filled with a liquid (such as mercury), or a vapour (such as freon), or a gas (such as helium), under pressure. As the temperature changes so does the pressure of the liquid inside the system. The change in pressure alters the bourdon tube. The movement is arranged to be directly proportional to the change in temperature therefore the gauge is calibrated as a thermometer.

Bi-metallic thermometer

When a metal is heated it expands and the expansion rate of metals is different from one to the other. The principle of operation of this type of thermometer is that of differential expansion between two different materials rigidly joined together. If the two bonded metals are subjected to the same heat they will want to expand at different amounts. However, as they are held together the bonded bar will bend. If that bar is arranged in a helix to start with then the resultant heating or cooling will result in the helix trying to open out or close into a tighter helix.

Figure 11.1 illustrates a typical design employed between –40°C and 320°C. The helix coils or uncoils with temperature variation and as one end is fixed the movement rotates both the shaft and pointer. Invar (36% Ni, 64% Fe) has a low coefficient of expansion and when welded to a Ni–Mo alloy gives a good bi-metallic strip. These are very robust instruments and if looked after will give many years of trouble-free service.

Electrical thermometry includes resistance thermometers and thermocouples.

Resistance thermometer

This type of thermometer works because the electrical resistance of a metal varies with temperature. Therefore if a known voltage is placed across a circuit where the resistance is also known then the current drawn by the circuit will be of a set value. If the resistance changes and the voltage is kept the same then the current required will change. The actual arrangement to detect this change is shown below.

Figure 11.2 illustrates a resistance thermometer utilising the Wheatstone bridge principle. A variable resistance $r_1 r_2$ is used for balance purposes; R_1, R_2 and R_3 are fixed resistors. At balance the equation is:

$$\frac{R_1 + r_2}{R_2 + r_1} = \frac{R_3 + r}{R_4 + r}$$

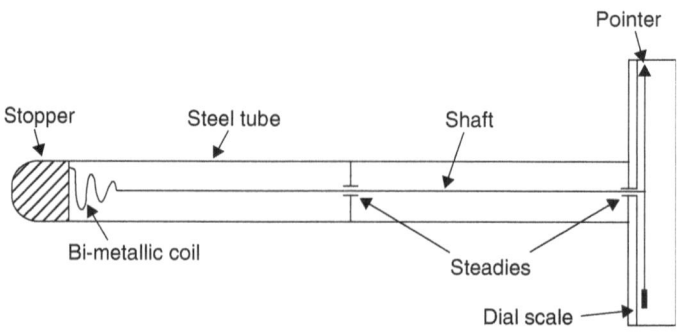

▲ **Figure 11.1** *Bi-metallic thermometer*

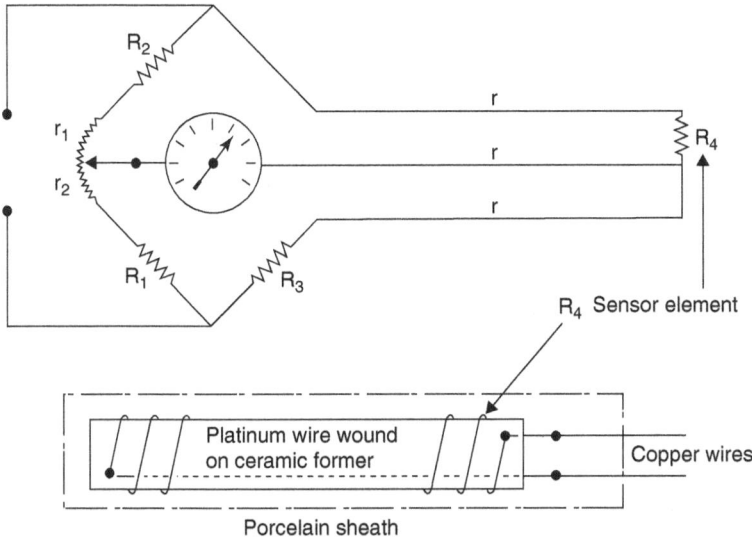

▲ **Figure 11.2** *Resistance thermometer*

The equal resistance of each of the wires is r. Temperature alteration causes change in resistance and electrical unbalance. By use of the variable resistor r_1, r_2 balance can be restored (i.e. the meter reading returned to zero) and while this is being done another pointer can be moved simultaneously and automatically to give the temperature – this is a null balance method. Alternatively, the meter can be calibrated directly in temperature units, thus giving the temperature reading directly. In this case r_1, r_2 is not required. Platinum is the most suitable sensing wire element but copper and nickel wire are used in the range –100°C to 200°C. Tungsten, molybdenum and tantalum are used in a high-temperature pyrometer, up to 1,200°C, and in protective atmospheres. Constantan can be used for the other resistances, if required, as its resistance varies negligibly with temperature variation. When the resistances utilise semi-conducting material, whose resistance decreases with temperature increase, the device is called a thermistor – it is an accurate instrument and there are many advantages in using it in control circuits.

Thermocouple

Whenever the junctions formed of two dissimilar homogeneous materials are exposed to a temperature difference, an emf will be generated that is dependent on that temperature difference, the temperature level and the materials involved. This causes a current to flow in the circuit (Seebeck effect) and the two materials, usually metals, form the thermocouple. Figure 11.3 (top diagram) shows a thermocouple consisting of two wires, one iron and one constantan, with a millivoltmeter coupled to the iron wire. If junction A is heated to a higher temperature than junction B current will flow since the emf at one junction will be greater than the opposing emf at the other junction. The difference is measured by a millivoltmeter, which is calibrated directly in temperature values. A third wire can be introduced (middle diagram), where AB and AC form the couple wires. A will be the hot junction and B with C will form the cold junction. Provided the junctions B and C are maintained at the same temperature, the introduction of the third wire BC will not affect the emf generated. A copper (+) constantan (−) couple is used up to 350°C (constantan 40% Ni 60% Cu). An iron-constantan couple is used up

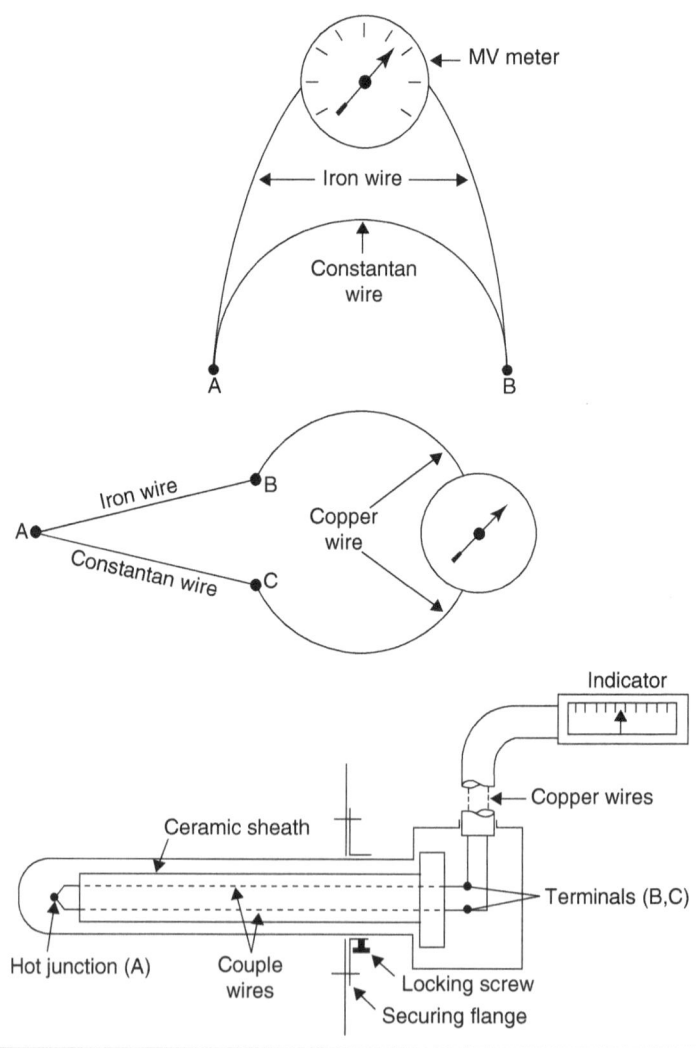

▲ **Figure 11.3** *Thermocouple*

to 850°C and a chromel (90% Ni 10% Cr) – alumel (94%Ni 2% Al) couple up to 1,200°C. Platinum – platinum plus 10% rhodium couples have been used to 1,400°C.

Pressure measurement

When measuring pressure, students should remember that the atmosphere could, in some cases, have an effect on values measured. There is a requirement for some pressure measurement to be much more sensitive than others. The manometer is used for low pressures, the pressure gauge for high pressures and the DP cell for differential pressures.

Manometer

Essentially this instrument is a U tube, one limb of which is connected to the system whose pressure is to be measured and the other limb is open to the atmosphere, which is able to act as a reference. For low pressures, such as a ventilation fan pressure etc., fresh water could be used in the tube.

- 1 m³ of fresh water has a mass of 1 Mg and weighs 9.81×10^3 N.
- 1 m head of fresh water exerts a pressure of 9.81×10^3 N/m².
- 1 mm head of fresh water exerts a pressure of 9.81 N/m².

Hence a difference in level of 20 mm between water levels in the two limbs indicates a pressure of 0.1862 kN/m². For higher pressures, such as the scavenge air pressure for a diesel engine, the fluid used would be mercury, which has a relative density of 13.6. Hence a level difference of say 20 mm between mercury levels in the two limbs indicates a pressure of 3.532 kN/m².

A well-type mercury manometer is shown in Figure 11.4. This instrument has a uniform bore glass tube that has a small internal diameter and when mercury is displaced from the well into the tube, the fall in level of the mercury in the well is very small, due to the large difference in cross-sectional area, and can be ignored. The pressure reading can therefore be taken directly from the level of mercury in the tube. The volume displaced in the well equals the volume displaced in the tube:

$$\text{Then } A \times x = a \times h$$

It can be seen that by altering the ratio of the well and tube areas ($A : a$), the value of h (level variation in tube) can be made any multiple of x (level variation in well). For

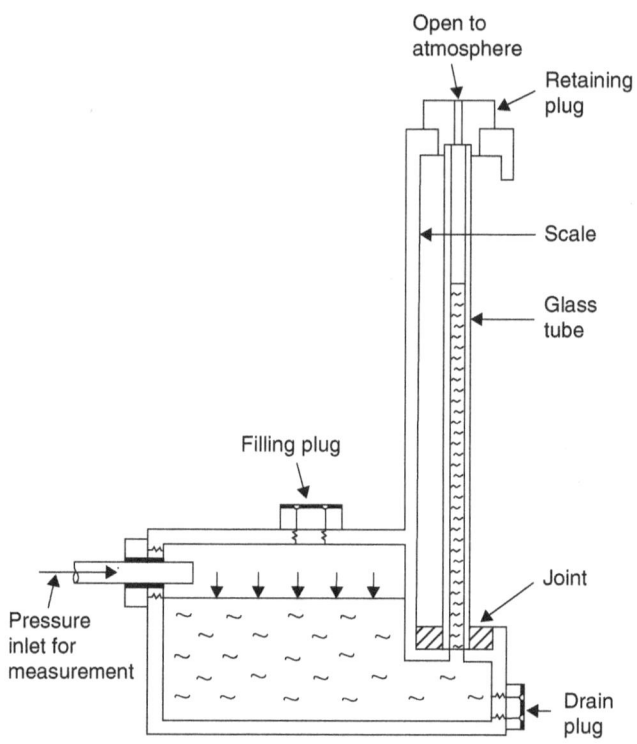

Open to
atmosphere

Retaining
plug

Scale

Glass
tube

Filling plug

Joint

Pressure
inlet for
measurement

Drain
plug

▲ **Figure 11.4** *Well-manometer*

example the simple mercury barometer is essentially the same as the well manometer
of Figure 11.4 but with the top of the tube sealed. A vacuum space down the tube
to the mercury level is formed and the change in atmospheric pressure is measured
from the normal atmospheric pressure of about 760 mm mercury (102 kN/m²,
1.02 bar).

Pressure gauge (Bourdon)

The high-pressure gauges rely on the fact that if a metal is deflected from its original state
then, within its elastic limit, the deflection will be proportional to the force applied. The
sensing element in the bourdon tube gauge is the relay tube, which is semi-elliptical in
cross-section and is arranged in a circular form. When this tube is subject to increased
pressure it tends to unwind (straighten out) and this motion is transmitted to the gauge
pointer via the linkage, quadrant and gear. If pressure is reduced the tube tends to
wind (curl) up. This gauge is therefore suitable for measuring pressures above or below
atmospheric. A diagrammatic sketch is shown in Figure 11.5. The tube is generally made
of phosphor bronze or stainless steel, as are other components except the case, which

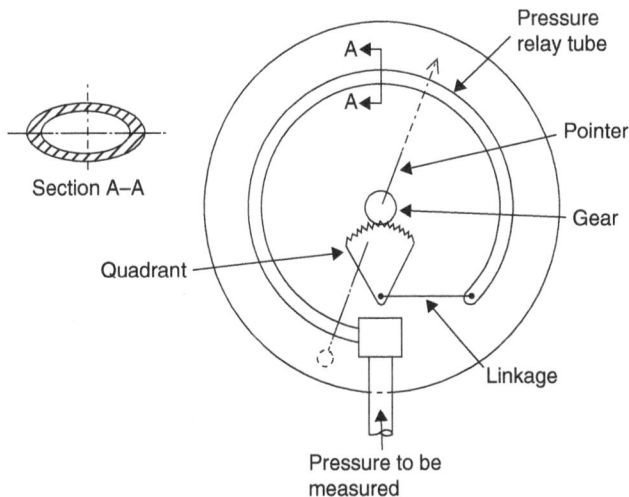

▲ **Figure 11.5** *Pressure and vacuum gauge*

is usually brass or plastic material. The Bourdon sensor is often used as a transducer device in pneumatic control systems.

Differential pressure cell

Figure 11.6 illustrates a twin-membrane sealed capsule secured in the cell body with different pressures applied at each side. The capsule is filled with a constant viscosity fluid (such as silicone), which also dampens any oscillation. Mechanical movement of the capsule is proportional to differential pressure. The DP cell is also used for flow and level measurements.

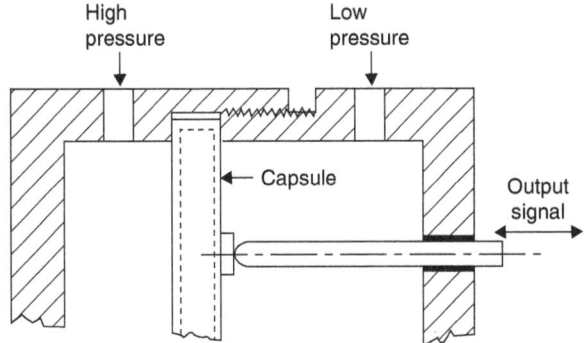

▲ **Figure 11.6** *Differential pressure cell*

Level measurement

Direct methods include sight glasses, float devices and electrical probe elements (or photo-cells). Inferential methods involve any of the sensing devices used for pressure measurement such as manometers, pressure gauges, diaphragms, capsules, etc. Chapter 3 details various boiler water level indicators and shows how level may be transmitted to a remote position. The level of liquid in a tank can be measured from two different perspectives. The measurement can be taken from the bottom of the tank to the level of the liquid. The manual method of achieving this is called a 'sounding' – a sounding tape is a measuring tape that has a weight on the end. The tape is lowered down a tube that is fixed to the top of the tank, called a sounding pipe. It must go through the liquid to the bottom of the tank and as the tape is raised again the level measurement can be read on the tape. This has some drawbacks, for example, if the level in a water tank is required the level might be difficult to see on the tape or if an oil tank is to be checked then the oil left in the pipe from a previous sounding might be left on the clean tape to give a false reading.

Another method of reading a level is to measure the distance from the top of the tank (or sounding pipe) down to the level of the liquid. This level is called a ullage and is used extensively in tankers to measure the level of oil in cargo tanks. If the quantity of liquid in the tank is required then reference must be made to a set of tables that give the corresponding values required.

Although direct level measurement through sight glasses is fairly straightforward the watchkeeper must be aware that tank sight glasses can become blocked and give false readings.

Purge system

Air is supplied at a measured flow rate of about one bubble per second. This keeps the dip tube inside the tank full of air and therefore a variable pressure, equal to the level of liquid in the tank, will be applied to the indicator, as shown in Figure 11.7. The bubbler device is similar to the *pneumercator* as used for determination of tank liquid levels in land-based applications. Air supply to the open-ended pipe in the tank will, in the steady state, have a pressure that is directly proportional to the depth level of liquid in the tank.

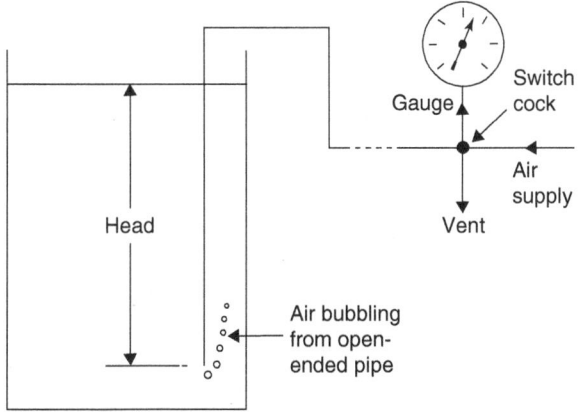

▲ **Figure 11.7** *Level sensor*

Flow measurement

Quantity meters do not include time while flowmeters involve rate of flow; the latter are inferential, that is, volume inferred from velocity. One type of flowmeter will now be described.

Inferential-differential pressure

Consider Figure 11.8. The Bernoulli equation, incompressible flow for fluid of density ϱ, is:

$$KE\ at\ 1 + PE\ at\ 1 = KE\ at\ 2 + PE\ at\ 2$$
$$\tfrac{1}{2}v_1^2 + p_1/\varrho = \tfrac{1}{2}v_2^2 + p_2/\varrho$$

11.1

where KE is kinetic and PE potential energy. This also assumes unit mass, negligible friction and shock losses. The continuity equation is:

$$v_1 A_1 = v_2 A_2$$

11.2

By substituting for v_2 from (11.2) in (11.1) and using mass flow rate RH as equal to $v_1 A_1$ then:

$$\dot{m} = k\sqrt{p}$$

where p is the pressure difference $(p_1 - p_2)$ and k is a meter constant. The Venturi sensor, as sketched in Figure 11.8, is the primary element and the pressure-measuring device (such as a DP cell) is the secondary element. The measure scale will be non-linear for direct recorders, due to the square root relation, and telemetering control will not be satisfactory unless a square root eliminator correcting device is fitted.

We have seen in Chapter 2 that the bunker fuel supplied to a vessel should be to the standard specification ISO 8217:2010. The bunker fuel supplied to ships is currently residual fuel oil that has been left over from the refining process. Therefore the potential exists for 'off spec' bunkers to be delivered to the vessel. Samples should be taken to check the composition of the fuel but analysis takes time. Also, the fuel delivered might have a proportion of other liquids, such as water, mixed in with the initial supply, and if the invoicing is paid by mass and the delivery measurement is by volumetric means there is room for error and dispute over the bill.

To avoid this, another type of meter has started to appear on board ships. This is the mass flow meter. This device measures the mass of the liquid flowing through a pipe and not the volume.

Figure 11.8 shows the flow of liquid through a pipe shaped in the form of a 'U'. The mass of the liquid flowing through the pipe causes a deflection of the pipe, which develops into a vibration at right angles to the flow and in proportion to the mass of the liquid. The deflection is known as the Coriolis effect. The mass flow is calculated by measuring the asymmetric displacement of the U tube. The mass flow meter from the Emerson process management, for example, will also record the temperature to give the volume, density

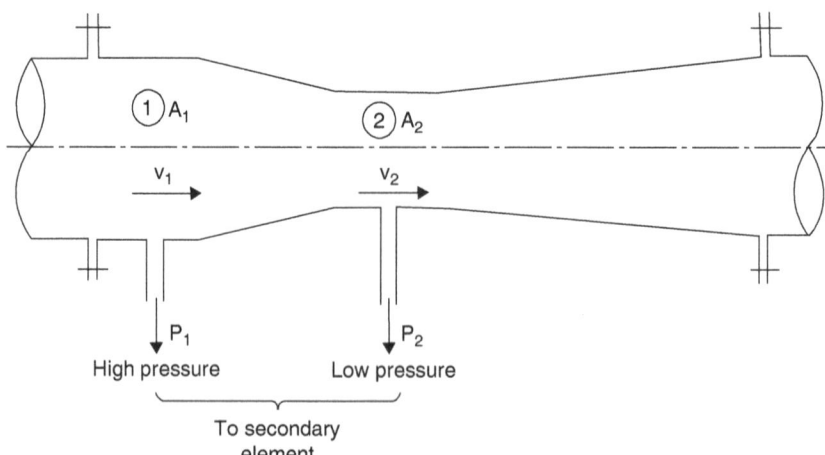

▲ **Figure 11.8** *Flow sensor (Venturi)*

and the viscosity of the fuel being loaded. This gives high-quality accurate information without the need for expensive precision instruments and accurate look-up tables, as required when using a volumetric system of measurement.

Other measurements

It has been appropriate to include details of some instruments in appropriate places elsewhere in this volume, for example, the viscometer and CO_2 recorder in Chapter 2 and the torsionmeter in Chapter 6, etc. Please refer to the index to find details about the instrument that you are looking for. Advancement in control technology has brought with it the need for additional instruments. However, the use of microcontrollers, programmable logic controllers (PLC) and computers has meant that measurement instruments have become smaller and have electronic circuits included enabling communication with the controllers. These small instruments are called sensors and are described on pages 494–495. There is still a need for the marine engineer to be informed about traditional forms of instrumentation as these will still be in use today on some ships. It is also useful to understand the operation of the more traditional instruments because some of the modern sensors will be derived from these devices.

Tachometer

The traditional tacho-generator is a small, precision, direct current generator driven by the shaft whose rotational speed is to be measured. The output voltage is directly proportional to speed therefore the output is the same as a conventional voltmeter but the read-out is usually arranged to be calibrated in terms of rotational speed. A digital counter can also be used (Figure 11.9).

The DC version is known as the Hall effect sensor. As the (ferrous) toothed wheel rotates each tooth alters the air gap and flux in a pick-up coil (P) whose output pulses are amplified (A). Pulses pass through a timing gate (G), say 1 second opening period, and are counted on a digital counter (D) that calculates the rotational speed by relating to the number of teeth per revolution, and displays the result as revolutions per second, or alternatively rev/min readings can be arranged with different gate or scale settings.

Photo-cell

Photo-conductive cells are constructed with a thin layer of semi-conductor material that has the ability to alter its resistance in proportion to the amount of light energy falling upon the material. They are used in some temperature sensors and flame failure devices.

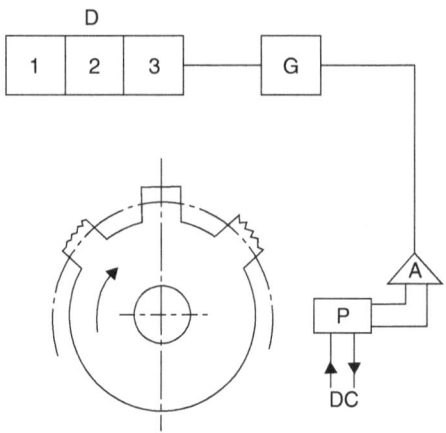

▲ **Figure 11.9** *Digital-Tachometer*

Photo-emissive cells use the light energy to release electrons from a metallic cathode. If visible light, which is radiation and hence energy, falls upon certain alkali metals – such as caesium – electrons will be emitted from the surface of the metal. Metals in general exhibit this characteristic but for most materials, the light required has threshold wavelengths in the ultraviolet region so that visible light does not cause electron emission.

Light energy is in particles called *photons*. The energy of the photons is used to remove the electrons and then to give them kinetic energy after they escape from the metal. Figure 11.10 shows a simple photo-cell where visible light falls on the metal cathode from which electrons are emitted. They collect at the anode and in this way create a potential difference (*voltage*) that can then be amplified and used for alarms and/or measurement for control, etc.

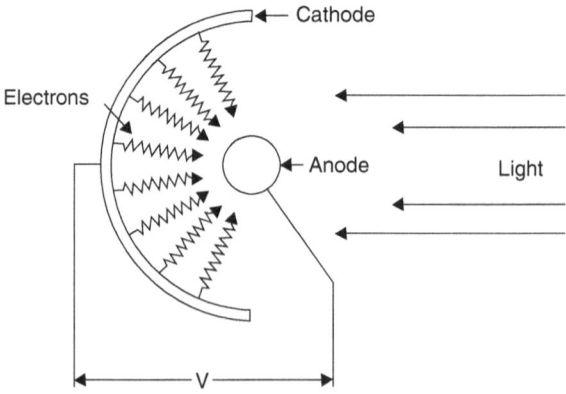

▲ **Figure 11.10** *Photo-cell*

In the vacuum cell all current is carried by photo-electrons to the positive anode. In the gas-filled cell emitted electrons ionise the gas, producing further electrons, so giving amplification. Secondary-emission (photo-multiplier) cells utilise a series of increasingly positive anodes and give high amplification.

Photo-transistors exhibit similar characteristics and small size and high amplification make their use particularly attractive especially when applied to counting systems, that is, digital tachometry.

Calibration

Instrument calibration and testing is a specialised task. Pneumatic instruments are tested by reference to a master gauge, standard manometer or hydraulic deadweight tester. Electrical instruments are tested against standard resistors and potentiometers.

Using a Bourdon calibration as example the correct procedure would be:

1. Start with the zero (error) adjustment, which changes the base point without changing the slope or shape of the calibration curve. It is usually achieved by rotating the indicator pointer relative to the movement, linkage and element.
2. Multiplication (magnification) adjustment alters the slope without changing base point or shape. This is completed by altering the drive linkage length ratios between primary element and indicator pointer.
3. Angularity adjustment changes the curve shape without altering base point and alters scale calibration at the ends. This error is minimised by ensuring that link arms are perpendicular with the pointer at mid-scale.

Over the design range the pointer movement has a linear relationship to pressure, and the scale is calibrated accordingly. Hysteresis – a vibration phenomena – is best eliminated by correctly meshed gearing and fitted pivots to reduce backlash, etc. Hysteresis of an instrument is the maximum difference between readings at given points moving up the scale, to those taken when moving down the scale, that is, hysteresis curves are plotted from up-scale and down-scale readings.

Figure 11.11 shows calibration curves and adjustment for the Bourdon link type of instrument mechanism. Instrument readings (I), true values (T), desired result (D). Zero error and adjustment (Z), multiplication error and adjustment (M), angularity error and adjustment (A) – error curves or lines for actual values.

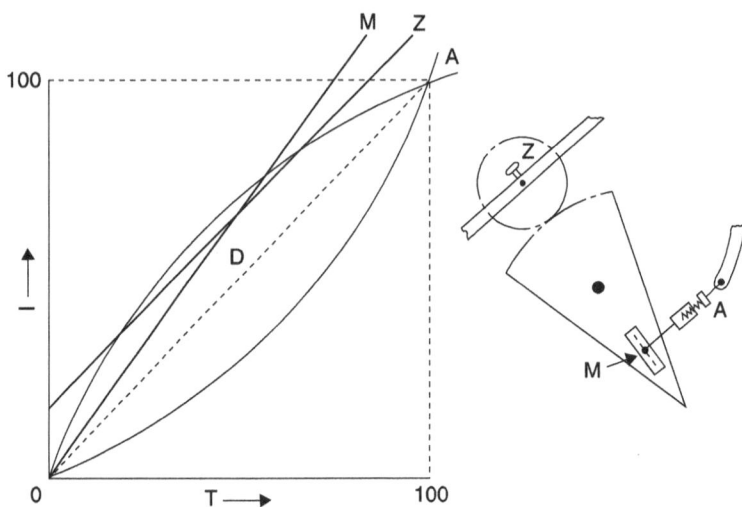

▲ **Figure 11.11** *Instrument calibration*

Sensors

In modern control technology, sensors can make or break the system. They combine the measuring and transducing function into one small component, which means that they are often required to work in hostile environments. The difference between the sensor and more traditional measuring devises coupled to transducers is that the sensor will produce smaller voltages for detection by micro-controllers and other types of electronic processors. The older transducers will have amplifiers and be required to send information to pneumatic controllers that require more force to drive the internal components.

The list of different types of sensors is growing longer by the day and as the maritime industry embraces the lessons of condition-based monitoring/maintenance, the types of sensors that are being used will also grow.

Some of the more common types of sensors currently available are:

- pressure sensors (also measuring differential pressures);
- temperature sensors;
- RPM sensors;
- force and torque sensors;
- magnetically inductive sensors;

- position sensors;
- magnetostatic sensors (Hall effect devices);
- optoelectronic sensors;
- compass sensors;
- vibration sensors;
- gas sensors;
- voltage and current.

Telemetry

Telemetry is the process of measuring process characteristics and transmitting the results to a remote monitoring station over a considerable distance. The transmission medium could be air and space for aircraft and satellites or wires for power stations and ships. The name has become popular in the mind of the general public due to its use by Formula One motor sport for the real-time transmission of data to and from the race cars as they circulate around the track. The device at the measuring point is called a transmitter with the receiver located at the recording or control centre. Telemetry on board a modern ship is also involved with centralised instrumentation, that is, display, alarm scanning, data logging, etc. or with remote-control devices, or both.

Staff working on superyachts will recognise this technology as some owners will like to have a read-out of the measured data on their own yacht even when they are sitting in their office many miles away.

Students familiar with 'drones' and underwater autonomous vehicles will recognise the advances made in this area and some manufacturers are working on systems for unmanned ships where 'telemetry' will be at the heart of the technology.

Centralised instrumentation and control

Display panel and controls

The centralised information and control is carried out in an air conditioned machinery control room (MCR). The reason for this is that the electronic alarm and monitoring systems are temperature sensitive and need to be kept cool. However, with the increase in the volume of information it is also necessary for the engineers to discuss plant operational issues in a quieter environment than is available in the main engine room.

The data logger has been a beneficial addition to the recording of information about the operation of the plant. Virtually all components are electronic and are fitted with standard printed circuit boards. Faults will be located by an analysis of a faulty board and replacement of the card rather than on-the-job repair.

In selecting alarm circuits great care must be taken in the preference choice utilised. Important circuits should be fitted with distinctive alarm indications and a quick and easy position location. Less important circuits can be fitted with a secondary importance alarm and isolating-locating system. The provision of too many alarms, not easily discriminated from each other, can cause confusion. Similar remarks apply to remote-control room gauge boards where only essential measurements should be *frequently* scanned. Manufacturers of modern machinery have a tendency to wire all their alarms into the shut-down sequence when there may be no need for it.

The control room itself requires careful design with reference to comfort, lack of lighting glare, selective positioning of instruments for rapid viewing, correct placing of on-off and position and variable quantity indicators, improved instrument indication techniques, rapid control fault location and replacement, etc.

Various types of indicators and recorders are in use, for example: lights, dial gauges with pointer, colour strip movements, flat screen displays, counters, charts, etc. There is a trend toward using Windows-based PCs and care must be taken to ensure that all the engineers are familiar with the 'back-up' proceedings for when the system crashes.

It is also very important that software updates are completed when the new versions are produced. The current versions should be recorded in a central position where all the engineers know the information can be found. If service engineers come on board to repair almost any equipment, they will undoubtedly want to know the version of software that the control systems are using.

Alarm scanning, data logging, terminology

The very latest ships can have up to 15,000 alarm points and the modern scanner can monitor between 1 and 400 channels per second. Each measuring point is selected in turn by the electronic circuitry and the software logic. Alarm conditions can be set close to the operation condition giving almost instantaneous responses to potentially dangerous conditions (see Figure 11.12).

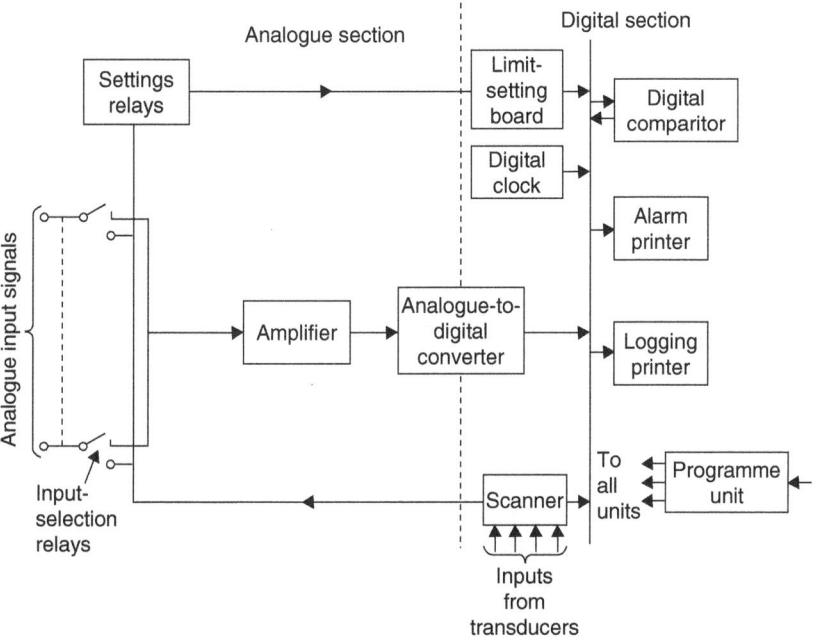

▲ **Figure 11.12** *Alarm scanning and data logging*

Measurement

Analogue inputs are amplified from the low voltages produced by the instruments, and the information as a voltage representation of the measured value is translated in the analogue-to-digital converter using the pulse code modulation (PCM) technique. The advantages of the digital signal are given below:

1. The transmission and receiving units are only detecting the presence of a pulse. Therefore if there are stray voltages or electrical interference or 'noise' these are largely ignored. But with the analog systems there is a distortion of the signal making it more difficult to interpret accurately.

2. Digital signals can be matched to a more diverse range of modern systems, such as logic controllers and microcomputers. They are also an absolute requirement for the digital data 'BUS' technology that is creating another step change in the capability of modern control and systems engineering.

Displays

The results of the signals from the control circuits can be displayed in several ways. The primary method on modern ships is in the form of actual values overlaid on graphical displays. The older displays would be cathode ray tubes but the modern ones would be LCD/plasma flat screen panels.

A second function is to compare digitally the analogue inputs with pre-set limit values and to display an alarm condition on the visual display. However, due to the complicated nature of the systems it is essential to also print out or display the alarm condition as it occurs. This system will then become the initial source of 'primary cause' fault diagnosis as the alarms that were activated first will appear first on the list.

Programme

The strength of modern systems depends upon the algorithms that are designed into the software. The latest systems have many thousands of lines of code and will initiate action to a predetermined logical sequence and scanning routine that gives storage, print-outs and allows manual control by the engineers. The main log print-out is timed by the special digital clock. This does introduce an additional 'planned maintenance' procedure, which is to update the software to ensure that the latest drivers are on the system.

Electronic design

Modern practice is for solid-state silicon components to be surface mounted on insulated boards as printed circuits. Any relays are hermetically sealed on plug-in cards. Test boards are provided for fault detection and repair is by the replacement of faulty boards. Data logging will be completed by printing information out of a robust printer. This could be a simple dot-matrix printer in the engine room or a high-quality inkjet printer in the Chief Engineer's accommodation.

It should be possible to download data to a spreadsheet so that the senior staff can complete any 'trend analysis' necessary for identifying faults as they develop. The most sophisticated systems will complete this in real time and have the ability to transmit the information back to the company's main office for further analysis.

Analogue representation of information

Analogue representation is the conversion of measured quantity into another physical quantity *in a continuous way*, and it may retain an analogue display such as a pointer that moves in unison with the measured value. For example, temperature converted into DC voltage by a thermocouple where the voltage is proportional to the temperature measured. The analogue display is useful if there were readings from several sources that should all be the same. A quick look along a row would enable the operator to spot any that are not in line faster that studying a row of digital readings.

Digital representation

In digital representation, measured quantity is represented by repeated individual increments *at given intervals* and displayed in a digital form. For example, a revolution counter that trips to alter the reading after each engine revolution. Which is useful for long-term presentation, for example, 'full away' watchkeeping readings.

The digital/analogue debate has very much swung away from the analogue systems as the digital equipment has become so fast that as far as humans are concerned the digital measurement, transmission and interpretation is instantaneous and therefore the digital equipment will give a continuous reading where it is necessary.

Scaling unit

The older mechanical registers could record about two pulses per second (maximum) without slip although the last designs could reach 50 pulses per second. Electronic GM tubes can record 5,000 pulses per second so that the scalar functions to reduce output pulses to the register in the ratio 5,000:50.

Advantages of electronic control systems and data recording are that it:

1. Reduces staff time and the number of separate systems.
2. Provides continuous observation and fault alarm indication.
3. Provides accurate and regular operational data records.
4. Increases plant efficiency due to close operational margins.

Telemetry components

Amplifier

The amplifier is used to step up the sensor low-power signal for use in a high-power actuator element. Two designs are given, namely electronic and pneumatic (relay).

The upper diagram of Figure 11.13 shows a transistor amplifier of the common emitter type. A small change in input current signal produces a larger amplified change in load current. B, C and E refer to base, collector and emitter, respectively.

The lower diagram of Figure 11.13 illustrates a reverse-acting pneumatic relay amplifier. Increase in the magnitude of the input signal air pressure reduces air flow from the pressure energy source and output pressure to load falls to a corresponding equilibrium value.

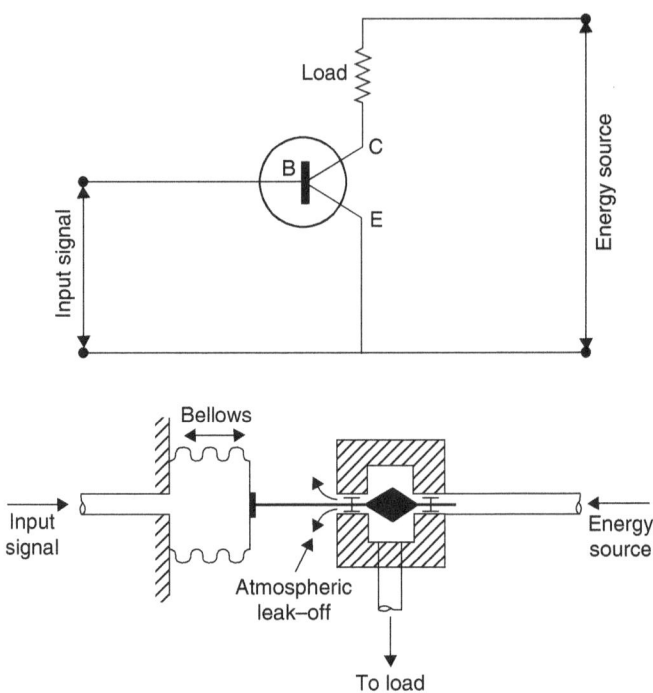

▲ **Figure 11.13** *Amplifiers*

Transducer

A transducer converts the small sensing signal into a readily amplified output, usually in a different form. Designs can generally be simplified into three basic reversible types:

Mechanical displacement ↔ Pneumatic
Mechanical displacement ↔ Electrical
Pneumatic ↔ Electrical

A mechanical displacement → pneumatic type of transducer is the flapper nozzle device as shown in Figure 11.16.

An electrical → pneumatic transducer is illustrated in Figure 11.14. For an increase in current the created S pole to the left will be attracted up to the N pole of the magnet so giving a clockwise rotation because the moment arm is greater than that caused by the created N pole being attracted down on the S pole of the magnet. This action closes in on the nozzle so giving a higher output air pressure and increasing the feedback bellows force until equilibrium is achieved.

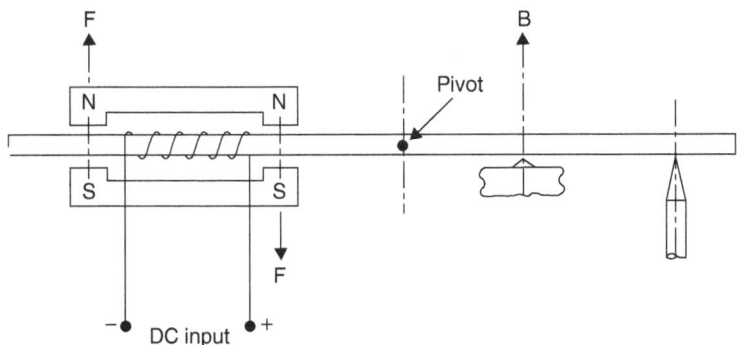

▲ **Figure 11.14** *Electro-pneumatic transducer*

Signal media

Traditional marine control systems used pneumatics or electrics as preferred systems although electro-hydraulic systems were used in many steering gears (still retained on modern tonnage). Pneumatic systems generally had the advantages of lowest first cost, inherent safety and proven reliability but they exhibited time lags and required clean air supplies.

Electrical-electronic systems have advantages of small component size, low power consumption and rapid response but generate some heat and are susceptible to variations in power supply. These traditional systems will be seen on board ships for many years to come and therefore the engineer presenting himself/herself for examination will need to be familiar with the designs of these systems.

Modern tonnage is moving rapidly towards full digital monitoring and control as is described in the next section.

Mechatronics and digital control systems

Mechatronics is the technology that is starting to transform the world of control engineering. The electronic control of diesel engines has taken them to new levels of performance. This has largely been due to the developments in manufacturing, which has enabled the production of reliable and robust sensors.

These sensors can detect pressure and temperature to a much finer tolerance and act infinitely faster than any mechanical system. The power of microprocessors and the development of the algorithms required to interpret the signals from thousands of sensors has brought everything together at a time when the industry needs a step forwards in efficiency.

Mechatronics is the synergistic integration of mechanical components, electronic devices, computer and software engineering along with embedded control features, coming together to keep machinery operating as close to its design condition as possible. It has been described as 'mechanical engineering for the 21st century'.

During the past 30 years or so, as we have seen, the technology consisted of pneumatic control equipment where precision-made mechanical instruments and other devices interpreted the condition of machinery and associated systems and transmitted pneumatic signals in an effort to control processes such as the temperature of cooling water or lubricating oil. These control systems were usually localised at the point of use, although some had remote indicators such as that for boiler water level. However, on the whole, control systems were independent of the main machinery alarm systems and the measurement of processes inside hostile environments, such as the engine combustion space, were just not possible.

As discussed, we are already seeing the electronic control of diesel engine combustion increasing engine flexibility and lowering emissions. Now, other marine machinery is also being produced with the ability to link into the central control and management system, and as algorithms become ever more sophisticated so will the efficient use of the equipment.

All communication systems need a common language and the focus for the development of mechatronics on board is now on the topology and protocols being used with Modbus and CAN-bus being the top runners. However, some manufacturers are producing equipment that uses the protocol called PROFIBUS.

The developments of these standards into a unified system or one protocol is important so that manufacturers can move forward rather than having to cover different systems or having to change standards halfway through their production run, which is a possibility at present.

Mechatronics also has the potential to provide the industry with an increased quantity of information relating to the condition of machinery thereby giving owners the evidence they require to ensure their vessels are complying with legislation. In addition, such information will enable managers to increase efficiency thus decreasing fuel consumption and maximising profits. The capability of mechatronics to assist the industry with its move toward higher efficiencies, lower energy consumption and more eco-friendly vessels is on the horizon. Given the right development, more sophisticated and so-called intelligent systems will be capable of linking the control of machinery with the production of machinery condition data. This will drive efficient management systems, enabling ship and head office staff to work together by improving not only the vessel's voyage planning but also the efficient use of machinery. Progression will increasingly mean different types of fuel and/or propulsion types and will also include more efficient maintenance systems and changes in the general management of vessels. The integration of embedded systems into the control of marine machinery will once again bring to the fore the importance of crew competence and focused staff development.

We now have in place the 'Manila amendments' to the STCW convention. Specific outcomes from this update require a focus on the operation of pollution prevention equipment and, more generally, additional emphasis being given to environment management. There will inevitably be a concentration on professional development including diagnostic techniques for ship staff since they will be in the frontline in the event of any emergency, for example, if the alarm monitoring system is indicating that the ship is not producing the correct exhaust emissions as the vessel is approaching an Emission Control Area (ECA). This issue would clearly have to be resolved urgently so the vessel does not risk incurring a hefty fine. Integrated approach to design has potential for monitor efficiency gains. With sophisticated control systems in place, crew familiarisation will be essential, therefore continuity of staff, efficient handovers and the use of integrated management information systems will be required in running the modern fleet effectively.

Communication systems

The increased use of sensors and electronic control systems will undoubtedly bring a steep change in both the accuracy and the amount of information that will be available to the engineers in charge of a marine power plant. The sensors are linked to programmable logic controllers (PLC), microprocessors or full computer systems. However, as with any communication system they all have to be speaking the same language and using the same transportation system. Computer systems can be networked in a number of different ways, the most popular configurations are:

- ring;
- star;
- bus.

A computer controlled system employing a universal bus system instead of individually wired circuits is the modern trend for control engineering.

Compared to a control system with conventional wiring a computer-controlled bus system has the following advantages:

- cost, weight and construction savings due to the reduction in wiring (valves, pumps and actuators, etc. are operated by data signals travelling along one communication bus instead of electrical signals travelling down individual circuits;

- greater redundancy and operational reliability due to a much lower number of plug-in connectors and easy use of multiple transmission paths;
- much simpler and reduced construction time;
- reduced commissioning time due to much simpler connecting procedures;
- multiple use of sensor signals;
- simple upgrade procedures.

One of the features of this system that promises to deliver so much is the production of real-time data. Machinery control, performance monitoring and maintenance systems can all be updated with real-time data. This will have substantial benefits for the efficiency of ship machinery and for energy management. Remote monitoring will be made so much easier and therefore there will be knock-on effects for management systems, training and professional development of engineering officers.

Automation and Control

This section discusses only the basic requirement and is deliberately brief and simple. A more detailed treatment is given in Volume 10 of the Reeds series. In Volume 10 there are illustrations on pneumatic systems to give practical examples of the control actions taken. The modern systems are electronic control systems and as such cannot be seen operating.

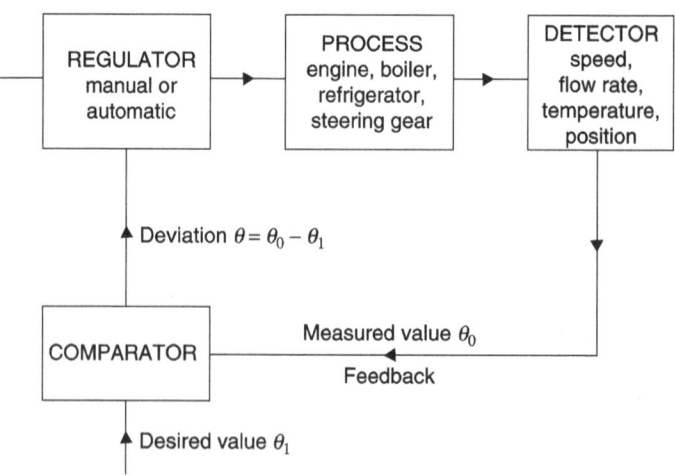

▲ **Figure 11.15** *Closed-loop control system*

Terminology

The correct terminology is given in the relevant British and International Standards. These are the International Electrical Commission (IEC) – BS IEC 60092-504:2016 standards for electrical installations in ships.

A few simplified terms, related to Figure 11.15, are now given.

Closed loop control system

In this system control action is dependent upon the output. The system may be manually or automatically controlled. Figure 11.15 shows the basic elements in a closed-loop control system.

The measured value of the output is continually being fed back to the controller, which compares this value with the desired value for the controlled condition and produces an output to alter the controlled condition if there is any deviation between the values.

Measured value is the actual value of the controlled condition (symbol θ_0).

Desired value is the value of the controlled condition that the operator desires to obtain. Examples: 2 rev/s, 25° off helm, 55 bar, –5°C, etc. (symbol θ_1).

Set value is the value of the controlled condition to which the controller is set – this should normally be the desired value and for simplicity no distinction will be made between them.

Deviation (or error) from the set value is the difference between measured and desired values (symbol θ). Hence $\theta = \theta_1 - \theta_0$. This signal, probably converted into some suitable form such as voltage to hydraulic output or voltage to pneumatic output, etc., would be used to instigate corrective action with the objective to reduce the error to zero.

Offset is sustained deviation.

Feedback is the property of a closed-loop control system that permits the output to be compared with the input to the system. Feedback will increase accuracy and reduce sensitivity.

Control actions

Three basic actions will be described: (i) proportional; (ii) integral; (iii) derivative.

Proportional control (P)

A pneumatic type of proportional controller is shown diagrammatically in Figure 11.16. A set value of pressure P_s is established in one bellows and the measured value of pressure P_m is fed into the opposing bellows (the measured value of pressure could be proportional to some measured variable such as temperature, flow, etc.) Any difference in these two pressures causes movement of the lower end of the flapper, alteration in air flow out of the nozzle and hence variation in output pressure P_o to the control system.

If the upper end of the flapper were fixed, that is, there are no proportional action bellows, then a slight deviation would cause output pressure P_o to go from one extreme of its range to the other. This is simple proportional control (output pressure change P_o and deviation $[P_m - P_s]$) with a very narrow proportional bandwidth and high gain (gain = controller output change/deviation). Moving the nozzle down relative to the flapper increases the sensitivity and gain, and further narrows the proportional band width.

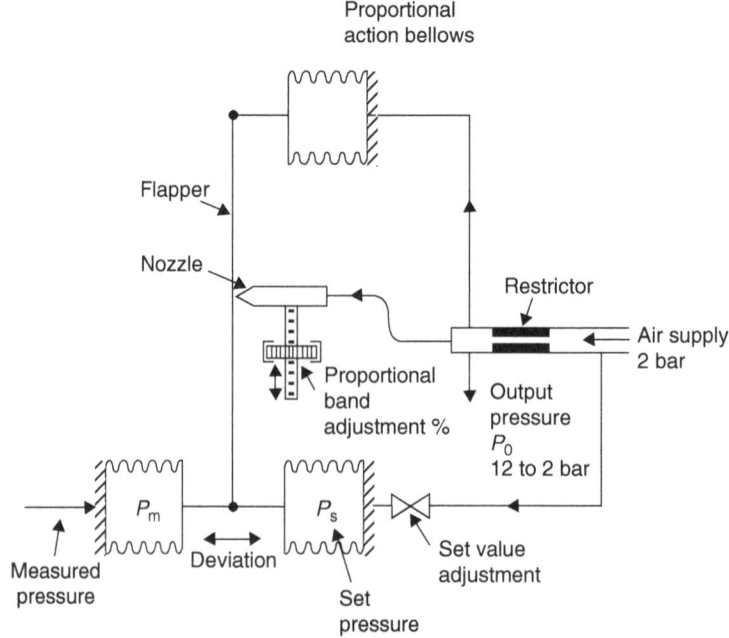

▲ **Figure 11.16** *Pneumatic proportional controller*

With output pressure P_o acting in the proportional action bellows, the top end of the flapper will always move in the opposite direction to the lower end, which reduces the sensitivity and widens the proportional band. When adjusting the controller to the plant, the object would be to have minimum offset with stability, that is, no hunting. Commencing with maximum proportional band setting (200%) the set value control is moved away from and back to the desired setting (step input) and the effect on the controlled variable noted. Using incremental reductions of proportional band and for each reduction a step input, a point will be reached when oscillations of the controlled variable do not cease, and a slight increase in proportional band setting to eliminate the oscillations gives optimum setting.

Proportional plus integral control

Proportional control will arrest a change and hold it steady but at a different point from the desired value. The difference between these values is called *offset*, which is different at each load; this is the shortcoming of proportional control. Offset can be reduced by increasing sensitivity (i.e. narrowing proportional band) but this can lead to hunting and instability. Integral (reset) action addition (P + I) gives arrest of the change and a reset to the desired value irrespective of load. Integral action will always be occurring while deviation exists. Controllers of this type have an adjustment to vary reset time: the shorter the time setting the greater the integral action. Too great an integral effect will cause *overshoot* past the desired value.

Proportional plus integral plus derivative control

This can be represented as (P + I + D). Derivative action may be added as a damping action to reduce overshoot. Derivative action is also anticipatory where the sensed rate of change of deviation corrects to reduce likely deviation – it opposes the motion of the variable. Derivative action time, usually adjustable at the controller, has the effect that the longer the time setting the greater the derivative action.

Summary

- (P) Proportional control: action of a controller whose output signal is proportional to the deviation. That is, correction signal $\propto$ deviation.
- (I) Integral control: action of a controller whose output signal changes at a rate that is proportional to the deviation. That is, velocity of correction signal $\propto$ deviation. Objective: to reduce offset to zero.
- (D) Derivative control: action of a controller whose output signal is proportional to the rate at which the deviation is changing. That is, correction signal $\propto$ velocity of deviation. Objective: to give quicker response and better damping.

- (P) Single-term controller.
- (P + I) or (P + D) Two-term controller.
- (P + I + D) Three-term controller.

Pneumatic controller

Figure 11.17 shows in diagrammatic form a three-term controller (P + I + D). Set value control and proportional band adjustment have been omitted for simplicity (see Figure 11.16). Often, controller manufacturers produce a standard three-term controller and the installer can adjust for type of control action necessary, that is, either single-, two- or three-term as required.

Approximate sizes are restrictor about 0.2 mm bore, nozzle about 0.75 mm bore and flapper travel at nozzle about 0.05 mm. To ensure exact proportionality and linearity the effective flapper travel is reduced to near 0.02 mm, giving less sensitivity and wider

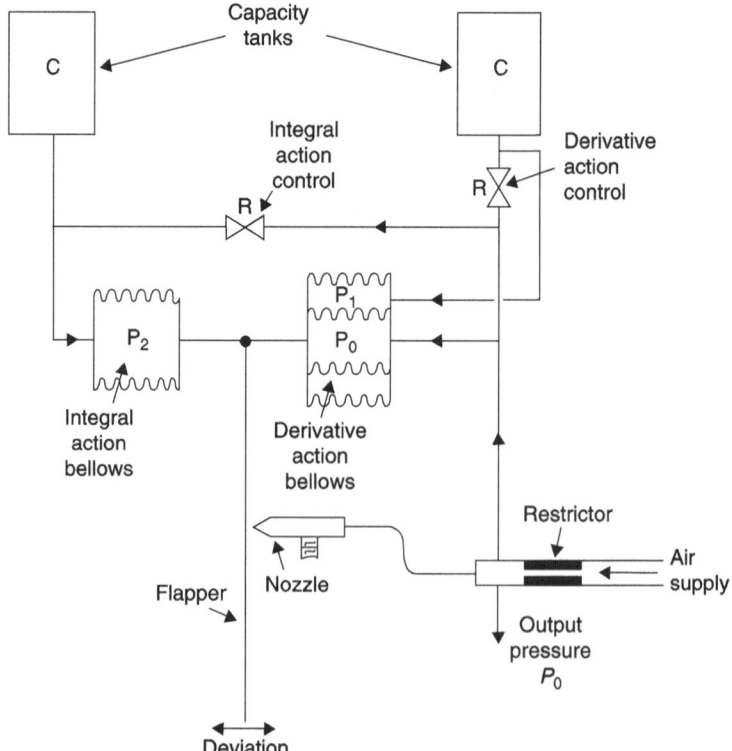

▲ **Figure 11.17** *Pneumatic P + I + D controller*

proportional band, with negative feedback on the flapper due to the inner bellows and pressure P_0 acting on it. Note that whichever way the bottom of the flapper is moved, by the deviation, if the top is moved in the opposite sense this is negative feedback, if in the same sense it is positive feedback.

Integral added

This is applied by adding positive feedback with pressure P_2 acting on the integral action bellows. Integral action time is the product of the capacity C and the resistance of the integral action control R, that is, RC (note the similarity with electrical circuits). Increasing R by closing in the integral action control increases integral action time.

Derivative added

This is applied with further negative feedback with pressure P_1 acting on the derivative action bellows. Derivative action time is the product of the capacity C and the derivative action control resistance R. Increasing R by closing in the derivative action control increases derivative action time.

Note that integral action is very rarely applied on its own. Derivative action is never applied on its own.

Electrical-electronic controller

Figure 11.18 shows the compound controller, which should be compared with its pneumatic counterpart (Figure 11.17). The upper part of the diagram illustrates grouping to controller (note the summer and the potentiometer gain adjustment) while the lower diagram is basic operational amplifier configuration. On the lower diagram after summing of measured and desired value, the input voltage is V_i. The gain adjustment is the factor n related to resistances R. The rate (derivative D) circuit is at input and the reset (integral I) on the feedback line.

Control Systems

Microprocessors, sensors, electrical power systems and instrumentation all networked together using digital CAN-bus transfer protocols, make up the structure of the monitoring and control engineering systems on board modern ships.

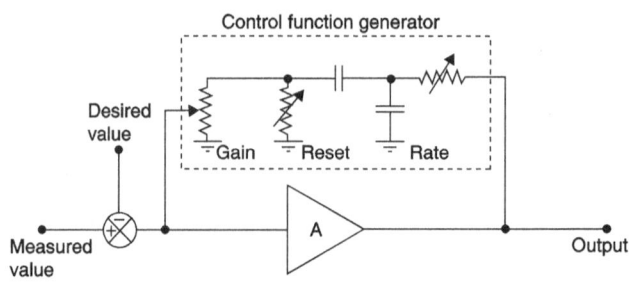

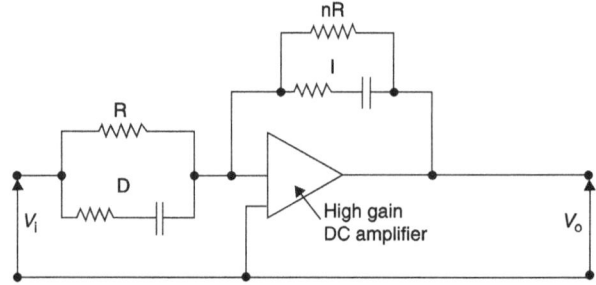

▲ **Figure 11.18** *Electrical-electronic controller (P + I + D)*

These systems are fast and much more flexible than the older electro/pneumatic/ hydraulic control engineering systems that they replace. However, some of the older equipment is still retained and using these to illustrate the principles involved makes it easier for the student to learn the basics and concepts. For more information about the PLC based systems please see Volume 10 of the Reeds series.

Therefore, the following control systems have been retained here to allow the students to have a clear vision about the overall design philosophy of the control systems. These are;

- one final controlling element, the diaphragm valve;
- one telemetry system, the electric telegraph;
- three control circuits
  - o fluid temperature control
  - o automatic boiler control
  - o bridge control of the different IC engine configurations.

Diaphragm valve

The diagram (Figure 11.19) illustrates such a valve used for controlling fuel quantity to the burners of a boiler.

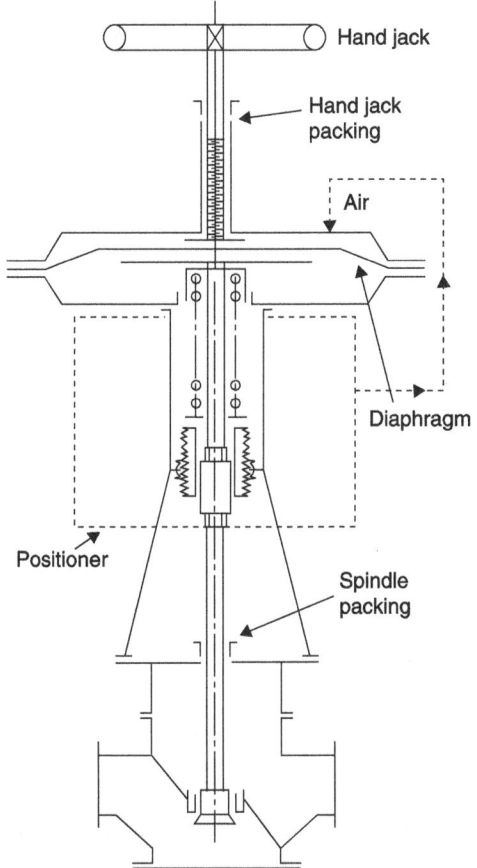

▲ **Figure 11.19** *Diaphragm valve*

Control air acts on top of the synthetic rubber diaphragm, and increasing air pressure causes the valve to move down, permitting increased fuel supply to the burners.

If air supply to the diaphragm should cease then the valve will 'fail safe', that is, it will close against the flow of the oil (right to left in Figure 11.19) and the burners will be extinguished. Hand regulation could be used by operating the hand jack.

A positioner would be used if:

- the valve is remote from the controller;
- there is high-pressure difference across the valve;
- the medium under control is viscous;
- the pressure on the gland is high.

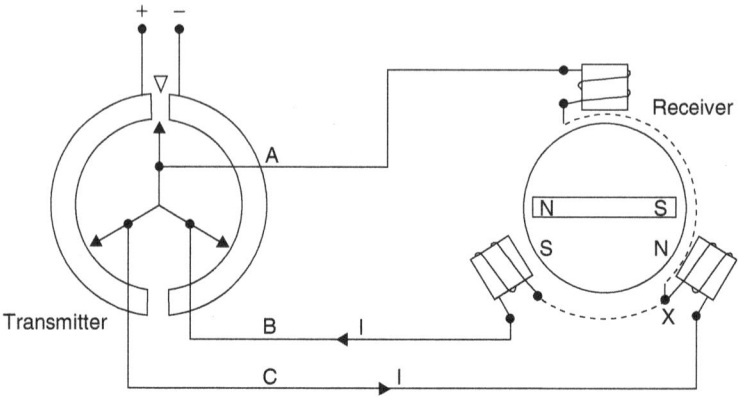

▲ **Figure 11.20** *Position indicator (electric telegraph)*

In this case a flapper is fastened to the valve stem at one end and the other end operates against the nozzle whose air supply pressure acts on top of the diaphragm. The control air signal acts, via bellows and spring, between the two extremities of the flapper and fixes its position relative to the nozzle, dependent on magnitude of control pressure signal.

Position indicator (electric telegraph)

This device is inherently telemetering rather than a control system as such. Figure 11.20 shows the system in equilibrium with equal DC currents (I) in lines B and C and zero current in line A. The receiver rotor is locked by equal and opposite torques from the attractions on unlike pole faces. Assume the transmitter to be moved 30° clockwise. Current flows to the receiver from line C subdivides at point X and equal currents return through lines A and B of magnitude 1/2. This creates a strong N pole at fixed magnet X and two weak S poles at the other two fixed magnets.

The receiver indicator will therefore turn to the corresponding position, that is, 30° clockwise. Energy supply failure alarms are fitted.

As a telegraph, a bell can be activated until the hand lever on the receiver is moved to correspond with receiver position.

For clockwise (say ahead) rotation of the transmitter, line A will be carrying return current. If the engine-driven speed tacho-generator is rotating in the correct direction, its output can be arranged to be in the same sense as line A. A summed signal (additive) will maintain a 'wrong way alarm' in the isolated condition. Incorrect rotation of the engine will create signals in opposition, which causes the alarm to be activated. The same applies for anti-clockwise (say astern) rotation of the transmitter when line A will be carrying supply current.

Fluid temperature control

The arrangement sketched in Figure 11.21 is single element for fresh water coolant temperature control applied to an auxiliary IC engine. It would be equally suitable for oil coolant or lubricant (engine or gearbox) temperature control. Full flow of sea water is arranged through the cooler. A three-way valve (two inlets, one outlet) operates to mix quantities of coolant, bypassing or going through the cooler, dependent on coolant return temperature.

With correct analysis of parameters, and careful valve selection, a simple single-element control system can be utilised for most duties. A wax-element-activated valve gives a simple but fixed control (over say 10°C), by adjusting one outlet for the return coolant to the cooler with the other bypass to the engine. For large thermal variations, such as manoeuvring

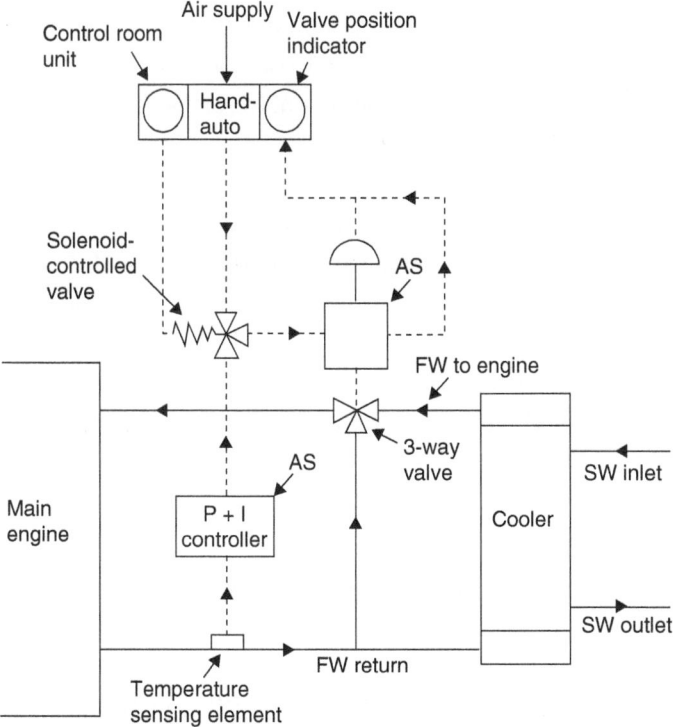

▲ **Figure 11.21** *Fluid temperature control*

conditions, single-element control may not meet requirements. 'Cascade control', involving one controller (the master) amending the set value of another controller (the slave) could then be used. The slave controller, sensing temperature of coolant leaving the cooler, would detect variations of sea temperature and adjust sea water flow accordingly. The master controller, sensing return temperature of coolant from engine or gearbox, would if necessary amend the set value of the slave controller and so alter sea water flow.

For warming through, or operation with low sea temperatures, the refinement of 'split level (range) control' could be added. Over a range of low temperatures of coolant, sensed by the slave controller, a lower-magnitude signal would operate heat input to a coolant heater. At a given point, when coolant temperature had increased, the signal would close the heat control and for the rest of the range would operate to control sea water flow to the cooler in the usual way.

Automatic boiler control system

Refer to Figure 11.22 for the lighting sequence and emergency devices.

1. The pressure switch initiates the start of the cycle. The switch is often arranged to cut in at about 1 bar below the working pressure and cut out at about 1/5 bar above the working pressure (this differential is adjustable).
2. The master initiating relay now allows 'air-on'. The air feedback confirms 'air-on' and allows a 30-second time delay to proceed.
3. The master now allows the arc to be struck by the electrode relay. The 'arc made' feedback signal allows a 3-second time delay to proceed.
4. The master now allows the fuel-initiating signal to proceed. The solenoid valve allows fuel on to the burner. The 'fuel on' feedback signal allows a 5-second time delay to proceed (this may be preceded by a fuel heating sequence for boiler oils).
5. The master now examines the photo-electric cell. If in order the cycle is complete, if not then fuel is shut off, an alarm bell rings and the cycle is repeated.

Obviously failure of any item in the above cycle causes shutdown and alarm operation. In addition the following apply:

1. High or low water levels initiate alarms and allow the master to interrupt and shut down the sequential system.
2. Water level is controlled by an Electroflo type of feed regulator and controller. Sequential level resistors are immersed in conducting mercury or non-conducting fluid, so deciding pump speed by variable limbs level. The fixed limb level passes over a weir in the feed box.

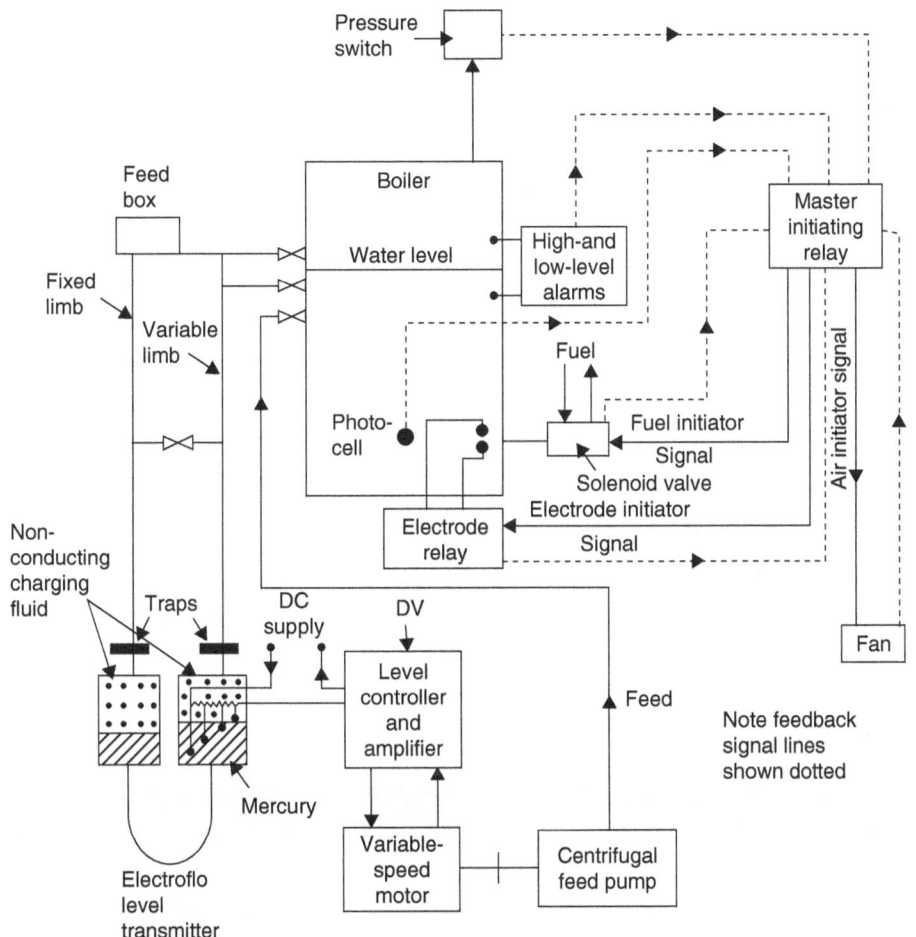

▲ **Figure 11.22** *Automatic boiler control system*

Unmanned machinery spaces operation

In the past an unmanned machinery space (UMS) may have been known under other names such as 'Unattended' Machinery Space, but IMO has chosen the wording 'unmanned' and this is the terminology used in the STCW document as well as M Notices in the United Kingdom and by all participating flag states.

In vessels that have a machinery space designated as unmanned, it does not mean that the engine room is unattended or unsupervised; in fact quite the opposite. STCW actually describes such an engine room as 'periodically unmanned engine-room', which is the key description to the system of watchkeeping practised on ships built to operate in such a way.

The watchkeeping engineering officer is supervising, controlling, monitoring and working closely with the machinery just as s/he would in a fully manned engine room. The difference is that some of the routine monitoring is undertaken by the control systems that will activate a machinery alarm if any of the controlled conditions move outside of the pre-set values. This means that the watchkeeper does not have to carry out routine adjustments to the machinery systems and his/her time is freed up to carry out other tasks. It also means of course that if the watchkeeper completes all the checks and duties that the machinery should operate for up to 8 hours without the watchkeeper being physically present in the engine room.

The watchkeeper is, however, still on duty and responsible for the supervision of the machinery space but to be effective s/he can monitor the alarm system remotely from the vessel's accommodation block. In the event of a machinery alarm the watchkeeper must be present in the machinery control room (MCR) and respond to the machinery alarm within 90 seconds of the alarm first sounding.

All modern ships are built with a sophisticated alarm and monitoring control system and they will all have the ability to run with UMS; however, not all vessels will be operated in this way. Passenger ships, for example, have 15,000 alarm points but they still have an engineering watchkeeper in the MCR at all times due to the reassuring message that this sends to the customers.

Cargo ships on ocean passage will be able to operate with the machinery space unmanned if the vessel has the appropriate certificate to do so. To get the approval the vessel must have the following essential requirements for operating with 'unattended machinery spaces' (UMS), that is, particularly for unmanned engine rooms during the night:

1. Bridge control of propulsion machinery. The bridge watchkeeper must be able to take emergency engine control action and the control and instrumentation on the bridge, for the engine, should be as simple as possible.

2. Centralised control and instruments are required in the machinery space. Engineers may be called to the machinery space in an emergency and controls must be easily reached and fully comprehensive at that central point (usually the Machinery Control Room – MCR).

3. Automatic fire detection system. Alarm and detection system must operate very rapidly. Numerous well-sited and quick response detectors (sensors) must be fitted (see Chapter 8 of this volume).

4. Fire extinguishing system. In addition to conventional hand extinguishers a control fire station remote from the machinery space is essential. The station

must provide control of emergency pumps, generators, valves, ventilators, extinguishing media, etc.

5. Alarm system. A comprehensive machinery alarm system must be provided for control and the alarms repeated in accommodation areas.

6. Automatic bilge high-level fluid alarms and pumping units. Sensing devices in bilges with alarms and hand or automatic pump cut-in devices must be provided.

7. Automatic start emergency generator. Such a generator is best connected to separate emergency bus bars. The primary function is to give protection from electrical blackout conditions.

8. Local hand control of essential machinery.

9. Automatic start of standby pumps (suitable alarm activated – slow-down of engine if system is a critical one (oil, cooling, exhaust or fuel)).

10. Adequate settling tank storage capacity.

11. Regular testing and maintenance of instrumentation, including engine emergency shutdowns, generator overload trips, bilge alarms, loan working alarms and fire alarms.

Bridge control where the vessel has an Internal Combustion (IC) engine

Ships are built with types of IC engines arranged in different configurations. These are the:

- large two-stroke, low speed, direct drive, reversing engine;
- smaller, four-stroke, more compact, medium speed, indirect drive constant speed engines;
- higher speed, four-stroke diesel generators driving electric propulsion motors.

To enable a direct drive, a reversing engine relies on its size to be able to generate the power required to drive the ship and as a consequence there is a significant amount of heat produced, which is not all changed into effective work. Therefore, the heat must be absorbed by the components of the engine, creating considerable 'thermal' stresses.

To protect the engine from being used outside of its design envelope and to be controlled from the bridge, basic procedures and safeguards must be built into the control system.

The following shows some of the checks that may be incorporated into the algorithms of the control programs to protect the engine during starting and running.

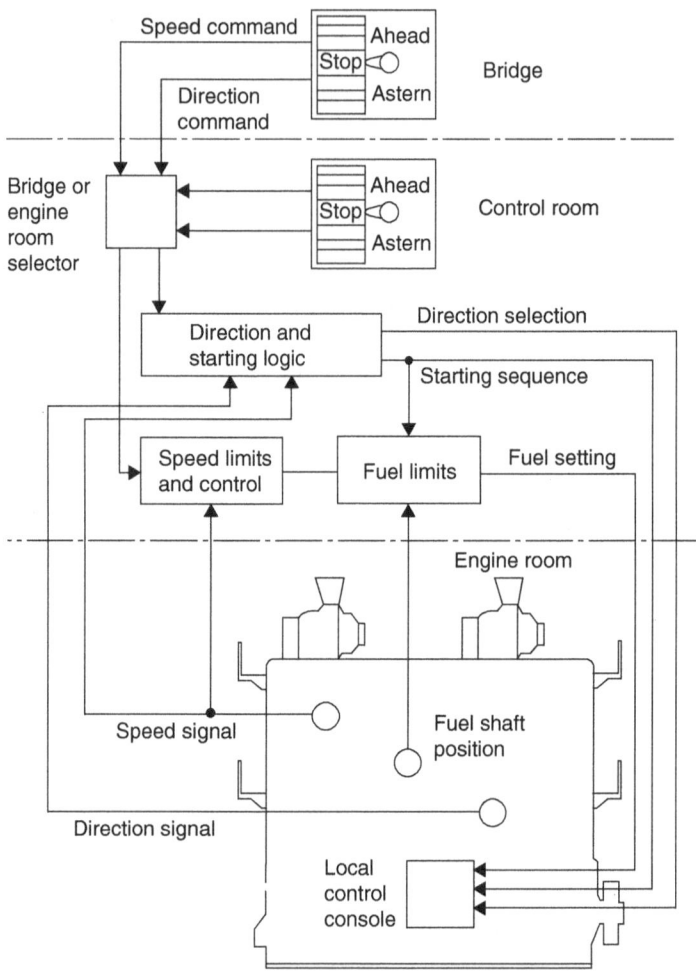

▲ **Figure 11.23** *Bridge control: block diagram*

1. Confirmation that the turning gear is disconnected – all engine types driving a propeller directly or through a gearbox (this is a major point of safety for both the staff and the machinery as considerable damage and danger would be present if the engine was able to start with the turning gear engaged).

2. Confirmation that the engine is running in the correct direction on air, before the fuel is applied – low speed, direct drive, reversible engine ONLY.

3. Confirmation that the engine is accelerating on fuel before the starting air is cut off – low speed, direct drive, reversible engine ONLY.

4. Alarm if a start is not confirmed within a reasonable period of time – low speed, direct drive, reversible engine ONLY.

5. Speed limitation, that is, avoidance of critical speed ranges or limits imposed by excessive jacket temperatures or vibration ranges – low speed, direct drive, reversible engine ONLY.

6. Acceleration limits – this will be controlling the rate at which the fuel is applied to give a safe torque or to prevent the fuel exceeding the air available from the turbo-blower or to prevent excessive thermal stressing of the engine.

7. Automatic rundown to half speed if, for example, the cylinder jacket temperature is too high.

8. Automatic stop if the lubricating oil pressure fails.

Other checks or alarms may be fitted if required. A typical system is shown in block diagram form in Figure 11.23.

The various signals may be electrical or pneumatic. Final connections to the engine are usually pneumatic cylinders that operate the engine controls. However, on the modern 'common rail' or electronically controlled engines the bridge control systems will be Programmable Logic Controller based and will be integrated with the engine software.

Selection of bridge or engine room control is made in the engine room in consultation with the bridge, thus enabling the engineer to take control at any time, if required. When an electronic governor is fitted, the speed signal is generated by a tacho-generator and the fuel quantity is measured by a position transducer fitted to the fuel control rack. When a hydraulic governor is fitted, speed is measured by a watt-type governor and hydraulically amplified. In this case, the blower and torque limits are usually incorporated into the governor. To protect the engine, fuel limits are applied after speed limits. For the auto-pilot steering control, see Chapter 5 in this volume.

With these systems in place it is vital that the engineering staff work closely with their bridge team counterparts. It would not be safe, for example, to suddenly change over from bridge control without informing the navigator, especially if altering the engine output would endanger the vessel itself.

12

MANAGEMENT AND LEADERSHIP

The Regulation of Shipping

It is very important that the modern engineer presenting himself/herself for examination has a working knowledge about the regulatory framework that is responsible for guiding international shipping.

International shipping transcends national boundaries and therefore national laws. The law relating to one nation is never exactly the same as the law that has jurisdiction over another. Therefore for sovereign states that wish to engage in international trade there must be some overarching organisation that has the authority to discuss the problems of international shipping and make binding rules for the nation's administration to follow.

The International Maritime Organisation (IMO) is a self-governing, autonomously funded department of the United Nations. It is headquartered in London and has the authority to provide the forum for governments to meet and decide on collective action to formulate the rules by which ship owners can then conduct their business (Figure 12.1).

The assembly of the IMO members carry out their work through the council and five main committees. All of the members (170 plus 3 associate members during 2013) participate in the assembly meetings and elect the 40 member states that will make up the council. All the flag state members participate in the work of the five main committees. Their work is assisted by the nine technical sub-committees.

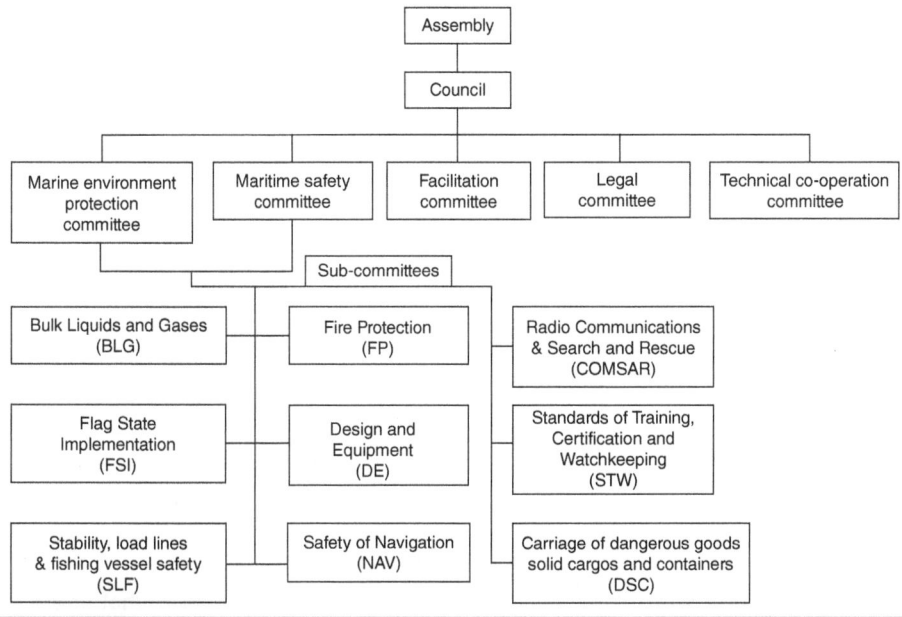

▲ **Figure 12.1** *International Maritime Organisation structure of IMO bodies*

As well as the governments that make up the full membership there are 62 non-government organisations with observer status and 78 non-government organisations that have consultative status. These organisations provide invaluable technical support to the member states.

Once the rules have been agreed on at IMO by a 'threshold' number of nations then each of the member governments must pass legislation bringing the new convention into their own domestic law. In the United Kingdom, the Maritime and Coastguard Agency is the government department that is responsible for representing the United Kingdom at IMO and is also tasked with implementing any new law into UK statute.

The MCA is the UK government's department responsible for maritime issues and therefore it is the custodian of the names and details of vessels registered to operate under the rules and administration of the United Kingdom.

Port State Control

Port State Control (PSC) is the inspectorate responsible for checking foreign ships in a nation's port. The action is to verify that the visiting vessel is complying with all the regulations set out by IMO. In the United Kingdom, the MCA is responsible for PSC inspections. Flag states that are geographically close have organised themselves into groups. The groups are bound by memorandums of understanding (MoUs). One example is the Paris MoU, which binds 27 different countries such as the United Kingdom, Greece and Norway. Some members are not in the EU such as Canada. Another MoU is the Tokyo MoU, which has 18 members including Japan and Australia.

M Notices

The Marine or M Notice system is the method by which the UK government has official contact with the owners, managers and operators of the ships under its administration. Other flag states will have similar systems. The Statutory Instrument or SI is the actual document of legislation, however, M Notices can still relay mandatory information. There are three different types of M Notices, which publicise to the shipping and fishing industries important safety, pollution prevention and other relevant information.

Merchant Shipping Notices (MSNs) convey mandatory information that must be followed under UK legislation. These MSNs relate to Statutory Instruments and contain the technical details of these regulations.

Marine Guidance Notes (MGNs) give detailed advice and guidance relating to:

- improving the safety of shipping;
- improving the safety of life at sea;
- preventing or minimising pollution from shipping.

Marine Information Notes (MINs) are intended for a more limited audience, for example, training establishments or equipment manufacturers. They may also contain information that will only be useful for a short period of time, such as timetables for MCA examinations. Within each series of M Notices suffixes are used to indicate whether documents relate to merchant ships and/or fishing vessels. The suffixes following the number are as follows:

    (M) for merchant ship

    (F) for fishing vessels

    (M+F) for both merchant ships and fishing vessels.

These are the most important IMO conventions for the marine engineer:

- Safety of Life at Sea (SOLAS).
- MARPOL (MARine POLlution convention) and its Annexures.

- Standards of Training, Certification and Watchkeeping for seafarers (STCW).
- Ballast Water Management convention (BWM).

The SOLAS convention was the first and remains the most important IMO convention. Set up in response to the *Titanic* disaster it not only lays out the minimum requirements for lifesaving equipment but also covers:

- ship construction'
- fire prevention and firefighting;
- radio communication;
- navigation aids;
- rules to prevent collision.

SOLAS 1974 was the last total rewrite of the convention; however, the rules have changed as the convention has grown by adding amendments and using the system of 'tacit acceptance procedure', which allows changes to be agreed upon more easily than under the older system in place before 1974.

MARPOL is a newer convention, adopted in 1973 in response this time to a series of accidents with tankers that caused considerable pollution. The convention is set out in a general section and six annexures. The general section covers definitions of the terms used and the types of ships. Of interest to the marine engineer preparing for examinations will be the annexures to the convention. These cover the following:

- Annexure I covers the regulations for the Prevention of Pollution by Oil (entered into force on 2 October 1983). It deals with both the prevention of pollution by oil from operational measures and accidental discharges.
- Annexure II contains Regulations for the Control of Pollution by Noxious Liquid Substances in Bulk (entered into force on 2 October 1983).
- Annexure III details Prevention of Pollution by Harmful Substances Carried by Sea in Packaged Form (entered into force on 1 July 1992). This provides the requirements for standards on packing, marking, labelling, documentation, stowage, quantity limitations, exceptions and notifications.

Within the regulations 'harmful substances' are those substances that are identified as marine pollutants in the International Maritime Dangerous Goods Code (IMDG Code).

- Annexure IV details Prevention of Pollution by Sewage from Ships (entered into force on 27 September 2003).

The discharge of sewage into the sea is prohibited, except when the ship has in operation an approved sewage treatment plant or when the ship is discharging comminuted and disinfected sewage (MSN1807) using an approved system at a distance of more than 3 nautical miles from the nearest land; sewage that is not comminuted or disinfected has to be discharged at a distance of more than 12 nautical miles from the nearest land.

- Annexure V contains Prevention of Pollution by Garbage from Ships (entered into force on 31 December 1988).
- In July 2011, IMO adopted extensive amendments to Annexure V, which prohibits the discharge of all garbage into the sea, except as provided otherwise, under specific circumstances.
- Annexure VI contains Prevention of Air Pollution from Ships (entered into force on 19 May 2005).

This convention is the latest and sets limits on sulphur oxide and nitrogen oxide emissions from ship exhausts and prohibits deliberate emissions of ozone-depleting substances; designated emission control areas set more stringent standards for SO_x, NO_x and particulate matter.

In 2011, IMO also adopted mandatory technical and operational energy efficiency measures that will significantly reduce the amount of greenhouse gas emissions from ships; these measures were included in Annexure VI.

STCW sets out minimum standards for training and competence of watchkeepers. The standards set out the content and extent of the training courses that make up the different levels of operational qualifications of the seafarers serving on board.

All seafarers should now be familiar with personal survival techniques and this is a requirement for all crew members on board ship. Everyone working in the engine room on board will need basic training before they can start in a junior capacity. Basic safety training now contains five parts:

1. Personal Survival Techniques.
2. Fire Prevention and Firefighting.
3. Elementary First Aid.
4. Personal Safety and Social Responsibilities (including Security) and Environmental Awareness.
5. Security Awareness.

Following on from the basic training, the engineering officers in charge on the watch will need to hold the Officer of the Watch (OOW) qualification. The Second Engineer (sometimes called the First Officer Engineering) will need the Second Engineering Officer qualification and the Chief Engineer will need to hold the Chief Engineering Officer qualification. Following the Manila Amendments there will also be the qualification of Able Seafarer (Engineering), which will be considered as the highest-ranking engine room rating on board the ship.

The STCW code is set out in two parts: part A is mandatory and part B is advisory. Clause 53 in part A of the code states: 'The officer in charge of the engineering watch is the Chief Engineer officer's representative and is primarily responsible, at all times, for the safe and efficient operation and upkeep of machinery affecting the safety of the ship and is responsible for the inspection, operation and testing, as required, of all machinery under the responsibility of the engineering watch.' Therefore, IMO is being quite specific in that the senior watchkeeping engineer is in *full* charge of the engine room even when the Chief Engineer is in the engine room unless the Chief Engineer makes it quite clear that he is taking over the watch and that everyone understands that change.

The 2010 (Manila Amendments) revision of the STCW convention has formally adopted Leadership and Management competence at both the management and operational levels for Unlimited Certificates of Competency and at management level for the limited certificates of competency. It has also clarified the guidelines for refresher training. Since the 1995 revision of STCW there has been the need for refreshing the basic safety training. However, flag states were allowed to accept on-board drills as satisfying the requirements for refresher training. Manila Amendments stated the activities (such as live fire practice) that cannot be carried out on board. Therefore a refresher course completed ashore will be required for the revalidation of the Certificates of Competency as OOW, Second Engineer and Chief Engineer. Engineers will also need to complete a high-voltage training course to be able to sail on vessels with high voltage on board.

Human element

Ships are designed, built and operated by humans and yet the focus seems to be always on the technical and financial side of the business and the common theme of the human element seems to take a back seat.

Most ships are owned and/or operated by the owner to give a return on his/her investment and the people can be viewed as a 'cost' or as an 'asset'. If the view taken is that the crew of a ship is a cost then the aim will be to reduce this 'cost' to the lowest value possible.

If, however, the crew are seen as an asset the aim will be to make this 'asset' as cost-effective as possible and this will mean investing in the crew to enhance their performance.

At the meeting in Manila, IMO recognised that some formal, human-element subjects were required for the education and training of officers, as part of the Certificate of Competency licencing structure for the sea-going officers.

The subject 'Human Element Leadership and Management' (HELM) is split into two areas: HELM (Operational) and HELM (Management).

Both are education and training subject areas that are related to management and operation of a ship. At the operational level the learning objectives relate more to the management of physical resources and the 'human' side of working in and operating a ship. This is completed as part of the Officer of the Watch (OOW) Certificate of Competency (CoC) and is aimed at enhancing the duties of the watchkeepers.

At the management level the focus is very much upon the management of people and the wider aspects of managing the vessel. The HELM Management is completed at the 'Chief Officer'/Second Engineering Officer level.

Time and again studies are finding that accidents happen as a result of people not acting correctly, usually in a combination of ways and not just due to a single factor.

Technical Issues that Require Careful Management

Ballast water

Ballast water has been used since the introduction of steel hulls as a method of stabilising a vessel, especially when the vessel has little or no cargo. Unfortunately, along with the water came everything that was living within it. The species travelled with the ship and were then discharged at the port of call where they were released into an environment that could be completely different from their origin. Some will not have survived but others have taken over to the detriment of the indigenous species. In some cases the invasive species have had a devastating effect on the local aquatic wildlife. This has led to the development of the Ballast Water Management Convention, which requires every ship to manage their ballast water and sediments to a predetermined standard according to their ship-specific ballast water management plan; all ships are required to carry a ballast water record book.

During 2017 the IMO's International Convention for the Control and Management of Ship's Ballast Water and Sediments (BWM) came into force. The convention is designed to slow down and stop the transfer of harmful aquatic organisms and pathogens via the ballast water that is carried in ships.

This problem has presented the industry with a significant technical challenge. Just the effective sampling of thousands of tons of water as it is discharged is in itself a problem and then there is the 'effective' treatment of the water.

Every vessel of 400 gt and above must have an 'approved' ship specific ballast water management plan and a ballast water management record book. The plan will act as:

- a guide for the on-board management team about complying with the regulations;
- information for the Port State Control Officer about the way that the vessel meets its obligations;
- a 'dual language' document including the working language of the ship as well as one of the following:
  - o English
  - o French
  - o Spanish
- an indication to the name of the ship's BTM Officer;
- a record of the training that has been completed by the on-board team.

Ballast water exchange was an early alternative to implementing one of the full Ballast Water Treatment systems. Generally the treatment process is made up of filtration/separation of the solid matter from the water and then 'treatment' of the solid matter.

The filtration/separation could be done by using fine-mesh mechanical filters or with a process using centrifugal force (hydrocyclone).

The treatment/disinfection process could be chemical, using for example chlorination or electrolysis; or physical, using one or a combination of processes that employs things such as ultraviolet, ultrasonic, heat or pressure to complete the process.

Classification of ships

It is taken for granted these days that ships are built to be strong enough to complete the tasks that are required of them. This was not always the case and in the past many ships have sunk because they were not built well enough. Modern ships, however, are built to a high standard, which ensures that the ship's crew, passengers and cargo are kept safe while the vessel is sailing across the ocean. It is up to the skill of the officers and ratings to ensure that the vessel is operated correctly. This means that the vessel must be built, maintained and operated to the required standards established by industry experts.

The rules and regulations – which determine the size and strength of the materials and machinery that go to make a ship strong enough to sail the ocean safely – are determined by the classification societies and to ensure that ships are kept in a good, safe condition they are checked at regular intervals. This is required to keep them 'in class' or in compliance with the flag state's requirements. The process of checking ships is called surveying and is carried out by marine surveyors. Ships are surveyed for a number of reasons and by a number of different organisations but the overarching principles are that they are inspected by an independent, technically competent, reliable and honest person or organisation.

Ships over 500 gt operating commercially will be required to conform to the Safety of Life at Sea (SOLAS) convention detailed by the International Maritime Organisation (IMO). As a result, these ships will need a survey from the flag state to ensure compliance. Approximately 80% of the survey work required for statutory purposes is delegated to non-governmental organisations that act as certifying authorities on behalf of the MCA. For surveys required by international conventions the certifying authorities must be 'recognised organisations', in accordance with EU Directive 94/57/EC. In the United Kingdom, the recognised organisations are authorised by written agreement, which follows the model developed by IMO Resolution A.739 (18). The MCA has written agreements with the other certifying authorities, which are available on the Master List of Documents.

The seven Class Societies that are recognised organisations authorised by the United Kingdom are:

- Lloyds Register (LR);
- American Bureau of Shipping (ABS);
- Bureau Veritas (BV);
- Det Norske Veritas (DNV);
- Germanischer Lloyd (GL);
- Registro Italiano Navale (RINa); and
- Nippon Kaiji Kyokai (Class NK).

Administrations might also require vessels to be registered, which fall outside the framework of the IMO conventions due to type, size or service limits – for example, superyachts or near-coastal vessels. This then allows them to comply with the conventions or other regulations and to be issued with a statement of compliance and/or a certificate of measurement on a case-by-case basis. Similarly, where there is no requirement by the administration, owners may require compliance to ensure an acceptable level of safety is provided on their vessels or to meet the requirements of third parties such as insurance companies.

The National Audit Office report 'Ship Surveys and Inspections', published in March 2001, contains the following definitions:

Surveys: May be undertaken by the agency's surveyors or by surveyors from the classification societies or other certifying authorities, such as the British Waterways Board. They cover specific items depending on the type of survey. Surveys require close examination of the construction and/or equipment or operations on board a ship to ensure that the requirements of the relevant regulations are complied with. Although it may not be practical to examine every component of a ship, it should be an examination of sufficient depth to ensure the vessel complies with each requirement.

Inspections: Usually unscheduled, are intended to check on vessels in between surveys and also on aspects that are not covered by the survey. They also cover vessels that are not subject to a mandatory survey regime, such as fishing vessels under 12 m. An inspection may look at the whole or specific parts of the vessel, its structure, equipment or operation. It gives a measure of the safety and pollution-prevention standard of the vessel concerned; the scope of the inspection is chiefly determined by the professional judgement of the surveyor. If problems are found in a general inspection, the surveyor may focus on areas in depth.

Generally, a survey differs from an inspection in the level of detail observed as well as the fact that a survey leads to the issue of some form of statutory certification.

Any survey is a report of the facts that are found within the remit of the survey undertaken. For example, a condition survey of a complete vessel might not include an accurate assessment of the hull if the vessel is in the water and was not inspected by divers.

The performance of ships is coming under closer scrutiny all the time and as the number of different checks by regulatory bodies increases, there is a greater need to work intelligently to reduce the workload in gathering the information required to satisfy the different needs of surveys.

The two ways to achieve substantial savings on the surveying of the hull and machinery is to have a comprehensive planned maintenance scheme that has the ability to alert the technical management to the survey requirements as well as to the ordinary maintenance requirements.

This means having a complete 'asset' register of all the equipment and machinery that is in need of maintenance or survey. On modern vessels this would mean a computerised system of some kind. A UK-registered vessel can then also adopt the MCA's Harmonised System of Survey and Certification or HSSC (for more details, access www.mcga.gov.uk/mca/msn1751.pdf).

The second action to be taken that will reduce the need for an outside surveyor is to have the Chief Engineer registered so that s/he can survey in a limited capacity on behalf of the classification society. The classification society such as Lloyd's Register can provide a list of the machinery that can be surveyed by the Chief Engineer and a list of the machinery that must be surveyed by class surveyors.

Surveys and inspections will be required in the following broad categories:

- hull and machinery surveys;
- flag state inspections;
- environmental survey;
- safety survey;
- ISM audit;
- cargo survey;
- seaworthiness survey;
- pre-purchase survey;
- condition survey.

All these surveys will be required by SOLAS-approved vessels at some stage during their life depending on the system that they are operated under and the length of ownership. In addition, depending upon the class of vessel, the following will also need to be issued as part of the flag state requirements:

- Passenger Certificate (PC);
- Cargo Ship Safety Radio Certificate (Radio);
- International Load line Certificate (Load Line);
- Cargo Ship Safety Equipment Certificate (SEC);
- Cargo Ship Safety Construction Certificate (SAFCON);
- International Certificate of Fitness for the Carriage of Liquefied Gases in Bulk (IGC/GC);
- International Certificate of Fitness for the Carriage of Dangerous Chemicals in Bulk (IBC/BCH);
- International Oil Pollution Prevention Certificate (IOPPC);
- International Oil Pollution Prevention Certificate for the Carriage of Noxious Liquid Substances in Bulk (MARPOL Annex II).

Harmonised System of Survey and Certification

The Harmonised System of Survey and Certification (HSSC) seeks to standardise the period of validity and the intervals between surveys for the nine main SOLAS-convention certificates to a maximum period of validly for all certificates, except a passenger ship safety certificate, to five years. In so doing it aims to simplify the survey and certification process. Following a period of transition, the streamlined format of the HSSC will bring benefits to the industry in terms of flexibility of survey schedule, reduced numbers of surveyors, survey time and paperwork, all therefore reducing costs.

Under the HSSC, there are seven types of survey:

1. Initial Survey.
2. Renewal Survey.
3. Periodical Survey.
4. Intermediate Survey.
5. Annual Survey.
6. Inspection of the outside of the ship's bottom.
7. Additional Survey.

Details of the HSSC can be found within MSN 1751 issued by the MCA. The special codes relating to chemical and gas vessels and Marpol 73/78 Annex. II apply to all ships, with no lower limit to size. Similarly, SOLAS applies to all passenger ships that carry more than 12 passengers engaged on international voyages. The 1972 Regulations for the Prevention of Collisions at Sea also applies to all vessels without a lower limit on the size of the vessel.

Leadership and management training

The Manila Amendments to the STCW code has added the requirement for leadership and management training to be included in the flag state examinations. One of the problems of having people appointed to a rank and given authority to command others just on their technical understanding is that they may not know much about leading the people under their command.

The military place a lot of emphasis on the leadership qualities of the officers and as a consequence there is a considerable amount of leadership training within a military officer's career. Traditionally the merchant service has not followed the same practice, placing more emphasis on practical and technical skills and less on personnel management and leadership skills.

A well-known writer of books about management, Peter Drucker, said that the difference between leadership and management is that management is about 'doing things right' and leadership is about 'doing the right things', and as we progress through this chapter it may be worth reflecting upon this statement to see if you agree with it.

Collaboration between individuals for a common objective, with division of labour under a recognised leader, has been practised for centuries. Social and organisational facets within the work environment were recognised at the beginning of the Industrial Revolution and have evolved in this century. However, application of the scientific method – observation, data collection, analysis, classification, hypothesis, experimental verification, formulation of laws and use for prediction – to the work situation is more recent. Such applications have resulted in a systematic approach leading to the recognised discipline of management.

Management is about putting into practice knowledge of the (five) processes of planning, organising, directing, co-ordinating and controlling. This relates to technical resources, human resources, materials, method and money. Management by objectives, with targets and accountability, through line management and staff functions is established practice.

Leadership is not necessarily about being 'The Boss'. People are promoted to positions of authority for many reasons, not always because they are good leaders. Leaders are made not by themselves or by their bosses, they are made by the people they lead. The manager of a team may be able to tell people what to do while they are at work but they don't necessarily accomplish the team's goals through leadership. Leadership is earned through the behaviour demonstrated to the followers.

Leadership

So much of the important messages and information contained within this chapter are fundamental to the successful operation of the ship. The lessons need to be practised by all engineering officers. Sympathetic leadership showing care and empathy for the team will result in an efficient and safe group of people who want to work for each other as well as for the company, vessel and Chief Engineer.

In the United Kingdom, the flag administration is working hard on research and leading discussion groups relating to the human element aspect of ship operation. They identify four key areas for ship's officers to work on:

1. Confidence and Authority.
2. Empathy and Understanding.

3. Motivation and Commitment.
4. Openness and Clarity.

The theory about development of leadership skills has changed, especially over the past 70–80 years. Prior to the 1940s it was accepted, in the United Kingdom that leadership was inherent from birth or instilled from an early age. This thinking even influenced the development of the UK education system up to the 1940s. Gradually, it became apparent that all manner of people from different backgrounds could become leaders and since then the approaches to the problems of leadership have fallen under one of three general headings:

- trait theories;
- style or situational theories; and
- contingency theories.

Each of these theories seem to contain elements of truth but in the long term each have failed individually to convince researchers of the difference between effective and ineffective leadership to be useful as a complete indicator of good leadership. However, these theories do give useful insights on leadership as a whole. Each one is briefly described below.

Trait theory

If we describe a person as 'punctual' we mean that he or she has a consistent tendency to be at a certain place on time when s/he has arranged or given an undertaking to be at that place and time. We do not change our mind dramatically if on the odd occasion that person is held up for some reason and is, therefore, not on time on that occasion. Therefore, we can see that a trait is a description of a regular situation response but it is not cast in stone, it is a description of what is most likely to happen. So the most important aspect here is that when we consider traits in people we do not mean that they will display those traits 100% of the time but that they will be predisposed towards the descriptions attributed to them.

The first studies into leadership concentrated on the study of a leader's personality hoping to find the 'magic' ingredients that go to make up a leader. Set against the background of a belief that leaders were born, researchers attempted to demonstrate that the following traits were the perceived qualities of a leader: height, responsibility, social skills, integrity, knowledge, popularity, physical powers, self-confidence, judgement, prestige, appearance, mood control, insight, co-operation, speech, socio-economic status, originality, initiative, intelligence, social activity, adaptability and dominance.

Possession of all these traits by one person, however, would be very unusual, if not impossible, and the following questions were not considered, or were overlooked, when considering leaders and/or their training:

- Can these qualities be identified?
- How do we measure them?
- Do effective leaders need all of these qualities?
- Can training develop the required qualities?

The biggest departure from traditional thinking came when the question was asked:

Are the 'core' traits the same for all situations?

There are, however, some interestingly important traits when considering leadership potential within people. These are:

- ambition (high levels of energy and a show of imagination);
- motivated to lead others;
- honesty and integrity;
- self-confidence;
- cognitive ability (ability to make sound judgements and show an analytical ability);
- technical knowledge – about the area of responsibility;
- emotional maturity.

Style theory

This is based upon the findings that different circumstances require different characteristics, approaches and interactions. Indeed a group, free from the constraints of having an appointed leader, may have different leaders emerge under different circumstances. Therefore, in a group or organisation that has an appointed leader, it follows that flexibility of response by the leader according to situation should be the leader's approach. This is the emphasis of the style or situational theory. The leader's style changes from the authoritarian/autocratic to achieve results at the one extreme, to the total delegation of responsibility and authority to the group in order to achieve results at the other extreme.

The basic problem with situational theory is that most people work within a narrow band of behaviour. Very few people can use a wide repertoire of behaviour and we normally have a preferred 'best set' of behaviours, which makes it difficult to change because the preferred 'best set' of behaviour is due to a person's personality, for example, some people are basically 'task' oriented and others 'people' oriented.

The identification of these problems led to the development of the contingency theory, which is centred on the approach that the appropriate behaviour is determined by the situation.

Four variables, however, stand out as interesting when considering leadership roles:

1. The personality cluster of the leader and his or her learned behaviour.
2. The predispositions, expectations, skills and personality clusters of the followers.
3. The organisation, its structure, function, tasks, and the situation it faces (leadership in a matrix organisation, for instance, might be different from that in a hierarchical structure, or the leadership required at the growth stage is different from that of a decline stage in the organisation or group).
4. The milieu (environment and social surroundings), that is the climate, ethos and values expressed within the institution.

Contingency theory

This theory centres not only on leadership style but also upon the situation in which the leader finds himself or herself. The site or situation is referred to as the structure. Examples of a highly structured site would be a production line and an unstructured site might be the research and development (new ideas) laboratory.

Charles Handy (1974 and 1993) has his 'best fit' approach in which the four areas of leader personality, subordinates, task and environment are brought together in the most effective and efficient way. Given that all four are held within a narrow band of established characteristics he describes how a given task is best accomplished by the four variables coming together in the most efficient manner.

He goes on to suggest different factors that would influence the structured or unstructured nature of each of the four areas, such as the leader's value system and/or his or her confidence in subordinates or the subordinates' estimate of their own confidence.

This highlights the complex nature of the role of the leader. He has to assess factors such as his own value system, to what degree he or she is prepared to involve subordinates in the planning, decision-making and controlling processes of running a task/business.

Qualities of a leader	Method of achievement
Creative insight	Asking the right questions
Sensitivity	Empathising with the other person's position
Vision	About creating the future
Versatility	Anticipating change
Focus	Concentrating on the key decisions without distraction
Patience	Looking for long-term rather than short-term gains

He or she must also assess the subordinates' preferred styles of leadership and how this 'fits' in with his or her own style.

The efficient leader must assess the movement towards the most suitable styles of both themselves and that of their subordinates and how this is related to the task and the environment. How many times have we heard, for instance, that an interview panel is looking not for the most qualified person but for the person who will 'fit' in with an established team.

Leadership should be to do with lifting vision, raising performance and building personality, motivating the group to accomplish tasks it would not otherwise accomplish.

The broad qualities needed to carry out this feat and the method of achievement can be summarised as below:

Leadership must be seen in terms of the performance of the group that is being led. The leader must accept responsibility for the interaction of the group's members and for the ability of the group to complete the task that it is required to undertake. The leader who automatically thinks 'we the group' and not 'I the leader' will have the advantage of being part of the group and not standing outside. The most effective leaders seem to be the people who concentrate on the members of the group more than on the tasks that have to be completed.

The leader should be building a structure that can handle the tasks required and if the leader is removed for a short time, the operation should carry on as if nothing had changed. The worst situation is if the system collapses because the leader is removed for a short period. This does not mean that the leader becomes redundant, on the contrary, the leader should be concentrating on the overview of the group's activities or on the next task still to be undertaken and not on the day-to-day details of individuals' activities.

The success of the leader can be judged by the success of his or her team and how well the team can:

- cope with conflict;
- communicate with each other;
- understand its goals.

All these functions of the team are a reflection of the leader's ability to have the team working together towards a common aim. It must be remembered that a team working well together will perform better than both individuals and teams that are not working together.

The art of delegation

Humans work together in teams, groups and organisations to accomplish all manner of activities. The team will accomplish more than individuals if they all work for the common goal and are directed in the appropriate way. When the size of the task or tasks becomes too much for one person, the workload and responsibility has to be distributed in some way. Delegation is more than a useful tool, it is essential to the success of an organisation.

With the development of flatter 'matrix'-style management structures it is now more common to delegate authority to employee teams than it has been in the past. This can lead to employees having more input into the management and running of an organisation by virtue of taking responsibility for delegated tasks.

The manager must choose carefully the person who is to have the work or task delegated to them, as the manager is actually delegating part of his or her responsibility to subordinates/team members. However, the overall responsibility for the wider project/business remains with the manager. Therefore, the manager must be able to trust their team. There must be clear guidelines for each party as to the nature and extent of the task. Discussions should take place beforehand to ensure that the person taking on the task is conversant with the task and has the necessary qualifications and experience to undertake such a project. If this is not the case, some staff development might be required.

Delegation with empowerment is vital. It is important that you delegate not only the work but also the authority to get the work done. Try also to delegate the good and the bad. Do not just push your dirty work down to subordinates for they will soon become fed up and demoralised.

I have given tasks and sometimes whole projects to colleagues; they have the day-to-day responsibility and they use their own resourcefulness to overcome difficulties. However, there should also be some perks that go with the work. For example, they might be given the freedom to set their own meeting structure, and determine their own agendas and timescales within overall constraints. This work might be away from their normal place of work, which will require resources if the staff are to visit customers on their own site.

Do not delegate and then overrule. Give your staff the chance to progress and flourish. They will not do this is if they think they are to be taken off the job for the smallest mistake. One of the most demotivating situations is where a subordinate is asked to carry out a job or task and is then chastised because it is not done the same way as the supervisor would have completed the task, even though the end result is satisfactory. There is nothing more demoralising than receiving negative comments because your way of working is different. A manager/leader can certainly point out different ways of operation but this should come after the praise for completing a successful job.

Where organisations have to respond to customer's needs and react rapidly, then delegation of appropriate responsibility and authority to people that are nearer the day-to-day operation of the business is highly beneficial. Decisions can then be taken when and where they matter most without reference to more senior managers who will undoubtedly be further from the operation.

Motivating factors

Motivation is a term that we hear often but information about motivating factors is generally not incorporated into formal training courses. The word is associated with human behaviour, meaning, conscious thought that moves us to action. It is one of those characteristics of life that seems to fit the old adage:

'I know it when I see it.'

As we move towards more matrix styles of management and the use of employee teams continues to grow, one question that is taking on great importance is how to keep the team motivated over the longer term. What is the nature of teams that seem to sustain high levels of motivation?

First let us take a look at some of the things that motivate individuals:

- What makes us do anything?
- Why did we join the maritime industry?
- Why did some people choose navigation and others engineering?

Within our daily lives each day brings with it an endless list of decisions to be made. The process of making those decisions is generally driven by the hope of a benefit or the fear of a consequence. For example, many of us enjoy chocolate but we pay for the benefit of enjoying the taste. Therefore, it is prudent to limit our intake for fear of the consequences of too much sugar and fat in an individual's diet, or the thought of the extra time on the treadmill to burn the calories.

Psychologists have taken this further by defining these consequences as needs. Our needs for sustenance, safety, security, belonging, recognition and a sense of growth and achievement become strong drivers (motivators) of behaviour.

So in an attempt to explain motivation, we need to appreciate the subtleties that exist in human behaviour, and focus our attention on general principles of motivation that have wider application. At least if we can understand some of these principles, we might be better prepared as managers or team leaders to lead a highly professional team in the long term.

Motivation can be intrinsic or extrinsic to the individuals concerned. Intrinsic motivational factors have been familiar as common sense long before being confirmed by more precise investigation by psychologists. The intrinsic motivational factors include the desire to find out things and explore the environment, the desire to manipulate things, the sense of competence and the achievement motive. If these motivational factors can be channelled in the right direction, they can be very powerful in helping people to perform to the best of their ability for the good of the organisation.

We would not be able to introduce the concept of motivation without a mention of the work of important researchers in this field. The most famous is of course Maslow. Maslow's studies into motivation have influenced the theories of management development for years partly due to the identification of the higher-level needs of people, which will have an influence on their performance in the workplace.

Maslow's hierarchy of needs can be summarised as follows:

1. Physiological needs – food, sex, sleep.
2. Safety needs – the need for an environment relatively free from threats.
3. Love needs – the needs relating to affection and relationships within a group.
4. Esteem needs – self-respect, self-esteem.
5. Self-actualisation needs – the need for self-fulfilment.

A committed and highly motivated workforce might show characteristics such as:

- loyalty;
- conscientiousness;

- responsibility; and
- reliability.

This ideal can only be approached if the workforce has a sense of identification with the organisation and a feeling of ownership of the decision-making process within the company. It has been argued in the past that the rising strength of Japanese industrial organisations was due in part to the mutual trust and loyalty between management and employees, as well as decision-making being shared at all levels. It is much better to be open with staff and involve them in the making of decisions by asking their opinion from the start of the process.

The most common extrinsic motivator is financial, increased pay through a bonus scheme or the incentive of earning commission. Some companies offer preferential share options to employees. This has the effect of linking the profitability of the company with the efforts of individuals' work practices. Other extrinsic motivators could be enhanced status within an organisation or enhanced status outside the organisation and within the wider industry or community. Pleasant working conditions and other fringe benefits are also examples of motivational factors that can be used as incentives in return for commitment to work from the staff.

Being part of a team may be voluntary. For example, when you decide to take up a new position within a company you will invariably be expected to become part of a team working for the common goals of the organisation. Sometimes participation in a team will not be open to discussion. If, for example, a ship's Superintendent or Chief Engineer is to be the owner's representative at the dry docking of a ship, or at the new build stage, then he or she will become part of the team that has responsibility to work toward the successful completion of the work, to the correct timescale and within the budget constraints. Whether being part of the team is voluntary or not, successful performance will depend upon many different factors. Studies have, however, come up with the result that 65% of employees would rather have a 'better boss' than more pay.

In the first instance becoming or intending to become part of a team will inevitably give rise to questions such as these:

- What is the purpose of the team?
- Is it a topic that interests me?
- Who will be on the team with me?
- What kind of authority will we have?
- Is it important to management?
- What is the reward for participating?

- What is the risk (perceived as punishment) for not participating?
- How long will it run?
- Will I be better off as a result of my participation?

Important aspects of team motivation

Earlier in this chapter, we outlined the importance of the need to give clear direction for the team to focus upon. This is not only a need for the team to be able to complete the tasks before them; it is also necessary to motivate the team members to perform well. Studies into team behaviour and motivation put a clear purpose, focus, or mission at the top of almost everyone's list. However, for the longer term, people must ensure that the purpose or mission of the team or organisation aligns with their personal wants and needs. When a team is brought together for a specific task, if the mission is clear and the subject of sufficient importance, then he or she might be able to sustain motivation for the duration. However, if it is a topic that is not in line with the person's wants and needs, motivation may diminish as time goes by.

One strategy with a lethargic team might be to stop the process, revisit the team's purpose or mission and see if there is alignment on it. Even with a team that seems well-motivated, it still is a good strategy to recheck once in a while.

The second important aspect of team motivation is for the members to feel stretched. Many people will say that their most rewarding team experiences resulted from some sort of challenge. However, in the workplace, these challenges occur infrequently and teams are not presented with stimulating challenges every day. So the question for the team leader becomes how to provide challenges for the team at frequent intervals.

If a challenge has a motivating effect, then there are caveats. If the challenge is too difficult, perhaps perceived as impossible, then team members may give up before they start. However, the same result may occur if the members perceive the challenge as too easy. Little energy is required to accomplish something so easily obtained.

So, for ongoing teams, periodic stimulation in the form of a worthy challenge is a method of maintaining motivation.

A shipping company building a team of managers to oversee the successful commercial operation and technical maintenance of its ships will require more than just attending to the professional requirements of the task. Building interpersonal relationships is just as important. Teams that get on well and like each other consistently perform better than those teams that do not have good interpersonal relationships.

There are times when companies have appointed staff to a particular post more for the fact that they will 'fit in' better with the existing team than for their technical qualifications and skills. This is one reason why industry sometimes recruits graduates that have a degree in an unrelated subject to the job that they are expected to carry out. The company wishes an employee educated to a certain standard that they can mould into the corporate way.

It follows that if members of a team genuinely like each other and there is a spirit of camaraderie, meaning comradeship, fellowship and loyalty, then the people on these teams work hard to develop and maintain their relationships. They carry out activities instinctively that lead to effective teamwork, which is open and direct communication, frequent praising of each others' contributions and mutual support.

If you have an existing team that needs building but do not have the camaraderie that you would like then you will have to arrange a better understanding between the team members. For most of the time, our like or dislike of someone is related, in the main, to how well we understand them. Since our formal training has not addressed this, most of us enter adulthood ill-equipped to deal with the myriad personalities, temperaments, cultures, values, beliefs, ideologies, religions and idiosyncratic behaviours of those we meet and work with.

Do not overlook simple solutions to this problem such as designing an off-site activity for the team. This is the reason for the increase in corporate adventure weekend breaks, but sometimes it is beneficial just to have strategic planning days away from the office as a way of building camaraderie.

As stated earlier in this chapter, people are motivated by having responsibility for carrying out and completing a task. Having ownership of an identifiable block of work is also recognised as providing motivation for teams. Implied in this concept is the understanding that the responsibility comes along with authority to make the necessary changes. Teams that have both the responsibility and authority tend to maintain motivation over longer periods of time.

Responsibility can be demotivating if the consequences of error or failure are too great. If the organisation, for example, has a history of punishing mistakes, then the giving of responsibility is viewed more as a negative. The short-term performance may be good (remember fear is only a short-term motivator), but long-term motivation will suffer. It is difficult to sustain high performance when energy is being sapped by fear.

Personal and team growth can provide another basis for long-term motivation. When people feel they are moving forward, learning new concepts, adding to their skill base and stretching their minds, motivation tends to remain high. Personal growth adds value to the individual, enhancing self-esteem and self-worth. Appraisal systems are a good opportunity to focus on the development of individuals in the context of the organisation's own goals and strategic plans.

Objectives or goals

Some motivational theorists suggest that staff are motivated by working towards goals or objectives and that this, in itself, is a motivational condition. It is certainly true that the best way to tackle large projects is to work towards smaller objectives. However, if staff have too many objectives or work to carry out this will be a big demotivator. Motivation is also enhanced when individuals received feedback on their performance but always remember 'Politeness oils the wheels of conflict' and disagreements are so much easier to resolve if people are polite to each other. The objective approach can have a detrimental effect on business especially if the objectives are imposed instead of being negotiated.

It can be seen from this brief description of motivation that the appraisal system, if used well, can be such an important tool for the manager in the motivation of his or her staff.

Communicating with staff

Meetings can be one of the most time-consuming and costly activities that an organisation can undertake. If you add up the hourly rate of the people involved then the cost of meetings become apparent. It is up to individual managers to ensure that the meetings within their control are necessary, focused and to the point.

However, most organisations need meetings. They are good for communication between teams, making group decisions and are useful for pooling knowledge and skills. However, if they are not focused or the meetings are too frequent then frustration, tension, hostility and conflict can be the result.

This is particularly true in a modern company where time is of the essence and people need to be focused or they will not complete the work for which they are responsible.

The composition of the meeting is also critical. If people are requested to sit through meetings that are not relevant to them, then a negative effect will be the result.

So how can we ensure that meetings are an asset to the company instead of being a liability?

The first thing to consider is why meetings are needed and is it necessary to hold a meeting under the circumstances given. Meetings are usually required to solve a problem or accomplish a goal or purpose by bringing together a group of people who have the relevant mix of knowledge and skills to bring about a successful outcome.

Given this purpose, the priority should be to choose the composition and set the time and venue to ensure that the members can attend. They should be informed well in advance and there really should be a good reason for calling a meeting at short notice if it is not just going to be a waste of time.

The agenda is the primary source of information and this together with any supporting documentation should be circulated within a timescale that allows the members to read and digest the information. It is important to allow time for people to gather any additional information that may be required as a consequence of seeing the agenda and any supporting documentation.

The job of the chair is most important if the meeting is to run smoothly and on time. The chair must be firm and even if necessary set the end time for the meeting and not allow it to overrun.

The area most likely to cause problems is 'Any Other Business' (AOB) usually placed at the end of the meeting. This has the effect of being a dumping ground or a talking shop for all the different issues that may or may not be relevant.

One of the ways to limit the effect of over running due to AOB is to take the items that are proposed under AOB and set them as agenda items for the next meeting. Another way to focus the activity of the group is to have items on the agenda that are for information only and not to allow any discussion under this section. The next section can then be limited and discussion only takes place on these items.

Communicating with others, especially in multicultural situations

The formal forms of communication in businesses and organisations can be divided in three ways:

1. First is the mechanics of communication – memos, telephones, e-mails and other computer systems such as intranets.
2. Second is the interpersonal skill of communication, such as giving instructions, conducting interviews and holding meetings.

3. Third is the organisation of the lines of communication, management structures, committee structures and project teams.

There is also the informal system of communication – the so-called jungle/galley telegraph where rumours spread quickly. It is very important that if there is information to give to your team that it is you that tells them. It is very demoralising for members of a team to hear important information from sources other than their leader.

One of the worst habits of leaders is to receive information and then assume that everyone else knows the same information and to the same level. Another important factor when setting up lines of information is to have a positive reporting system. For example:

- 'you must report project progress at regular intervals'; and not
- 'you will report when you have made progress'.

The maritime industry knows to its cost the importance of positive reporting but individuals will also have to decide which form of communication is preferable in different circumstances. Writing things down will have the added bonus of accuracy whereas picking up the phone and having a brief conversation can be more beneficial since it allows interpretation of meaning from the voice tones that are not present in the written word, especially e-mail.

Managing cultural differences has always featured heavily in the successful operation of ships. Seafarers have learnt to accommodate the needs of others and tolerate peoples of different nationalities but there are two important aspects to prepare for in the modern setting:

1. The first is that during an emergency ship's staff will revert to their native language and if this is not the same as the other crew members, then there will be a loss of communication
2. The second issue is that some cultures will not question the boss even if they know that the boss is wrong. Furthermore, staff will also say 'yes I understand' when in fact they do not.

Ship Management

Until the 1980s the common model of ship management was for the technical, operational and human resource (HR) management to be carried out by people working directly for the ship owner. The vast majority of the staff were ex-seafarers who had taken a job ashore. This meant that the staff ashore had a very good working knowledge of the ship's operational requirements.

As the ships moved under the ownership of the banks and the accountants started looking for ways to reduce costs, there was an opportunity to remove liabilities from balance sheets by outsourcing the staffing. Ship management companies were the result. Some just undertook to supply the ship's staff while others provided the technical back-up as well.

Today, the picture is mixed as some companies, especially oil companies, have taken back the responsibility for providing the staff due to their concerns about maintaining quality, loyalty and general all-round standards. The shipping industry has found to its cost that it might be able to delegate service but it can't delegate responsibility. An oil company where one of its vessels pollutes the ocean or shore line will still find its share price reduced and will still receive the fine despite the ship being managed by a third-party company.

Having said that, the rise of ship management companies has been a feature of the industry that has produced some very good professional companies. Another feature of the industry in recent times has been the move away from 'nationalised' crew towards 'international' crews. There are now very few ships where both the officers and ratings are from one country. It is usual to source unskilled and semi-skilled labour from the cheapest 'global' area, which tends to be the country that has the lowest cost of living at the time of need.

Company Management Structures

The traditional organisational structure of a typical shipping company's management used to be established around the hierarchical 'tree'-type divisional format, with the operational managers working independently of HR or the accounts. The lines of communication were via line managers and then across divisions via the management.

The modern company set-up may be shown in a similar way on paper but a 'networking' style of operation is now encouraged with the lines of communication as shown in Figure 12.2.

Individual ships in the fleet can be set up as an independent unit with special management conditions and recently with the advances in modern communications, the ship can, if needed, keep in constant touch with the shore staff. This has led to a change in the way that ships are managed and again this change is being approached in different ways.

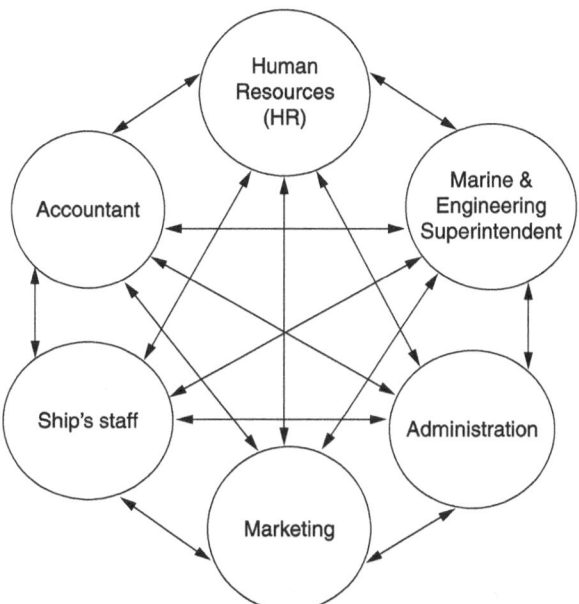

▲ **Figure 12.2** *Lines of communication in a modern company*

Some companies emphasise high-level expertise in their shore staff and only employ staff on board who have the minimum qualifications and experience determined by international agreements. Other companies have a different approach and use the education and training systems to enhance expertise on board. This may be a more expensive short-term approach, although, companies with this approach have found a reduction in 'total life cycle' costs. This is especially true where the move to condition-based monitoring/maintenance is concerned.

The increased automation, decreased crew sizes and the trend to cost consciousness adds to the pressure on all staff. Planning is still usually the responsibility of a small team of senior ship's officers, in liaison with company management, which requires the application of:

- planned maintenance systems;
- replacement policies;
- financial considerations.

There is now slightly less emphasis on departmental segregation and much more emphasis on co-operation between all the staff on board liaising closely with the staff based ashore. The Chief Engineering Officer now tends to take responsibility for all the technical equipment on board independent of where it is situated (on deck or in the engine room).

Responsibility for maintenance of the machinery plant lies ultimately with the Chief Engineer officer although a significant part is delegated to the Second Engineer officer and certain other sub-departmental heads. The Chief Engineering Officer must ensure that links with the head office, between departments on ship and within his own department are effective and maintained. The Second Engineer Officer has an important role in delegation of engine room duties and responsibilities, control and recording of spares, maintenance schedules, etc. and, not least, effective personnel relations. The responsibility for efficient operation, firm leadership and good communications is an inherent requirement for all ship's officers. The 'networking' management system is used extensively, using committee structures, involving key officers and ratings, so as to improve decision-making operations. This also allows more effective communications as the central 'hub' can relate closely to the peripheral 'wheel' with signal transmission via 'spokes' as well as around the 'rim'. 'On the job' training, to clearly defined objectives, has assumed an increased importance on-board ship.

Machinery Maintenance

Introduction to maintenance

The advancement and better understanding of material science is improving the reliability of machinery. This is particularly true within the marine industry where the quest for saving fuel, improving energy efficiency and reducing the environmental impact is driving the research and development within the sector. However, these improvements will only be realised if the machinery is operated and maintained correctly over its entire life cycle.

Engine manufacturers are working to increase the time between overhauls (TBO) but this will only happen if the supporting components are also kept in good condition, for example, the piston and piston rings will only remain in good condition if the fuel injection equipment is kept in good condition and on a large two-stroke engine the good condition of the piston rings will also depend very much on the correct operation of the cylinder lubricating oil system.

Breakdown maintenance or 'run to failure' is a perfectly legitimate maintenance strategy if used in the correct way. It is not correct to apply this method to all machinery or to machinery that is critical to the operation of the vessel. However, it may be cost-effective to allow a small transfer pump to run until it is showing visible signs of failure and then replace the whole pump.

Ships form a vital link in global transport systems and any failure in this link will have a substantial effect on the supply chain, customer costs and time schedules, which will continue after the technical problems have been resolved. Unplanned loss of critical machinery will not only cause immediate financial damage, but it also erodes customer confidence, extending the effects of that damage far beyond the actual event.

The size and complexity of engineering management within the industry is infinitely variable, ranging from that required for a small coastal vessel to the large management structures found on a complex state-of-the-art cruise or transatlantic passenger ships. The successful management of this highly complex, 24/7 operation requires a fully informed, proactive and co-ordinated management team, with the fullest possible understanding of their own and other team members' objectives within a common framework.

In the past it was fully acceptable to operate the manual, paper-based, maintenance systems in such a way that it did not impede the core business of carrying cargos around the world. The ship's schedules would allow sufficient time for maintenance to be carried out in accordance with the manufacturer's recommendations and the staff on board were sufficiently experienced to be able to complete any work necessary to keep the vessel sailing. Sometimes the equipment would be over maintained because time constraints were not an issue it was better to have the machinery in top condition than to risk a breakdown at the worst possible moment.

However, with the introduction of containerisation the efficiency gain in carrying cargo was enormous, mostly at the expense of time in port or non-operating time. This operational requirement placed totally different constraints on the maintenance of ship's machinery. Other factors were also creeping into the system, such as the lowering of the experiential time in the qualification structure of engineering officers, increases in automation and a reduction in the number of staff carried on board. We now hear of phrases such as 'maintenance-induced failure' and 'increases in main engine failure' 'main engine failed to run astern'. All of these changes have meant that the industry has had to embrace a new way of looking at the maintenance of ship's machinery.

Planned maintenance and stock control are an essential part of that process and increasingly technical departments are embracing the lessons learnt from other branches of engineering and incorporating the ideas of condition-based maintenance.

Planned maintenance

One function of higher-level strategic management is about creating policy and controlling future investment and direction. Today's requirement for accountability in safety matters and where only short periods of inattention can cause major financial loss means that senior managers must also have access to current 'real-time' information at all levels.

It is not acceptable for the maintenance department to be continually involved with breakdown maintenance. The worst situation is for the department to be staggering from repair to repair and not having enough time to be proactive and carry out maintenance that will help to prevent breakdowns from happening.

All the engineers on board must become involved with helping the company's technical services department in their quest to reduce the down time of equipment and help to eliminate unscheduled breakdown of machinery. In today's climate of corporate accountability, it is also vital that senior management have an appreciative understanding of the abilities of the staff within the ships under their jurisdiction and are able to exercise a degree of informed real-time control over the whole operation.

To achieve an overall picture, an effective recording and identification system will need to be put in place. On modern ships this 'asset register' will be computerised and the efficient systems such as the Amos (maintenance and procurement) will assist with the stock control and also link to the planned maintenance systems.

Traditional maintenance systems have evolved from breakdown maintenance to time-based planned maintenance systems. These systems could be calendar based, running hours based, or a combination of the two. However, it is vital that the two are not mixed in such a way that could promote a machinery failure.

For example, it is so important to keep a record or the running hours of machinery. It is also important that a running total is kept for each piece of equipment. Often these days this function is not carried out. This could so easily lead to the unexpected failure of machinery literally because the staff do not know when machinery should be maintained. If the running hours are not kept and the maintenance is carried out on a calendar basis then machinery could be over or undermaintained and sometimes dangerously so.

Condition-based maintenance

The condition-based maintenance (CBM) system is gaining favour as a credible alternative to the traditional systems of maintenance. With CBM it is essential to install a system to provide comprehensive real-time information that will allow all levels of management to monitor performance in relation to the optimum requirement, and to constantly ensure the availability of equipment in relation to the maintenance requirement.

Manufacturers have to ensure that their recommendations for maintenance intervals are set for a worst-case scenario. Therefore, naturally if the operating conditions were light then the maintenance interval could be extended. Operators would be reluctant to do this, as it would leave them open to the criticism of not completing the maintenance correctly.

If condition-based maintenance systems are adopted then a whole new set of circumstances will exist. For example, an operator could claim that the machinery was running well just before a failure, in an attempt to justify an insurance claim. In the past the maintenance would have been completed at set time intervals or based on the machinery's operating hours. Under a condition-based system much more rigorous testing and monitoring systems must be employed to justify extending the maintenance periods.

It must also be remembered that maintenance periods might need to be shortened if the machinery is subjected to adverse conditions for a period of time and therefore it is vital that the marine engineering officer pays attention to the maintenance system that is used aboard any vessel that they are to work on.

Companies such as WÄRTSILÄ and MAN Turbo & Diesel are looking to take the system further and have all the data about the operation of one of their engines sent back to them on a regular basis. In this way they can ensure correct operation and also advise about the maintenance required.

IACS produce a very good 'Guide to Managing Maintenance' that gives a clear statement about the objectives of maintenance, the link to International Safety Management (ISM), class and flag state. It makes sense to collect data about the machinery and place it into one system. The system should then produce the reports to satisfy the following requirements:

1. Safe operation of the machinery.
2. ISM system.
3. Classification society inspection schedules.

The maintenance system is set up initially around the maintenance schedule recommended by the manufacturers. The intervals can be modified by the following condition monitoring techniques such as:

- lubricating oil analysis;
- vibration analysis.

Condition-based monitoring, when applied to planned maintenance is sometimes referred to as 'predictive maintenance' and can be used for the purposes of initiating maintenance procedures that are intimately linked to 'performance monitoring'.

Condition-based monitoring means any measurable condition that may be used as part of a maintenance routine. However, there are practical limitations. For a condition-based system to work, the condition being monitored must be convenient to read with repeatable results being achieved. For example, it might seem to be a good idea to measure the surface roughness of a main bearing journal and use the result as a means of regulating the planned maintenance on that bearing. However, at the moment, given the current techniques available, the bearing would have to be dismantled before the reading of roughness could be made, thus making 'roughness' an impracticable 'real-time' condition upon which to make the decision about whether maintenance is required.

However, it may be possible to embed a small wire in the material of the shell bearing below the surface, so that when and if the bearing material wears away then the wire surface is worn and an open circuit made that triggers an alarm. This is similar to the warning that is built into the braking systems of motor vehicles. The MAN system shown in Volume 12 of the Reeds series involves a wear sensor fitted to the two-stroke engine that measures the distance to the cross-head slide, which gives an indication of wear in the running bearings associated with that sensor.

Condition-based monitoring can apply to items such as filters. If the condition is poor, a maintenance routine is executed. However, this is also difficult to quantify so as to be useful in everyday situations. Engineers would normally say that a filter with debris in it means that the associated machinery needs some sort of examination. How do we judge when the filter is in a 'poor' condition so that a maintenance routine is carried out? Do we extract and weigh the debris? Can we separate out rubber debris from any other material? Or do we just measure the differential pressure across the filter.

Usually, condition-based monitoring is a term that is applied to monitoring in real time, that is, some property is monitored as the machine is operating or running. This is what the operator would ideally want. A condition-monitoring record can also be used to present to class surveyors. If experience with the equipment is positive, a good record can be used to show that a particular item of machinery (e.g. pumps) is in good order and does not need to be dismantled for the survey.

This approach can have considerable savings in terms of spares and down time, as well as a saving in terms of the human resource necessary to carry out the stripping down and re-assembly. Some manufacturers of machinery will offer such systems as bolt-on enhancements.

A very dramatic influence in this area of monitoring will be wireless technology. For example, at present it is difficult to monitor accurately the temperature of the bearings in a diesel engine due to the need to transmit a reading from the moving bearing to the static casing of the engine where it can be read or sent to a remote gauge. Some efforts have been made to measure the temperature of the oil as it sprays from the bearing but by using wireless technology the manufacturers will be able to send the bearing temperature information to a remote pick-up via a wireless signal.

One of the world leaders in this technology is Parker Kittiwake. They offer a range of real-time, on-line technologies such as the following.

Acoustic emission

Acoustic emission (AE) monitors the very high frequency shocks from a bearing arc measured by a sensor located as close to the monitored surface as possible.

Microscopic impacts from the moving machinery become more frequent as the equipment deteriorates over time due to increased wear and tear and an electronic circuit is used to filter out the background machinery 'noise' and translates the shock frequency into a meter reading. This reading can be measured either by a hand-held unit, or by means of a hard-wired continuous monitoring installation where comparisons can be made with previous readings and automatic alarm systems incorporated for use on modern ships with sophisticated control systems.

One of the disadvantages of the system is that the detector is very sensitive and can be prone to spurious signals (loose pipe clamps being one source). This is particularly important if used in on-line systems where alarm functions are included. Therefore, interpretation must be carried out with care, and a visual inspection is necessary before any action is taken.

Vibration/noise analysis

Vibration/noise analysis is similar to acoustic emission analysis but here the vibration of the machinery (as a whole) is measured and monitored, as opposed to filtering out some of the unwanted 'noise'. On-line monitoring and alarm conditions are available as with the acoustic emission analysis.

With a hand-held device as opposed to a permanent sensor, the position for taking the reading is indicated by a distinctive marking that ensures that the reader is applied as close as possible to the same point on the machine each time. However, it is difficult to replace the instrument precisely in the same position for every reading; there is a consequent loss of accuracy when comparing the trends in the readings.

The analysis does require sophisticated software but the harmonic vibrations set up with engines running mean that care has to be taken about when the readings are taken. Advancements in this area are taking place all the time; for example, it is now possible for measurements to be made with accelerometers, which gives a two-dimensional graph of machine movement. The measurements are achieved with two channel simultaneous vibration measurements and two transducers placed at an angle of 90° to each other, plus a trigger signal from a tachometer probe. The vibration analysis possible is very sophisticated and can give early warning of potential defects.

Oil analysis

These techniques include measuring and monitoring:

- viscosity;
- water content in oil;
- total base number (TBN);
- insolubles;
- wear debris.

Monitoring these properties of oil can give valuable insight into the condition of machinery, especially when combined with information from temperature, pressure and vibrations sensors.

Oil mist detector

The oil mist detector has seen some serious development in recent times, after many years of a single producer's standard instrument. The equipment is used to detect the presence of oil mist that might be in explosive/flammable mixtures with air. This is done by comparing the optical properties of one compartment within the engine with the optical properties of the other compartments.

For many years Graviner were the sole manufacturer in this area and their products are still in use and in demand (see Figure 12.3). However, there are alternatives available such as the highly compact Heinzmann Oil Mist Detection micro-sensor and the fast-acting oil mist detector from Quality Monitoring Instruments (QMI).

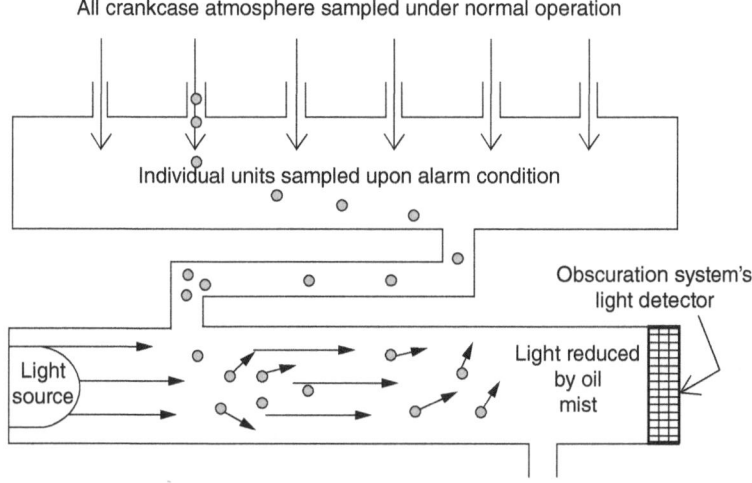

All crankcase atmosphere sampled under normal operation

Individual units sampled upon alarm condition

Obscuration system's light detector

Light source

Light reduced by oil mist

Some detectors detect 'lightscattered' by oil mist

▲ **Figure 12.3** *Oil mist detector*

Machinery condition/performance monitoring

The sensing equipment used for determining machinery performance can also be used for condition monitoring (CM). However the difference being that with CM we are looking for faults in the machinery/plant, whereas for performance monitoring (PM) we are looking to measure the efficiency of the machinery.

The objective of CM is to determine the maintenance required and the reasons for PM is more to do with making adjustments to improve the efficiency of the machinery or supplement the CM information to help with the maintenance choices.

Engine/machine performance monitoring relies on real-time data to monitor the actual performance of the machinery against a predetermined specification. This area of development is growing rapidly helped by the progress of 'mechatronics' (see Chapter 11 pages 501–503).

Most modern manufacturing processes are using mechatronics to inform:

- the process control;
- machinery performance;
- maintenance systems.

Managers can view directly up-to-date readings of plant performance, determine any lack of performance that leads to a reduction in plant efficiency and instigate or actually make any remedial action straightaway. Alternatively, the machinery can be left to operate, but within pre-set boundary conditions.

When a particular set of measured conditions (e.g. pressure, temperature, flow, power consumption or speed changes) reaches pre-defined limits then maintenance action can be considered to return the plant to full efficiency. This system of monitoring and maintenance is very efficient and is in widespread use in other industries. The development of solid-state electronics has allowed electronic control to take over from pneumatic systems. The new methods are very reliable, introduce more redundancy and require fewer back-up systems such as hydraulic or pneumatic power systems.

Non-destructive testing

Non-destructive testing is an important part of the safe operation of the ship and is mostly to do with surveying by the classification society either as part of general maintenance or as a result of mechanical or structural failure.

Detection of surface cracking would generally be by:

- visual inspection;
- magnetic particle testing;
- liquid penetrant testing.

The International Association of Classification Society's (IACS) rules state that any personnel carrying out a visual inspection of damaged parts must have sufficient knowledge and experience and any person undertaking magnetic particle testing or liquid penetrant testing must be qualified to IACS standards. Such persons will have certification to prove their qualifications as IACS requires a more onerous system for the testing and repair of crankshafts, propeller shafts and rudder stocks.

Magnetic particle inspection works on the principle that a defect in the metal, which could be below the surface of the metal, causes a distortion in a magnetic flux, at the point of the defect, which extends to the surface of the metal. This leakage of magnetic flux can then be used to attract coloured iron particles suspended in a solution that is spread over the surface of the metal.

Liquid penetrant testing is a low-cost method of testing for cracks that break the surface of a non-porous material. It relies on the penetrating ability of a low surface tension fluid such as paraffin to carry a dye into a surface crack that may not be visible under normal circumstances.

Ultrasonic testing is commonly associated with discussions about NDT. However, this is more a technique for detecting the thickness of metal than to determine the extent of any corrosion taking place. However, by checking the reduction in thickness over time, informed decisions can be made about the continued reduction in the future. This method of testing has the added advantage of being open to a high degree of automation and, therefore, slightly less training is needed to carry out the tests successfully.

Engine room record books

All ships should keep proper records about their engineering operations. These will include records such as the engine room main log, the working log and the oil record book.

The main log should record the:

- performance of the main and auxiliary machinery;
- time of the stop/start and the running hours of machinery;
- transfer of fuel;
- use of any lubricating oil;
- operating state of the main propulsion motors.

The oil record book must be kept on-board for at least three years and, therefore, should be available for the surveyor to examine. If the ship has recently changed ownership then the original book might be with the previous owners. However, a 'certified' copy covering the previous six months should be retained on-board. The book must record all movements of oil or oily liquids, whether the movement is by gravity, pump or by hand.

MARPOL 73/78 (consolidated editions 2002 and 2006) give a list of the operations that should be recorded. IMO have recently brought out (November 2010) a new MEPC Circular 736 in an attempt to clarify and standardise ORB entries with the revision of the ORB under Resolution MEPC.187(59), which came into force on 1 January 2011.

Collecting the information and compiling the engineering log is a very important task because it not only forms a record that a diligent watch is being keep on the machinery but it also provides a 'standard' set of readings. This standard set of information can then be analysed by the Chief Engineer for any adverse trends in the machinery performance.

The flag state examiner will be very interested to know if the student understands that the engineering log is the responsibility of the watchkeeping engineer. This is true even though the 'data logger' might complete the task on a modern ship. The watchkeeper MUST still check the readings to identify possible mistakes or adverse conditions that might affect the machinery.

Strategic approach to maintenance

Technical departments should also take a 'strategic' approach to the planning of their maintenance system. With modern computerised systems this is an achievable task even for highly complicated vessels such as cruise ships. With the introduction of field bus technology and the soon-to-be implemented wireless collection of data, a lot of the time-consuming tasks of data collecting and recording is already undertaken automatically, shifting the emphasis towards analysis and validation of information.

The exact system that is put in place will differ depending upon the circumstances of each company. There are three different scenarios that can be used to illustrate the point of a differing approach to maintenance.

Case 1

The first case is about the ship owner who has purchased a vessel for a budget price and intends to operate the vessel for one charter period and sell the vessel upon completion, making a profit on the project.

The strategy would be to run the machinery using only the spares that were provided with the vessel. The survey records would be up to date at the time of the purchase and the intention would be to sell the vessel before the next major survey is due.

Adopting this approach the ship owner would have allocated a small budget for the maintenance of the vessel and there would be very little left to implement a complicated planned maintenance system. In fact, most of the machinery might be operated on a 'run to failure' basis. Critical machinery and components would be identified and budgets allocated to ensure that these items do not fail and cause a breakdown and subsequent fine due to failure to complete the charter agreement.

Case 2

This is where a ship owner might purchase a new or second-hand vessel with the view of operating it over a reasonably long period but with a crew that only hold the basic relevant STCW qualifications necessary to operate the vessel.

Here, the ship managers or management company will need to ensure that the vessel is maintained to a good standard for the extended period of time. This will include an effective planned maintenance or condition-based maintenance system.

Having to work with a crew that only have a basic understanding might mean that a planned maintenance system operated on calendar hours or running hours would be the best system to adopt.

As the crew would be less able to operate the CM equipment or complete the trend analysis required, a CBM system might prove costly in the long run. The analysis work can be contracted out to one of the OEMs or to an independent company who will interpret the data and give advice about a maintenance strategy.

Case 3

Increasingly, the well-established, big name companies are turning away from instances such as Case 2 and moving towards a different mode of operation. This is especially so where the reputation of an important brand is at stake.

A sophisticated and efficient maintenance strategy can be installed if the company has sufficient well-qualified and motivated staff having access to an appropriate staff development scheme.

Adopting this system means that savings can be gained from buying high-quality machinery and maintaining it in the appropriate way. The strategy will take into account the operating conditions and matches them to the most efficient maintenance methods, depending upon the circumstances of each piece of equipment.

A full CBM system could be run for most of the major items such as the main engine and generators, pumps and steering gear. This is especially true for the machinery where it is a simple procedure to ascertain its condition using any of the techniques explained earlier in the section.

Where it is difficult to obtain any meaningful data but the machinery is still important then a maintenance system based on running hours could be used. For equipment of less importance or infrequent use, the 'run to failure' method might be the most cost-effective. It may even be that the air conditioning compressor for example is operated on this basis.

This system has a high initial set-up cost but the payback is with reduced running and operating costs. The system can be operated either in partnership with an OEM or just by using the company in-house expertise. The data collected can be analysed on board or sent to the company's office for analysis by the Superintendent. Copies can also be sent to the manufacturers for a third avenue of analysis.

One of the most important aspects creeping into the modern ship manager's life is one of producing the evidence to protect oneself against accusations of having caused an infringement of the pollution regulations. Proof of having planned maintenance up to date will be vital evidence in a court of law where the ship owners and/or managers will be required to provide evidence of a well-run ship.

Owners are now being advised to place tamper-proof devices on board that will record any emissions from the vessel. These systems are linked to the ship's GPS giving a precise time, date and position of the vessel.

Report Writing

There is an important and increasing need for junior and middle managers to provide reports for consideration by the top management and to write technical letters to external organisations. This becomes more vital at sea where senior ship management is remote from the central organisation of the company.

A common question in the MCA Chief Engineering Knowledge examination paper requires a report to be written. This could be about a report compiled by the Chief Engineer Officer for the technical staff in the administration office on a selected technical topic.

Second Engineering Officer questions are more likely to be directly concerned with on-board operational management. Some of the questions are not specifically directed to report writing but do require a knowledge of operational management principles – these are often related to safe working practices.

Most students seem to have difficulty in writing formal letters that would be sent with a report as there is no 'model' answer to such examination questions. The following notes covering general use of English, examination requirements and specimen question-answer examples that should be studied. The test examples at the end of this chapter should then be attempted and critically evaluated.

Using the English Language

1. The student could collect his/her thoughts by asking: 'what am I trying to say?' and 'what are the words that will best express what I want to say?' The final letter should be worded clearly and briefly.

2. A simple skeleton plan, of sequential ideas, may then be listed.

3. Technical wording and information can be used provided this is not overdone.

4. The 'shape' of a letter is important. A clear opening sentence, or short paragraph, is preferable to catch the immediate interest of the reader and to indicate what is to follow. The 'body' of the letter must follow a logical thought sequence – there is no objection to the use of a, b, c, or bullet points to define specific items, provided this makes the content clearer. The end of the letter should indicate some positive summary or conclusion.

5. Sentences should be short and well punctuated with compact paragraphs relating the main points.

6. Simple words should be used in place of more complicated words or phrases – 'join' for 'integrate', 'send' for 'despatch', 'about' for 'respecting', 'walk' for 'capable of locomotion', etc.

7. Standard references such as 'it is regretted', 'for your information', 'I am further to point out', etc. should not be used.

8. Verbosity, especially with adjectives (serious danger, unfilled vacancy), adverbs (to risk unduly, to enhance markedly) and prepositions (in terms of – about, until such time – until), are to be avoided.

9. The use of 'buzz words', for example, maximise, optional, orientated, etc., is inclined to be showy.

10. Clichés ('my grateful thanks', 'in this day and age'), similes ('as good as gold', 'works like a horse') and metaphors ('hit for six', 'backed a winner') are best avoided.

11. A positive statement is preferable to a double negative one ('not unnaturally', 'not unoily!').

12. Letter endings are best made 'Yours faithfully' if the method of address is 'Dear Sir' and 'Yours sincerely' to a 'Dear Mr Smith' address. Subservient ('Your obedient servant') and somewhat ridiculous (at your convenience) remarks are not to be used.

Examination requirements

1. The examiners look for clarity of expression, good punctuation, paragraphs, etc.– that is, English presentation is being assessed.

2. Technical accuracy is not so important, within reasonable limits.

3. Major details, without minor technical points, are required.

4. The experience of the candidate in matters of management, personnel relations, work study, etc. is part of the assessment.

5. The examiners attach particular importance to the following:

   Machinery surveys: arranging, preparing and recording.

   Safety equipment: certificates.

   Planned maintenance: schedules, surveys.

   Testing: machinery space lifting gear.

   Oil in navigable waters act: instructions to staff.

   Requirements of Marpol: instructions to staff.

   Firefighting: instructions and training for ER staff.

   Firefighting: co-operation between deck and ER staff.

   Training: engineer cadet training schedules.

   Training: instructions to new junior officers.

   Inspection: essential tests etc., 24 hours before sailing on a strange vessel.

   Performance: assessment of voyage records and test data.

   New ships: improved ventilation and equipment, suggestions.

   Safety schedules: day-to-day safety training.

   Bunkering: information on bunker chits, stability during bunkering.

   Crews: duties of staff, general-purpose duties.

   Ship maintenance: overall maintenance criteria for whole vessel.

   Emergency conditions: machinery failures, operation.

A full report on the other hand will be different and should be set out in a clear manner as shown below.

Report writing

A report is quite different from, say, an essay. Essays do not usually have headings or sub-headings. A report is full of headings, sections, sub-headings and sub-sections. It is important that they are all numbered and follow a logical sequence that is clear and concise. Depending on the type of report that you have been asked to write, you should consider the following sections. The headings might change to suit the local circumstances, but should be used as a framework to guide the surveyor when writing his or her report.

Title page

Should include the name of the report (what is it about), name and affiliation or title of the author, release date and any additional useful information, such as for whom the report has been prepared.

Declaration

That the material in the report is the author's own work.

Acknowledgement

To those who have helped or influenced your work (used only for academic or research work where you have had significant help from other people).

Contents page

If the report is longer than a few pages, then the contents listing is essential. It should list the structure of the report with appropriate page or section references:

1.0 Abstract or executive summary
2.0 Introduction
    2.1 The purpose of the report
    2.2 The scope of the report
    2.3 Limitations, assumptions or background
3.0 Glossary
    3.1 Essential for technical reports or reports of a specialised nature

4.0 Main technical chapters

    4.1 Survey work

    4.2 Accident investigation

    4.3 Legal reports

5.0 Conclusions

    5.1 Should be referenced back to the main body of the report

    5.2 Conclusions should present a well-reasoned argument based on fact

6.0 References

    6.1 State the source of material from books or other places such as the internet, journals or films and videos

7.0 Appendices

    7.1 Additional detailed information that is relevant to the report

If you are producing a report longer than a few pages, then it will be essential to use page numbers within the contents page as shown on the following page.

The report will be most useful for producing a report from the survey of equipment. The Chief Engineer is allowed to survey some machinery in the engine room and this may well help to reduce the cost of classification society support (please see page 529).

Specimen question

Write a letter to your company's Superintendent Engineer concerning the circumstances attending a fire in the boiler or machinery spaces. The letter should state the probable cause, action taken and suggest preventive measures.

The candidate should first ask himself the following questions and then answer accordingly (the same technique is applicable to most examination questions in the subject):

1. What happened . . . when?
2. What was the cause . . . effect?
3. How was the condition dealt with?
4. How can a re-occurrence be prevented?
5. Any other relevant comment?

Specimen answer

MV Eastern Glory,
c/o Foster Johns (Managers) Ltd,
'Ocean View',
Brisbane,
Queensland,
AUSTRALIA   10 March 2018

Chief Engineer Superintendent,
The Moss Line Ltd,
Star House,
Leadenhall St.,
LONDON, W2 5MK

Engine Room Fire, 25 February 2018, at Sea

Dear Sir,

Further to my email of 26 February, I wish to confirm that the above vessel was stopped from 15:00 hours to 23:00 hours on that date, because of an engine room fire that required evacuation of this space for about five hours.

Due to an overflow when filling a settling tank, oil escaped on to the hot engine exhaust manifold causing a serious fire. The general fire alarm was sounded and the bridge informed but within two minutes the engine room was untenable and I ordered immediate evacuation.

At 15:07 hours all engine room staff were accounted for, the engine room was sealed off and inert smothering gas injected. At 18:00 hours an attempt was made to re-enter the engine room via the tunnel but without success. At 20:00 hours the engine room was entered and small fires still burning were put out with portable extinguishers. No serious permanent damage was noted but the space was severely blackened. The machinery was prepared for sea and, before getting under way, all lagging was stripped from the manifolds.

The cause was established as a faulty tank indicating float and overflow gooseneck whose outflow was directed near the manifold. It is suggested that a mercury-type level and alarm switch be fitted as a replacement and that the gooseneck be replaced by an overflow pipe (with sight glass) to an overflow tank. I ask for approval to put this work in hand immediately. In the meantime, special care is being exercised in tank filling. A detailed damage report will be sent in the near future.

I would like to record the excellent behaviour of the engine room staff during the whole incident and the efficient communication between deck and engine departments. No injury occurred to personnel.

Yours faithfully,
William J. Hall (Chief Engineer)

Test examples – technique

An outline *framework,* as suggested method of answer, to three of the test examples at the end of this chapter is now presented for consideration before the reader proceeds to attempt the remaining test examples.

Class 3: Test example no. 1

1. Fire location – accommodation, deck, engine (need to vary).
2. Advance warning (notice) and practice alarm (method).
3. Fire stations – assembly, numbers, roll call.
4. Communications – central and remote.
5. Checking responsibilities – (individual and collective).
6. Testing alarms and equipment.
7. Simulated attack on fire.
8. Emergency sealing arrangements – check.
9. Short seminar on effectiveness – feedback.

Class 2: Test example no. 1

1. Increased ignition potential – burning, welding, sparks, smoking.
2. Increased combustible potential – waste, dirt, spillage.
3. Increased air potential – draughts, openings, circulation.
4. Increased space potential – open tanks, holds, stores.

Easier air and combustible access, less efficient sealing arrangements, more people/less co-ordination, reduced availability of immediate ship firefighting services, etc.

Reduced a, b, c, d. Improved emergency arrangements, close liaison ship-shore personnel, ready access to sealing/opening facilities, fire patrol (24 hours), etc.

Class 1: Test example no. 1

1. What major repairs?
   Precise nature, voyage effects, difficulties reaching port.

2. Why necessary after major refit?
   Possible cause, blame, deterioration on voyage.

3. Justification of shore labour?
   Scale of repair, time available, ship labour resources/facilities.

4. Justification of cost?
   Typical cost figures, estimates obtained, agent's advice.

5. Balance costing?
   Survey in lieu of port time, cargo, voyage schedules.

6. Inspection?
   Classification society, ship staff, work standards, expected outcomes.

Note

In cases where a report is on a subject likely to be used in subsequent legal actions (safety, law breaking, etc.), it is vitally important to present information very accurately with respect to time, date, names of personnel involved, etc.

TEST EXAMPLES

Test Examples 1

Chief Engineering Officer examinations

1. Name four copper alloys associated with marine engineering, in each case giving its constituents, physical properties and a practical example of its use.

2. Explain why a material may fracture when stressed below its yield point. Give examples of components that might fracture in this way if suitable precautions are not taken. Explain how such fractures can be avoided with reference to the materials chosen, careful design and workmanship.

3. Give the approximate composition and the properties of the following metals:

    (a) manganese bronze

    (b) cupro-nickel

    (c) Babbitt metals

    In each case give two examples of the metals in use on board a ship, and explain why the metal is chosen for the applications you mention.

4. Give properties, uses and constituents of:

    (a) phosphor bronze

    (b) blackheart malleable iron

    (c) monel metal

5. Describe the following:

    (a) case hardening

    (b) flame hardening

    (c) nitriding

    (d) induction hardening

Second Engineering Officer examinations

1. Briefly describe the tests made on a piece of metal to determine its suitability for use in engineering. Explain clearly what is meant by any *four* of the following metallurgical terms:

   (a) work hardening

   (b) case hardening

   (c) annealing

   (d) normalising

   (e) yield point

   (f) creep

2. Sketch graphically the load–extension diagram for a mild steel test piece. Would you expect a similar diagram if you tested a non-ferrous metal? Explain yield point, elastic limit, limit of proportionality and proof stress.

3. State the approximate proportions of carbon contained in (a) cast iron and (b) cast steel and mention the forms in which the carbon may occur therein.

   Compare the physical properties of these two metals and name some of the more important parts of machinery for which these materials are respectively suitable.

4. Explain the difference between 'strength' and 'stiffness' of steel. Discuss the importance of these properties in shipboard structural members and machinery components.

5. Describe the effects of varying the percentages of the following constituents on the physical properties of steel:

   (a) carbon

   (b) phosphorus

   (c) manganese

   (d) molybdenum

Engineering Officer of the Watch examinations

1. Explain the following terms:

   (a) elastic limit

   (b) UTS

   (c) safety factor

2. What is the advantage of case hardening? How is it done? Give an example of a component that may have this treatment.

3. Explain the essential differences between the properties of cast iron and mild steel.

Test Examples 2

Chief Engineering Officer examinations

1. With reference to fire or explosion explain the significance of the following properties of a flammable gas:

   (a) vapour pressure

   (b) explosive limits

   (c) flash point

   (d) density

   (e) fire point

2. Define each of the following terms in relation to lubricating oil:

   (a) pour point

   (b) cracking point

   (c) flash point

   (d) auto-ignition point

   State, with reasons, when each of these characteristics will be of primary importance.

3. Explain the effects of differences in chemical composition, calorific value and viscosity of the fuel on engine performance.

   Describe how engine operation and maintenance may need changing in order to burn heavy-distillate instead of light-distillate fuel.

4. For a carbon dioxide recorder explain:

   (a) the principle of operation

   (b) the action to be taken if the reading is unacceptably low

   (c) the normal maintenance required

   (d) how the accuracy of the recorder is checked and adjusted

Second Engineering Officer examinations

1. Sketch the apparatus used and describe the test to determine the following properties of oil:

   (a) viscosity

   (b) calorific value

2. (a) Suggest, with reasons, which of the following data is relevant and significant to the quality of fuel oil:

Viscosity	Conradson number
Pour point	Total base number
Closed flash point	Octane number
Open flash point	Specific gravity

   (b) Define the significance of lower and higher calorific value in assessing the standard of liquid fuel.

3. (a) Specify, with reasons, where test samples should be drawn from a main lubricating oil system.

   (b) Describe shipboard tests to determine:

   (i) water content

   (ii) acidity

   (iii) suspended solids

4. (a) Describe, with sketches, an instrument for indicating the carbon dioxide content of the gases in the uptake.

   (b) Explain the meaning and importance of the readings obtained.

Engineering Officer of the Watch examinations

1. What is meant by the term 'calorific value' and how does the calorific value of fuel oil compare with that of coal?

2. Why is the flash point an important criteria with regard to lubricating oil? State how it is determined, and give one reason why the flash point of a lubricating oil sample from an engine might be lower than expected.

3. When referring to fuel oil, the terms cetane number and Conradson number are used frequently. Explain these two terms.

4. Complete the following combustion equations:

   (a) $C + O_2 \rightarrow$

   (b) $C + \dfrac{1}{2}O_2 \rightarrow$

   (c) $H_2 + \dfrac{1}{2}O_2 \rightarrow$

Test Examples 3

Chief Engineering Officer examinations

1. Assess the value of regular systematic inspection of auxiliary boilers and ancillary equipment. With particular reference to a vertical, smoke tube, hemispherical furnace boiler, identify three common faults on the water side and two common faults on the fire side. Describe the remedial action in each case to retard development.

2. With reference to main safety valves for handling steam at 60 bar, 500°C explain the following:

   (a) Why is precise, rapid and ample valve movement essential during opening and closing?

   (b) How are the characteristics in (a) achieved in practice?

3. Discuss the merits of fitting a low-pressure exhaust gas boiler into the uptakes of a diesel-engined installation. Sketch a boiler suitable for this purpose.

4. What are the essential differences between an ordinary single spring-loaded safety valve and an improved high-lift safety valve? How would you set the safety valves for a multi-boilered installation? What is meant by accumulation of pressure and how would you conduct a test to determine if the safety valves are of correct capacity?

5. Sketch and describe two types of remote boiler water level indicator. To what defects are the indicators of your choice liable and how would you remedy these defects?

Second Engineering Officer examinations

1. Describe the essential steps in the structural examination of an auxiliary boiler. State, with reasons, where wastage is likely to be found. Explain why it is equally important to examine both the fire side as well as the water side.

2. With reference to auxiliary boiler safety valves:

   (a) State, with reasons, what clearances need checking when lapping in valves and seats.

   (b) Why must the drain be kept clear?

   (c) How is setting done under steam?

   (d) Why should the opening gear be kept in good working order at all times?

3. Sketch and describe a sea water evaporation plant using engine coolant as the heating medium. State how the distillate is rendered fit for drinking.

4. Describe how to 'blow down' and 'open up' an auxiliary boiler for inspection. Identify with reasons those parts that normally require especially close examination during internal inspection.

5. Sketch and fully describe the operation and construction of a remote boiler water level indicator. To what defects is the instrument liable and how you would remedy these defects?

Engineering Officer of the Watch examinations

1. With the aid of a simple sketch, explain how a water gauge fitted directly to a boiler is tested for accuracy when the boiler is steaming.

2. Describe the start-up sequence of an automatic auxiliary boiler.

3. Describe how fire tubes are attached to the tube plate of a fire tube boiler.

Test Examples 4

Chief Engineering Officer examinations

1. Describe how you would make a quantitative test of boiler water for:

   (a) alkalinity

   (b) chlorinate

   (c) hardness

   State the values obtained from the above tests that you would consider suitable for a water tube boiler. Describe in each case the action you would take if a test gave an unsatisfactory result.

2. Why is it necessary to keep oxygen out of the boiler? Describe how this is done mechanically and chemically.

   State the procedure for laying up a boiler:

   (a) for a considerable period

   (b) for a few days

3. Enumerate the scale-forming solids in fresh and sea water, respectively. How would you steam a boiler on contaminated feed? Give reasons for your action.

4. Describe the boiler water tests carried out for boilers, and the results expected from:

   (a) low-pressure boilers

   (b) high-pressure boilers

5. Specify, with reasons, those parts requiring particularly close scrutiny during internal and external examination of independently fired auxiliary boilers. With reference to those examinations distinguish between metal fatigue due to caustic embrittlement, corrosion fatigue, overheating (plastic flow) and direct overpressure.

Second Engineering Officer examinations

1. Give an analysis of the dissolved solids in an average sample of:

   (a) sea water

   (b) fresh water

   Which of these solids can form scale and which can cause corrosion?

2. Discuss the contamination of boiler feed water. What action should be taken in the event of such contamination to prevent damage to boilers and machinery? What tests are made?

3. What are the causes of corrosion in boilers? What precautions would you take to prevent corrosion in the following instances:

   (a) when the boiler is steaming?

   (b) when the boiler is idle?

   How would you test the boiler water for acidity and alkalinity?

4. Suggest with reasons which four of the following impurities in the feed water of a 'package' boiler operating at 7 bar, dry saturated are likely to contribute most to scale formation:

   (a) silica

   (b) iron compounds

   (c) sodium chloride

   (d) magnesium bicarbonate

   (e) calcium bicarbonate

   (f) calcium sulphate

   (g) sodium sulphate

   (h) magnesium chloride

5. Give a reason why sodium phosphate, sodium hydroxide and hydrazine are each used in boiler water treatment. Describe any three of the analytical tests normally applied to boiler water. Explain how the results influence further treatment. State two precautions to be observed when storing and handling these chemicals.

Engineering Officer of the Watch examinations

1. Why is oil in boiler water considered dangerous? Where does it usually come from? How can it be removed?

2. Water for boilers is usually kept as pure as reasonably possible. Give reasons why this is so.

3. Briefly describe why boiler water needs to be tested periodically and state two of the tests.

Test Examples 5

Chief Engineering Officer examinations

1. With reference to stabiliser fins that either fold or retract into hull apertures:
   (a) make a simplified sketch of the essential features of the activating gear for both fin extension and altitude
   (b) explain how it operates

2. With reference to steering gears explain the following:
   (a) Why are multi-piston variable stroke pumps used rather than rotary positive displacement pumps with controlled recirculation or delivery?
   (b) Why are independent, widely separated power supplies to the electrically driven pumps provided together with duplication of pumps in many instances, yet the hydraulic telemotor system has usually only one run of double piping from the bridge to the receiver?
   (c) Why is pump, piston and cylinder wear of considerable consequence?

3. State the effects of the following faults in steering gear telemotor systems:
   (a) low liquid level in replenishment tanks
   (b) weak receiver springs
   (c) worn cup leathers or rings in transmitters or receivers
   (d) leaking pump connections
   (e) specify with reasons the nature and properties of the fluid generally used in such systems

4. With reference to steering gears, explain the following:
   (a) Why are four rams commonly employed on large vessels?
   (b) How is the steering function maintained despite loss of hydraulic fluid from the telemotor system?

Second Engineering Officer examinations

1. (a) Sketch a hunting gear as fitted to a hydraulic steering gear, labelling the principal items.

    (b) Explain the purpose of the hunting gear.

    (c) State how worn pins in the hunting gear effect steering gear operation.

2. With reference to hydraulic steering gears explain the following:

    (a) Why are relief valves provided as well as shock valves?

    (b) Why is the pump of constant speed, variable stroke?

    (c) Why are the ram glands filled with soft or simple moulded packing?

3. Describe a simple test to ensure that steering gear hydraulic telemotor systems are 'air free'.

    (a) Define two ways whereby air enters such systems.

    (b) Give reasons why it is essential that such systems be 'air free'.

4. With reference to electro-hydraulic steering gears explain how the ship can be steered in each of the following circumstances:

    (a) destruction by fire of primary supply cables

    (b) destruction by fire of telemotor lines

    (c) bearing failure in running pump

Engineering Officer of the Watch examinations

1. Itemise the tests you would carry out on a steering gear before leaving port.

2. If the main steering gear failed on a small coaster describe how you would rig an emergency steering system to enable the ship to get to a port.

3. Briefly describe how the torque is transmitted to the rudder stock in a rotary vane steering gear.

4. Explain how the following are achieved for a ship's electro-hydraulic steering gear:

    (a) relief of over pressure in the rams' hydraulic circuit

    (b) replenishment of hydraulic oil to make good losses caused by minor leakages from glands

Test Examples 6

Chief Engineering Officer examinations

1. Describe how alignment of shafting between engine and propeller shaft is checked with the vessel afloat. State how alignment is corrected in the case of appreciable hull deflection.

2. (a) Make a simplified sketch of the operating mechanism for a controllable pitch propeller.

   (b) Describe briefly how pitch is altered in accordance with telemotor signal.

   (c) State what 'fail-safe' feature is incorporated into the logic of the mechanism.

3. Identify the defects to which propeller shafts are commonly susceptible. Explain how propeller shafts are surveyed in order to detect these defects.

4. (a) Sketch a 'muff' (flangeless or sleeve) coupling for connecting adjacent lengths of main transmission shafting.

   (b) Describe the manner in which the coupling is mounted on and transmits torque between the adjacent lengths of shafting.

   (c) State how astern thrust is accommodated by the coupling.

Second Engineering Officer examinations

1. With reference to keyless propellers explain the following:

   (a) Why have keys and key ways been eliminated?

   (b) How is angular slip avoided?

   (c) Why does mounting upon and removal from a propeller shaft require a different technique than that employed for propellers with keys?

2. (a) Sketch a coupling enabling external withdrawal of propeller shafts.

   (b) Give a general description of the coupling.

   (c) Give one advantage and one disadvantage of this coupling compared to the solid flange type.

3. With reference to controllable pitch propellers answer the following:

   (a) Explain why the blade attitude assumed upon control failure is considered safe.

(b) Describe how the 'fail-safe' feature operates.

(c) State how the ship can be manoeuvred when the bridge control is out of action.

4. (a) Describe how unequal loading of main transmission shaft bearings may be partially corrected at sea.

(b) Suggest, with reasons, what remedial action should be taken upon arrival in port.

(c) State the indications, while at sea, that unequal loading of such bearings exists.

Engineering Officer of the Watch examinations

1. How is alignment of a crankshaft checked? Does the loaded condition of the ship have any effect on alignment?

2. What are the forces on a diesel engine crankshaft and are these forces uniform along the length of the crankshaft? Give reasons for your answer.

3. What is the purpose of putting a thrust bearing between the main engine and the propeller? How is the thrust bearing cooled?

4. Explain how a variable pitch propeller operates.

Test Examples 7

Chief Engineering Officer examinations

1. (a) Draw a line diagram of an accommodation air conditioning plant, labelling the principal components and showing the direction of air flows in all ducts.

(b) Explain why humidity control is essential for comfort.

(c) State how ambient temperature affects humidity control.

(d) Give a reason why compensation for air losses is necessary and how it is accomplished.

2. (a) Identify, with reasons, those properties of Freon that makes it such an attractive refrigerant.

(b) Give reasons why each of the following gases has fallen into disfavour as a refrigerant:

(c) carbon dioxide

(d) ammonia

(e) methyl chloride

3. Give a reasoned opinion as to the validity of the following references to accommodation air conditioning:

   (a) 'rule of thumb' method whereby rate of air change is directly related to cubical capacity of the compartment concerned is quite satisfactory for all practical purposes

   (b) mechanical ventilation with air heating is inadequate for comfort in ships operating within a wide range of ambient air temperature and humidity

   (c) humidity control is absolutely essential for long-term comfort of personnel

4. In refrigeration define the following:

   (a) specific heat capacity

   (b) specific enthalpy of evaporation

   (c) specific volume

   (d) critical temperature

   Give typical values for each of the above for three refrigerants.

Second Engineering Officer examinations

1. (a) Draw a line diagram of a refrigeration system for servicing a large number of insulated containers, labelling the principal items and showing the direction of flow in all lines and ducts.

   (b) Explain how the system works in order to maintain containers at different temperatures.

2. Suggest, with reasons, the most likely cause of the trouble if the suction to a multi-cylinder refrigerant compressor is subject to considerable icing under the following simultaneously prevailing conditions:

   (a) compressor in good condition and running at normal speed

   (b) throttling regulator valve open more than usual

   (c) no detectable loss of refrigerant

   (d) brine temperature rising

3. With reference to refrigeration plants answer the following:

   (a) How are very low evaporator temperatures achieved?

   (b) How are automatic expansion valves in direct expansion plants adjusted?

   (c) How are compressors protected against appreciable 'carry-over' of liquid refrigerant?

   (d) How is air in the system detected?

(e) How is overcharge of refrigerant indicated?

4. Briefly describe the following in main refrigeration plants:

(a) How can sea temperature restrict plant operation?

(b) How are the limitations in (a) overcome?

(c) How does short cycling occur?

(d) How is short cycling avoided?

Engineering Officer of the Watch examinations

1. Sketch a simple refrigerant cycle of compression type and on the sketch show a position in the cycle where you would expect the refrigerant to be:

(a) a liquid

(b) a gas

2. Describe the basic refrigerant circuit for a compression-type plant. What kind of gas is commonly used in this type of plant?

3. State the reasons why Freon 12 is a popular refrigerant gas.

4. Explain why the refrigerant gas in a compression-type domestic refrigeration plant is passed through a condenser after being compressed.

Test Examples 8

Chief Engineering Officer examinations

1. Describe, with sketches, an inert gas system using gas from the main uptakes. Explain:

(a) the scrubbing process and its purpose

(b) safety devices

2. Compare the advantages and the disadvantages of the following fixed fire extinguishing systems:

(a) high-pressure water spray

(b) carbon dioxide smothering

(c) chemical foam smothering

3. Differentiate between fixed temperature and rate of rise types of fire detector.

   Sketch and describe a fire detector of the rate of rise type and explain how a gradual rise of ambient temperature is accommodated.

4. With reference to fire or explosion explain the significance of the following properties of a flammable gas:

   (a) combustion pressure

   (b) explosive limits

   (c) flash point

   (d) density

5. Sketch the construction and describe the operation of the following types of fire detector:

   (a) vapour products (ionisation)

   (b) flame sensor (infrared)

   (c) heat sensor (rate of rise)

   Explain why use of all three types together is to be preferred to the use of one of these types alone.

Second Engineering Officer examinations

1. Compare, with reasons, the merits and demerits of the following permanent fire extinguishing systems installed in machinery spaces:

   (a) high-pressure water spray

   (b) carbon dioxide smothering

   (c) chemical foam smothering

2. Describe with sketches how the following portable fire extinguishers are operated:

   (a) chemical foam

   (b) carbon dioxide

   Explain how they extinguish fire. State with reasons what type of fire each is most suited for.

3. Sketch and describe a self-contained breathing apparatus. Give two advantages and two disadvantages of this equipment compared to the smoke helmet. State the signal system used when wearing breathing apparatus. Describe with line diagrams a fixed carbon dioxide fire-smothering system for an engine room.

4. Explain the need for an action alarm stating when and how it operates. Give a reason for gang release and explain how this is achieved.

5. If a fire broke out in the engine room, explain the following:

   (a) How can the fuel supply be shut off?

   (b) How can the supply of air be shut off?

   (c) How can the fire be dealt with from outside the engine room? Give a summary of all the facilities available for this purpose.

Engineering Officer of the Watch examinations

1. Sketch a cross-section through a portable fire extinguisher suitable for use on oil fires. Identify the components.

2. State the component parts and associated equipment of a bellows-type breathing apparatus.

3. Sketch and describe a smoke detector of the type fitted in an engine room. How is it tested?

Test Examples 9

Chief Engineering Officer examinations

1. Describe with a line diagram a hydrophore system, that is, a fresh water system incorporating an air reservoir.

   Describe how a drop in pressure actuates the fresh water pump.

   State two advantages of this system over the gravity head system.

2. Sketch a tubular oil cooler and explain the following:

   (a) How is differential expansion between the tubes and shell accommodated?

   (b) How is corrosion controlled?

   (c) How is automatic control of the oil temperatures effected?

   State how by construction, operation and maintenance tube failure can be minimised.

3. Give a simple explanation of the nature and effect of cavitation in rotodynamic pumps.

   Describe with sketches a supercavitating pump. State the purpose of such a pump and give an instance of current shipboard application.

4. Describe with sketches a centralised cooling system incorporating plate-type heat exchangers.

   Explain the purpose of this system and how it is achieved.

5. Describe with sketches an axial flow pump.

   Explain its principle of operation.

   State what important advantage and serious disadvantage it possesses compared to other pumps.

   Explain the effects of throttling either the suction or discharge valve.

Second Engineering Officer examinations

1. Sketch and describe a plate-type heat exchanger. State one advantage and one disadvantage of this design compared to the tubular type.

2. Sketch and describe a pump other than the reciprocating or centrifugal type. Explain how it works.

   State with reasons the duty for which it is most suited.

3. Sketch and describe a centrifugal bilge pump.

   Explain the need for a priming pump.

   Give one advantage and one disadvantage of the centrifugal pump compared to the direct acting pump for bilge duties.

4. Sketch the construction of an oil–water separator.

   Explain how it functions in service.

   State the purpose and operation of all alarms or safety devices fitted to the separator.

5. Make a line diagram of a bilge pumping system for a dry-cargo ship.

   Indicate the type and position of each valve fitted.

   In view of the possibility of collision, explain how the integrity of the bilge pumping system is ensured as far as possible.

Engineering Officer of the Watch examinations

1. Make a diagrammatic sketch of a bilge pumping system, itemising the main components.

2. Describe the passage of water through a centrifugal pump.

3. Describe an oil–water separator and state how you would check that it was working satisfactorily.

Test Examples 10

Chief Engineering Officer examinations

1. Describe a centrifugal oil purifier and explain in detail how separation occurs. Should a purifier be worked at its rated capacity if the maximum efficiency of separation is desired?

2. Sketch and describe a purifier and explain how the separation of dirt and water from oil is accomplished. What factors increase the speed and efficiency of the separation? Comment on the advantages and disadvantages of water washing lubricating oils.

3. Discuss the effects of water in lubricating oils and fuel oils. Explain how the presence of water can be detected and explain how water can be separated from oil by virtue of their different densities.

4. Sketch and describe three types of lubricating oil filters used aboard ship. Give reasons for using different types and relative positions in the system.

5. Explain the necessity for regular laboratory analysis of lubricating oils and the importance of obtaining a truly representative sample for this purpose. Without specialist equipment explain how to test for the following:

    (a) sludge

    (b) water

    (c) acidity

    State what effect each of these has on bearings and how it is countered.

Second Engineering Officer examinations

1. Sketch a self-cleaning filter and describe its operation. Explain the function of magnetic filters. State why magnetic filters frequently complement self-cleaning filters in lubricating oil systems and why centrifuges do not render static filters redundant.

2. With reference to centrifugal separators answer the following:

    (a) Differentiate between the purpose and operation of purifiers and clarifiers.

    (b) Explain how these different roles are achieved.

    (c) Describe with sketches a self-cleaning arrangement.

3. Sketch and describe an oil centrifuge. Give a general explanation of the principles involved in the function of a centrifuge. State the adjustments made to the machine between handling oils of different densities.

4. Describe the operation of a differential pressure alarm fitted across an oil filter. Explain how such a device would be tested while the filter is in service.

5. Give two reasons why regular laboratory analysis of both main engine and auxiliary engine lubricating oil is desirable. State where and how representative samples of lubricating oil would be obtained from the systems. Describe the simple shipboard tests you could apply to determine:

    (a) insoluble content

    (b) water content

    (c) acidity

Engineering Officer of the Watch examinations

1. Explain why it is usually necessary to purify crankcase lubricating oil and how the purification is carried out.

2. Apart from providing lubrication to engine bearings, what other important function does lubricating oil perform?

3. If during passage you had reason to think that the lubricating oil in the main engine was contaminated, can you state any simple checks that would help you come to a conclusion about the contamination?

Test Examples 11

Chief Engineering Officer examinations

1. Sketch diagrammatically an auxiliary boiler automatic combustion control system. Explain how it operates. Specify how 'fail-safe' conditions are ensured.

2. With reference to automatic combustion control in auxiliary or main boilers answer the following:

    (a) How does the master controller follow steam pressure variations?

    (b) Why is pressure drop across air registers measured?

    (c) How is air fuel ratio adjusted?

3. **(a)** Sketch diagrammatically the arrangements to control lubricating oil temperature at cooler outlet by either of the following means: (i) diaphragm-actuated control valve as part of a closed-loop system or (ii) self-operated valve of wax element type.

   (b) Describe how the selected arrangement operates.

   (c) Define, with reasons, where a control valve of the wax element type should be positioned to be effective.

4. With respect to shipboard control equipments, state the following:

   (a) Where is an electronic system essential?

   (b) What are the advantages of the electronic system over the other systems?

   (c) What are the disadvantages of the electronic system?

Second Engineering Officer examinations

1. Complete the following:

   (a) Sketch a diaphragm-operated control valve of any design.

   (b) State how flow changes are sensed.

   (c) State how command signals are transmitted to actuators.

   (d) Explain why a pneumatic control system requires clean, dry air. Explain how the following pollutants are dealt with: water, oil, dust and dirt.

   (e) Describe, with sketches, a bridge/engine room telegraph interconnecting gear.

   (f) Explain how the system may operate a 'wrong way' alarm.

2. Describe, with sketches, instruments used for measuring the ambient temperature in the following spaces:

   (a) refrigerated compartment

   (b) main machinery exhaust gas uptakes

Engineering Officer of the Watch examinations

1. Describe the start-up sequence of an automatic auxiliary boiler.

2. Describe a device that would automatically activate an alarm when the lubricating oil supply to an engine fails.

3. Explain the principle of operation of a pneumatic diaphragm activator.

4. List three automatically controlled systems in a ship's machinery space that need to be provided with alarms. State the consequences of an alarm failure in each system.

Test Examples 12

Chief Engineering Officer examinations

1. The ship's crew have conducted a series of combined lifeboat and fire drills over the period of the voyage.

   As Chief Engineer, make a brief report to the ship's management to explain the lessons learnt from these drills, proposing ways of improving the effectiveness of these emergency procedures.

2. Draft a Chief Engineer's report to Head Office outlining a three-month practical training programme on-board ship for engineer cadets who have not been to sea before.

3. Your ship sustained a sudden and irrevocable loss of main propulsion power while entering port. As Chief Engineer, make a report to Head Office explaining the cause of the failure.

4. Most foreign-going cargo ships are required to possess a valid Safety Equipment Certificate, renewed at intervals after survey of the safety equipment.

   (a) Identify those items covered by the Safety Equipment Certificate that are usually the responsibility of the Chief Engineer.

   (b) Suggest how the survey should be organised in order that it be completed with the least trouble and delay.

   (c) Suggest how it can be ensured that this safety equipment is in a full state of readiness at all times.

Second Engineering Officer examinations

1. Give four common sources of fire in a vessel under repair in a yard.

   Explain why fire may spread more rapidly in a ship under repair in a yard than when at sea.

   Describe the precautions taken to minimise the possibility and effect of fire in a repair yard.

2. Explain what is meant by 'planned maintenance'.

   Give details of its application to the main lubricating oil pumps.

3. List the safe practices to be observed when personnel are:

   (a) using lifting tackle

   (b) working beneath the floor plates

   (c) overhauling valves or renewing jointing in steam lines

   (d) dismantling machinery during rough weather

4. (a) Explain the role of the Safety Officer in relation to the ship's Safety Committee.

   (b) Outline the specific duties of this officer on board ship.

Engineering Officer of the Watch examinations

1. Describe briefly how a practice fire drill should be carried out.

2. If a large item of machinery is to be lifted out of the engine room, state what precautions should be taken to:

   (a) prevent injury to any personnel

   (b) prevent damage to either ship or the item being lifted

3. Briefly describe a system aboard ship by which the spare gear can be monitored and an adequate supply be ensured for machinery repairs.

4. Explain what is meant by planned maintenance with respect to ship's machinery.

INDEX

REEDS MARINE ENGINEERING AND TECHNOLOGY SERIES

Vol. 1 Mathematics for Marine Engineers
Kevin Corner, Leslie Jackson and William Embleton
ISBN 9781408175552

Vol. 2 Applied Mechanics for Marine Engineers
Paul A Russell, Leslie Jackson and William Embleton
ISBN 9781472910561

Vol. 3 Applied Thermodynamics for Marine Engineers
Leslie Jackson, William Embleton and Paul A Russell
ISBN 9781408160749

Vol. 4 Naval Architecture for Marine Engineers
Richard Pemberton, E A Stokoe
ISBN 9781472947826

Vol. 5 Ship Construction for Marine Engineers
Paul A Russell, E A Stokoe
ISBN 9781472924285

Vol. 6 Basic Electrotechnology for Marine Engineers
Christopher Lavers, Edmund G R Kraal and Stanley Buyers
ISBN 9781408176061

Vol. 7 Advanced Electrotechnology for Marine Engineers
Christopher Lavers and Edmund G R Kraal
ISBN 9781408176030

Vol. 8 General Engineering Knowledge for Marine Engineers
Paul A Russell, Leslie Jackson and Thomas D Morton
ISBN 9781472952738

Vol. 9 Steam Engineering Knowledge for Marine Engineers
Thomas D Morton
ISBN 9780713667363

Vol. 10 Instrumentation and Control Systems
Gordon Boyd and Leslie Jackson
ISBN 9781408175590

Vol. 11 Engineering Drawings for Marine Engineers
H G Beck
ISBN 9780713678574

Vol. 12 Motor Engineering Knowledge for Marine Engineers
Paul A Russell, Leslie Jackson and Thomas D Morton
ISBN 9781472953445

Vol. 13 Ship Stability, Resistance and Powering
Christopher Patterson and Jonathan Ridley
ISBN 9781408176122

Vol. 14 Stealth Warship Technology
Christopher Lavers
ISBN 9781408175255

Vol. 15 Electronics, Navigational Aids and Radio Theory for Electrotechnical Officers
Steve Richards
ISBN 9781408176092

Reeds Introductions: Essential Sensing and Telecommunications
Christopher Lavers
ISBN 9781472922182

Reeds Introductions: Physics Wave Concepts
Christopher Lavers & Sara-Kate Lavers
ISBN 9781472922151